Werner Vogel and Henry Kalb

Large-Scale Solar Thermal Power

Related Titles

Quaschning, V.

Renewable Energies and Climate Change

344 pages
Hardcover
ISBN: 978-0-470-74707-0

Freris, L., Infield, D.

Renewable Energy in Power Systems

300 pages
Hardcover
ISBN: 978-0-470-01749-4

De Vos, A.

Thermodynamics of Solar Energy Conversion

205 pages with 80 figures
Hardcover
ISBN: 978-3-527-40841-2

Eicker, U.

Solar Technologies for Buildings

336 pages
2003
E-Book
ISBN: 978-0-470-86506-4

Ackermann, T. (ed.)

Wind Power in Power Systems

742 pages
2005
Hardcover
ISBN: 978-0-470-85508-9

Werner Vogel and Henry Kalb

Large-Scale Solar Thermal Power

Technologies, Costs and Development

WILEY-VCH Verlag GmbH & Co. KGaA

The Authors

Werner Vogel
Königsbronn, Germany
werner.h.vogel@web.de

Henry Kalb
Karlsruhe, Germany
Kalb-Heinz@t-online.de

The Translator
Prof. William D. Brewer, Berlin, Germany

Cover picture
Solar tower power plants

Artist's concept based on Solar One (pilot plant)
Barstow, California, 1993

PIX Number 00036
Sandia National Laboratories
National Renewable Energy Laboratory
U.S. Department of Energy
Copyright U.S. Department of Energy

All books published by **Wiley-VCH** are carefully produced. Nevertheless, authors, editors, and publisher do not warrant the information contained in these books, including this book, to be free of errors. Readers are advised to keep in mind that statements, data, illustrations, procedural details or other items may inadvertently be inaccurate.

Library of Congress Card No.: applied for

British Library Cataloguing-in-Publication Data
A catalogue record for this book is available from the British Library.

Bibliographic information published by the Deutsche Nationalbibliothek
The Deutsche Nationalbibliothek lists this publication in the Deutsche Nationalbibliografie; detailed bibliographic data are available on the Internet at <http://dnb.d-nb.de>.

© 2010 WILEY-VCH Verlag GmbH & Co. KGaA, Weinheim

All rights reserved (including those of translation into other languages). No part of this book may be reproduced in any form – by photoprinting, microfilm, or any other means – nor transmitted or translated into a machine language without written permission from the publishers. Registered names, trademarks, etc. used in this book, even when not specifically marked as such, are not to be considered unprotected by law.

Cover Design: Grafik-Design Schulz, Fußgönheim
Typesetting: Toppan Best-set Premedia Limited, Hong Kong
Printing and Binding: Strauss GmbH, Mörlenbach

Printed in the Federal Republic of Germany
Printed on acid-free paper

ISBN: 978-3-527-40515-2

In Memoriam: Werner Buckel

We owe very special thanks to Professor Werner Buckel (1920–2003), previously the Director of the Physics Institute at the University of Karlsruhe, Germany, and former president of the European Physical Society. From 1985 until his death in 2003, he gave support to our work with an exceptional dedication, and helped to put forward the concept of a large-scale energy supply based on solar thermal power at the scientific and the political levels. Our work in the last years and, in particular, the present book would certainly not have come about without the steady encouragement and support of Professor Buckel.

Contents

Preface *XV*
Preliminary Remarks and Summary *XVII*
The Significance of the Rapid Deployment of Solar Thermal Power Plants for Energy Policy *XVII*
Acknowledgments *XXV*
List of Tables *XXVII*

1   **Introduction** *1*
1.1   Historical Background *1*
1.2   Formulating the Problem *1*

2   **The Salient Facts** *7*
2.1   Solar Tower Power Plants as the Basis for Cost Estimates: Cost Analyses *7*
2.2   The Combined System of Solar and Backup Power Plants ("Solar Power System") *9*
2.2.1   Solar Base-Load Plants *10*
2.3   How Much Does Solar Power Cost? *11*
2.3.1   Introductory Remarks *11*
2.3.2   Investments and Power Costs *12*
2.3.3   Are the Additional Costs Compared to Nuclear Plants Affordable? *18*
2.3.3.1   Burden on the Economy Due to Higher Power Costs (The Cost Difference Solar Energy–Nuclear Energy) *19*
2.3.4   Possibly Lower Cost Differences, Potential for Further Development *21*
2.3.5   "Hidden" Costs of Conventional Power Plants *24*
2.3.5.1   Nuclear Power Plants *25*
2.3.5.2   Coal-Fired Power Plants *26*
2.3.5.3   Fossil-Fuel Backup Power Plants for the Solar Power System *28*

Large-Scale Solar Thermal Power. Werner Vogel and Henry Kalb
© 2010 WILEY-VCH Verlag GmbH & Co. KGaA, Weinheim
ISBN: 978-3-527-40515-2

2.4	Possible Time Scales for the Operational Readiness of Solar Thermal Power Plants and the Comprehensive Replacement of Current Power Plants *29*	
2.4.1	Special Aspects of Solar Power-Plant Development *29*	
2.4.2	The Simplest Technology – Consequences for Development and Construction on a Large Scale *32*	
2.4.3	The Basic Development Tasks for Heliostats *43*	
2.4.3.1	Stability *43*	
2.4.3.2	Cost Predictions *44*	
2.4.4	The Most Important Single Point: A Cost Study for the Standard Heliostat *47*	
2.4.5	The Interdisciplinary Character of Solar-Plant Development *48*	
2.4.6	Consequences for the Organization of Research *49*	
2.4.7	Industrial Initiatives and Start-up Funding *49*	
3	**Solar Technologies – An Overview** *51*	
3.1	Dish Plants *52*	
3.2	Tower Power Plants *55*	
3.3	Parabolic Troughs *61*	
3.4	Linear Fresnel Plants *64*	
3.5	Updraft (Chimney) and Downdraft Power Plants *67*	
4	**Some Additional Economic Factors** *71*	
4.1	Detailed Treatment of the Costs of the Solar Power System – Comparison with Competing Types of Power Plants – Discussion *71*	
4.1.1	Solar Power Systems with Coal-Fired Backup Power Plants (Instead of Natural Gas Plants) *71*	
4.1.2	Overview of Costs *74*	
4.1.3	Coal-Fired Base-Load Power Plants with CO_2 Sequestration *82*	
4.1.4	Coal-Fired Power Plants without CO_2 Sequestration *84*	
4.1.5	Nuclear Power Plants *85*	
4.1.6	Weighing Cost Differences *89*	
4.1.7	Separate Considerations of Solar and Backup Power Supplies *91*	
4.1.8	Solar Power at the Plant Site *92*	
4.1.9	Hydrogen Production *93*	
4.2	Comparison with the Study of Sargent and Lundy *94*	
4.2.1	Costs from Various Studies *98*	
4.2.1.1	Investment Costs *98*	
4.2.1.2	Operating and Maintenance Costs *99*	
4.2.2	Response of the NRC to the S&L Study *103*	
4.2.2.1	The Research-Political Context of the S&L Study and the Criticism of the NRC *106*	
4.2.2.2	Conclusions Based on the Current Preliminary State of Knowledge *108*	

4.2.2.3	Conclusions Regarding the NRC Report	*110*
4.3	Some Special Points Concerning Cost Estimates	*110*
4.3.1	The Effect of Mass Production on the Indirect Costs	*111*
4.3.2	Solar Multiple/"24-h Design Insolation"	*112*
4.3.2.1	Recalculation for a "Base-Load" Power Plant	*113*
4.3.3	Land Prices in Spain	*114*
4.3.4	Political Costs – North African Solar Energy as a "Relative" Alternative for Europe	*115*
4.3.4.1	European Alternatives in Negotiations with North African Countries for Potential Power Plant Sites	*115*
4.3.5	Specific Land-Area Requirements	*117*
4.3.6	Horizontal Salt Circuits	*120*
4.3.6.1	Costs	*120*
4.3.7	Dry Cooling	*121*
4.3.7.1	Literature References to Dry Cooling for Solar Power Plants	*123*
4.3.7.2	Literature References to Dry Cooling for Conventional Power Plants	*125*
4.3.8	Technical Reliability	*126*
4.3.9	Power Transmission via Overhead Power Lines	*127*
4.4	Calculating the Power Costs	*129*
4.4.1	Capital Costs, Nominal or Real Interest, Operating Lifetimes	*130*
4.4.1.1	Note on the Technical Operating Lifetime	*132*
4.4.2	Interest Rates	*133*
4.4.3	Equity Capital and Outside Capital	*134*
4.4.3.1	Conclusions	*138*
5	**The Potential of Solar Thermal Power Plants for the Energy Supply: Capacity Factor, Availability of Solar Energy, and Land Availability**	***141***
5.1	Overview	*141*
5.2	Spain: Capacity Utilization and Insolation	*147*
5.3	The USA	*152*
5.4	Solar Tower Plants – Permissible Slope of the Terrain	*155*
5.5	Spain: Availability of Sites	*157*
5.6	Morocco/Sahara	*160*
5.7	China, India, and Potential Sites in Tibet – Inaccuracy of the Available Maps	*164*
5.7.1	Conclusions	*169*
5.8	Insufficient Accuracy of the Insolation Data; Measurement Program	*171*
6	**Heliostats**	***181***
6.1	Estimating the Heliostat Costs	*181*
6.1.1	Examples	*182*
6.1.2	Preliminary Conclusions	*184*

6.2	Necessary Measures for the Precise Determination of Costs in Mass Production *185*
6.3	Stretched-Membrane Heliostats *186*
6.3.1	Technology *186*
6.3.2	Development Aspects *191*
6.4	Installations for Operational Testing of the Heliostats *194*
6.5	Comparison of the Cost Assumptions with Those of Other Studies *196*
6.5.1	Heliostat Costs in the S&L Study *196*
6.5.2	The Sandia Heliostat Study *197*
7	**Receivers** *209*
7.1	SOLAR TWO: Development Requirements for the "Advanced Receiver" *209*
7.1.1	Costs and Basic Technology *209*
7.1.1.1	Costs *209*
7.1.1.2	Design and Function *210*
7.1.1.3	Developmental Requirements *211*
7.1.2	System Development: Molten-Salt Circuits and Receivers *214*
7.1.2.1	Molten-Salt Circuits *215*
7.1.2.2	The Development of Hybrid Boilers *217*
7.1.2.3	A Test Installation for Receiver Development *217*
7.2	Air Receivers *218*
7.2.1	Technology *218*
7.2.2	Development *225*
7.2.2.1	Airflow Piping *225*
7.2.2.2	Heat Storage Systems *226*
7.2.2.3	Air-Recovery System *226*
7.2.2.4	Test Installation for Receiver Development *227*
8	**Parabolic-Trough Power Plants** *229*
8.1	Basic Facts *229*
8.2	Costs *232*
8.2.1	Preliminary Remarks *232*
8.2.2	Investment Costs *234*
8.2.3	Operating and Maintenance Costs *237*
8.2.4	Power Costs *238*
8.3	Development Program and Cost Estimates for Mass Production *240*
8.3.1	Test Plants *241*
8.4	Heat-Storage Systems for Parabolic-Trough Power Plants *241*
8.4.1	Preliminary Remarks *241*
8.4.2	Molten-Salt Heat-Storage System *243*
8.4.3	Heat-Storage Systems Based on Concrete *246*
8.4.4	Test Facilities for Solid and Thermocline Heat-Storage Systems *248*

9	**Solar Updraft Power Plants** *251*	
9.1	Introductory Remarks *251*	
9.2	The Principle *252*	
9.3	Investment and Power Costs *256*	
9.4	Development Program *259*	
9.4.1	The Development of Components *259*	
9.4.1.1	The Chimney *260*	
9.4.1.2	Heat Storage *261*	
9.4.2	A Demonstration Plant *262*	
9.4.3	Detailed Cost Estimates *263*	
9.4.4	Development Costs *264*	

10	**Fossil-Fuel Power Plants** *265*	
10.1	Natural Gas Plants *266*	
10.1.1	Investment Costs *266*	
10.1.2	Gas Costs *267*	
10.1.3	Operating and Maintenance Costs *268*	
10.2	Conventional Coal-Fired Plants *269*	
10.2.1	Investment Costs *269*	
10.2.2	The Price of Coal *270*	
10.2.3	Plant Efficiencies/Contribution of Coal Price to Power Costs *272*	
10.2.4	Operating and Maintenance Costs *275*	
10.3	Coal-Fired Plants with CO_2 Sequestration *275*	
10.3.1	Cost Estimates According to EIA AEO 2007 (Without Storage Costs): The Cost of Power *276*	
10.3.2	The Cost of Storing the Separated CO_2 (Including CO_2 Transport) *277*	
10.3.2.1	Storage on Land *277*	
10.3.2.2	The Cost of CO_2 Storage at Sea *277*	

11	**Other Technologies for Backup Power Generation and Alternatives for Future Energy Supplies** *281*
11.1	Generating Backup Power Without Natural Gas and Coal-Fired Power Plants *281*
11.1.1	Overview *281*
11.1.2	Gas from Coal Gasification for Backup Power Plants *283*
11.1.3	Smaller Coal-Fired Installations in the Solar Plants – Solar-Coal Hybrid Power Plants *284*
11.1.4	The Combination of Solar Thermal and Offshore Wind Plants – Offshore Wind Power as a Conditional Alternative to Solar Energy for Europe *290*
11.2	Coal Gasification as a Gas Source for Backup Power Plants and as an Important Component of the Future Energy Supply *292*
11.2.1	Gasification versus Direct Power Generation Using Coal – Solar Energy for Coal Replacement in Power Generation and for Hydrogen Production *292*

XII Contents

11.2.2 The Cost of Coal Gasification (for H_2 Production) 293
11.2.2.1 Conventional and Advanced Gasification 297
11.2.2.2 Operation and Maintenance (O&M) Costs 298
11.2.3 The Assumed Cost of CO_2 Storage 298
11.2.4 Syngas as a Particularly Inexpensive Substitute 300
11.2.5 Backup Power Plants as Consumers of Gas – Gas Transport and Storage Costs 302
11.2.6 Backup Power Plants: Switching to Other Fuels When Gas is in Great Demand – Development of Combustion Chambers 304
11.2.7 Development of "Advanced Technology" with a View to a General Gas Supply and IGCC Power Plants – Barriers to Development 305
11.2.7.1 Gas Purification and Separation 305
11.2.7.2 Advanced Technology for IGCC Power Plants 307
11.2.7.3 Development of Gasification Facilities – The Higher Efficiency of the Shell Process 307
11.2.8 Preconditions for the Substitution of Natural Gas by H_2 or Syngas: Modification of the End-User Appliances and the Transport Networks 310
11.2.9 The Possible Extent of Coal Gasification Using Substitutable Power-Plant Coal 312
11.2.9.1 Gas Quantities Made Available by the Substitution of Current Coal-Fired and Gas Power Plants 315
11.2.9.2 Limitations of the Natural-Gas Reserves in the USA 318
11.3 Coal as the Only Major Alternative to Oil and Gas? – The Scope of the Coal Resources for Power Generation and Gasification on a Large Scale – the Potential for Sequestration of CO_2 318
11.3.1 Coal Reserves 319
11.3.2 The Future Consumption of Coal – Depletion Time of Resources 324
11.3.3 The Potentially Limited Capacity for Economical Storage of CO_2 329
11.4 Solar Hydrogen 332
11.4.1 Hydrogen Production from Electrolysis 332
11.4.2 Transporting Hydrogen 341
11.4.3 Sun Methanol for Around 90 $/Barrel Oil Equivalent – An Effective Brake on the Oil Price. The USA as a Future Sun-Coal-Fuel World Power. "OPIC" as the Answer to OPEC 344
11.4.3.1 Costs 346
11.4.3.2 Coal Consumption 352
11.4.3.3 A Price Brake on Petroleum – The Potential of Sun Methanol in the USA 355
11.4.3.4 Liquid-Fuel Production from Coal Alone? – Sun Methanol to Conserve US Coal Reserves 357
11.4.3.5 Methanol Production using Nuclear Hydrogen 358
11.4.3.6 OPIC 359
11.4.3.7 The CO_2 Balance 360

11.4.4	Hydrogen and Coal for Liquid Energy Carriers in a Future Solar-Hydrogen Energy System *362*	
12	**The Large-Scale Use of Nuclear Energy** *367*	
12.1	The Costs of Nuclear Power – Results *367*	
12.2	Investment Costs under Mass Production *367*	
12.2.1	Estimates According to the "Chicago Study" *367*	
12.2.1.1	Conclusions from Table 12.1 *376*	
12.2.2	A Problem: The Lack of Competition among System Manufacturers – The Contrast to Solar Energy *381*	
12.3	Operation and Maintenance Costs; Fuel Costs *383*	
12.3.1	Operation and Maintenance (O&M) Costs *383*	
12.3.2	Enrichment and Other Fuel Costs, Not Including the Cost of Natural Uranium *384*	
12.4	Consumption and Cost of Natural Uranium per kWh_{el} *386*	
12.5	The Problems Associated with Nuclear Energy *387*	
12.5.1	Consequences of the Development of Centrifuge Technology *387*	
12.5.2	General Problems of Nuclear Power Generation *387*	
12.6	Uranium Reserves *390*	
12.6.1	Lifetime of the Reserves in the Case of a Massive Increase in Nuclear Power Production *390*	
12.6.1.1	Lifetime *390*	
12.6.1.2	Classification of Ores According to Their Uranium Content *394*	
12.6.1.3	Unconventional Uranium Reserves *395*	
12.6.1.4	Thorium Reserves *397*	
12.6.2	The Present and Future Price of Uranium – Geographical Distribution of the Uranium Reserves *398*	
	Appendix A	**Solar Tower Power Plants: Comparison of Kolb (1996), Kalb/Vogel, SunLab, S&L** *403*
	Appendix B	**Inflation, Purchasing Power Parities** *439*
	Appendix C	**Energy Statistics** *443*
	Appendix D	**Comments on the Earlier Study (Kalb and Vogel 1986a)** *455*
	References	*461*

Preface

This book is based on the work done by the authors that began in the mid-1970s on the topics of solar thermal power plants, long-distance power transmission, and solar hydrogen – in particular, on a study for the European Association for Renewable Energy (Eurosolar) carried out in 1996–1998. In the face of considerable resistance from the scientific and especially from the political communities, the authors have attempted since the early 1980s to introduce the concept – of importing solar-thermally generated electrical energy from sunny regions – into the public debate on energy. Over many years, this concept, in spite of support from well-known public figures, including Professor Werner Buckel (former president of the German Physical Society) and Hermann Scheer (president of Eurosolar, Member of the German Parliament), has been almost completely ignored. A gradual change in the political perception (initially in Germany) began to make itself felt when in the year 1995, the German Physical Society took up the topic of "Solar thermal power plants and imported electric power" in an energy memorandum and continued to publicize it with increasing emphasis in the following years. At the beginning of the new millennium, the German Federal Ministry for the Environment then recognized the significance of solar thermal power plants and also of the concept of importing solar thermal power; in the year 2003, this concept was taken up by the Club of Rome in cooperation with the German Aerospace Center (DLR) (under the new appellations "TREC" or "Desertec"). This then opened the way to a broader political acceptance worldwide and also to the initiation of concrete projects. In this phase, it seemed expedient to us to describe the whole topic of solar thermal power plants and a future world energy supply based largely on them in a comprehensive and thorough manner. In particular, it is our aim to present to a broad spectrum of readers the enormous but still underestimated potential of solar thermal power generation for the general energy supply, as well as the developments required to make this vision a reality.

Large-Scale Solar Thermal Power. Werner Vogel and Henry Kalb
© 2010 WILEY-VCH Verlag GmbH & Co. KGaA, Weinheim
ISBN: 978-3-527-40515-2

Preliminary Remarks and Summary

The Significance of the Rapid Deployment of Solar Thermal Power Plants for Energy Policy

Solar thermal power plants have been barely considered by a wider public until a few years ago. This is all the more surprising since they not only offer the promise of relatively low power costs (under mass production), but also have a notable advantage over other large-scale energy technologies: owing to their rather simple structure, consisting of conventional, straightforward components such as mirrors, systems of piping, insulated containers, and steam power plant blocks, they could be produced in large numbers within a fairly short time. If necessary, available production capacities from many branches of industry could be utilized for the fabrication of the individual components. After the completion of the required further development program, which if carried out rapidly could be finished within around 4 years, the replacement of today's coal- and natural gas-fired power plants for the base-load power supply could be started. With a "crash program" (maximum speed with strong time pressure), this would take 10–15 years. If the necessary preconditions for such a rapid implementation were met, the whole "energy turnaround" including the development program could be completed within ca. 15–20 years.

The energy carriers which would then be freed up and thus far have been used in fossil fuel power plants (in particular coal) could then make an important contribution to the substitution of the present imported energy carriers outside the electrical power generating sector. Coal can be converted into fuel gas at a relatively low cost. Worldwide, including the USA, the amounts of coal burned in power plants, and thus the potential amounts of gas which could be produced, are enormous. Solar energy would thus make an *indirect* contribution to the substitution of oil and natural gas. Such a reasonably priced alternative to natural gas is significant both for Europe, with its high proportion of imported gas, as well as for the USA, where the gas reserves are limited. Since the supply of gas would then be increased, oil could be substituted as well.

Solar hydrogen and gas from coal gasification could furthermore together form the basis for the large-scale manufacture of liquid fuel ("sun methanol"). In view

Large-Scale Solar Thermal Power. Werner Vogel and Henry Kalb
© 2010 WILEY-VCH Verlag GmbH & Co. KGaA, Weinheim
ISBN: 978-3-527-40515-2

of the nearly unlimited production potential for such a substitute fuel in the USA in terms of the solar regions (hydrogen) and a sufficient supply of coal, this fuel could become a major *direct* alternative to petroleum. With a successful development of solar technology, sun methanol made with US coal should cost about 90 $/barrel of oil equivalent. Given the enormous potential capacities – in principle, the world's oil consumption could be supplied from the USA alone – in the medium term the price of crude oil could even be limited to parity with the cost of this fuel (a "price brake" for crude oil).

Next, we give a brief explanation of the following aspects:

1. Costs

Solar thermal power plants offer favorable conditions for economical power generation. Using heat-storage systems, they can deliver power 24h per day. Transmission of the power over a distance of, for example, 3000 km is possible with only minor losses (11.5%) using present-day modern transmission technology (800 kV direct current, HVDC). The power plants could thus be located in regions with a high and uniform insolation, for example, in Spain, North Africa, or in the southwest USA (providing power to the East Coast). As backup power plants, the substituted natural gas and coal-fired plants, or also new, relatively low-cost gas plants would be available. They would perform the task of bridging over gaps in the solar power supply due to weather conditions. In Morocco and the USA, this would correspond to about 20% of the overall power generated; in Spain, it would make up 25–30%.

As we describe in detail in this book, based on current knowledge, solar power from mass-produced plants would not be much more expensive than the present-day power, which is generated mainly in fossil fuel plants: about the same as from natural-gas CCGT plants at today's gas prices. For the power supply of Europe from Spain, the cost including backup power (from new gas-fired plants) would be about 5.2 ¢/kWh, and in the USA, it would be around 4.7 ¢/kWh (all prices quoted in US cents at the monetary value of the year 2002). This can be compared with the cost of energy from natural-gas CCGT base-load plants (4.8 ¢/kWh) or from newly constructed nuclear plants (3.1 ¢/kWh). The latter value would decrease in the case of large-scale deployment, possibly to as little as 2.4 ¢/kWh (without including the societal costs). The increased costs to the national economies for solar energy as compared to nuclear energy would be readily tolerable, even with a very large-scale deployment of solar plants, as we shall show.

2. The time required – The need for a special development program

Solar power plants, due to their extremely simple technology, can not only be produced rapidly and in large numbers, but also, for the same reason, they can be quickly developed and optimized. It must, however, be considered that not just a *single* type of solar power plant, but rather several families of plant types, and within them, a multiplicity of technological branches will have to be developed. For it is not yet clear which variant will achieve the lowest costs under mass production. However, nearly all the individual technologies represent relatively simple development tasks. Insofar as all the different branches are developed in *parallel*, the time required will not be increased. In each case, the economic potential under mass production must be explored; within the overall development program, this represents a special task for each case.

Thus, although the individual development problems are simple as a rule, the large number and wide variety of process steps make a broadly conceived and tightly enforced development program essential, if we wish to reach our goal as quickly as possible. This, in turn, presupposes a suitable organizational structure, which is adapted to these particular goals for the planning and execution of the program. Thus, it must be guaranteed that each new problem that arises, in whatever technical field it may lie, can be countered by a rapid and flexible reaction within the development program.

As we shall show in the discussion of the individual technologies, the greatest portion of the development tasks could be accomplished within about 4 years, insofar as the organizational preconditions are met. This will require not only an efficient organization but also an unhampered access to the necessary resources. The rapid development of solar energy thus requires a similarly structured and optimized approach ("crash program") as, for example, the Apollo space program, although with a much more modest financial effort.

Some of the developmental tasks will require more time. With a correspondingly intensive program, the last of these should be completed within around 8 years. Thus, if rising (or even exploding) oil prices force us to act as quickly as possible, for example, already after 6 years (assuming completion of the main phase of the development after 4 years), the mass production of solar plants could be started. In the case of a few particular components, one would then begin with suboptimal versions and would allow further improvements in the course of development to enter successively into the ongoing production process; for solar tower plants, this applies, for example, to receivers optimized for maximum efficiency. A comparable program for nuclear energy would require at least twice this time for completion: in the case of this complex and *security-relevant* technology, a rapid increase in the production capacity would be incomparably more difficult, especially in terms of obtaining the necessary highly qualified personnel capacities. This aspect has always been emphasized in connection with the nuclear energy debate. Even with the greatest possible haste, it would presumably require 30–40 years for the completion of a full conversion to nuclear power.

While the USA has a practically unlimited potential for solar energy at its disposal, the resources in Europe are more scant. Precisely for Europe, however, a conceivable combination with other renewable energy sources should be considered. Most probably, offshore wind energy, for which there are favorable conditions and a great potential in the North Sea, would also lead to low-cost power. This renewable energy source could also be relatively quickly developed and deployed if necessary. It has, however, the disadvantage that power generation is less uniform so that by itself, it does not represent an alternative to solar energy for generating base-load power. In combination, solar and wind energy would complement each other in terms of seasonal variations. If the expected costs for wind energy prove correct, Europe's future power supply could be mainly based on renewable sources, consisting of one-third solar power from Spain, one-third solar power from Morocco, and one-third offshore wind power. (Two-thirds of the power would then be generated on European territory.) If one wished to replace today's total base-load power consumption (EU-25), then in Spain, a solar power plant capacity of ca. $100\,GW_{el}$ would have to be deployed. If necessary, that is, accepting somewhat less favorable plant sites, an area for up to four times this capacity should in any case be available in southern Spain. Without wind energy, thus presuming half of the solar capacity to be in Spain and half in Morocco, each location would require a generating capacity of $150\,GW_{el}$.

An even more rapid reaction to increasing energy prices is possible only through energy-saving measures. A combination of energy conservation with solar plants, which would be rapidly available on the energy-economic timescale (possibly combined with wind energy), thus probably represents the quickest path to a restructuring of the energy supply.

3. The importance of substituted power plant coal for the supply of gas

Coal can be converted into gas (syngas) at a very favorable price using conventional technology, insofar as one dispenses with CO_2 sequestration, that is, the separation and deposition of CO_2. However, also the production of pure hydrogen from coal – including sequestration – would appear to be possible in the future at a relatively favorable price. According to the report of the US National Energy Technology Laboratory, on which the major American hydrogen study of 2004 is also based, hydrogen from cheap American coal should cost around 2.5 ¢/kWh, and in Europe, from imported coal, 3.4 ¢/kWh; this, however, refers to a future advanced technology. Syngas made with current technology would cost ca. 2 ¢/kWh in the USA, and in Europe, from imported coal, ca. 2.8 ¢/kWh; utilizing low-cost German lignite, it would cost only about 2 ¢/kWh, as in the USA. This gas would thus be cheaper in the USA and in Germany (from lignite) than the natural gas as a fuel for power plants (USA, Germany: 2.5 ¢/kWh in 2007); its cost corresponds to that of imported natural gas in Europe (2.0 ¢/kWh in 2007).

Both syngas as well as hydrogen are thus considerably cheaper than the cost of oil in the year 2008. If we take, for example, 100 $/barrel as a benchmark for the future increased crude oil price (in 2008-$), this corresponds (in 2002-$, the reference year used in this book) to 5.3 ¢/kWh. Given the low cost of gas from coal, the latter thus appears to offer an important alternative to oil and natural gas; indeed either using syngas as produced today, or in the future (with improved technology) using "CO_2-free" hydrogen. The barriers to development of this advanced technology will, however, not be negligible. This applies to the same extent also to coal power plants with integrated gasification and CO_2 sequestration, which are under lively discussion at present. (Syngas, the raw (desulfurized) product gas of coal gasification – a mixture of CO and H_2 – has similar technical characteristics to those of pure hydrogen. Both gases differ in these characteristics from natural gas, and therefore require conversion of consumer appliances, in particular of gas burners and meters.)

With the quantities of coal which will be consumed in power plants in the coming years if the current power-generating strategy utilizing coal-fired plants is continued (including the expected worldwide increased power consumption from coal-fired plants by the year 2030), and taking into account the amounts of coal required for backup power generation in the case of a theoretically complete and worldwide substitution of the coal-fired base-load power plants, a quantity of substitute gas equivalent to 3000 GWa could be produced. This is more than the total gas consumption at present (2900 GWa), and that is 60% of today's worldwide petroleum consumption. Starting from the *current* coal consumption, that is, without considering the expected strong increase in coal consumption for power generation, and including the natural gas from substituted gas-fired power plants, ca. 1200 GWa of gas could be produced or replaced; this corresponds to 55% of the natural gas consumption at present outside power plants, or 25% of the petroleum consumed. In the USA, it would correspond to ca. 65% of today's gas consumption (outside power plants) or ca. 30% of the petroleum consumed. In Europe, comparatively little coal is employed for electric power generation; the substituted quantities would, therefore, be smaller. The gas that could be produced from coal corresponds there to ca. 40% of the current natural gas consumption (outside power plants) or ca. 20% of the petroleum consumed. Worldwide, but also in the USA, the replaceable coal thus represents significant quantities in terms of the energy economy. This also holds for the corresponding contribution to the CO_2 emissions. Even without separation of the CO_2 (syngas), the CO_2 emissions would be reduced by the substitution of oil and natural gas; utilizing hydrogen (with CO_2 sequestration), they would be completely avoided.

Petroleum thus far represents the only alternative to imported natural gas. For this reason, the gas price is currently tied to the oil price. With the use of coal gas, however, the possibility would open up of producing a replacement gas in very large quantities. In price negotiations with gas exporters, this replacement gas would then represent a significant competitor to natural gas. The new gas would then define the upper limit for possible price demands. This presumes, as stated, that coal thus far used in power plants be substituted; only then could the new gas be manufactured in large quantities without a massive increase in coal production (and thereby even in the case of syngas without a major increase in the CO_2 emissions).

In principle, coal gasification *without* substitution of the coal used in power plants can be imagined. This, however, would require an increase in coal production. Such an expansion of coal mining would probably require a similar time as the deployment of solar energy plants. And this strategy would not be acceptable for the future, given the high coal consumption that it would entail. Because of the CO_2 problem, it would also force sequestration of the resulting CO_2, a technology which likewise is still to be developed. The power costs from CO_2-free coal-fired plants cannot be expected from today's standpoint to be lower (not even with future "advanced technology") than the cost of solar power. There is thus no economic motive for the deployment of such power plants. In terms of the price of coal, this strategy would probably also be risky: in view of the expected worldwide increase in power consumption (by 2030) and – without substitution of the coal-fired power plants – of the corresponding increase in coal demand, the coal price would certainly rise, leading to high costs in particular for coal-importing countries. Furthermore, it must be considered that the development of CO_2-free coal-fired power plants ("future technology") may require considerable time, even with an intensified development program, in any case longer than the development time for the technically simpler solar power plants.

If natural gas and coal-fired power plants are to be substituted, the only alternatives for the generation of base-load electric power on a large scale are solar and nuclear energy (with a smaller contribution from wind energy). Since nuclear technology offers no comparable possibility of large-scale, rapid deployment, a rapid response program must focus on solar thermal power plants.

4. Sun methanol production in the USA

Sun methanol is manufactured from equal parts (in terms of energy content) of solar hydrogen and coal gas. For the production of hydrogen using solar power, we assume here that an efficient high-temperature electrolysis process will be developed. With the cost of solar power quoted earlier and the resulting hydrogen price, and assuming the price of coal gas from current technology, we find for the USA a methanol cost of ca. 90 \$/barrel of oil equivalent (2008-\$); without the new electrolysis process, the cost would be around 100 \$/barrel.

Sun methanol from the USA could replace the entire world consumption of oil. Once the technical preconditions for its production were fulfilled and a rapid build-up of the manufacturing capacity thus would be possible, one would no longer accept crude-oil prices higher than those of methanol. Most probably, the establishment of a relatively "limited" manufacturing capacity corresponding to ca. 10–20% of the world's petroleum production would suffice to provide an effective limit to the oil price.

The decisive point for price negotiations is the ability to make a believable threat of strongly increasing methanol production capacity should the negotiations fail to yield an agreement. The coal reserves represent a certain limitation to such a potential expansion of methanol production capacity. The USA has indeed large, but not unlimited reserves of low-cost coal. Without solar hydrogen, methanol production would require twice the amount of coal. (Even in this case, production with nearly no CO_2 emissions would be possible if the CO_2 produced were to be sequestered.) With a methanol production from coal alone corresponding to the worldwide annual consumption of oil, the low-cost coal reserves in the USA would be exhausted after only 20 years, so that the threat of oil substitution by the USA itself would not be convincing. In the case of sun methanol (utilizing solar hydrogen), the lifetime of the coal reserves would be 40 years. In terms of a methanol production rate (still very large) corresponding to only 40% of the world's petroleum consumption, the lifetime of the US coal reserves would be 50 years without solar hydrogen, and 100 years with solar hydrogen.

With a view to a conceivable methanol production using nuclear energy, we must consider the situation regarding uranium reserves – along with the fundamental questions raised by a large-scale expansion of nuclear power generation. In the case of a massive application of nuclear energy, electric power generation would have the first priority. To supply the future worldwide power demand, the uranium reserves, allowing for the acceptance of correspondingly higher uranium mining and extraction costs (given our present knowledge of low-cost reserves and a speculative estimate of those with higher extraction costs) would probably suffice for only *one* generation of power plants. The production of an additional, comparable amount of power for hydrogen manufacture would probably not be possible without resorting to very expensive uranium reserves. Their extraction would furthermore be accompanied by still greater environmental damage owing to the need to mine ever more ore with lower and lower uranium content. Just to meet future demands for electric power (and even utilizing future uranium-conserving technologies), annually ca. six times more uranium would be required than at present.

Near the end of the year 2008, the oil price again decreased. Since then, new hope has sprung up that the energy prices will remain at a moderate level for a certain time. That, however, does nothing to relieve the necessity of rapidly developing new energy systems. On the contrary, this renewed price decrease offers a chance to prepare for the future "emergency situation," which will occur sooner or later. This means not only the full technical development of solar power plants and the additional required technologies, but also – even though initially on a small scale – the substitution of gas and oil by the new energy carriers to provide a practical demonstration of this alternative.

In this book, all the variants of solar thermal technology are described. The main emphasis is, however, on the cost considerations relating to mass production, applied in particular to solar tower power plants, and to a lesser extent also to parabolic-trough and chimney plants. For each topic, still open questions and concrete research approaches are discussed.

With a view to the cost differences relative to other conceivable energy supply routes, we also treat the new CO_2-free coal-fired power plants as well as modern nuclear plants; in the case of the latter, in particular we discuss the costs to be expected under mass production. And to complete the discussion, we summarize the situation concerning uranium reserves. This book also contains information on coal gasification and methanol production. In the appendix, among other things, the relevant energy-statistical data (for the world, the USA, Europe, and Germany) are presented in a clear form. This book thus intends not only to provide the necessary knowledge for a comprehensive estimate of the economic outlook for solar thermal power plants and the related concrete developmental requirements and possible courses of action but also it provides the information needed to rank this new energy technology within the greater energy-political context. Thus, along with the specialized topics related to solar energy, the general question of the fastest possible conversion from oil to other, more secure future energy sources (and the associated costs) as a whole is discussed. In this connection, the most important elements of the necessary development program are described. The book thus sketches an overall plan for a rapid turnaround of the energy supply, beginning as soon as possible, and is therefore directed not only at readers interested in solar energy, but also at all those who are asking themselves what options are available in view of increasing oil prices and in the face of the increasingly pressing questions of environmental protection and climate change.

Acknowledgments

We owe sincere thanks to the many people who made our work on this book possible and who supported it.

Firstly, we wish to express our heartfelt thanks to the members of the private sponsor group which – beginning in the 1990s, when the topic of solar thermal power plants appeared to be practically "dead" – nevertheless generously supported our work.

Our special thanks for the support of this book project, in particular, go to the Ludwig Bölkow Foundation (Munich), and especially to Dr. Walter Kroy.

We also wish to thank Prof. Jörg Schlaich, Wolfgang Schiel, Dr. Gerhard Weinrebe (of the engineering company Schlaich, Bergermann und Partner), Dr. Reiner Buck (German Aerospace Center), and Prof. Bernhard Hoffschmidt (Solar Institute Jülich) for valuable information and discussions.

We owe sincere thanks to Wiley-VCH for their genial co-operation, and in particular we thank Mr. Ron Schulz-Rheinländer and Ms. Ulrike Werner, who made an important contribution to the completion of the manuscript through their friendly, helpful, and (in the face of many delays) patient disposition.

Finally, our special and heartfelt thanks go to Prof. William D. Brewer (Physics Department, Freie Universität Berlin), who took on the difficult task of translating the German manuscript into English. We consider it to be a stroke of fortune that the publishers were able to convince him to accept this assignment.

List of Tables

Table 2.1	Investment costs for solar tower plants with a molten-salt circuit (2002-$).	13
Table 2.2	Investments for the overall power plant system (solar power system).	13
Table 2.3	Power costs of the solar power system based on solar tower power plants, compared to gas-fired, coal-fired, and nuclear plants.	14
Table 2.4	Comparison of the electric energy costs from solar and nuclear power plants.	18
Table 4.1	Investment costs of the solar power system with natural gas (CCGT) or with coal-fired backup power plants.	72
Table 4.2	Energy cost from the solar power system with CCGT or coal-fired backup power plants; cost of solar power at the plant and of solar hydrogen.	75
Table 4.3	Energy cost from conventional power plants, cost of nuclear hydrogen, of gas from coal gasification and of crude oil at 100 $/barrel.	78
Table 4.4	The cost of solar power from Spain (including power transmission) and the cost of backup power from gas or coal-fired plants.	91
Table 4.5	The cost of solar power from Morocco or in the USA (including power transmission) and the cost of backup power from gas or coal-fired plants.	92
Table 4.6	Comparison of various cost studies – investment costs per 1000 MW of installed solar power generating capacity for locations in Morocco or in the USA (a more detailed comparison is given in Appendix A).	95
Table 4.7	Operation and maintenance (O&M) costs of solar tower power plants.	100
Table 4.8	Energy costs for the solar power system with natural-gas backup power plants (cf. Appendix A).	102

Large-Scale Solar Thermal Power. Werner Vogel and Henry Kalb
© 2010 WILEY-VCH Verlag GmbH & Co. KGaA, Weinheim
ISBN: 978-3-527-40515-2

Table 4.9	Annuity factors: constant annual rates (in % of the invested capital) for principal repayment and interest payments on the capital.	131
Table 5.1	Solar irradiation in Almeria (Spain): The direct normal irradiation (PSA 1997, 2008).	150
Table 5.2	Calculated values (Global Horizontal Insolation (GHI)) from NREL (2005) and DLR (Schillings et al., 2004a) in comparison to individual measured values in Tibet.	170
Table 6.1	Early heliostat cost prognoses (Sandia) for a production rate of 50 000 heliostats/a.	182
Table 6.2	Heliostat costs for different production rates (with today's technology, prices in 2006-$).	199
Table 6.3	Cost components for a conventional ATS heliostat (compare Kolb et al., 2007a, Table 3-14).	199
Table 6.4	Costs of the motion drive components for a heliostat (compare Kolb et al., 2007a, Table 3–13).	200
Table 6.5	Cost reduction as a result of mass production – the importance of the costs of the mechanical drive system (compare Kolb et al., 2007a, Tables 3-14 and 3-15).	200
Table 6.6	Cost comparison of conventional and stretched-membrane heliostats at a production rate of 50 000 heliostats/a (cf. Kolb et al., 2007a, Table 3-22).	202
Table 6.7	The cost of the mirror modules for stretched-membrane heliostats (cf. Kolb et al., 2007a, Table 3-22).	202
Table 6.8	The cost of the mirror modules for conventional heliostats (cf. Kolb et al., 2007a, Table 3-11).	202
Table 8.1	Technical data on which the "SunLab long-term" case is based (S&L, 2003, p. 4-3).	233
Table 8.2	Elements that contribute to the annual efficiency in the "SunLab long-term" case (S&L, 2003, p. 4-5).	233
Table 8.3	Investment costs (rounded values) for parabolic-trough power plants.	234
Table 8.4	Investment costs for the complete solar power system.	236
Table 8.5	Energy cost from the solar power system using parabolic-trough plants.	238
Table 9.1	The specific investment costs for updraft power plants, referred to an output power of 1000 MW, corresponding to 10 plants of 100 MW each.	258
Table 9.2	Investment costs for the complete solar power system (for a site in Morocco, transmission distance 3000 km, with transmission losses of 11.5%).	258
Table 9.3	The energy cost from the solar power system based on chimney power plants and CCGT backup plants.	258

Table 9.4	Estimated development costs for the updraft power plant.	264
Table 10.1	Coal prices in recent years.	271
Table 10.2	Coal-fired power plants.	273
Table 10.3	Cost of CO_2 transport and injection into the deep ocean according to WBGU (2006).	278
Table 11.1	Energy costs from the solar power system with coal-fired ancillary boilers and CCGT backup power plants.	287
Table 11.2	The cost of coal gasification (H_2 production); according to Stiegel and Ramezan (2006), but here as a real cost calculation and demonstration of the macroeconomic costs (without taxes, insurance. etc.).	294
Table 11.3	Operating and maintenance costs of coal gasification.	299
Table 11.4	Electric power generated (worldwide) in the year 2004, and predictions up to 2030 from (EIA – International Energy Outlook, 2007), along with our assumptions regarding possible future coal-fired power generation.	314
Table 11.5	Quantities of gas which could be produced from substitutable coal and substitutable gas from power plants (at today's power consumption).	316
Table 11.6	Coal reserves as of 2006 (BGR, 2007).	321
Table 12.1	Investment and energy costs of future nuclear power plants for large deployment series and for power plant pools. In the latter case, power transmission and backup plants are included.	368
Table 12.2	Nuclear power plants: specific investment costs for a certain number of reactors constructed by a particular manufacturer (according to the Chicago Study, 2004).	374
Table 12.3	Operating and maintenance costs for nuclear power plants according to the World Nuclear Association (WNA Report, 2005).	383
Table 12.4	Nuclear fuel costs without the cost of natural uranium (WNA Report, 2005, p. 11).	384
Table 12.5	Uranium reserves, ordered by production cost (source: Framatome, 2005).	391
Table 12.6	Uranium resources by country.	399
Table A.1	Basic data.	404
Table A.2	Efficiencies.	408
Table A.3	Investment costs: 200 MW, 220 MW.	411
Table A.4	Investment costs: 1000 MW.	417
Table A.5	Investment costs: solar power system (1000 MW).	423
Table A.6	Capital costs, gas costs.	428
Table A.7	O&M costs, electric power costs.	432
Table B.1	Inflation in the USA and Germany (OECD, 2008).	439
Table B.2	Inflation (USA, Germany): Price increases 1995–2002.	440

Table B.3	Purchasing power parities and exchange rates.	441
Table C.1	Energy consumption per year.	446
Table C.2	Electricity generation per year.	448
Table C.3	Power plant capacity.	450
Table C.4	CO_2 emissions per year.	452
Table C.5	US CO_2 emissions per year.	452
Table C.6	Emissions of greenhouse gases per year.	453

1
Introduction

1.1
Historical Background

In writing a book about solar power plants, one is initially tempted to start with a summary of their historical evolution. Even though knowledge of the first steps in the historical development is not always necessary to understand the current state of the art, it is nevertheless often instructive. Thus, although the historical development toward solar thermal power plants cannot offer much help in this connection, in particular in terms of finding a solution to today's developmental problems, it is still rather fascinating to learn about the first ideas and steps toward their implementation, which were taken by creative and truly far-sighted innovators in the past. In contrast to the visionary accomplishments of these early pioneers, today's suggestions and concepts for the application of solar energy often appear banal.

At the same time, it becomes clear that ideas can be implemented only when their whole technological framework is at a sufficiently advanced stage of development, and – incidentally – we see here again that war is not only the "father of all things,"[1] but it can also be the "death of very many things" (cf. the early solar power plant in Egypt, below).

However, in this book we cannot treat the historical aspects in detail, but instead we refer to the corresponding literature, for example, a comprehensive description of solar energy use in general by Butti and Perlin (1980), and a more specialized treatment of the subject of the technological maturity of solar power systems 100 years ago by Mener (1998). Smith (1995) offers in addition a very good account of the early technical development of solar thermal machines, which took place over a period of more or less 60 years (1850–1910) in the 19th century. This historical period of development ended in the year 1914 with the closure – and, according to Smith, later during the First World War the destruction – of a (for that time) "large"

---

1) This refers to a legend according to which in ancient times, the heliostat principle, that is, the solar power plant, was invented as a way for setting fire to enemy warships (Archimedes).

Large-Scale Solar Thermal Power. Werner Vogel and Henry Kalb
© 2010 WILEY-VCH Verlag GmbH & Co. KGaA, Weinheim
ISBN: 978-3-527-40515-2

solar power plant in Egypt.[2] A curtain closed on the technology, and it was raised again only nearly a half century later.

In this book, we also do not intend to describe systematically the developments of the past 30 years, with all their ups and downs, although this development, in particular in terms of its many missed chances for rapid progress, is hinted at over and over in the text. Instead, we emphasize the potential importance of solar thermal power plants in connection with today's energy problems. Humanity must solve these problems within the coming 30 years, and for their solution, the necessary strategic decisions must be taken already *today* (or as soon as possible).

Concentrating and Nonconcentrating Solar Thermal Power Plants – the Advantages of Heat Storage There are two fundamentally different types of solar thermal power plants:

1) The direct solar irradiation is concentrated by systems of mirrors, and with the resulting high-temperature heat (the usable temperature range, depending on the technology applied, is 300–1200 °C), one generates electrical energy by means of heat engines and electrical generators. These (optically) concentrating solar–thermal systems are often subsumed under the term "concentrating solar power" (CSP). (Sometimes, however, the term CSP is also used as a generic name for concentrating thermal *and* concentrating photovoltaic systems.)

2) Under a large area, glass roof at a height of a few meters, with a diameter of several kilometers, the direct *and* the diffuse solar irradiation (i.e., the global insolation) is used to heat the ground and thereby indirectly the air. This air, whose temperature is raised by ca. 30 K, rises in a central tower of 1000–1300 m height due to the chimney effect, while cooler air from the surroundings flows in through the open sides of the structure. This air flow drives turbines at the base of the tower. One refers to an *updraft power plant* ("solar chimney").[3]

---

2) It consisted of several parabolic troughs, about 50 m long. The energy conversion proceeded as in all plants of the time only up to mechanical energy (steam engines were used to operate pumps for irrigation). The output power of this plant, which, like all of its predecessors was thus not really a power plant (i.e., for the production of electrical energy), consisted of some 40 kW.

3) The principle of the *updraft power plant* was suggested already in 1903 in Spain by Cabanyes (1903) (cf. Lorenzo, 2002) and later, in the 1930s among others in Germany by Günther (1931). In the 1920s, there was supposedly a detailed suggestion ("with blueprints") for constructing a kind of updraft power plant at a cliff in Algeria.

"The principle was already clear: instead of water flowing downwards into a valley, air heated by solar energy would flow upwards and would perform work in the process." (Eisenbeiß 1995). A corresponding US patent was granted in 1981 to R. Lucier. The solar chimney has been promoted very energetically in the past 30 years by the German structural engineer J. Schlaich (Stuttgart), who has in particular developed it to a state of readiness for construction. He has shown such a strong dedication to the project that it seems justified to call him the "father of the updraft power plant." See for example (Schlaich, 1995; Schlaich *et al.*, 2005; Weinrebe *et al.*, 2006).

Conversely, evaporative wet cooling at the top of a high tower can be used to produce cooler air there; it flows downward within the tower and likewise drives turbines at the base. In this case, one refers to a *downdraft tower* or often simply to an "energy tower".[4]

The use of solar energy as heat for generating electrical energy has the advantage in principle that *thermal energy storage* is possible and can be implemented at a relatively low cost. In contrast, for direct conversion using photovoltaic cells, chemical energy storage using batteries or hydrogen technology would be required. This, in turn, is very expensive, owing to the high investment costs and the high loss rate (20–50%). Storage is, however, a decisive precondition for a regenerative energy source to compete seriously with conventional energy production sources (fossil and nuclear power), which themselves represent chemical or nuclear–chemical storage media.

At this point, we wish to make the following point regarding energy storage:

Although energy storage is such an important aspect, it is surprising that this topic has been flagrantly neglected over the years by many people – even professionals. This criticism applies in the main to European solar-energy research and in particular to research policies in Germany, which for many years regarded thermal solar power plants exclusively as an export product for "niche markets" and treated the topic of energy storage as a "niche area" of solar-energy research. Even R. Pitz-Paal, director of the Solar Research Division of the DLR,[5] stated clearly in 2004 in relation to solar thermal power plants that "the development of storage systems has long been neglected in Europe" (Pitz-Paal, 2004). He is correct in mentioning Europe, but also the US Department of Energy has thus far given too little support to the development of energy storage systems (thermal reservoirs). One can only hope that a basic rethinking is about to occur, when recently (2007) in the US Congress, targeted financial support for the development of thermal storage reservoirs was a topic of discussion and the subject of a funding proposal which would provide funding for several years (Committee on Science and Technology, 2007).

---

4) In 1975, Carlson obtained a US patent for a *downdraft tower* (Carlson, 1975), and in the past several years, this scheme has been promoted in particular by Zaslavsky; cf. for example (Zaslavsky, 1999, 2008; Altmann et al., 2006; Czisch, 2005). This type of power plant, however, is possibly not a "real" base-load plant, owing to seasonal and daily variations in its output, although it can generate a certain amount of power even at night. An excellent comparison of the two concepts was published in 2001 by Weinrebe and Schiel (2001). Here, for the present, the arguments favor the solar chimney principle. However, one should compare the newer investigations of Altmann et al., according to whom the downdraft principle (in a preliminary investigation) shows a highly interesting economic potential (Altmann et al., 2005 and 2006). We will consider this power-plant principle again briefly in Chapter 3, it might prove to be extremely attractive in terms of power-generating cost, and it could also be developed relatively quickly so that its real economic value could be tested.

5) The German Aerospace Center Deutsches Zentrum für Luft- und Raumfahrt (DLR) is the national research institution that carries out the main portion of research and development in the area of solar thermal power plants in Germany.

1.2
Formulating the Problem

The emphasis on energy storage is related to the central question that arises in the discussion about future world electrical energy supplies:

How can we replace as much fossil energy as possible by solar energy?

Or – if one is skeptical with regard to nuclear power for reasons which will be discussed later –

How can we dispense with nuclear energy by making use of as much solar energy as possible – and the least amount of fossil energy?

For a simple model calculation (to permit an economic comparison), the more specialized question mainly discussed in this book is, for example, appropriate:

How can we replace fossil-fuel and/or nuclear base-load power plants to the greatest possible extent by solar power plants at an acceptable cost?

This is in any case a question which in terms of energy policy plays a much more important role than the often (more or less explicitly) discussed and very general question, "How and for what can we make use of solar energy?"

A problem-oriented strategy has to take into account, along with the principal goal of finding an economically feasible concept, also the time aspect (especially in today's situation of massively increasing energy costs). In the past few years, many considerations of the subject of solar power plants in particular – and thus also of energy and research policy – have started from the question:

How long will it take, presuming a (worldwide) annual development budget of – let us say – US $50 million, until at least *one* of the various solar thermal technologies is roughly competitive with nuclear power and coal/gas power plants?

The question which corresponds to the problem at hand can, however, be formulated as follows:

What effort would be required to ensure that within about 4–8 years, at least one of the various solar thermal technologies could be implemented at a cost which lies in the economically affordable range?

Questions such as the second one are posed when one wants to solve a serious and pressing problem.[6] Questions such as the first one are posed when no acute problem with time pressure is at hand. Similarly, one could ask simply out of strong intellectual interest just how long it will take to land humans on Mars (or on Pluto) with an annual R&D investment of US $500 million.

---

6) In fact, this question should have been asked seriously 20 years ago; in Germany, in particular, at the latest in 2002, the year in which the schedule of abandonment of nuclear power was decided upon (in a so-called consensus with the electric power producers). However, the opposite was the case: the past roughly 15 years are correctly termed the "lost years," as a result of the blatant failure of responsibility of the political class in the rich industrial countries. This statement does not detract at all from the attainments of those workers who continued carrying out solar research under the given financial boundary conditions. On the contrary, they achieved far more than might have been expected.

The sums mentioned in fact indicate the orders of magnitude of the financial support within the past years for these different areas. In European countries (especially Germany, Spain, and Italy), the financial support for solar thermal energy consisted of a few million US dollars per year in each case. Similar amounts hold in the USA, where the financial support by the government in Washington was even canceled completely for a short time. In contrast, according to the "NASA Fiscal Year 2008 Budget Request," the American NASA budget in the years 2006–2012 for Mars missions ("Mars exploration") was around $600 million per year (NASA, 2007).[7]

---

7) Additional budget items, which can be allocated to the Mars missions, are included especially in that part of the budget that concerns the development of space vehicles ("Exploration Systems": $5000 million per year) and in the "Aeronautics Technology" part ($500 million per year).

2
The Salient Facts

2.1
Solar Tower Power Plants as the Basis for Cost Estimates: Cost Analyses

Although in this book all variants of solar thermal energy production are considered, we wish to emphasize the solar tower plant concept, since it has the best chance – according to most studies – of being implemented. To be sure, it is still much too early to make this prediction with certainty. We therefore point out that based on current knowledge, all the solar thermal and in particular all the light-concentrating technologies must be treated as being nearly equivalent, and thus they should all be developed with the same priority. Various studies in the past 30 years, such as for example that of Sargent and Lundy (2003), carried out for the US Department of Energy on the medium- and long-term costs of solar thermal power plants, arrived at the result that for electrical power generation on a large scale, tower plants have the greatest economic potential. This is related, among other things, to the high temperatures obtainable with this technology and thus their high efficiencies, and to the cost-effectiveness of heat storage in this case. In addition, they have relatively less stringent requirements in terms of the maximum slope of their location, which in particular could be important for sites in Spain.

In the USA, in 1996 a pilot solar tower plant of a type which can be used for base-load power generation was put into operation and tested through 1999: the "Solar TWO" power plant (Figure 2.1). This type of power plant uses molten salts as heat-transfer medium (molten-salt technology). The solar radiation that is reflected from a large number of movable mirror units (heliostats) is concentrated onto a so-called receiver, which is located at the top of a tower, and in which the molten salt is heated (solar tower plant). A portion of the hot salt is used immediately for steam generation for the turbine, while the larger portion is stored in a tank for nighttime operation. Thanks to this generously dimensioned heat storage system, the power plant can generate electrical energy around the clock, as long as the sun shines during the daylight hours. Figure 2.2 shows a schematic drawing of the molten-salt circuit.

Since during the day, solar energy must be stored for nighttime use, the mirror field and the receiver are correspondingly larger than for a power plant of similar

2 The Salient Facts

Figure 2.1 The solar tower power plant Solar TWO (Barstow, California) (SANDIA).

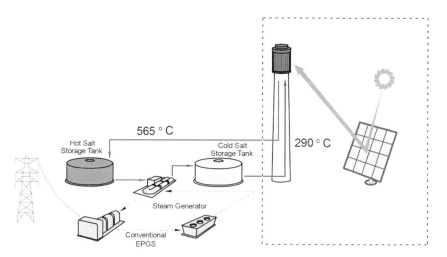

Figure 2.2 Schematic drawing of a solar tower plant with molten-salt technology (SANDIA).

output power without a heat storage system. The enlargement factor is called the "solar multiple" (SM).

The US national research institute Sandia National Laboratories in Albuquerque, New Mexico, carried out *cost estimates* in connection with the Solar TWO project, giving an indication of the costs of later, larger power plant units (Kolb, 1996a).

In these estimates, however, they considered only production series which were small in terms of the overall energy-economical scale. The results, however, already permit the clear-cut conclusion that for large-scale production, for example, the construction of 1000 MW of installed capacity each year, after intensive development work it would be possible to attain power costs of the same order as those from fossil-fuel power plants. (These results confirm, incidentally, several comprehensive studies dating from the 1970s and 1980s.[1]) The Sandia data form the *basis for the cost analyses* in this book. The expectations for large-scale production derived from them are compared here in detail with several newer studies based on higher production rates, which lead to practically the same conclusions (cf. Chapters 4 and 6).

2.2
The Combined System of Solar and Backup Power Plants ("Solar Power System")

One disadvantage of the concentrating solar power plant is its requirement of direct solar radiation, which in Central Europe and in most other centers of power consumption, such as the American East Coast, is available to only a limited extent.[2] The solution of this problem lies in shifting the site locations of the power plants to regions with a generous supply of solar radiation. (In the cloud- and haze-free desert regions, one not only has considerably more solar irradiation on the whole, but at the same time, only a small proportion of diffuse radiation.) The electrical energy must then be transported via transmission lines to the region where it is to be consumed. For the power supply to Europe, possible sites are located especially on the semiarid plains of southern Spain and in the desert regions of North Africa (Sahara). There are a certain number of possible sites in southern Italy and in Greece, but the available areas are not ample, so that the potential of these countries would be sufficient at most to supply their own energy demands. Furthermore, the annual number of hours of insolation is not as great as in southern Spain or especially in the Sahara.

A similar situation to that of Europe is also found in the USA. There lies a very large solar potential of the sunny Southwest within the country itself. The southwestern states of the USA have among the best sites worldwide. From here, the eastern part of the country could be supplied with solar electric power. The distances would be roughly the same as those between North Africa and Central Europe (of the order of 3000 km). The cost data, which are given below for locations

---

1) The most important previous study was the so-called Utility Study of 1988, whose import can also be seen from the fact that – as its name implies – it was carried out with the significant participation of American power-generating operators (Hillesland, 1988).

2) In contrast to the concentrating systems, a solar chimney power plant can make use of the diffuse irradiation, in addition to the direct insolation.

in *Morocco*, are therefore very similar to those for the *USA* (power supplies for the Eastern Seaboard).

The transmission losses and the required investments for transmission lines naturally enter into the economic considerations, but they are, at the current state of transmission technology, not a serious problem. With today's high-voltage direct-current (HVDC) transmission at a voltage of ±800 kV, the losses over the distance from southern Spain–Central Europe (2000 km) are ca. 8%. At a distance of 3000 km, they lie around 11.5%. The latter corresponds to the case of North Africa–Central Europe or the US-Southwest–East Coast. For example, the distance from the favorable sites in the south of the state of New Mexico to Washington, DC, is 3000 km, while the distance to Chicago is about 2300 km. High capacity utilization contributes to the low transport costs. Thus only base-load power, and not peak-load power, can be economically transported over these distances (at least not with comparable cost effectiveness). Since the large-scale power supplies we are considering here are mainly relevant to the base load, these desirable uniform energy transport rates can be taken for granted.

2.2.1
Solar Base-Load Plants

The day-night problem (base load) can be solved by using heat storage for solar thermal power plants. The problems of cloudy weather and the seasonal changes in the angle of the sun cannot, however, be efficiently solved by energy storage, since storage of heat over many days or even weeks is too expensive.[3] For this reason, in the present investigation, we compare nuclear and fossil-fuel base-load power plants with the model of a *combined system* (for solar base-load): it consists of solar thermal power plants (including energy transmission) and of fossil-fuel backup power plants, and will be referred to in the following for brevity as a "solar power system."

As reserve power plants (also called the *backup system*), one could envisage for example natural-gas fired gas- and steam-turbine power plants (combined-cycle gas-turbine (CCGT) plants, often briefly referred to simply as CC plants), or also coal-fired power plants. If these are located in the region of high power consumption, for example, in Germany, they guarantee power supplies in the event of an interruption of the transmission lines or even in the case of a hypothetical very long shutdown of the solar power plants (e.g., due to natural catastrophes). CC power plants can be operated both with natural gas as well as with domestic fuel oil. Even in the hypothetical case of a simultaneous failure of the solar plants and the gas supply, the continued provision of electric power could be guaranteed. Furthermore, gas turbines have a very short startup time and would be available

---

3) There are plans, for example in connection with the Italian solar thermal development program, to bridge over cloudy periods of several days by using very large thermal storage systems (ENEA, 2001). The heat losses of the storage reservoir are not a problem here; instead, investment costs are the limiting factor. In the end, the question of economic optimization must be dealt with.

within minutes in the event of a sudden failure of the long-distance electrical transmission.[4]

If today's coal-fired plants were all replaced by solar power plants (to release the coal supplies for other uses), the coal plants would remain available as backup systems at practically no cost.[5] This is also true for the current base-load gas-fired plants, if they were to be replaced by solar power plants.

2.3
How Much Does Solar Power Cost?

In the following section, we present an overview of the cost estimates for solar tower plants with molten-salt cycles. In the later sections, we then give more details for solar tower plants and also for the other types of solar thermal power plants.

2.3.1
Introductory Remarks

Predictions of the production costs of various components of solar thermal power plants in mass production are necessarily limited in their accuracy. From the present viewpoint, we can thus not be certain that the costs estimated by Sandia in connection with the Solar TWO project, and the resulting derived costs for mass-produced plants, will apply in practice. The same holds for the prognoses related to the other solar thermal variants. Further research and development will, however, clarify this point. The general perspective that solar thermal power plants have the potential of low, economically feasible costs, is however already plausible today: there are, as we have pointed out, other solar thermal technologies in addition to solar tower plants, which offer in principle a similar economic perspective, although their chances of success may not be quite as favorable. Each of them has, considered from today's level of development, the potential of acceptable costs for mass-produced series. Since these systems are basically different, the R&D tasks

---

4) A CC power plant consists of a gas turbine and a steam turbine, which follows it in the power train. About two-thirds of the output power of the plant is produced by the gas turbine and one-third by the steam turbine. Gas turbines can be started up on a routine basis in 15 min, and even more quickly in an emergency. The steam turbine part of the plant previously required ca. 1 h for a normal startup (not an emergency start). In newer plants, it can be started up in 30 min (cf. Chapter 10). In order to ensure the rapid switch-on time of the gas turbines (only a few minutes in an emergency) for the whole backup power plant, the present cost estimates for the solar power system include *additional* gas turbines in the CCGT backup plants, corresponding to the power output of the steam-turbine part, that is, one-third of the overall power output. Thus the power plant can provide its total rated output power within a few minutes after start-up.

5) Coal-fired power plants must be supplied with supplemental facilities in order to be able to start up rapidly; using either gas turbines or diesel engines (with the full output power of the coal-fired plant). If superseded coal-fired plants are to be maintained at low cost as backup plants, these additional facilities lead to tolerable supplemental costs for the overall backup plant system.

with the goal of reduced costs for mass-produced series are also different. The inability to meet these goals for one or another of these technological paths would not be a decisive failure. Owing to the great degree of independence of these paths from one to other, the probability that at least one of the technologies would attain the predicted cost for power produced is relatively high. We can illustrate this point with a numerical example: if we assume for four equivalent individual technologies that the probability of success for each one is 33%, then the probability that at least one of them will succeed is 80%. If the individual success probability is 50%, then the overall probability of success with at least one is 94%. (To be sure, such "probability considerations" must generally be taken with caution due to the difficulty in defining the input values precisely.)

2.3.2
Investments and Power Costs

In Tables 2.1–2.3, the costs which are to be expected based on current information for a large-scale solar energy-supply system, including transmission and the backup systems, are summarized in compact form. A more detailed treatment is given in Chapter 4. Tables 2.1 and 2.2 give an overview of the investment costs, while Table 2.3 shows the resulting power costs (the latter in comparison to gas-fired, coal-fired, and nuclear power plants). As mentioned, for the solar plant, the cost estimates are based upon those found in connection with the project *Solar TWO*; there, however, for relatively small production series (Kolb (Sandia) 1996a). They were adapted to mass-production series under assumptions which will be discussed later. The basis of these cost estimates will be explained in detail in later chapters, see especially Section 4.2.

The "capital costs" per kWh associated with the investments are calculated using the *real interest rate* throughout this book. This is the nominal interest rate reduced by the inflation rate. These costs are thus given in "inflation corrected" form (for details see Section 4.4). The amortization time is assumed throughout to be 45 years, for gas-fired, coal-fired, and nuclear power plants as well. This time period corresponds to the technical life expectancy of the power plants, in contrast to financial or tax amortization times which are in general taken to be considerably shorter. The costs are – with a few exceptions – quoted in *US dollars as valued in the year 2002*. Sources which give costs in Euros (€) are usually recalculated and quoted directly in 2002-US$ without mentioning this specifically in each case.

The conversion of EUR into dollars was not carried out using the exchange rate (from the year in question), but instead using *purchasing-power parity*. The exchange rate is subject to wide fluctuations, which have nothing (or very little) to do with the manufacturing costs of a product within a country. It, therefore, reflects the factual production costs in another country only in a very limited fashion. Purchasing-power parity is the preferred quantity for this purpose. Using the figures of the OECD (see Appendix B), it was in the range of 1–1.1 in the past 15 years; that is, 1 EUR (or the equivalent in the German predecessor currency DM) corresponded over this period of time with respect to its factual buying (purchasing)

Table 2.1 Investment costs for solar tower plants with a molten-salt circuit (2002-$).

Investment costs	Solar TWO advanced technology	Spain	Morocco/USA
Solar multiple	2.7	4.4	3.7
Heliostat costs	138 $/m²	83 $/m²	83 $/m²
	Million dollars (2002) per 1000 MW installed		
Heliostat field	1710	1670	1400
Receiver + tower	295	480	405
Horizontal salt circuit	–	180	150
Heat storage reservoir	355 (13 h)	435 (16 h)	435 (16 h)
Conventional components	590	590	590
Land preparation costs	60	95	80
Land purchase costs		140	
Sum ("direct costs")	3010	3590	3060
Indirect costs[a]			
Interest during construction	240	145	125
Owner's cost	180	105	90
Planning and project management	270	145	125
Miscellaneous/unexpected costs	210	–	–
Overall investment costs (wet cooling)	3910	3985	3400
Dry cooling (+8.7%)			295
Total investments	3910	3985	3695

The cost data from Sandia (Kolb, 1996a), based on the Solar TWO project (but assuming a receiver with "advanced technology") were adapted with certain assumptions to a future large-scale, mass production scenario with mature technology; this concerns especially the heliostat costs and the indirect costs.

a) The indirect costs according to Kolb are given in Section 4.3.1.

Table 2.2 Investments for the overall power plant system (solar power system).

Investment costs	Spain	Morocco/USA
	Millions of dollars (2002) per 1000 MW	
(Solar power plants per 1000 MW at the plant site)	(3985)	(3695)
Solar power plants per 1000 MW after power transmission[a]	4335	4175
Transmission lines	500	665
Backup power plants (natural gas CC power plants)	715	715
Total investment costs	5550	5555

a) 1000 MW at the end of the transmission line requires 1090 MW at the plant location for Spain (2000 km, 8.1% losses), and for Morocco or USA (3000 km, 11.5% losses), 1130 MW would be required.

2 The Salient Facts

Table 2.3 Power costs of the solar power system based on solar tower power plants, compared to gas-fired, coal-fired, and nuclear plants.

Power costs			Solar power system		Gas-fired[a]	Coal-fired[a]		Nuclear power[b]
			Spain	Morocco/USA				
			US-cents per kWh (2002)					
Capital costs[c]			3.1	3.1	0.3	0.7		0.7
Operation and maintenance	Solar		0.7	0.7	0.3	0.7		1.0
	Backup		0.1	0.1				
Gas[d]			1.3	0.8	4.1	–	–	–
						EU	USA	
Coal[e]			–	–		2.5	1.3	–
Uranium	Fuel cycle[f]							0.5
	Natural uranium[g]							0.2
Power cost			5.2	4.7	4.8	3.9	2.7	2.4

a) Gas-fired power plants: 615 M$/GW, 8000 operational hours/a, efficiency 60%. Coal-fired power plants (without CO_2 sequestration): 1200 M$/GW, 8000 operational hours/a, efficiency 45%.
b) Nuclear power plants in mass production (new American type): 1100 M$/GW, 8000 operational hours/a.
c) Real cost estimate: 4% real interest, 45 years operating life.
d) Gas price (2002-$): 2.5 US-¢/kWh$_{gas}$ = 40 $/barrel oil = 6.6 $/MMBTU (HHV) = 0.68 ¢/ft³ = 22 US-¢/Nm³ (8.8 kWh LHV) = (assumed: €1 = $1.25) EUR 2 ¢/kWh.
   In the case of price increases for natural gas, the backup power plants of the solar power system could be supplied with gas from coal gasification, which is available at roughly the same cost (for large-scale users with a separate gas line) (see Section 11.2). The proportion of backup power from gas-fired plants would be 30% for solar plants in Spain, and 20% for solar plants in Morocco or the USA.
e) Coal price (2002-$): Europe (imported coal) 90 $/tce (= 1.1 ¢/kWh$_{coal}$); US 45 $/tce (= 0.55 ¢/kWh$_{coal}$).
f) Fuel cycle not including the cost of natural uranium (but including waste-disposal costs according to WNA (2005) for the USA).
g) Natural uranium costs:
   - Uranium price assumed: 130 $/kg U (= 50 $/pound U_3O_8). This is in the upper cost range for disposal of the uranium reserves. (For comparison, the maximum uranium price up to now (in mid-2007) was 350 $/kg U; in mid-2008, the price was 115 $/kg U, in August 2008 it was 135 $/kg U.)
   - Assuming a *reduced future natural uranium consumption* of 14.5 kg U/GWh$_{el}$. With today's reactors, the world average consumption is 25.5 kg U/GWh$_{el}$. (At this consumption rate, the cost factor due to natural uranium in the power price range would be 0.33 ¢/kWh.)

power to $1–1.1. In the year 2002, the year on which the costs estimates in this book are based, it was given by $1 = €0.96.

In the first column of Table 2.1, those costs are set out which Sandia (Kolb, 1996a) estimated for a 200-MW solar tower plant (converted to 2002-$ instead of the original 1995-$, inflation factor 1.18). For clarity, the costs were extrapolated to an output power of 1000 MW.[6] The 200 MW plant corresponds in principle to the Solar TWO technology (10 MW), but with a receiver of the design denoted by Kolb as "advanced technology."

In the two right-hand columns, the costs are set out for a power plant optimized for base load (with a high annual total operating time, i.e., a large solar multiple (SM)). Which annual total operating time a solar power plant (with a thermal storage system) can attain at a particular site depends significantly on the size of its mirror field. If the mirror field is made only large enough to allow the power plant to just attain 24 hours of full-output power on a cloudless winter day, then even a small decrease in solar radiation would permit operation at only less than nominal output power. With a larger field, the power plant would operate at full capacity even with a minor decrease in solar radiation, as often occurs. The size of the mirror field is thus an important aspect of the economic optimization. (We have already mentioned that the relative size in relation to a power plant of the same electrical output power without a heat-storage reservoir is termed SM.) If the conditions of solar radiation are similar to those in southern Spain, and an annual capacity factor of 70–75% is to be achieved, one requires an SM of 4.4. In Morocco or in the USA, where the solar radiation is more even and stronger, an SM of 3.7 suffices in order to achieve ca. 80% annual capacity factor. The mirror field in the two right-hand columns was, therefore, increased relative to the initial data from Sandia (SM = 2.7) for locations in Spain by 63% (SM = 4.4) and for locations in Morocco by 37% (SM = 3.7) (and at the same time the capacity of the heat-storage reservoir was increased from 13 to 16 hours). The proportion of the power required from the backup power plants for uninterrupted service from the solar power system (at 100% of nominal output power) is then for Spain only ca. 25–30%, and for Morocco or the USA only 20%.

Extrapolating to a *mass-production scenario* was carried out as follows (cf. also Chapter 6 and Section 4.3.1):

- Regarding the costs of the heliostat field (conventional glass-mirror heliostats), estimated by Kolb for a production rate of only 2500 heliostats per year (each

---

6) A 200 MW$_{el}$ plant at a solar multiple of 2.7 (receiver: 1400 MW$_{thermal}$) as described by Kolb corresponds for a solar multiple of 4.4 (in Spain) – that is, with a larger mirror field relative to the output power of the plant – to only 125 MW$_{el}$. A solar plant with a nominal output power of 1000 MW$_{el}$ (and an SM of 4.4) would thus consist of eight such solar tower installations. The tower and mirror field of a solar tower plant cannot be enlarged arbitrarily; the optimum size for an individual tower installation lies roughly in the range of 1400 MW$_{thermal}$ (at the receiver), as described above. Making use of thermally insulated molten-salt piping (i.e., a *horizontal salt circuit*), however, allows several individual tower installations to be connected to a single central steam power plant (power block); that is, they can be "interconnected" to form a larger solar power plant. Ideally, this interconnection would comprise five to six tower installations of the size mentioned (in the case of smaller installations, correspondingly more of them must be interconnected). This yields a base-load solar power plant (at an SM of 4.4) of ca. 700 MW$_{el}$ output power, with the advantage that efficient, reasonably priced, and reliable steam turbines in the widely used 700 MW class, as in current coal-fired power plants, could be employed.

with 150 m²) to be 138 $/m², we assume that with a mature technology and a production rate on the order of 100 000 heliostats per year, a reduction to 83 $/m² could be realized. This corresponds to a construction rate for power plants of roughly 1 GW/a installed output power.[7]

- An additional assumption refers to the so-called indirect costs. These include for example, the interest which accrues during the period of construction and the costs for the infrastructure. These costs are also strongly subject to economies of scale, the construction time for a power plant decreases for higher production rates, since the fabrication of the parts, for example, the heliostats, is the limiting factor for construction time. Therefore, assumptions are made that generally correspond to mass-production series as required in the case of a large-scale system. This point will be discussed in more detail later.

- In the two columns for Spain and Morocco in Table 2.1, the costs for a "horizontal" molten-salt circuit are shown. They include the coupling of several tower installations to a larger overall power plant by means of connecting pipes for the molten-salt heat-transport medium (horizontal molten-salt piping – in contrast to the vertical pipes which connect to the receiver at the top of the tower).

- Since at desert locations such as in Morocco, wet cooling does not usually appear to be practicable, the investments for a solar power plant in this case were increased by 8.7% to take into account the additional expense of dry cooling.[8] This also holds for solar power plants in the USA.

In Table 2.2, the investments for the overall "solar power system" are set out. They refer to a capacity of 1000 MW at the end of the transmission line (thus e.g., in Germany). The required power plant capacity at the location of the solar plants is larger, owing to the transmission losses.

Table 2.3 shows the estimated costs for electric power generated by the solar-fossil combined system (the *solar power system*), compared with the costs of power from gas-fired, coal-fired, and nuclear power plants (in 2002 US $). In the case of "Spain," solar power would cost 5.2 ¢/kWh; in the case of "Morocco/USA," 4.7 ¢/kWh. This corresponds roughly to the cost of power from natural-gas-fired power plants at the current price for gas.

With the assumptions mentioned above (including 4% real interest and 45 years operating lifetime for all the power plants compared), nuclear power from light-water reactors (future reactor types constructed on a *very large scale*) would cost 2.4 ¢/kWh, and power from coal-fired plants (without CO_2 sequestration) in Europe using imported coal at current prices would cost 3.9 ¢/kWh, while in the USA at current coal prices, it would cost 2.7 ¢/kWh.

For solar power plants, the *capital costs* are the most important determining factor. They are computed from the investment costs and the operating lifetime by

---

7) A base-load power plant of 1000 MW with a solar multiple of 4.4 consists of 135 000 heliostats of 150 m² each. A construction rate of 1 GW/a then corresponds to the production of 135 000 heliostats each year.

8) It is assumed here that the power output of a solar plant would be 8% lower in the case of dry cooling, which corresponds to an increase in the specific investment cost of 8.7%; cf. Section 4.3.7.

multiplying the investment costs (per GW) by the so-called annuity factor of 0.0483 or 4.83% (corresponding to 4% real interest and 45 years operating life). This yields the annual costs resulting from *interest and repayment of the principal* of the invested capital. (The annuity factor quoted above assumes the debt to be completely repaid in 45 years.) These annual capital costs are then divided by the amount of energy generated per year at an overall power capacity of 1 GW so that the *capital costs per kWh* are obtained. The amount of energy produced annually is found from the capacity factor of the power plant and its power output capacity.

For the solar power system – a double system consisting of solar and backup power plants – the capacity factor is 100%. This yields an annual energy production of 8760 GWh (1 GW × 365 d × 24 h/d = 8760 GWh). In the case of gas-fired, coal-fired, and nuclear power plants, we assume 8000 hours of full-capacity production per year (corresponding to a capacity factor of 91%), that is, 8000 GWh of electrical energy per year.

The operation and maintenance costs in Table 2.3 are shown separately for the solar and the backup power plants. An additional point in the cost estimate is the fuel cost. In the solar power system, this refers to the natural-gas consumption of the backup power plants. Given the poorer insolation conditions in Spain, the expected proportion of backup power is 30% of the overall power production; in the more favorable desert sites in Morocco and in the USA, a proportion of ca. 20% is reasonable. This is reflected in the difference in natural gas costs shown in Table 2.3.

The costs for power from coal-fired plants shown in Table 2.3 refer to today's coal plants *without* separation and storage of CO_2 (so-called sequestration). It is, however, clear that the use of coal for energy production in the mid- to long term must be accompanied for the most part by sequestration, which will make the electrical energy more expensive. In these costs, one must distinguish on the one hand the cost of separating the CO_2 at the power plant, and on the other, that of transporting the CO_2 to the storage point and of storing it in a depot (former gas-field or a so-called aquifer, or even passing it into the ocean). Employing the data from (EIA AEO, 2007) for computing the costs of power plants, the cost of power would increase due to separation of the CO_2 alone (without transport and storage), for power plants with integrated coal gasification (IGCC) and assuming the cheaper American coal costs (45 $/tce), from 2.7 ¢/kWh (conventional power plants) to 3.4 ¢/kWh; for the case of the more expensive imported coal in Europe (90 $/tce), it would increase from 3.9 ¢/kWh (conventional) to 4.7 ¢/kWh (see Tables 4.3 and 10.2).

Regarding the costs of transport and disposal of the separated CO_2, there are widely divergent estimates. The literature available to the present authors quotes costs ranging from ca. 5 $/t coal (tce) up to 70 $/t coal (tce) (recalculated as an equivalent increase in the coal price).[9] The former would increase the price of

---

9) In the literature, the costs are quoted in $/t CO_2; they vary between 2.7 and $25 /t CO_2. Here, we have recalculated these costs as $/t coal (tce), which has the advantage that then, in the cost table in Chapter 10 "Fossil-fuel power plants", the influence of these costs on the price of electric power can be seen directly. (1 t coal (tce) = 0.75 t C (carbon); 1 t C = 3.66 t CO_2; 1 tce = 2.75 t CO_2.)

Table 2.4 Comparison of the electric energy costs from solar and nuclear power plants.

	Solar plants	Nuclear plants 1100 $/kW at the plant	Nuclear-plant pools 1100$/kW power transmission over 1000 km	Difference solar–nuclear
Operating life	45 a	45 a	45 a	
Power transmission	±800 kV		±800 kV	
	US-¢/kWh (2002)			
Morocco/USA	Solar power system			
4% Real interest	4.7	2.4	2.9	2.3/1.8
2% Real interest	3.8	2.2	2.6	1.6/1.2
	At the solar plant			
4% Real interest	3.3	2.4		0.9
2% Real interest	2.5	2.2		0.3
Spain	Solar power system			
4% Real interest	5.2	2.4	2.9	2.8/2.3
2% Real interest	4.3	2.2	2.6	2.1/1.7
	At the solar plant			
4% Real interest	4.1	2.4		1.7
2% Real interest	3.1	2.2		0.9

power from coal-fired plants by an additional 0.14 ¢/kWh, the latter by an additional 2.0 ¢/kWh[10] (for more details of the cost estimates, see Section 10.3). The overall costs of power from coal-fired plants would thus lie within the range of 3.6 to possibly 5.4 ¢/kWh in the USA; in Europe from ca. 5.0 up to possibly 6.7 ¢/kWh. If, for example, one assumes disposal costs of 10 US $/t CO_2 (27 $/tce), the resulting energy cost would be 4.2 ¢/kWh in the USA or 5.5 ¢/kWh in the EU (cf. Table 4.3). In comparison with these coal-fired power plants, solar power would then not be much more expensive in the USA (4.7 ¢/kWh) or (depending on the development of sequestration costs) possibly even cheaper; in Europe (at 5.2 ¢/kWh) it would at least cost the same, but in fact would probably be even cheaper.

2.3.3
Are the Additional Costs Compared to Nuclear Plants Affordable?

In Table 2.4, the cost difference between solar and nuclear electric power is displayed. For locations in Morocco and the USA, this difference is 2.3 ¢/kWh; for sites in Spain, it is 2.8 ¢/kWh. In Table 2.4, in addition to the power costs as shown in Table 2.3

---

10) Passing the CO_2 into the oceans would lead to additional costs; for the case of a transport distance of 1000 km on land (pipeline) plus a distance over the ocean of 2000 km, they would correspond to additional effective coal costs of $30 to $130 per tce (i.e., an increase in electric power cost of 0.7 to 3.6 ¢/kWh).

(i.e., the capital costs assuming 4% real interest), also the expected power cost at 2% real interest and furthermore the power costs at the solar power plant and – for nuclear power plants – the costs for the case of nuclear power-plant pools located far from the large consumer centers are given. These three points are discussed only briefly here. (For more details concerning lower interest rates, see Tables 4.2 and 4.3 and Section 4.4; regarding nuclear power-plant pools, cf. Table 4.3).

2.3.3.1 Burden on the Economy Due to Higher Power Costs (The Cost Difference Solar Energy – Nuclear Energy)

The burden on the economy due to possibly higher costs of solar power will be summarized in the following with a list in outline form, using as examples the USA and Europe. In spite of the assumed massive substitution of fossil energy sources, the economic burdens would in both cases be tolerable (USA 1.8%, EU 1.9% of the annual gross domestic product (GDP)).

Example: USA

- Assumption: *Difference in power cost = 2.3 ¢/kWh*
- Assumption: *1000 GWa$_{el}$ solar or nuclear power generation per year**

   * This is 11 times today's annual power production from nuclear power plants. (For this production, with a capacity factor of 91% (2004), a nuclear power plant capacity of *1100 GW$_{el}$* would be required in the USA; currently, it is 105 GW$_{el}$)
   Compare USA (2004):
   Electric power generation: in total *450 GWa$_{el}$*, of this 90 GWa$_{el}$ from nuclear energy; coal-fired plants 230 GWa$_{el}$, gas-fired plants 80 GWa$_{el}$.
   Primary energy consumption: in total 3100 GWa (coal 750 GWa); without primary energy sources for electric power generation (coal 670, gas 190, oil 35, nuclear 250, hydro 80 (cf. Appendix C) = 1225 GWa), this amounts to ca. *1900 GWa*.

The assumed 1000 GWa$_{el}$ corresponds, for example, to 350 GWa for the electric power supply (this should correspond roughly to the *total base-load* portion of the electric power generated in the USA) and 650 GWa for the substitution of primary energy sources in other areas (assumption: substitution by electric power) corresponding to *33% of the primary energy* consumption (1900 GWa). In evaluating the scope of this substitution scenario, the production of gas from the coal which would be substituted in the power plants (280 GWa of gas from coal gasification) must be considered, as well as ca. 80 GWa of natural gas which would be conserved in the gas-fired plants; cf. Section 11.2.9. All together, around 1000 GWa, out of 1900 GWa primary energy (mainly oil and natural gas), could thus be replaced outside the power plants, i.e., 53%.

With the assumptions given above, the additional costs to the economy amount to *$200 billion annually*.

   (2.3 ¢/kWh × 0.0876 billion $/GWa* × 1000 GWa/a = 202 billion $/a).
   *1 ¢/kWh = 0.01 million $/GWh = 87.6 million $/GWa = 0.0876 billion $/GWa

> Comparison: *Gross Domestic Product* (GDP) USA 2003: 11 000 billion $.
> The additional costs (200 bill. $/a) thus correspond to *1.8%* of the GDP
>
> Comparison: *Defense spending* (USA)
> 2000 (before 11th Sept. 2001):   in total 390 billion $*
> 2006:   in total 590 billion $*
>   (Defense Department budget: 410 billion $**)
>
> * Source: World Military Spending 2007   ** Amadeo 2007
>
> The additional costs estimated above (200 billion $/a) thus correspond to:
>
> - *50%* of the total defense spending in the year 2000 (before 11th Sept. 01)
> - *33%* "   "   " in the year 2006
> - *100% of the increase* in defense spending from 2000 to 2006

Example: Europe (EU-25) – plant sites in Spain

- Assumption: *Difference in power cost* = 2.8¢/kWh
- Assumption: *800 GWa$_{el}$* solar or nuclear power generated per year*

> * This is seven times today's annual power production from nuclear power plants. (For this production, with a capacity factor of 91% (2004), a nuclear power plant capacity of *880 GW$_{el}$* would be required in Europe; currently, it is 113 GW$_{el}$.)
> Compare EU-25 (2004):
> *Electric power generation*: in total 360 GWa$_{el}$, of this 113 GWa$_{el}$ from nuclear energy; coal-fired plants 110 GWa$_{el}$, gas-fired plants 70 GWa$_{el}$.
> *Primary energy consumption*: in total 2280 GWa (coal 410 GWa). The primary energy sources for electric power generation are not listed in the EU statistics. Making use of assumed average efficiencies for electric power generation (efficiency: coal 38%, natural gas 50%, oil 50%), a rough estimate yields the following amounts: coal 280, gas 140, oil 30, nuclear 290, hydro 100 = 840 GWa. The primary energy consumption without energy sources for electric power generation thus amounts to roughly 1440 GWa.
>
> The assumed 800 GWa$_{el}$ corresponds, for example, to 280 GWa for the electric power supply (this should correspond roughly to the *total base-load* electric power production in Europe) and 520 GWa for the substitution of primary energy sources in other areas (assumption: substitution by electric power); this is 35% of the *primary energy* consumption in Europe (1440 GWa). In debating this substitution scenario, the possible gas production from the substituted power-plant coal (110 GWa gas from coal gasification) should be considered, as well as ca. 50 GWa of natural gas which would be conserved in the gas-fired plants; cf. Section 11.2.9). All together, around 680 GWa, out of 1440 GWa of primary energy (mainly oil and natural gas), could thus be replaced outside the power plants, i.e., 47%.

With the assumptions listed above, the additional burden on the economy amounts to *$200 billion annually* (the same total as found for the USA)

(2.8 ¢/kWh × 0.0876 billion $/GWa* × 800 GWa/a = 196 billion $/a).
*1 ¢/kWh = 0.01 million $/GWh = 87.6 million $/GWa = 0.0876 billion $/GWa

Comparison: *Gross Domestic Product* (GDP) EU-25 2002: 9900 billion €. Converted to US $ at purchase power parity according to the OECD (2002 1 $ = 0.96 €), this amounts to 10300 billion $. The additional costs estimated above (200 billion $/a) then correspond to *1.9% of the European GDP.*

Comparison: *Defense spending* (European NATO countries 2002) 161 billion US $ (IMI 2002). (This is only 27% of the defense budget of the USA in the year 2006). The additional costs estimated above (200 billion $/a) then correspond roughly to the current defense spending in Western Europe (only 25% more).

2.3.4
Possibly Lower Cost Differences, Potential for Further Development

Regarding the comparison of power-generating costs in Table 2.4 and the resulting burden on the national economy, several aspects should still be considered that might change the estimated costs noticeably:

- The cost difference of 2.3 ¢/kWh (USA) or 2.8 ¢/kWh (Europe) (Table 2.4) is based on the cost of nuclear power from plants in urban areas. If the plants were to be constructed far away from populated zones (nuclear-plant pools), which is reasonable in view of the large number of nuclear power plants needed, and if the electric power must, therefore, be transported over a distance of, for example, 1000 km via transmission lines, then the resulting power costs are greater. The cost difference would then be reduced in both cases (EU/US) by about 20% and the additional burden on the economy with solar power would then be correspondingly lower.

- A noticeable decrease in the cost difference would also result in the case of lower interest rates, for example, for 2% instead of 4% real interest.[11] This, of

---

11) Many economists now believe that in future, when economic growth in the industrialized countries has stagnated, it will be possible to avoid a high proportion of unemployment only by maintaining a relatively low real interest rate. Considering the long amortization times for power plants (operating lifetime 45 a), it is thus quite plausible that this changeover to an employment-oriented finance policy might occur relatively soon, for example, within 10 or 20 years. The greatest part of the investment financing for future power plants would then take place at low interest rates. Under these economic boundary conditions, solar power, which is capital intensive, becomes more economically favorable than with today's comparatively high real interest rates.

course, reduces the costs of power both from solar plants and from nuclear plants, but not to the same extent. At a real interest rate of 2%, and in comparison to nuclear power-plant pools (with power transmission to consumer centers), the cost of nuclear energy in the USA would be 2.6¢/kWh, and the cost from the solar power system only 3.8¢/kWh. The difference in costs of 1.2¢/kWh is then only half as great as assumed in the previous considerations (2.3¢/kWh). In Europe (solar sites in Spain), the difference would be only 1.7¢/kWh as compared to the value of 2.8¢/kWh assumed above.

- In this calculation, the construction of *new* gas-fired backup power plants is assumed. In the case of a strategy involving the rapid replacement of the present, mostly relatively new coal- and gas-fired base-load plants, the plants replaced would be available at quasi "zero cost" as backup plants for the solar power system. (If the fossil-fuel power plants were to be replaced by nuclear plants, on the other hand, they would be shut down.) The power cost from the solar power system would be reduced effectively by ca. 0.5¢/kWh in comparison to nuclear power plants (cf. Section 4.1).

- The expected improvements in power-transmission technology, especially with superconducting transmission lines, would also shift the cost balance in favor of solar power plants. A decrease in transmission costs by more than half is not unthinkable.

- The true costs of solar power plants, by the way, could be considerably lower than assumed above, considering the *very large* production scenarios (and the accompanying intensive development of the power plants), which would be required. The numbers quoted above for solar power plants are based on a construction rate corresponding to less than 1 GW of new generating capacity per year (at an SM of 3.7), while the construction rate assumed for the nuclear power plants is a factor of 3 to 12 higher (see Chapter 12). Under the assumption of such a very large production scenario, it can be expected that simply the optimization of the power plants (making use of the overall *innovation potential*[12]) would give rise to a cost reduction compared to the costs assumed

---

12) Making use of the innovation potential is, however, in part already included in the above cost estimates, namely in the heliostat costs. A new large-scale heliostat cost study by Sandia (see the chapter on heliostats) yielded nearly the same heliostat costs for large production series (i.e., 80 \$/m² in 2002-\$) as assumed by the present authors in 1998 (70 \$/m² in 1995-\$, corresponding to 83 \$/m² in 2002-\$). These costs (Kalb/Vogel 1998) were, therefore, adopted in this book without change (and simply recalculated to 2002-\$). The value of 80 \$/m² quoted in the Sandia study, however, already includes a cost reduction through further technical development; it amounts to ca. 15% cost reduction. Fulfilling the innovation potential for the heliostats is thus, at least partially, already taken into account in the cost estimates for solar power given above. For all the remaining components of the solar power plant, this is, however, not the case. Furthermore, the cost-reducing effect of improvements in fabrication processes for the various power plant components (the so-called "learning curve") is also not included in the cost estimates. Here, and to a limited extent even for the heliostats, the large-scale application of solar energy should lead to further cost reductions.

above. The precondition for this is that the previously assumed costs, based on relatively low construction rates, are correct. This could be determined after only a short time in the course of the required intensive development phase (within ca. 2 to 3 years).

- Concerning the innovation potential, we should mention a particular point: simply the introduction of supercritical steam circuits might well yield a considerable cost reduction. This is shown in Table 4.6 and in Appendix A for the "SunLab 220 MW" power plant. These advanced steam circuits for coal-fired power plants are just coming on the market. (However, for "SunLab 220," along with these steam circuits, lower heliostat costs were also assumed, namely $76 instead of 96 \$/m^2). In the USA or in Morocco, the advanced steam circuits would lead to energy costs "at the solar plant" of only 2.5 ¢/kWh (in spite of the lack of wet cooling), compared to 3.3 ¢/kWh (Table 2.4) and 2.4 ¢/kWh for nuclear power plants. (The energy costs "at the power plant" are relevant both in view of the provision of electrical energy for regions in the immediate neighborhood of the solar power plants, and for hydrogen production.) Supercritical steam circuits are, however, a special topic in connection with the development of solar power plants. Since there are no concrete statements on this point in the literature (and thus also not in S&L 2002!), we shall not consider them in more detail in this book. In order to achieve the higher steam temperatures required, the temperature of the molten-salt circuit (receiver, heat-storage reservoir) would have to be accordingly increased by ca. 100 °C. This, however, approaches the temperature range in which the currently used salts become unstable. What advantages such high-temperature steam circuits might in fact hold for solar power plants can thus not be readily judged. It is, however, clear that this possibly important opportunity has to be intensively investigated in a research program dedicated to the improvement of solar power plants.

If the present cost basis is thus confirmed, the costs in the later development stages of solar power plants will be lowered. Taking the considerably higher production rates into account in addition, we can expect correspondingly lower power costs.

- In contrast, for nuclear power plants it is to be feared that with strongly increased construction rates and accompanying limited construction capacity, a wide margin in the price calculation of the producers would be present; this could lead to prices that might be *substantially higher* than those quoted above, which follow closely the projected production costs based on current conditions and represent the lower limit of the cost range considered possible. Nuclear power plants, *in contrast* to solar power plants, are manufactured by only a few system suppliers owing to their complexity so that here a supplier cartel would be possible. This will be discussed in more detail in Chapter 12. A similar cost-increase effect is possible for the price of natural uranium.

"At the power plant," the cost difference between solar energy and nuclear energy is relatively minor (see Table 2.4). Thus for *site-proximate power-consuming centers*, the cost balance looks somewhat different from the above example (where the distance from the power-plant site was assumed to be 3000 km). Power transmission is unnecessary, or at most it is required over short distances. This holds, for example, for the West Coast of the USA, but also for Spain or the North African countries. With a view to the provision of base-load power, a backup system is indispensable. Insofar as the solar plants replace operating fossil-fueled power plants (which then, as mentioned, would be available "at zero cost" as backup plants), the energy costs would be roughly those given in Table 2.4 "at the power plant."

Energy costs "at the plant" are also relevant to solar H_2 production (electrolysis). Compared to nuclear H_2 production, however, somewhat higher costs for the transport and storage of the hydrogen gas must be assumed. In the USA, a difference of 1.5 ¢/kWh$_{H_2}$ would then result (see Section 4.1). In the case of truly large-scale production series, and with full realization of the development potential, the difference would probably be negligible.

Regarding the cost problem, we must remember that it is much more important to avoid high oil and gas import prices than to prevent such (tolerable) cost increases as described above. No one can predict the developments in the Near East with certainty. A scenario involving a sudden increase in the oil price to 200 \$/barrel sometime within the next 10 years is not unthinkable. (100 \$/barrel (in 2008-\$) corresponds to 6.3 ¢/kWh (2008-\$) or to 5.3 ¢/kWh (in 2002-\$).) If this should indeed happen (accompanied by the inevitable economic disturbances), it would not be important whether the replacement energy were especially cheap, but rather that it be available as quickly as possible. The question of whether, for example, solar hydrogen would cost 4.7 ¢/kWh (in 2002-\$) (= 90 \$/barrel of oil in 2008-\$), or 3.2 ¢/kWh (= 60 \$/barrel of oil), as for nuclear power (compare Tables 4.2 and 4.3), would then be irrelevant.

By the way, we should mention that the costs given here for nuclear power are based on future advanced reactors (generation III and III+) and adopted from a study carried out at the University of Chicago for the US Department of Energy (Chicago Study, 2004). The nuclear-power experts expect that these new reactor series will yield clear-cut cost reductions compared to the current reactors of generation II, in addition to improved security. Also, the costs assumed in the above estimates are based on very large production series for the nuclear reactors. The current reactors were considerably more expensive. These higher costs were accepted without complaint, and this would have been the case even if the reactors had been used on a much larger scale than was in fact the case.

2.3.5
"Hidden" Costs of Conventional Power Plants

In the cost comparisons given above, only the microeconomic costs of power generation were taken into account. This holds both for nuclear power plants and for coal-fired plants. However, power generation also gives rise to economic burdens for the general public, which are not contained in the construction and operating costs of the plants ("external" or "social" costs).

2.3.5.1 Nuclear Power Plants

In the case of nuclear power plants, these social costs refer especially to the risk of a major nuclear accident to the environment of the power plant. (Only the loss of the power plant itself as a result of the accident is insured, not damage to its environment.) This nuclear risk is naturally not easy to evaluate, and the various studies which have been carried out in the past years have arrived at quite different conclusions. The order of magnitude of the costs which can be expected becomes clear when one considers, for example, the study carried out by the noted Swiss Prognos Institute in 1994. It was prepared for a committee of the German parliament with the goal of estimating the costs associated with the risk of major nuclear catastrophes (Enquete Commission, 1994). The result was a price increase amounting to 3¢/kWh (2002)[13] for nuclear-generated electric power. Other studies have arrived at results lying notably lower, but some also higher. If we use the Prognos result as a working value, the costs of nuclear power would thus increase by 3¢/kWh (corresponding to the risk represented by currently operating reactors, on which the study was based); this means an increase in the value from 2.4¢/kWh[14] given in Table 2.3 to 5.4¢/kWh. This can be compared to the energy cost of power from the solar power system of 5.2¢/kWh (Spain) or 4.7¢/kWh (Morocco/USA).

For future reactor series with improved security – in particular for inherently completely safe reactors (the so-called generation IV) – this risk supplement would be negligible (however, these reactors are more expensive, and furthermore one expects for them a development time of the order of 30 years). The Prognos value is, as mentioned, not to be taken as a "scientifically verified" number, and there are estimates that lead to much higher and much lower costs. Nevertheless, we can use it to give a rough orientation. A decisive aspect for the comparison with solar energy is in any case that those who make policy decisions must completely accept these external costs, which has thus far been the case; supplementary costs of this order (or the associated economic burden) were thus in their opinion economically tolerable. Therefore, similar costs for solar energy should also be tolerable. Furthermore, the prognoses estimates include only those supplemental costs resulting directly from a possible nuclear catastrophe, but not the associated human and social tragedies. Another aspect that cannot be quantified is the long-term problem of nuclear wastes, which is being passed on to future generations, while the possible consequences associated with the proliferation of nuclear weapons-grade fissionable material cannot even be assessed.

One more general remark on such risk-assessment studies is as follows: they are in principle always based upon the calculation of "damage from a major catastrophe multiplied by the probability that such a catastrophe will occur." In a lecture ("On the responsibility of scientists to the public") at the Spring Meeting

---

13) Prognos: 1992, 0.046 DM/kWh (= 2.3 Eurocent/kWh 1992) = 3.0 US-¢/kWh (2002) (inflation in Germany: × 1.20; €1 = 1.96 DM; purchasing power parity (OECD) 2002: €1 = US $1.043).

14) It should in fact be taken into account that the value of 2.4¢/kWh is already based on a new generation of cheaper reactors. All in all, this simplified consideration should still be valid.

of the German Physical Society in 1994, Werner Buckel[15] (the former president of the European Physical Society) made the following statement about such calculations:

> "... However, I hold it to be unscientific to try to quantify the risk of a serious accident in a nuclear power plant in terms of probabilities. For the computation of a probability in the scientific sense, one needs a large number of events; or, if one wishes only to extrapolate, a complete set of parameters and their mutual interdependencies. We have neither for the nuclear power plants." (Buckel, 1994) Considering the above-mentioned possible level of damage resulting from a single nuclear accident, he spoke on another occasion of the "lack of sense made by risk calculations of the type 'multiplication of zero by infinity'." (Buckel, 1996)

If in fact it is possible to develop inherently safe nuclear power plants with a catastrophe probability of zero, this argument would become irrelevant as applied to reactor safety. Considering the no less relevant problem areas of permanent waste storage and proliferation, as well as environmental damage in the uranium mining regions, it however maintains its validity. For high-temperature reactors (inherent design), however, the reprocessing of fuel elements has yet to be perfected (and it is evidently more difficult than for current reactor types). Therefore, in terms of the supply situation for uranium alone, using high-temperature reactors as sole energy supply is not feasible (as is also true of the currently used light water reactors), so long as uranium cannot be extracted from seawater. Instead, fast breeder reactors would have to be employed. These sodium-cooled reactors are, however, generally considered to be anything but "inherently safe."

2.3.5.2 Coal-Fired Power Plants

For coal-fired power plants, in terms of the external costs we must distinguish between power plants with or without CO_2 sequestration. In the case of sequestration, that is, separation and absolutely secure storage of the CO_2 for very long times, the external costs are reduced (i.e., those related to the effects of the CO_2 emissions on the climate), but power generation is more costly. We have already discussed these supplemental costs in Section 2.3.2. With typical disposal costs of 10 \$/t CO_2, power from coal-fired plants in the USA would cost 4.2 ¢/kWh, and in Europe 5.5 ¢/kWh.

In the long term – according to the most recent climate studies even in the medium term – generation of base-load power with coal (and also gasification of coal) on a large scale will be permissible only with accompanying CO_2 sequestration. For large amounts of CO_2 on a worldwide scale, the terrestrial depots are

---

15) Professor Werner Buckel (1920–2003), previously director of the Physics Institute at the University of Karlsruhe, was president of the German Physical Society from 1971 to 1973 and president of the European Physical Society from 1986 to 1988.

probably too small (although in some countries they might be sufficient). Then, only disposal in deep ocean regions would be adequate. Whether this can be carried out in practice, whether it would guarantee secure and permanent entrapment, and what it would cost, are all questions which remain to be addressed. Investigations into CO_2 disposal in the oceans are still at the stage of "basic research."

The costs of terrestrial disposal are also still highly uncertain, since no one as yet knows the sites of the depots which would be suitable for this purpose (in the future, they would require a very large total volume). Whether or not CO_2 storage in geological formations can in fact be carried out with absolute reliability will have to be demonstrated by further research. The potential storage capacity appears to be sufficiently great in some countries (e.g., in the USA, with its large land area). In many countries, this is however uncertain (for more details see Section 11.3.3).

Concerning storage in the oceans, Jochem (2004) – quoting Tzima and Peteves (2003) and Mazzotti et al. (2004) – states that "The possibility of discharging CO_2 into deep ocean regions is dismissed by the majority of experts owing to its considerable risks (authors' note: the risk that the stored CO_2 would again escape into the sea and rise to the surface) and the associated ecological damage." Three German ministries in a joint report to the German Federal government likewise rejected ocean storage (BMWi, 2007).

If CO_2 is not sequestered, as is the case for current coal-fired power plants, the external costs due to the effects of CO_2 on climate change must be taken into account. There have been many scientific investigations into this extremely complex subject in the past 20 years. The general tendency of the results is that the significance of fossil-fuel power plants on *public health* is not as serious as considered in the mid-1990s, at least not in terms of its monetary effects[16]. Krewitt (2002a) quotes a value in 2002 for bituminous-coal power plants of 0.7 ¢/kWh, likewise (apparently using the same database) the European Commission (2003). In a new expertise for the German government in the year 2006, as a mean value for Europe (EU-25) in the case of bituminous-coal power plants, a value of 0.3 ¢/kWh is given (Krewitt and Schlomann, 2006). In the same expertise, in contrast, referring to the *greenhouse effect* of the CO_2 emissions (likewise from bituminous-coal-burning power plants), external costs of 70 $/t CO_2 are quoted; recalculated in terms of the amount of carbon in the coal, this corresponds to an increase in the coal price by 190 $/tce[17]; that is, a cost rise for power from coal-fired plants by over 5 ¢/kWh$_{el}$. The costs of power from modern coal-fired plants (without CO_2 sequestration) in Europe (using imported coal at 90 $/tce) of nearly 4 ¢/kWh would thus increase by more than 100% when these external costs are

---

16) These changes relate not only to potential technical improvements in the power plants (SO_2, NO_x), but also to new scientific findings (e.g., the effects of particulate emissions) and simply to a changeover in methods, in particular for the monetary evaluation. We refer the reader to the excellent analysis by Krewitt (2002b).

17) 1 t coal (tce) = 0.75 t C (carbon); 1 t C = 3.66 t CO_2; assumed efficiency of the coal-fired plants: 45% for the computation of ¢/kWh$_{el}$.

taken into account. A remarkable fact in this connection is that in spite of intensive research, the bandwidth of the cost estimates ranges from \$15 up to around 300 \$/t CO_2.

In the German expertise mentioned above, the following statement on this topic is made: "The effects of a global climate change are various and possibly very great. The interactions between the global climate system, the ecosystem, and the socio-economic system are extremely complex. A study by the English Environmental Ministry (UK Department for Environment, Food and Rural Affairs – Defra) comes to the conclusion that the costs for damage resulting from climate change are with a high probability greater than a lower limit of 15 €/t CO_2. Model calculations using the integrated assessment model FUND in this study show that with plausible assumptions, damage costs of up to 300€/t CO_2 result. After evaluation of the relevant literature and taking special account of the results of the Defra study, we recommend that as "best estimate" for computing the external costs due to CO_2 emissions, a value of *70€/t CO_2* should be used (lower limit:15 €/t; estimated upper limit: 280 €/t)."

2.3.5.3 Fossil-Fuel Backup Power Plants for the Solar Power System

The problem mentioned above basically affects a solar power system as well, if the backup power were to be generated by coal-fired plants. In that case, however, the CO_2 emissions would be comparatively small owing to the small percentage of backup power within the overall power generated. For solar power plants in Morocco or in the USA, only ca. 20% of the power would need to be generated by the backup plants. This should be compared with the remaining ca. 10% CO_2 emissions from IGCC gas turbine plants with sequestration (CO_2 separation efficiency 90%). For the case "USA/Morocco," the reduction of the CO_2 emissions would thus not be much less than for an IGCC base-load power plant (80% instead of 90%), but for the case "Spain" (with a backup power fraction of 30%), the CO_2 emission would be reduced to a lesser extent.

As mentioned above, gas-fired CCGT backup plants could be fueled with gas from coal gasification instead of natural gas; either with syngas (without CO_2 separation) or with hydrogen (with CO_2 separation, insofar as in the future, CO_2 disposal could in fact be carried out at the costs given). Hydrogen should be available through "advanced technology" both in USA as well as in Germany (there using cheap lignite coal) at a cost which *roughly* corresponds to the current price of natural gas (2.5 ¢/kWh$_{gas}$) (see Section 11.2). For hydrogen production, CO_2 is separated at the end of the gasification process and then disposed of. With this type of backup power generation, there would thus be no more CO_2 emissions from the solar power system (except for the remainder of ca. 10% which results from coal gasification).

We can summarize as follows: (1) for backup power generation using gas from coal gasification in CCGT power plants, CO_2 emissions from the solar power system can in principle be avoided. (2) Even in the case of backup power generation using conventional coal-fired plants, the emissions would be limited and future carbon penalties would play only a minor role in the final power cost.

2.4
Possible Time Scales for the Operational Readiness of Solar Thermal Power Plants and the Comprehensive Replacement of Current Power Plants

Some *fundamental* remarks on the question of time scales for development, which is discussed in detail below, and on the necessary research program were already made in the "Preliminary Remarks and Summary" and are presupposed here.

2.4.1
Special Aspects of Solar Power-Plant Development

For the substitution of oil and natural gas, the major options of nuclear energy, coal, and solar thermal power plants can be considered. The characteristics and problems of nuclear and coal-fired power plants are generally well known as a result of lengthy public debates on their relative merits. In contrast, those aspects typical of solar power plants have hardly been present in the public consciousness or that of decision makers. This applies especially to the question of how quickly and comprehensively this alternative energy source can be mobilized.

For solar power plants, several aspects are important which distinguish them *fundamentally* from conventional power plants. These will be discussed here using the example of solar tower power plants:

1) **The simplest technology (the solar field)**
   This applies in particular to the mirror systems, the main cost item of solar power plants. For solar tower plants, these are the *heliostats* which can be rotated about two axes. The consequence of their technical simplicity is that their development can be carried out very rapidly. It is indeed true that the heliostat field of a solar tower plant is exceedingly large. However, since this field is completely modular – consisting of many simple, identical structural elements – we can see that "large size and high cost" are by no means synonymous with "great development effort."

2) **Construction from mass-produced components**
   This has the consequence that the development tasks (regarding the solar field) can be concentrated more strongly on production aspects than on the technology of the heliostats themselves. The reliable predictability of mass-production costs is here an important – and for power-plant construction unique – element of the required research and development.

3) **Mainly conventional technology for the remaining power-plant components**
   A solar power plant has a completely conventional *electric power generating system* (power block), as in a coal-fired power plant.
   The *heat-storage system* also consists simply of insulated containers filled with molten salt. This salt is a mixture of sodium and potassium nitrates, two materials which have been produced by the fertilizer industry in large quantities for many years. Molten nitrate salts have been in use for some time also in the chemical industry as a heat transfer medium, although not at the composi-

tion required for solar plants. Therefore, the *heat-transfer circuit (molten-salt circuit)* of a tower power plant is in fact nothing new. The salt piping, the pumps, the associated control facilities, the construction materials, etc. need only to be optimized for the application at hand. The main difficulty will be overcoming the initial "teething problems"; this holds also for the steam generator, which is heated using the molten salts.

Only the *receiver* can be regarded as an essentially new component. There, the concentrated solar radiation is employed to heat the molten salt flowing through its tubing. The goal is to use tubes with walls as thin as possible, and solidification of the molten salts within the receiver tubing must be avoided at all cost. Along with a suitable choice of materials, careful dimensioning and a precise control of the salt flow are of great importance. However, even in the case of the receivers, one is in principle still in the realm of conventional technology (heat exchangers).

4) **A limited complexity of the overall power plant – the separate development of the components is possible**

In contrast to nuclear power plants or modern coal-fired plants (e.g., with integrated coal gasification), solar power plants involve a relatively simple technology even for the parts outside the mirror field. A similar conclusion holds for the interactions of these plant components with each other and with the mirror field. There is no complex overall process (such as in particular in a nuclear power plant with its many safety systems, redundancy of components, and the associated intricate control facilities), but rather the individual subsystems (mirror field, tower circuits, heat-storage systems, molten-salt piping, steam circuit with its cooling system) are essentially simply connected in series, without complex feedback effects. The result is that the individual components of the plant can be operated during the developmental phase essentially independently, that is, they can be developed and tested individually. For this purpose, only certain ancillary facilities are required, which replace the remaining power-plant components for the purposes of operational testing.

For example, the development of the heliostats does not require a molten-salt circuit (receiver) and can thus be carried out completely independently of the receiver development. For testing the optical quality of the heliostats, one merely requires a beam characterization system including a target area.

A similar conclusion holds for the remaining components of the power plant (and thus with reservations also for the receiver), that is, for the *thermal systems* of the solar power plant. These include all those components involved in heat transport or the conversion of heat energy into electrical energy, that is, heat storage media, steam generators, turbines, etc. One requires no solar field and no receiver to perfect all these components; instead, the molten salt can be heated using a *gas-fired* test facility. In this manner, the same temperatures, temperature variations, and salt flows can be obtained as in operation with solar heat. We are thus dealing with a complete *fossil-fuel power plant* with *liquid-salt technology*, consisting of a gas-fired heater (instead of the receiver), molten-salt piping, heat-storage system, and steam power block. With the gas-fired heater, temperature variations such as those that occur when clouds pass over the solar plant or during start-up and shut-down can be simulated precisely. All these tests and the complete development of the molten-salt components can thus be carried out with such a facility. The planning and construction of such a fossil-fuel test facility could be begun immediately, without the need to wait for the completion of a solar field and a tower installation.

For the receiver development, the situation is basically similar. Here, too – very probably – test installations will be possible in which the concentrated solar radiation can be simulated: namely by using the thermal radiation from a combustion chamber (natural-gas-fired radiation chamber). In such a chamber, similar radiation densities should be attainable as from a real solar field. (Such a relatively simple gas-fired installation could be planned and constructed within a short time.) If the required radiation values cannot quite be achieved, the receiver components would have to be tested additionally in the established test centers at Almeria or in Albuquerque (Sandia), or possibly in the available (relatively small) solar tower power plants (Solar TWO in the USA (if it can be reactivated) or PS 10 in Spain). If the existing tower power plants are – contrary to expectations – not sufficiently large for this purpose, one of them could be expanded relatively quickly into a complete large receiver test installation[18]. The optimization of the technology for the receiver could thus be begun comparatively quickly. The receiver will, however, most likely be the component which requires the longest development time (but also not an extremely long time).

In summary, for the development of the individual components, a large solar test power plant is *not* required; on the contrary, this would even be disadvantageous.[19] After the individual parts have been tested in the manner suggested above and optimized for mass production, one can begin with their fabrication on a large scale.

5) **The interdisciplinary character of the development program**
Development of solar energy cannot be limited – in contrast to the development of nuclear or coal-fired plants – to a particular special subject area. Nearly

---

[18] Since currently available heliostat designs could be utilized for this expansion (even though they are not optimized in terms of cost), an enlargement of the mirror fields could be begun immediately and the expansion program could be completed very rapidly, if needed (at an increased cost and under time pressure).

[19] In the case of nuclear or coal-fired (IGCC) power plants, such large research and development facilities were and are indispensable. There, several stages of improved and increasingly larger test plants were constructed in order to profit from the experience gained with the previous stage (scaling-up procedure).

In the area of renewable energy sources, within the research support and planning in the past one could discern a tendency toward the construction of large, expensive, but not very innovative test facilities (public "justification" of the R&D activities). A new, larger solar power plant (using a single technology, e.g., for the heliostats) would yield only minor advances in the technological development, but would cause an immense loss of time.

all the topics for research require a broad-based, interdisciplinary approach. By this, we mean that only a very small portion of the tasks lies within the field of development of solar technology in the narrow sense. Most of the tasks are situated in other areas; an example is the determination of the costs for mass production of the heliostats.

6) **Special requirements for the organization of the research and development program**
The characteristics mentioned above have serious consequences for the organization of the required research program. Given their special nature, an orientation toward a research organization such as is applied, for example, to the development of nuclear power plants (in particular the establishment of new, large central research institutions) would not be expedient.

In the following sections, some of these special characteristics will be discussed in more detail.

2.4.2
The Simplest Technology – Consequences for Development and Construction on a Large Scale

The decisive factor, namely the *simplicity* of solar thermal power plants, will be demonstrated in more detail in this section using the example of the heliostats. The reader is expressly challenged to form his or her *own* opinion in this connection. This is readily possible even for nonprofessionals, given the clarity of the situation. Only then will he or she be sufficiently prepared to judge the numerous inaccurate predictions regarding the necessary development time which are currently circulating, that is, regarding the time when solar thermal power plants could be operational on a scale relevant to the energy economy. At present, the date when a substantial contribution from solar thermal technology to the overall energy supply can be expected is frequently estimated to be in the years 2040–2050. Such estimates are due solely to unthinkingly equating solar energy to the classical energy technologies in terms of the aspects discussed here. With nuclear or coal-fired plants, the technological complexity indeed determines the speed of development, and it sets a limit to an increased annual rate of deployment (in particular in the case of nuclear power plants). While a nuclear power plant counts among the most complex facilities yet constructed (even in terms of its security systems alone), in the case of solar power plants, we are dealing with – one is tempted to say – a primitive technology. This becomes most obviously clear on considering their major cost item, the heliostats. Their technology is by the way also very much simpler than, for example, that of an automobile, which is a typical mass-produced item.

Figures 2.3–2.14 show various designs for glass-mirror heliostats. Their simplicity can already be seen in the fact that – apart from two movable joints and the associated drive units – they contain only static structures, namely the supporting frame for the mirrors and the pedestal, which is anchored to the base. The sup-

Figure 2.3 ATS heliostat (front side, as in Figure 2.4) (SANDIA).

Figure 2.4 ATS fourth-generation prototype (148 m^2) from the USDOE large-area heliostat development program (1985–1986) (SANDIA).

Figure 2.5 The solar tower power plant PS 10 near Seville (Spain): 624 Heliostats (121 m^2) (Photo: DLR).

Figure 2.6 The heliostat field of PS-10 (Spain) (Photo DLR).

Figure 2.7 A heliostat of PS-10 front side (SANDIA).

Figure 2.8 The heliostat field in front of the CESA 1 Tower (Almeria, Spain) (Photo DLR).

porting frame must be able to hold the mirror in the required position even against wind forces, and must not bend too much under load. In spite of the different shapes and sizes which are shown in the figures, all these various frames are constructed merely from elementary components such as sheet metal, angle irons, tubes, or rods. The frame rests with the two rotating joints on the pedestal. The

Figure 2.9 SAIC heliostat design (145 m^2), with 3-m diameter stretched membrane mirror modules (SANDIA). Note the "focal point" of the heliostat (the image of the sun) on the front face of the tower.

Figure 2.10 The back side of the heliostat in Figure 2.9 (NREL).

Figure 2.11 ESCOSolar 20 (Photo DLR).

latter is usually a large steel tube with branches at its lower end to connect it to the base.

For the heliostat designs which have thus far been favored, the drive motors and the gear drives must indeed meet high standards in terms of precision and stiffness under wind load, and at the same time they must be able to withstand strong forces during storms. But they remain entirely within the realm of conventional technology (electric motors, gear drives, linear actuators). The actual (movable) power train of a heliostat thus consists of only two motors and two gear drives.

In contrast, an automobile contains a large number of power trains and components such as motors with gear boxes, heating and air conditioning units, power steering, braking systems, indicators, and numerous electrical drive systems, quite apart from the complex body with its many preformed sheet-metal parts. All together, it contains several thousand different individual parts. The development of an automobile and the associated production system thus involves an immense package of technologies.

A heliostat, however, consists of only a very few *different* individual parts. This becomes especially clear in Figure 2.12 (development of a small-scale heliostat) and Figure 2.14. Large heliostats naturally contain more and larger (but mostly identical) parts; the number of "different" parts increases only slightly (compare Figures 2.4 and 2.5); this becomes still clearer in the case of the collector for a

Figure 2.12 Design of a "small heliostat" (8 m^2) (SHP Australia, now AUSRA) (Photo Solar-Institut Jülich).

parabolic-trough power plant; see Figures 2.15–2.18. The development of a heliostat thus does not correspond to that of a complete automobile, but to only one small portion of one.[20]

This simplicity can already be seen in the relative costs of the technologies. A 150-m^2 heliostat with mass production will cost ca. \$13 000 (83 \$/m^2). The first prototypes are of course more expensive. But even if their *fabrication* costs 10 times as much, this would still be only \$130 000 per heliostat. For the development of nuclear power plants, one is faced with the task of constructing a prototype reactor

---

20) Somewhat exaggeratedly, but descriptively, one might say: The development of a heliostat (with only rotational motions, but around two axes) corresponds more nearly to that of an "electrically operated trunk lid" than to a whole automobile.

Figure 2.13 Heliostats similar to those in Figure 2.12, here in the solar test field at Jülich, Germany (Photo Kraftanlagen München).

Figure 2.14 Solar Energy Development Center of BrightSource Energy, Negev Desert, Israel (BrightSource Energy, Oakland, CA).

with an output power of 1300 MW at a cost of ca. $3 billion, that is, at a production cost and effort roughly 20 000 times greater. Although, of course, such a simple cost comparison cannot be taken too seriously in terms of its information value, and although heliostat development in reality involves much more than just the construction of *one* heliostat prototype – namely the construction of many different

Figure 2.15 A parabolic trough (Photo DLR).

Figure 2.16 The supporting frame for the mirrors of a parabolic trough power plant (SBP).

Figure 2.17 Mounting of a parabolic trough collector in the field (SBP).

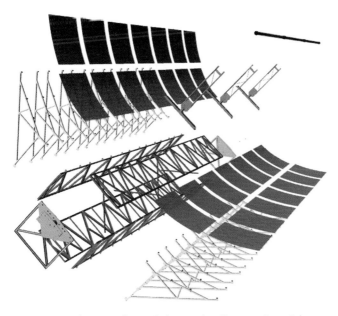

Figure 2.18 Elements of a parabolic trough collector (schematic) (SBP).

types and especially the preparation of cost predictions for each type – this comparison still makes the fundamental situation clear: the *completely* different order of magnitude of the development tasks.

The simplicity of the technology affects not only its development, but also the later stage of *mass production* to a great extent. Let us consider the support frame, which as mentioned consists of elementary steel parts that have to be welded together. These parts could be fabricated in almost any factory in the metalworking industry. They could and will presumably be delivered by a number of different manufacturers. If necessary, the production of even one particular part could be carried out by several different firms. The available production capacity could thus be utilized to its full breadth. In an emergency situation, all the free metalworking resources of a country could be mobilized for this production. This sort of distribution of the fabrication is of course not necessary, but the decisive point is that it would be *possible* if required for a *very rapid* increase in high production rates.

The individual parts, which were fabricated all over the country, would be transported to a central plant for *assembly*, where they would be combined into support frames (or also into a complete pedestal including attachment and mounting parts). In these assembly plants, the relatively large support frames would thus be produced and aligned. Due to the enormous dimensions of large heliostats, the assembly plants will certainly be set up in the neighborhood of the solar power regions.

In view of the very large numbers of such components and the simplicity of their assembly from only a few different parts, practically all the assembly steps could be carried out automatically, including feeding of the parts within the plant. Since a single assembly line would not be sufficient, the production would have to be spread over several parallel lines. As these would be identical (once the production technology had been tested and optimized), they could also be constructed quasi under "mass production" (worldwide, e.g., a total of 10 or 20 assembly lines), which would permit relatively short planning and construction times.

Furthermore, these plants, in contrast to automobile assembly plants, would remain practically unchanged over many years once they were set up. For here, unlike automobile factories, one would not have to contend with ever new models or variations. The small number of different parts (and thus different assembly procedures) simplifies the automation in general insofar as only a few types of assembly robots would be needed.

The complete large support frame, with mirrors and motion drives, will be mounted on the previously delivered and anchored pedestal in the solar field itself. This will also not be done "by hand," but instead using mobile automatic or semiautomatic machines. The same is true of the attachment and concrete embedding of the pedestal. Here, also, heavy machinery with manipulators will be used, which carry out at least the major portion of the work automatically. The material flow within the solar field will probably be managed by automatic transport systems.

Heliostats thus can be produced and installed rapidly in very large numbers.

This technical simplicity leads by the way to an additional important difference in comparison to conventional power plants: in spite of large-scale mass production, the fabrication of the solar components will not lead to oligopolies or cartels. A development of this type is to be feared in the case of nuclear power plants, in particular if their deployment rate were to be drastically increased. If a manufacturer of individual solar parts (sheet-metal parts, tubing, etc.) were to ask too high a price, one could switch readily to other suppliers without major problems. The setting up of assembly lines is likewise a comparatively clear-cut task (e.g., in comparison to an automobile plant). Here, also, many suppliers could compete for contracts to construct the production robots. The general precondition is of course that the *whole* mass production of the heliostats not be put into the hands of a single private supplier (or only a few); otherwise, one would be subject to their price diktat, in spite of the simple fabrication procedures. Instead, the production should remain under the control of the power-plant operator, possibly with the support of subcontractors (e.g., large planning agencies), who would organize the construction and operation of the assembly plants.

2.4.3
The Basic Development Tasks for Heliostats

As we have seen (regarding the cost of a prototype), the fabrication and construction of a single new heliostat initially represents only a marginal task. With the construction of *one single* variant, however, the task is not completed. Thus, the goal of the development program is to identify those heliostat designs that will be cheaper to deploy in mass production than currently existing types. This requires the development of many different types; and for each of these designs,

a) its stability must be tested, and
b) its costs under mass production must be determined *reliably*.

While the investigations of stability concern purely technical questions (mostly pertaining to the statics of the mounting system and frame) and the performance of tests, the determination of mass-production costs raises a multitude of questions for each of the designs. These encompass the entire production process for each type. A reliable determination of the costs thus includes a number of individual steps and is, therefore (in terms of its scope), one of the major tasks in the development of heliostats. Secure knowledge of the production costs is not only necessary for the identification of the *least expensive* solution among the various heliostat designs, but also it is indispensable for the estimation of the economic potential of the whole solar energy supply system. In the following, we consider both of these developmental tasks more closely.

2.4.3.1 Stability
This point will be described in detail in Section 6.4; in anticipation we mention here only the following: an investigation of the stability of heliostats can naturally not be carried out by first constructing the heliostats and then subjecting them to environmental influences and waiting until major stresses occur (storms, hail, sandstorms, earthquakes, and possibly snow and ice), in order to test their serviceability. That would take entirely too much time (and gaps in knowledge of their

resistance to certain environmental stresses would still remain).[21] A rapid and reliable investigation presumes the existence of *test installations*, in each of which a particular type of stress can be simulated:

- wind tunnels of a corresponding size (or other wind test installations for the large heliostats; cf. Section 6.4);
- sand blowers in combination with wind machines for testing leakage of the motion drives (and other parts of certain types of heliostats) in sandstorms;
- test beds for earthquake simulation;
- refrigerated chambers and snow machines for testing sensitivity toward snow and ice (for mountainous sites in parts of the USA and Spain), as well as hail machines, which likewise already exist (in the event of snowfall and hail, the mirrors are brought into their vertical positions);
- for plastic-foil mirrors, which are employed in a few heliostat types, aging effects must be investigated to estimate the operating lifetime of the components and how often replacement will be necessary.

We can see that with the exception of the more difficult questions regarding aging of plastics, the problems of stability with respect to environmental influences can be clarified reliably and quickly once the corresponding test installations are available. Thus, as soon as such a "test park" for heliostats has been set up, the actual test experiments are simple and can be carried out speedily, and represent on the whole a purely routine task.[22]

2.4.3.2 Cost Predictions

The "actual" task of heliostat development is the preparation of a reliable cost analysis for each design under consideration. This, as mentioned above, is a broad-based and multifaceted task. Not least due to this task, the "interdisciplinary" procedures are necessary. Such tasks are, however, in principle very ordinary in terms of the mass production of other items. Thus for the development of a new automobile model, such a cost analysis of the mass-production costs of new parts or assemblies is indispensable.

The costs need to be calculated for the following areas:

---

21) This "test procedure" was however employed in the past – owing to a lack of alternatives (i.e., lacking a systematic development program) – naturally complemented by calculations and simulations with models. Thus, Kolb *et al.* (2007a, p. 29) write: "The 148-m² ATS heliostat has successfully operated for the last 20 years at the NSTTF in Albuquerque. It has survived multiple high-wind events, some in excess of 90 mph ..." Better than such a waiting procedure, however, would be a realistic examination of the stability at high wind velocities in a test facility. Furthermore, the heliostat types being tested could also be set up in locations where the maximum design wind velocity occurs frequently.

22) Even aging effects in plastics can be at least roughly investigated using time-lapse procedures. This of course also requires the corresponding test facilities.

1) fabrication of the individual parts
2) assembly
3) field installation.

Fabrication of the individual parts. Regarding the costs of individual parts, we look once again at Figures 2.3–2.14. One can readily see how simple the individual components are. Their fabrication costs will thus be easy to determine. This applies in particular to the static components of the heliostats. The gear drives and motors are, however, likewise typical mass-produced assemblies, whose production costs can be estimated precisely and with relative simplicity.

Assembly. The situation regarding assembly is different. In order to determine its cost, the appropriate assembly line for mass production must first be designed. Insofar as the assembly process involves conventional handling steps such as welding, inserting screws etc., the costs can also be determined quickly and precisely (in view of the many comparable procedures in other applications of mass production). For new process steps or unusual dimensions of the parts, however, the corresponding assembly robots must first be designed, built, and tested, since there are no direct comparisons available. Although only a few steps in the overall assembly process would be in this category, this remains the most extensive and difficult task within the entire heliostat development program, in particular since it must be carried out anew and in a different manner for each heliostat design under consideration. This setting-up of assembly lines is thus (together with the construction of a test park as mentioned above) the *essential* development task. Here, incomparably more must be designed and constructed than for the heliostats themselves.

Thus far, there has been only a single investigation of production costs which deserves to be called *detailed* and which also (at least partially) includes the planning of production facilities and assembly lines: The *General Motors* study of 1981 (see Figure 2.19). Whether or not the *whole* production process was planned in this study to the degree of detail required today, and the corresponding investment costs were estimated, cannot be seen from the available reports.

What was said above concerning assembly naturally holds in particular for the "nonclassical" heliostat designs, such as membrane heliostats. There, considerably less experience can be drawn upon from conventional assembly processes. Several new process steps such as, for example, attaching the membrane to its mounting ring are involved. It must thus be assumed that more new assembly robots would have to be designed than for the other heliostat types. But it must still be kept in mind that here, too, only a few different individual parts are involved so that on the whole, the effort required for these developments should remain on a manageable scale.

Field installation. For field installation, the situation is basically similar. Here, too, the costs of the required automatic machines will have to be determined. In contrast to the assembly of the parts, these machines would however be rather similar for the various heliostat types. The installation includes mounting the

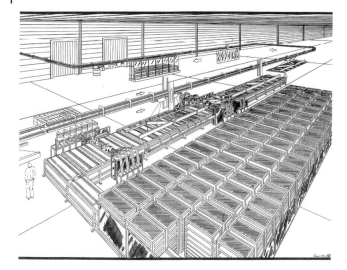

Figure 2.19 A figure from the detailed mass-production analysis performed by General Motors (1981), in which an assembly line was proposed that would be capable of producing 50 000 heliostats (McDonnell Douglas type) per year (SANDIA).

mirror frame onto the pedestal, attachment of the electrical connections, and adjusting the heliostat; furthermore constructing the base including attachment of the pedestal. Likewise, for all the heliostat types, the automatic or semiautomatic systems for material transport within the solar field are all practically the same. Nevertheless, at the beginning a considerable amount of development work will have to be carried out.

Thus, although detailed predictions of production costs lead to a whole series of development tasks and although various types of heliostats must be designed and tested, the costs of heliostat development will in the end be limited to a "modest" level. An estimate of these costs at present, without concrete examples, would be simply speculative. Nevertheless, a range from a few million dollars to some $10 million for the average development costs of each heliostat type would appear to be plausible. If we take $30 million per heliostat type and assume that 10 types would be considered, each differing substantially from the others (i.e., representing more than just minor variations of a particular type), this would yield an overall development cost of $300 million. Amounts of this order are, however, insignificant in comparison to the typical costs of energy research and development.

A certain indication of the price of investigations of production costs – namely the General-Motors study mentioned above – is given by Kolb et al. (2007a, p. 28): "The heyday of heliostat development in the United States occurred during the second-generation period ending in 1981. The DOE budget for heliostat development was $7.3 M, equivalent to *$19 M* in today's dollars. This budget level allowed for extensive optimization and cost studies, and more than 100 technical references can be found in Mavis [5]. An example is the detailed mass-manufacturing analysis performed by General Motors [6] in which assembly lines were proposed capable of producing 50,000 McDonnell Douglas heliostats per year. ..." ([5]: Mavis, 1989; [6]: McDonnell Douglas Astronautics Company, 1981.) The cost of the General Motors Study itself was not given. Since the entire development program cost only $19 million (in current monetary value), it can be assumed that this investigation cost at most $10 million. Whether it was sufficiently profound (i.e., sufficiently reliable in terms of current standards) is not clear.

Concerning the time required, as mentioned, one must always keep in mind that there would be no mutual hindrance due to the parallel development of different types of heliostats. With the correct organization of this parallel development program, the time for the overall development corresponds to the development time for a single type of heliostat. Nearly all the tasks of heliostat development (construction of the test park, planning of the assembly lines, design and construction of the plants for individual assembly steps) could be carried out in a short time in the framework of a program designed to be completed rapidly; probably within ca. 4 years. The most important intermediate results could be available even within 2 to 3 years.

In emerging countries without their own nuclear power plant construction industry, another important aspect of the development should be mentioned. Solar technology, owing to its simplicity, could namely be readily applied in these countries using *their own resources*. For nuclear power plant construction, these nations (if they wish to avoid an extremely time-consuming development program of their own) would have to rely on importation of plants, or at least on cooperation with internationally operating nuclear power-plant constructors.

These could, however, dictate their own conditions to a great extent. If either whole power plants or even parts of them must be imported, this would cost hard currency. Solar power plants, in contrast, could not only be manufactured within the country later, but also could be independently developed there. Having their own development program without the constraints of a cooperation would permit later plant construction without the involvement of other countries. In the present book, we in fact compare the economic characteristics of different power-plant technologies in the industrialized nations; there, this aspect plays no role. In the emerging industrial nations, however, a quite different cost relation in terms of different types of power plants might be obtained in the case of completely independent design and construction of the solar plants.

2.4.4
The Most Important Single Point: A Cost Study for the Standard Heliostat

As we have just discussed, an important part of heliostat development programs are cost analyses. Among these, the first would at the same time be the most important: a precise analysis of the mass-production costs of the standard heliostat, that is, complete and detailed comparisons with established processes for mass production, including the design of individual facilities for assembly and field installation. This first major investigation should be carried out with a high priority in order to obtain results as soon as possible. Then, we would finally have secure knowledge about the greatest cost factor – in the case of solar tower power plants – for the overall solar power supply system. Together with the analysis of the costs of molten-salt thermal circuits carried out in parallel, one would then have extensive information on the current developmental state of solar tower power plants. (Thereafter, one is dealing only with further developments and optimization.)

This investigation should, owing to its fundamental importance, even be carried out redundantly and in parallel by completely independent research groups. Differences in the results would then give indications of their reliability (i.e., reproducibility).

2.4.5
The Interdisciplinary Character of Solar-Plant Development

Nearly all the important questions will have to be treated in an "interdisciplinary" fashion. Some examples that we have already mentioned include the following:

- Cost estimates for heliostats by
  - comparison with costs in the automobile industry
  - planning of the production procedures and in the process
  - design of specialized automatic production equipment
- Construction of a test park for heliostats.

This interdisciplinary character, however, holds also for many other development tasks such as:

- development of the components for the molten-salt circuit, including;
- construction of simulation facilities for testing the plant components;
- a series of peripheral questions such as investigations of the insolation, of potential plant sites in Spain and other countries, of dry cooling at the corresponding locations, or of the infrastructure required by the solar-plant regions;
- further development of long-distance power transmission technologies (superconducting transmission lines, undersea cables);
- hydrogen production (among other things the development of high-temperature steam-phase electrolysis) and methanol production.

Nearly all these topics would have to be dealt with by "nonsolar" experts. Solar energy experts could thus carry out only a small portion of the required development program by themselves. Essentially, that portion consists of the technical development of the heliostats, the design of the solar field, and naturally, for example, studies of the overall project.

2.4.6
Consequences for the Organization of Research

The simplicity of the technology on the one hand, and the interdisciplinary character of the research and development required on the other, which goes well beyond solar technology itself, make the following suggestion regarding the organization of the development program attractive.

A large administrative apparatus with its own technical competence should *not* be set up![23] Instead, an *"innovation council"* should be established, which would in the end take all the important substantial decisions,[24] in which it would be supported by a *small* technical and administrative staff. This council should consist for the most part of "external" members, that is, not primarily of solar power-plant experts, but rather of engineers and scientists with proven *experience in innovation*, even though they might come from other (nonsolar) technical areas. Since the most important tasks lie outside the narrow area of solar technology and since many new concrete tasks would arise only in the course of the program, for these council members, a proven competence in innovations and their management would be more important than detailed knowledge in the field of solar technology. The relatively small amount of specifically solar-technical knowledge required could be made available to the council by solar experts. Furthermore, solar technology, owing to its relative simplicity, is readily accessible to anyone with a scientific-technical background.

In the case of the immediate initiation of a massive development program, a correspondingly large council would be required; it would be subdivided into a number of working groups. Such a collection of persons who were involved in the project only as "consultants," that is, with a limited time commitment (alongside their other occupations), could be quickly recruited in spite of its large size. Then the many required tasks could be rapidly carried out. After only a few years, the council could again be reduced considerably in size.

2.4.7
Industrial Initiatives and Start-up Funding

The Desertec initiative of the Club of Rome, which has been launched in Germany (see e.g., Desertec, 2009) and which beginning in the summer of 2009 is also being supported by large industrial companies and is currently entering a concrete planning phase, represents in the main an interesting *complement* to the systematic and in particular more comprehensive development program for an energy system

---

23) For nuclear energy, the situation is different: in that case, securing a particular technical competence within the project administration is necessary, that is, setting up a staff of professionals who have expertise on nuclear-technical and also legal questions, and who accompany the further development of a reactor series over decades after their introduction.

24) Above all, the council would have to award contracts for development work to companies working in the corresponding technical areas, and would be responsible for the evaluation of the results.

based on solar power plants, which we suggest here. This initiative can, however, not *replace* such a program. It appears essentially to be aimed at the large-scale realization of a version of the technology which is technically and economically still an intermediate phase. This strategy is expensive and time-consuming, insofar as one considers the (relatively low) power costs discussed in the present book to be a precondition for the large-scale application of solar technology. While the Desertec initiative thus presumes that the current technical state-of-the-art (with moderate further developments) is sufficient for a first major step, and that *governments* should make up the difference to full economic feasibility through corresponding subventions, from a long-term energy-political and national economic point of view, the primary goal must be to elucidate as quickly as possible just what the economic potential of solar energy in a state of advanced development could in fact be (especially considering the likely future increases in oil and gas prices). By initiating a comprehensive and immediate "crash program," cost-favorable technology must be made available as soon as possible, in order to begin in earnest with a true energy turnaround.

The new (industrial) Desertec plan has thus far (July 2009) still not been published in detail. But in any case, it appears – with an overall investment volume of around €400 billion (roughly $550 billion) – to be aimed, at least in its initial phase, toward the deployment of parabolic-trough solar plants in the Sahara, similar to those which have already been constructed in Spain.[25]

This initiative – if it indeed goes beyond the stage of "planning" or of a "memorandum of understanding" – could produce an enormous push forward for solar energy within the current political and economic framework. The commitment of very large firms makes it clear to everyone that this technology has great economic potential, and thus lays the political foundations for a large-scale governmental engagement.

Desertec is in particular welcome if the volume of its (subsidized) technical and economic intermediate stage remains relatively modest, and if its conceptual political framework is open to a rapid and broad-scale exploitation of the full economic potential of solar energy. As an "alternative" to the large-scale development program suggested here, however, the disadvantages of the Desertec plan would predominate. Unfortunately, there is reason to fear that just this might happen, since this private-economy initiative has a special "advantage": it takes the pressure off the governmental ministries that would be responsible for the planning and carrying out of a multifaceted R&D program, and frees them from this task and challenge. From the point of view of the responsible agencies, this is the simpler path. It requires only a law permitting subsidies, instead of the manifold measures necessary to carry out a full development program for which there are few role models. The higher overall costs and the time lost may appear to the bureaucracy to be of secondary importance.

---

25) Parabolic troughs require sites with flat terrain; these are available to the required extent only in North Africa, not in Spain. This is quite possibly one of the reasons for the fact that the Desertec plan concentrates on sites in North African countries.

3
Solar Technologies – An Overview

Before we continue discussing the economic aspects of solar energy, and enter into details in the following chapters concerning solar tower plants, parabolic-trough plants and chimney power plants as well as the requirements for their further development, we first present a systematic (but not exhaustive) overview in this chapter of the already existing variety of solar thermal concentration technologies.[1] In addition, at the end of the chapter we will briefly introduce the "downdraft tower." Chimney power plants, however, will be discussed in more detail later in Chapter 9.

Fundamentally, there are two methods of optical concentration of (direct) solar radiation: *point focusing* and *line focusing*. Both can be achieved either by using the paraboloid principle or the Fresnel principle.

The *paraboloid principle* is based on mirrors with a precisely parabolic cross-section. Mirrors with rotational symmetry, that is, with a dish shape similar to TV satellite antennas, provide point focusing (the *dish* principle). Mirrors in the shape of a long trough with a parabolic cross-section produce focusing onto a line (*parabolic trough* principle).

The *Fresnel principle* allows focusing with the aid of many planar mirrors which are set up on a level surface (e.g., on the ground) and are individually adjustable. When a large number of sufficiently small individual mirrors are used, focusing onto a point (the *tower* principle) can be achieved; if sufficiently narrow but relatively long mirrors are used, focusing onto a line is obtained (the *linear Fresnel* principle or simply the *Fresnel* principle).

The broad field of technologies is ordered in this chapter according to the principle by which the concentration of radiation is accomplished. Departing for the moment from the main focus of this book, which is on tower power plants, in the following section we begin by introducing the dish technology, since it represents a nearly ideal example of the concentration of solar radiation.

---

1) Readers who are primarily interested in the economic aspects of solar tower power plants, as discussed in Chapter 1, can skip over the present chapter.

Large-Scale Solar Thermal Power. Werner Vogel and Henry Kalb
© 2010 WILEY-VCH Verlag GmbH & Co. KGaA, Weinheim
ISBN: 978-3-527-40515-2

Figure 3.1 Dish plants with a continuous mirror surface (SANDIA).

3.1
Dish Plants

Using a paraboloid-shaped mirror (reflector), solar radiation can be most strongly concentrated, namely with a concentration factor in the range of 1000–3000. The system tracks the Sun so that the radiation is always incident parallel to the mirror symmetry axis. For the dish concentrator, there are essentially three different types of construction: it can be made with a continuous mirror surface (Figure 3.1), or composed of many closely spaced, slightly curved individual mirrors (facets) (Figure 3.2), or made with a few separate concentrators having a common focal point (Figure 3.3). (For the last method, the basic principle of a continuous parabolic mirror surface is no longer fulfilled.) The mirror surface areas are typically in the range of 50–100 m^2. In the radiation absorber ("receiver"), which is at the focal point of the mirrors, attached rigidly to the concentrator, temperatures of ca. 800 °C are produced. Electrical energy is then obtained from a thermal engine coupled to an electric generator. For this purpose, the first choice is a Stirling motor, which in principle is predestined to make use of an external heat source (in contrast to the more common internal-combustion engines), and which forms a unit together with the receiver and an electric generator. Since the Stirling motor, however, does not seem yet to be a completely mature technology, the microturbine is currently also under discussion; remarkable progress in its development has been made in recent years.[2]

---

2) The published data on the Stirling motor are somewhat contradictory. Often, a high degree of reliability is mentioned; on the other hand, statements such as "... availability has to be proven" are found.

Figure 3.2 A dish plant with mirror facets (SANDIA).

Thus in general, each dish installation produces electrical power independently (ca. 10–50 kW) so that its typical area of application was considered up to now to be in providing local energy supplies in sunny regions. This was held to be a renewable-energy alternative (and somewhat cheaper than photovoltaic cells) to diesel engines for local power generation, independent of the power grid, especially in developing countries.

For a long time, the dish technology appeared – even in the long term – to be the most expensive solar thermal technology and was, therefore, not of interest for the utilities industry. Furthermore, dish systems, in particular dish-Stirling engines, are based upon a very direct transfer of the heat collected in the receiver to the thermal engine, and therefore heat storage is out of the question for them, or at least it was not included in the systems built and tested in recent years. Their economic relevance for power production on a large scale thus in the current view remains limited to local power generation and peak-load support for power grids so that they are not at the focus of the considerations in the present book.

It is then surprising that recently (since 2005) in the USA, there is talk of plans to install large-scale systems with a great number of dish-Stirling units before 2010, for an output power of several hundred megawatts, corresponding to thou-

Figure 3.3 Dish Stirling solar power system with separate concentrators (SANDIA).

sands of individual installations[3] (Edison, 2005; Sempra Energy, 2005). Whether such plans can be put into practice in the near future is a fascinating question. Precise economic data are not yet available. But these plans must nevertheless be taken seriously, since there are already preliminary agreements between the producer (SES) and two energy suppliers (SCE and SDG&E).[4]

Initially, one would imagine a large number of self-contained small power units which could be electrically connected into a large power plant. There are, however, also plans to employ the dish installations only for concentration of the solar radiation, and to pass on the heat collected at high temperature in many receivers via heat-transfer fluids (HTFs) to larger central generating units. The question of heat storage would then need to be reconsidered.

At the beginning of the recent phase of development of the dish technology, that is, in the early 1980s, there were in parallel to the development of dish-Stirling

---

3) This is in fact surprising only at first sight. For example, as early as 1997 in a DOE report, the potential of the dish principle was emphasized with great clarity. In the chapter on the "Solar Dish Engine," it is stated that: "... by 2010 dish/engine technology is assumed to be approaching maturity. A typical plant may include several hundred to over a thousand systems." DOE, 1997).

4) Stirling Energy Systems (SES), Southern California Edison (SCE), San Diego Gas & Electric (SDG&E).

plants also some larger projects for central power generation: one in Kuwait and two in the USA. The largest of these projects was a 5-MW plant in California with 700 dish concentrators. These were conceptually and technologically not mature systems, however, and could not establish themselves successfully. In the case of the two smaller power plants, the use of a synthetic oil as cooling medium made it impossible to achieve the potentially high operating temperatures of the dish principle, and the simpler parabolic troughs are sufficient for lower temperature operation (see below). The plant in California operated at higher temperatures (using direct steam generation in the receiver), but in the end it was also not successful (Mohr et al., 1999).

Dish systems with central power generation could in the future again become interesting if it proves possible to adopt receiver cooling (and simultaneously heat storage) using molten salts as in the tower-receiver technology; this has been suggested by Murphy et al. (2002). But direct steam generation (DSG) in the receiver also remains an option; it has been employed in particular in Australia since the early 1990s, together with especially large dish reflectors ("Big Dish," $400\,m^2$) (Lovegrove et al., 2006). Finally, we mention also that in Australia, since the 1990s, a heat-storage concept based on ammonia is being investigated. In this process, solar heat at high temperature is used to split NH_3 catalytically into nitrogen and hydrogen and – after storage – it is induced to recombine catalytically, accompanied by the release of heat[5] (Lovegrove et al., 2004).

3.2
Tower Power Plants

For large-scale plants – in particular in the range of several hundred megawatts – it is probably more economically reasonable in the long term to focus the solar radiation onto a single fixed "point" by using a large-area field of movable, nearly planar mirrors (heliostats). At this central point, the absorber (or central receiver) is mounted on a high tower and in it, high temperatures are obtained (of the order of 600–1200 °C, depending on the cooling medium and the type of receiver[6]).

---

5) The second – exothermal – step is in principle simply the Haber–Bosch process, which has been in use for nearly 100 years in the chemical industry on a very large scale; one thus requires no "new chemistry." Whether chemical storage is superior to direct thermal storage, in the economic as well as in the technical sense, can be shown only by future development, in particular by the construction of larger demonstration plants. The fact that this approach is in principle of interest for "thermal" storage is in any case evident, since it offers the possibility of long-term storage, even though for 24 h storage, its efficiency is lower than that of the usual heat reservoir, namely 60–70% as opposed to 99%.

6) Apart from the early small steps taken roughly 100 years ago (see Smith, 1995), we should mention here that the first significant modern concepts and test plants for the application of the Fresnel principle to solar thermal installations were initiated by Baum et al. (Soviet Union) in the 1950s and by Francia (Italy) in the 1960s (Baum et al., 1957; Francia, 1968).

The main solar components are thus as follows:

- the heliostat field
- the receiver
- the storage reservoir.

With this concept, one needs only a single large turbogenerator (with a mature technology), instead of thousands of Stirling motors or microturbines. For such applications, there is – due to the developments of the past 20 years – a ramified system of individual technologies, of which we can treat here only a few of the most important.

In general, one can say that the economic optimum for the mirror area of a heliostat in a tower plant of the 100-MW class should lie in the range of 100–200 m^2, or possibly still larger. For the actual mirrors, there are (as in the case of dish reflectors, also) models with a continuous reflecting surface, but also models which consist of several (e.g., 10 or even only 2) individual mirrors. The mechanical support structure for the mirror is nearly always held in the middle by a post and can be turned around two axes through this support point. Two axes of rotation are required for combined mirror tracking of the local position of the Sun (cf. Chapter 6).

An interesting variant of the tower principle is the "beam-down" principle, an Israeli development suggested in 1976 by Rabl (1976). The receiver is installed not at the top of the tower, but instead at its base. At the top of the tower, a large reflector with a hyperbolic curvature is mounted; it directs the solar radiation coming from the heliostats downward to the receiver. Before entering the receiver, the radiation is further bundled by a funnel-shaped secondary concentrator. It is immediately apparent that a simplification in the installation and maintenance of the receiver is obtained by mounting it at the ground level without piping in the tower for the heat-transfer medium. One of the main problems is, however, the thermal loading of the tower reflector, and it cannot be judged with certainty at this point whether this component will be sufficiently inexpensive so that all together, a cost advantage would be obtained as compared to the classical tower principle. This concept has recently been revived in Japan, with a new reflector design and – instead of the previously planned high-temperature receiver – a special molten-salt receiver (Tamaura *et al.*, 2006).

For the tower receivers, there are three different systems:

1) The immediately obvious method is DSG in the receiver, where the production of either saturated steam or superheated steam is possible. For example, the American Solar ONE power plant (1982–1988, Barstow, California), for a long time the largest operating tower power plant worldwide, used a receiver which generated superheated steam. The associated technical problems were, however, one of the reasons that this technology was no longer pursued in the USA and worldwide. The first "commercial" European solar power plant, an 11-MW tower plant with the name PS 10, which went into operation in 2007 near Seville (Spain), had a saturated-steam receiver (Figure 3.4). Regarding

Figure 3.4 The tower power plants PS-10 and PS-20 with steam receivers near Seville, Spain (Wikimedia Commons).

this power plant, we refer the reader to a very instructive EU research report (Solúcar, 2006).

2) In the *volumetric air receiver*, the absorption of solar radiation takes place not only at the surface but also three-dimensionally within the volume of an open, porous structure (therefore "volumetric"). The material of the receiver, which is heated by the solar radiation, can be metallic (wire mesh) or (porous) ceramic, depending on the operating temperature level. The energy absorbed by this material is transferred to air which is pulled through the absorber structure toward the interior or (in the case of the so-called pressure receiver) is pushed through; the air exits the receiver at a high temperature.[7] Apart from the fact that the air is a relatively trouble-free material for heat transfer, volumetric absorption has the following advantage:

Since the absorption takes place essentially in the interior of the receiver material, its surface temperature remains relatively low. This, in turn, means that losses through convection and especially through infrared emission, which is proportional to the fourth power of the absolute temperature, are noticeably reduced.

There are two types of air receivers:

---

7) The principle of the volumetric receiver is closely connected with the name of Hans Fricker (Switzerland). He developed it to near technological maturity during the 1980s.

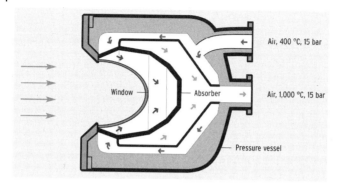

Figure 3.5 Principle of a pressurized volumetric air receiver (DLR).

- *The open or atmospheric receiver.* Air from the surroundings is sucked through the porous receiver. The air which exits at ca. 700 °C produces steam in a steam generator for driving a steam turbine.[8] This type of receiver is described in more detail in Chapter 7.

- *The pressurized receiver.* Air is pressed through the absorber material at a high pressure (e.g., 15 bar) within a closed system; it is heated to 800 °C or even to over 1200 °C, and then for example employed as combustion air in the gas turbine of a CC power plant. One thus obtains a gas/solar hybrid power plant (solar-hybrid combined cycle plant). The radiation enters through a (cooled) concave pressure-resistant quartz window into the actual absorber, which is located in the interior of the pressure system (Figure 3.5). Depending on the overall power output, many modules are connected together into a large-scale receiver by parallel and series hook-ups. Series circuits in the newer designs are triple: the first stage consists of simple receivers made of steel, in which the air is heated to ca. 600 °C by the solar radiation that is concentrated by the heliostat field. In a second stage, for example, consisting of special ceramic absorbers,[9] the hot air is further heated. Only the third stage is then constructed as a (ceramic) volumetric absorber, in which the air is finally heated to about 1200 °C. (However, according to Hoffschmidt (2007), in the latest designs even the second stage is as a rule a volumetric receiver.) In order to attain this high final temperature, the radiation entering from the heliostat field must be further concentrated. This takes place with the aid of funnel-shaped (and cooled) mirrors, the secondary concentrators, which are mounted in front of the quartz windows. They have a hexagonal shape so that when many individual absorber modules

---

8) The principle of the open air receiver also owes much to Hans Fricker, who developed it during the 1980s. In recent years, in particular Prof. Bernhard Hoffschmidt (director of the Solar Institute in Jülich, Germany (SLJ)) has made important contributions to the further development of the air receiver, making use of ceramic absorber materials.

9) The currently favored material is SiC ceramic (silicon carbide).

are combined to give a large-scale receiver, a densely packed honeycomb structure results. Due to the very high temperature, the pressurized receiver has an especially high average total solar efficiency (annual average well over 20%), which of course permits a corresponding reduction of the size of the (expensive) heliostat field. Since the total efficiency of the other receiver types is in the range of 16–18%, this means a reduction in the size of the mirror field by about one-fifth. A major disadvantage of the pressurized receivers, however, can be seen in the fact that a heat storage reservoir must be constructed as a pressure vessel. The pressurized receiver is still in an early developmental stage (by the way, it, like the open air receiver, is being developed almost exclusively in Europe) so that it is not yet clear which type will finally be successful. A certain (cautious) optimism is justified, as the development is proceeding continuously – in spite of enormous materials problems (Pitz-Paal and Hoffschmidt, 2003).

3) In the receiver, a liquid HTF is heated and is then used to produce superheated steam in a heat exchanger (steam generator). A substance suitable for this purpose must not have an exceedingly high melting temperature, and its vapor pressure must remain relatively low over a wide temperature range (with a high boiling point). Furthermore, the substance must remain chemically stable over long times at up to at least 600 °C, and it must not be chemically aggressive; it must also have a good thermal conductivity and – not least – it must be as inexpensive as possible. These properties can be found (with some restrictions) in molten alkali metals (e.g., sodium) and in certain molten salts. About 30 years ago, elaborate studies were carried out and even test installations were constructed using sodium receivers. For technical and economic reasons, today only the *molten-salt receiver* is still being developed for use in solar tower plants.[10] (Molten salts have, to be sure, only moderate thermal conductivities compared to sodium, but they are safer and, owing to their low material costs, they provide an inexpensive storage medium (typical: a eutectic nitrate salt mixture consisting of 40% KNO_3 and 60% $NaNO_3$; in the solar-energy industry, it is often called "solar salt").

In comparing these systems, one must always keep in mind along with the primary technical aspects also the question of heat storage, in particular when one

---

10) The sodium technology was established in the 1950s for the development of fast breeder nuclear reactors. Sodium was used in them as cooling medium, in part due to its extremely high thermal conductivity. For the construction of very compact receivers, this property is likewise very useful. In the 1980s, work on sodium receivers was discontinued, in Europe in 1986 following a sodium fire at the Spanish solar research center in Almeria. On a smaller scale, however, sodium receivers are used in dish systems ("heat-pipe receivers"). But the idea of a solar tower power plant with an alkali metal receiver may not be finally dead: in a recent US patent, a receiver concept using sodium cooling is mentioned (Litwin *et al.*, 2005). Furthermore, cooling with a eutectic mixture of potassium and sodium is considered there (solidification temperature: around −18 °C). It would eliminate the risk of freezing of the heat-transfer medium.

is considering power generation for base-load requirements, as in this book. The air receiver uses a working medium which is very readily manageable, and in the case of the high-pressure air receiver, one has furthermore a high efficiency. However, whether or not it will prove possible to develop large heat storage reservoirs for air receivers (even for open systems), which would permit 24-h operations, is still an open question, just as in the case of the steam receiver.[11] First and foremost, solid material reservoirs are being considered, in which heat energy is exchanged between the air and, for example, concrete, ceramics or sand, depending on the operating temperature. In the case of the open air receiver, up to now parallel channel monoliths (channel diameter: a few centimeters) or a packed bed with ceramic particles (spheres or saddle-shaped bodies with a diameter of a few centimeters) have been suggested, through which the hot or cool air is blown to release or gather heat. According to R. Pitz-Paal, however, such a storage vessel cannot be mounted on the tower if it becomes too large, while on the other hand the transport of large amounts of hot air to the ground is very problematic. "We can thus see that at present, 24-h operation of a solar tower plant using air receiver technology with a heat storage reservoir of this type does not appear to be competitive with respect to other solar thermal concepts …" (Pitz-Paal and Hoffschmidt, 2003).

For the open air receiver, operating at air temperatures up to around 800 °C, an apparently elegant storage method was suggested several years ago: the hot air coming from the receiver is passed through compact quartz sand, as in a cross-flow heat exchanger, while the sand is kept continuously (and controllably) in motion. Owing to the large surface area and small diameter of the sand grains, a very efficient heat transfer is achieved. The heated sand then falls into a simple silo reservoir, which is suitable for high temperatures. To withdraw energy from the reservoir, air is blown through the hot sand, allowing the resulting fluidized bed to transfer the heat in optimal fashion to tubes through which water is flowing, and thereby to generate steam[12] (Pitz-Paal and Hoffschmidt, 2003; Pitz-Paal, 2003; Hoffschmidt, 2007). The sand reservoir, apart from its technical advantages, has extremely low materials costs. Rapid further development of this high-temperature storage process, which is still at the stage of laboratory experiments, should likewise be one of the points of a goal-oriented R&D program for solar thermal power plants. The solar institute in Jülich, Germany, in cooperation with the German Aerospace Center (DLR), completed its investigations using a prototype on a laboratory scale "successfully" in the year 2008. The next step will be a test facility on the scale of several hundred kilowatts (Hoffschmidt, 2007).

In the case of the salt receiver, one can use the molten salt, as mentioned above, owing to the low cost of this material – simultaneously as heat-transfer medium

---

11) The Spanish plant PS 10 mentioned above, with a saturated steam receiver, has a steam reservoir which provides energy for only a half-hour of operation at full output power.

12) This second step in the operation of the storage system, that is, heat retrieval, thus corresponds to the fluidized-bed process which is well established, for example, in power-plant technology.

and as heat-storage medium; hot molten salt is simply stored in insulated tanks. For a thermal reservoir, one merely requires the corresponding amount of the relatively low-cost salt as well as an insulated tank for the heated and for the "cool" salt. Storage for 24 h is possible in this manner with less than 1% losses. If one is aiming at supplying base-load power using solar energy, this type of receiver has great advantages. (It will be treated in more detail in Section 7.1.) If – as mentioned – the topic of heat storage was long neglected, particularly in Europe, this neglect corresponds among other things directly to the neglect of the salt receiver over many years by European solar research.

Finally, we want to mention another excellent review of the tower technologies, in particular of air receivers, which was written by Romero *et al.* (2002), as well as the description of the (unrealized) precursor project of the PS-10 steam-receiver power plant mentioned above, given by Osuna *et al* (2000).

3.3
Parabolic Troughs

Here, the light is reflected by long concave mirrors with a parabolic cross-section onto a metal tube, which is located along the focus line of the mirrors. This tube is coated with a selective absorber material to guarantee maximum absorption of the solar radiation (at short wavelengths) and the least possible infrared radiation losses (at longer wavelengths). To avoid heat losses, the absorber tube is surrounded by an evacuated glass sleeve. The energy absorbed is transferred to an HTF, which is pumped through the tube. The hot fluid from many parabolic troughs is transported to a steam generator via collection piping; this then supplies the turbines. Each trough consists of a ca. 150-m long metal support structure with mirror segments mounted on it. The breadth of the troughs, that is, the aperture width, is typically about 6 m. The torques required to hold and rotate the mirrors are transferred to the trough from the drive unit via an extremely stiff metal construction (likewise 150 m long), which is mechanically connected to the mirror support structure over its entire length. Depending on the shape of the cross-section of this construction, one speaks of a "torque tube" or a "torque box."

Three different systems are under consideration:

1) Originally, mineral oil was used as HTF; currently, synthetic oil is in use, so-called *thermal oil* (e.g., a eutectic mixture of diphenyl oxide and biphenyl, "Therminol VP 1®"), in particular in the well-known Californian parabolic-trough power plants, and also in a newer power plant in Nevada ("Nevadav Solar One," 64 MW) as well as in the new Spanish power plants (e.g., "Andasol 1 and 2," 50 MW each). The problem, however, is that this oil remains chemically stable only up to ca. 400 °C. This limits the temperature of the steam generated, giving a correspondingly lower efficiency in the turbines.

    For heat storage, one can in principle store the hot oil in tanks for nighttime operation of the steam generator. For this purpose, however, this special oil is much too expensive. Instead, the nitrate salt mixture, which is used for solar tower plants, can also be employed here. One then requires an oil/salt heat exchanger and of course insulated storage tanks (one each for the "cool" and the hot molten salt, i.e., a "two-tank reservoir").

Since a molten-salt reservoir is relatively costly, especially for 24-h operation of trough power plants[13] (see Chapter 8), great expectations have been placed in solid-medium storage systems for some years, in particular using the low-cost storage medium concrete. It is, however, still to be determined whether the technical problems can be solved. They are clearly more difficult than initially expected. In the short term, it proved to be favorable in the case of the new Spanish parabolic-trough power plants (Andasol 1 and 2) to simply adopt the molten-salt technology already developed for SOLAR TWO and (since about the year 2000) also for the Spanish 15 MW tower project SOLAR TRES.

2) It is then altogether consistent that in the past few years in Italy, within the framework of a targeted, comprehensive and exemplary development program, the possibility is being investigated of using *nitrate salts directly as* HTF in the absorber tubes and not only as heat storage medium (ENEA, 2001). (In the USA, also, in recent years this option has been intensively discussed.) This would mean that the principle of SOLAR TWO would be adopted in its entirety. The oil/salt heat exchanger of the storage system would become unnecessary, and the temperature limitation mentioned above – due to the oil – would be eliminated, giving an increase in operating temperature to 500–550 °C, with a higher thermal efficiency. Note, however, that this temperature increase has a positive effect only on the efficiency of the turbines. The heat losses in the receiver also increase so that the overall efficiency would not increase so strongly. How great the overall economic advantage of this technology would be is not yet clear. Some misgivings are understandable, if one considers a system of piping many kilometers long through which hot, molten nitrate salts at up to 550 °C are being pumped. The main problem appears to be the more or less frequent solidification of the salt due to cooling below its melting point. Although the technical challenges of such a system are clearly serious, the experts place great hope in it as an important option for the cost reduction of parabolic-trough power plants (see Chapter 8). But skepticism is clear in the question posed (with a certain self-mockery) by D. Brosseau and G. Kolb (Sandia): "Are we crazy? Are we really serious about flowing molten salt through miles and kilometers of thin-walled steel tubes? Are we prepared for freeze-ups, freeze recovery, damage on expansion ..., human error?" (Brosseau and Kolb, 2007). To this we contrast the "Conclusions" of a presentation by A. Maccari (ENEA) at a seminar organized by the European Commission: "The 'crazy' idea to let a molten salt mixture flow in a trough field seems close to becoming a reality. A lot of work has been carried out to

---

13) Here, it must be kept in mind that for trough power plants, owing to the limited temperature range available for heat storage, namely ca. 100 K, the reservoir requires around three times more molten salt than in the case of solar tower plants (with a temperature range of 275 K).

address and solve most of the crucial questions regarding molten salt usage. ... There are still some points needing further research activities ..." Maccari's closing remark then illustrates the bottom line: "The next 7th European RTD Framework could represent a big opportunity to move further in the development of such very promising technology" (Maccari, 2006). One can only hope that this sentence will be read and taken seriously by the political leaders in the European Parliament.

3) Since the early 1990s, great expectations have been placed on the principle of DSG, which promises considerable cost savings compared to heat transfer from the receiver using heat-transfer media. High temperatures in comparison to oil cooling and the elimination of a separate steam generator are obvious advantages of this system. On the other hand, new problems arise: the absorber tubes must withstand high pressures, and two phases (vapor/liquid) are present, leading to correspondingly complex flow and heat-transfer properties.[14] For this reason, a considerable effort in basic research still needs to be invested here. However, in the course of the European DISS project,[15] which began in 1995, very encouraging progress has been made (Zarza et al., 2004) so that presently, an industrial consortium based on the intermediate project INDITEP[16] (2002–2005) has been organized to construct a small, complete demonstration power plant with an output power of 3 MW (PSA, 2006). (Up to the end of 2006, an output power of 5 MW was planned; cf. the detailed project description by Zarza et al. (2006).) Finally, we also mention that Eck and Zarza have investigated the question of whether one should initially work not with superheated steam, but instead with the technically simpler case of saturated steam. The construction of the collector field would be comparatively simple, safe operation has already been demonstrated, and the thermal efficiency of the collectors would also be higher (Eck and Zarza, 2006).

In recent years, it has been generally assumed that for a DSG system – elegant though it may be – thermal-energy storage would be rather difficult. This applies not so much to short-term storage, but rather to long-term storage for 24-h operation. This disadvantage has, however, not been considered as being very serious, as long as the power plants – as was frequently the case and still is – are mainly intended to supply only peak-load requirements. In this connection, we briefly mention a special strategy for heat storage, which has recently aroused some optimism: since the beginning of the modern development of solar thermal power plants, latent-heat storage schemes have

---

14) Problems with direct steam generation initially led to the use of heat-transfer media for all the planned solar thermal systems. Merely the avoidance of high pressures in the absorber seemed to justify or even to compel this step. In particular, for solar tower plants one wished to save on weight in the construction of the receiver. In the American SOLAR ONE plant, there were special problems with the generation of superheated steam. But technical progress can – as we see here – lead to a reversal of such conclusions.
15) DISS: Direct Solar Steam.
16) INDITEP: Integration of DSG Technology for Electricity Production.

been central to the considerations regarding heat storage systems, due to their high storage densities; these in turn are the result of the relatively high values of the specific heat of melting. The decisive problem up to now was that of the poor thermal conductivity of the substances under consideration (e.g., nitrate salts) in their solid phase so that the heat-transfer rates within the storage medium were too low to yield an economically feasible system. In more recent development work, an attempt has been made to combine the actual storage medium (phase-change material, PCM) with another material that has a good thermal conductivity (especially with so-called expanded graphite, which can – due to its extremely low density – be infiltrated with the PCM). If this proves to be feasible, and a low-cost latent-heat storage system can indeed be produced in this manner, the storage system would probably consist of three stages: on discharging, a storage medium, for example, for sensible heat (concrete) could provide preheating of the water (to a low temperature level). The latent-heat storage system, with a melting point adjusted to the operating temperature of the steam generator, would deliver the energy for vaporization of the water, since both solidification (of the PCM) and vaporization are constant-temperature processes. The second sensible heat storage unit (possibly also concrete) would deliver the energy for superheating the steam to a higher temperature range. For charging, the principle would be correspondingly reversed (Laing, 2007; Tamme et al., 2007).

3.4
Linear Fresnel Plants

In the "youngest" of the technologies based on this principle – at least in terms of its implementation[17] – the parabolic trough is replaced by a number of parallel planar (or nearly planar) mirror elements, which are all mounted at the same height near the ground. They follow the position of the Sun by rotating around their long axes so that they point to a focus line at a height of 10–15 m, which remains fixed over time (Figures 3.6 and 3.7). Along this line, an absorber tube up to 1000 m long is mounted, and in it – similarly to the development for the para-

---

17) First attempts, investigations and experimental installations were already established – as we mentioned in Section 3.2 – in the 1960s and 1970s. However, the current phase of development began about 10 or 15 years ago, when the concept was taken up again in Australia and in Belgium and introduced with great emphasis into the scientific discussion, particularly in Europe (on the basis of a test installation by the Solarmundo company in Belgium). In Australia in the year 2003, this led to the construction of a demonstration plant. In recent years, there were several theoretical investigations into the Solarmundo concept, in particular in Germany, for example, that of Haeberle et al. (2001) and Lerchenmueller et al. (2004a, 2004b). In 2007, these also culminated in the construction of a small demonstration plant, 100 m in length, at the Spanish Solar Research Center Almeria (ISE, 2007).

Figure 3.6 A linear Fresnel collector: in this foreground, the absorber pipe is visible; behind it is the piping with the secondary reflector (Photo DLR)

Figure 3.7 Here, one sees the support piping for the mirrors (Photo DLR).

bolic trough – water is directly vaporized. The steam from many parallel absorber tubes can operate a large turbine.

In the Solarmundo concept, which is the basic variant of this principle, the radiation reflected from the mirror field is focused by a secondary reflector onto the absorber tube mounted directly below it so that the concentration factor is sufficiently large in spite of the wide mirrors. Each absorber tube is served by a mirror field about 30 m wide located directly below it and containing many strip mirrors, each around 1 m wide, mounted close together. In Australia, a special variant of this basic concept is in favor; it is known as the compact linear Fresnel reflector (CLFR). In this concept, the individual mirrors are not associated with a particular absorber tube, but instead channel the radiation, depending on the position of the Sun, onto different parallel absorber tubes in a cost-optimized fashion, or else they alternate between one particular absorber tube and its neighbor. This strategy is intended to make optimal use of the land area. A secondary reflector is not used, but the primary mirrors are slightly curved by elastic deformation so that the focus lines of the mirror field are sufficiently sharp. For a project in Australia, a reflector length of 600 m was planned, consisting of individual segments of 200 m length (separate or coupled as parallel segments (Haeberle et al., 2001)), which could be directed toward the Sun by means of motor-driven units (University of Sydney, 2007). In connection with the CLFR project, a novel absorber concept was also developed, whereby steam generation takes place not in the long absorber tubes, but rather in many smaller tubes with lengths of about 1 m (Burbridge et al., 2000).

The Fresnel principle initially offers the advantage that the mirrors can be relatively inexpensive; they may be mounted with a slight curvature produced by elastic deformation. In addition, the absorber tube need not be moved to follow the Sun. This not only allows the use of simpler machinery, but also offers the advantage that the absorber tubes can be attached via fixed connections to the steam collection pipes. These connections still represent one of the major development problems for the parabolic-trough technology – no matter which heat-transfer medium is used.[18]

This system, which is relatively simple in terms of its components, has however the disadvantage that the concentration of the solar radiation is not as strong as with the trough technology so that lower temperatures and, therefore, lower efficiencies are obtained. In addition, owing to its flat construction, the system misses ca. 30% of the incident solar radiation for a given reflector area as compared to parabolic troughs (Hoffschmidt, 2007). The question is thus whether the components can in fact be produced, installed, and maintained so much more cheaply that the final cost of the output power is still equal to or even less than that from the parabolic trough systems. Depending on the location of the plant, if the costs were roughly the same, an advantage in principle of the Fresnel technology might tip the balance. For geometric reasons, that is, due to the reduction in mutual

---

18) For these connections, one makes use of either flexible tubes or ball joints. However, for parabolic trough plants, there are plans to construct the collector in such a way that its rotational axis is identical with the focus line, which would likewise permit fixed connections between the absorber and the steam distribution pipes. In this case, no rotation, but only the linear thermal expansion of the absorber would need to be mechanically compensated.

shadowing, its land-area requirements are, compared with parabolic troughs, modest: half the land area is required for a given mirror area, whereby, however, the above-mentioned performance reduction per unit of mirror area must be tolerated.

In the current state of the art, heat storage appears to be a problem (as with DSG in parabolic trough installations); it is however necessary in large plants. In general, the studies mentioned conclude that no storage concept is as yet available ("Another long-term option is thermal storage, which is not yet feasible for direct steam generating systems" (Lerchenmueller *et al.*, 2004a)). There was, however, an interesting suggestion in this direction, published by D. Mills in 2004. He suggested the storage of hot water or steam below ground (at a depth of several hundred meters) under high pressure. In a preliminary cost comparison, Mills showed why this method of storage might be considerably cheaper than the methods under discussion for the parabolic trough systems (Mills and Le Lièvre, 2004). (If it proves to be feasible, the same storage method could just as well be applied to DSG from parabolic trough plants).

All together, Mills reaches the conclusion that using Fresnel plants, lower power costs than those expected for future advanced parabolic trough systems could be achieved within a few years. It is hard to judge at present just how realistic this is since the first complete larger pilot plants are not yet in operation. However, the justification for a massive development effort of this solar thermal principle can hardly be questioned. A certain indicator of the great potential of this concept is the fact that at present (since 2007), European big industry (in the form of the MAN Ferrostal Corporation) is also investing in the technology. This is interesting to hear, since the technology used in the Californian parabolic-trough power plants in 1984 was based on the developments carried out in part by MAN (Munich) for the Spanish Solar Research Center Almeria a few years earlier. MAN thus started with parabolic troughs and ended up, roughly 30 years later, with the Fresnel system. It is, by the way, typical of alternative-energy research policy up to now that the early developments in the 1970s had to be terminated due to a lack of sufficient support funds (Burbridge *et al.*, 2000).

3.5
Updraft (Chimney) and Downdraft Power Plants

Chimney and downdraft power plants both use the "chimney effect" (the downdraft principle could be termed the "inverse chimney effect"): due to a temperature difference with respect to the ambient air, in a solar *chimney* power plant, *warmed* air rises within a large pipe (tower); the air is warmed under a large-area collector roof. In a *downdraft* power plant, *cooled* air sinks in a comparably dimensioned tower, the air being cooled by evaporation of seawater. In the first case, the air is lighter, and in the second case, it is heavier than the ambient air. In both cases, air turbines are driven by the flow. Downdraft power plants are also referred to as "energy towers." In order to avoid confusion (with "solar *tower* power plants," i.e.,

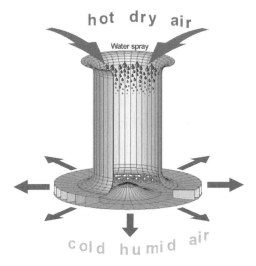

Figure 3.8 Downdraft tower (principle, according to Zaslavsky et al., 2001).

concentrating solar power plants), this term is not used in this book. Since the solar chimney plant will be treated in Chapter 9, we confine ourselves in the following to the second type.

In a *downdraft tower*, air is cooled at the upper end of the tower, which can be, for example, 1000 m in height and 400 m in diameter. The heavier air sinks into the pipe and drives turbines in passing (Figure 3.8). Because the temperature difference here is obtained purely by evaporative cooling, there is no need for a large collector roof, a major cost aspect of the updraft power plant. This in principle permits much lower construction costs. A precondition is the presence of warm and dry air (e.g., at the edge of a desert),[19] and at the same time the abundant availability of seawater for cooling.

As we have already emphasized in the introduction, this concept has been publicized in particular by Dan Zaslavsky (Technion Haifa). Zaslavsky (2008) himself quotes a possible (sensational!) attainable power cost of 2 US-¢/kWh; in (Frenkel, 2007), he gave a value of 2.5 ¢/kWh, and in earlier publications "ca. 3 US-¢/kWh," with 5% interest rate (Zaslavsky and Glubrecht, 2000). The feasibility of such low costs would not seem to be ruled out in principle, since the solar roof, which makes up a major part of the cost of a solar chimney plant, is not necessary. However, whether the basic assumptions about the well-defined nebulization of seawater at the top of the tower are realistic (this is probably the key problem of the technology – the size of the droplets must lie within a relatively limited range), and

---

19) In downdraft tower plants, solar energy is utilized in an *indirect* manner (use of the ambient air heated in the warm desert).

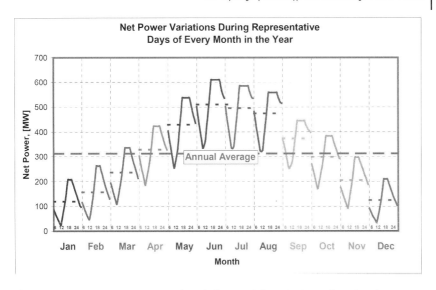

Figure 3.9 Power output variations (downdraft towers) for a site in southern Israel (Zaslavsky *et al.*, 2001).

whether all the other parameters which enter into the calculation can be achieved in practice, remains speculative for the outsider at this point. Not only the low costs are important, an additional advantage is the enormous application potential. On the Mediterranean coasts of North Africa and the Middle East, the coasts of the Indian Ocean, Australia, California, Mexico, etc., great amounts of energy could be generated (see Figure 5.9); Zaslavsky writes (in Frenkel, 2007): "We could easily produce between 15 to 20 times the total electricity the world uses today." Power generation can be carried out not only in the daytime but also at night, however, at a reduced rate. The decrease in power output at night and the seasonal output variations depend on the location: in *southern Israel*, a plant of a certain size would produce in the summer in daytime 600 MW, at night at least 300 MW; in the winter, however, only at most 200 MW in the daytime, and at night at least 50 MW (Figure 3.9). As the figure shows, an installation of this type could at least be termed as "semi-" base-load power plant (whereby one has to tolerate a certain surplus power production during the summer, which increases the average cost of the power by a small amount) – so long as the values attained in southern Israel can also be applied to other locations; but there may be many better locations in the world – for other regions cf. Figure 3.10.[20] Good descriptions (including detailed potential maps for various countries) can be found in Zaslavsky *et al.* (2001), Altmann *et al.* (2006 and 2008), and (in German) in Czisch (2005).

---

20) The seasonal variation is furthermore well matched to the high power requirements in the summer due to the use of air conditioning in such hot locations.

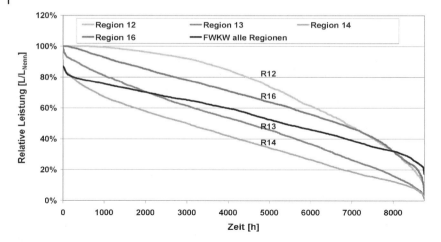

Figure 3.10 Relative power output (duration curves) of downdraft towers for various regions (Czisch, 2005). The curves refer to the annual trends in the regions indicated.

Ordinate: relative power; abscissa: operating time (hours).
Region 12: Saudi Arabia, Gulf Coast
Region 16: Senegal, Mauritania
Region 13: Morocco, Algeria
Region 14: Libya, Tunisia
FWKW: A power network combining all four regions

Given the low costs and the potential of this technology, it is self-evident that all the remaining open questions should be investigated in detail in the course of an *energy research program*, which we strongly support here. This would include not only the construction of a prototype but also preliminary research, which could be carried out rapidly, in particular into the efficiency of the water nebulization process. Zaslavsky thus proposes constructing a research installation on the upper floor of an existing high building. The comparison already cited in Chapter 1 (Weinrebe and Schiel 2001), however, suggests that there are still additional open questions. Since this technology is possibly not to be considered as a full-scale "base-load plant" concept, and since furthermore only relatively little information is available on this thoroughly noteworthy scheme, it will not be discussed further in this book. Research must first be carried out to determine more precisely just how great its potential is in fact. More detailed investigations, conducted not only by proponents or competitors as has thus far been the case, but also by neutral researchers, could provide clear indications within a short time. Furthermore, it is quite probable that Carlson (the inventor) and also Zaslavsky will have encountered the same treatment as have many others in the area of renewable energy (such as, e.g., the German research group of J. Schlaich und W. Schiel): their proposals were registered with a lack of interest or even active skepticism by the responsible energy research agencies, and never seriously tested, while at the same time, no funding was provided to the initiators so that they themselves could carry out the necessary proof of operational feasibility.

4
Some Additional Economic Factors

The subject of economic feasibility, which we discussed briefly in Chapter 2, will be treated in more detail in this chapter. In Section 4.1, we give some more information on the cost of the "solar power system" and of competing base-load power plant concepts; in particular, the possible role of these competing technologies for future energy supplies and their costs are discussed in comparison to those of the solar power system. In Section 4.2, we compare the cost data for solar power plants employed in this book with those given in some other studies, and in Section 4.3, various special aspects which enter into the cost calculation are elucidated.

4.1
Detailed Treatment of the Costs of the Solar Power System – Comparison with Competing Types of Power Plants – Discussion

For the solar power system, we consider here in particular the question of backup power plants and their influence on the overall power costs. The *investment costs* for solar power plants were already listed in Table 2.1 in outline form. We will consider them again in Section 4.2, where they will be compared with the results of other studies. (They are also listed in detail in Appendix A.) In the following discussion of power costs, we always refer to the solar power plant investment costs as given in Table 2.1.

4.1.1
Solar Power Systems with Coal-Fired Backup Power Plants (Instead of Natural Gas Plants)

Table 4.1 shows the investment costs for the solar power system at solar sites in Spain and in Morocco or the USA as a function of the type of backup power plant chosen. In the case of coal-fired backup plants, two variants are given: in the one case, new coal-fired plants must be constructed, and in the other, existing plants would be used:

Large-Scale Solar Thermal Power. Werner Vogel and Henry Kalb
© 2010 WILEY-VCH Verlag GmbH & Co. KGaA, Weinheim
ISBN: 978-3-527-40515-2

Table 4.1 Investment costs of the solar power system with natural gas (CCGT) or with coal-fired backup power plants.

	Spain (SM 4.4; 2000 km)				Morocco/USA (SM 3.7; 3000 km)			
	Million dollars (2002) per 1000 MW							
	Natural gas CC backup power plants	Coal-fired backup power plants – "annex" construction of plant capacity	Coal-fired backup power plants – "replacement" of base-load coal-fired plants		Natural gas CC backup power plants	Coal-fired backup power plants – "annex" construction of plant capacity	Coal-fired backup power plants – "replacement" of base-load coal-fired plants	
(Solar tower plants per 1000 MW at the plant site)	(3985)	(3985)	(3985)		(3695)	(3695)	(3695)	
Solar tower plants per 1000 MW after power transmission[a]	4335	4335	4335		4175	4175	4175	
Transmission lines	500	500	500		665	665	665	
CCGT gas power plants	615	–	–		615	–	–	
Coal-fired plants[b]	–	900	0		–	900	0	
Fast-start backup[c]	100	440	300		100	440	300	
Total investment costs	5550	6175	5135		5555	6180	5140	

a) Transmission losses: Spain 8.1%, Morocco/USA 11.5%
b) A mixture of 50% new coal-fired plants (1200 million $/GW) and 50% older coal-fired plants (600 million $/GW).
c) Fast start-up backup for *natural gas* (CCGT) power plants:
   Additional gas turbines (integrated into the power plants, 300 million $/GW), however, each with only 1/3 of the output power of the (CCGT) plant, corresponding to a power output of the steam turbines of 300 × 1/3 = $100 million per GW of backup capacity.
   Fast start-up backup for *coal-fired plants*:
   - Diesel engine plants (decentralized): 440 million $/GW; or
   - Gas turbines at the site of the coal-fired plant (partially integrated): 300 million $/GW (cf. Chapter 10).

1) Replacement

This case is based on the assumption that the modern coal-fired plants currently in operation, which would have a relatively long remaining operating life, are to be *replaced* as quickly as possible. If one is considering the principle alternative "solar power systems *versus* nuclear power," then the following situation arises: for the solar power systems – in terms of a cost comparison with nuclear power – the superseded base-load plants would be available nearly "for free" as backup power plants; in the case of substitution by nuclear plants, they would be shut down and their value would be lost. In determining the *absolute* costs of substitution, the average remaining value of the coal-fired plants would of course have to be included both for solar power systems as well as for nuclear power plants; this would, however, be rather involved[1] and here, where we are interested in the cost *difference* between the solar power system and nuclear power production, it is also superfluous.

2) Annex construction

In this case, we consider how to increase the generating capacity of the power-plant park, and thus in this sense we are dealing with *annex construction* of plants. Then, parallel to the solar plants, naturally also *new* coal-fired backup plants would be constructed. The corresponding investment costs are included in the columns labeled "annex" in Table 4.1.[2] Even in this case, a certain portion of the backup capacity for the solar power system could consist of used coal-fired plants. Indeed, new coal-fired backup plants must also be built to correspond to the new solar plant construction, but these would reasonably be exchanged for operating, older base-load plants, due to the higher efficiencies of the new plants. We make the assumption that the backup system consists of 50% new and 50% older coal-fired plants (average investment costs, 900 $/kW; average operating efficiency, 42%; cf. Section 10.2). When power generation from solar energy decreases and thus power is required from the backup plants, the newer coal-fired plants will first be brought into operation. The older plants will be used only if the solar energy supply drops below 50% of full capacity. They would thus have a relatively low annual capacity utilization so that their remaining operating life could be expected to be long, in spite of their many years of previous operation. (For solar plants in Spain, the capacity factor of these older backup plants would be only 15.7%, as found from an analysis of the daily insolation values (which are available over a period of 12 years), instead of the value of 28.6% for all the backup plants together, as quoted in Section 5.2.)

---

1) The current remaining value of the coal-fired plants is often not known and would likely be very difficult to determine precisely.
2) It is clear that the investment costs for the coal-fired plants have to be included in the case of "annex construction" even when they are not used to provide additional capacity, but instead represent a "replacement" (but not in the same sense as intended above) of coal-fired plants whose operating lifetimes have completely expired. This is the case for a slow, "natural" substitution of coal-fired plants.

Finally, in both the cases discussed above, the investment costs for fast start-up backup plants have to be considered. With coal-fired backup plants, either gas turbine plants or diesel engines could be used for this purpose (see Section 10.2). In Table 4.1, we calculated the costs in one case for gas turbines and in another for diesel engines; this was intended to make it clear that diesel aggregates, which can be switched on very rapidly (start-up time ca. 1 min), entail roughly the same costs as gas turbines.

4.1.2
Overview of Costs

Table 4.2 shows the costs for power from the solar power system, Table 4.3 those of competing types of power plants (all the entries are in 2002-US$; the values can also be interpreted as EUR (2002) with only minor errors[3]). The cost of nuclear power is 2.9 ¢/kWh for large-scale power plant parks under mass production, while the power cost for the solar power system in the USA (with "replacement") is 4.1 ¢/kWh. The cost difference is thus not very great.

If energy policy is intended to address the goal of replacing the many current *natural gas* base-load power plants (combined cycle gas-turbine (CCGT) plants or, for short, simply combined cycle (CC) plants) as rapidly as possible (and thus not just when their expected operating lifetimes have been completed), in order to make the natural gas they consume available to the general energy supply, then the same considerations as mentioned above for coal-fired plants also apply to the CCGT backup plants ("cost-free" backup plants). In order to keep the tables clear, we have included only the case of "annex construction" of CCGT backup power plants, in particular – again for simplicity – assuming 100% new CCGT plants. If

---

3) Based on the domestic purchasing power of the US dollar related to that of the Euro in Germany in the year 2002, according to the OECD: $1 = €0.96. In this book, all conversions from dollars to Euros or vice-versa correspond to the ratio of their domestic purchasing powers ("purchasing power parity"), not to the official exchange rates. The exchange rate depends upon noneconomic factors and is thus suitable to only a limited extent for comparisons of the production costs in different countries. The purchasing power, in contrast, reflects the real expenditure for the construction of plants in the different countries (here USA or Germany).

The exchange rate need be applied only for oil and other imported energy carriers, which are priced in dollars, especially imported coal. In the case of a relatively high Euro/dollar rate (e.g., in 2007: €1 = ca. US $1.40), the real costs of these imported energy carriers are lower for countries in the Euro zone than for USA. Thus, for example, a fictitious oil price on the world market of 100 $/barrel translates for the USA into 6.3 US ¢/kWh (2008-$; Table 4.3), but for Germany (as for the entire Euro zone) at the current exchange rate (19th December 2008) of €1 = $1.39, the price is only 72 Euro/barrel or 4.5 Euro-¢/kWh; recalculated using the purchasing power parity of (€1 = $1.136; this is the newest currently available value), this corresponds in Germany to a "real" price of 5.1 US-¢/kWh (2008-$). This would thus be ca. 20% lower than in the USA, not 30% – as would be suggested by the exchange rate. A corresponding situation holds for imported coal. We, however, need not take these effects into account in Tables 4.2 and 4.3, since the future relationship between exchange rate and purchasing power parity cannot be predicted.

4.1 Detailed Treatment of the Costs of the Solar Power System

Table 4.2 Energy cost from the solar power system with CCGT or coal-fired backup power plants; cost of solar power at the plant and of solar hydrogen.

2002-US$	Solar power plants								
	Solar power system (solar plus backup power plants)						Power at the plant		Solar H_2[d]
	Spain			Morocco/USA			Spain	Morocco/USA	Morocco/USA
Backup Plants	CCGT	Coal-fired "annex construction"	Coal-fired[c] "replacement"	CCGT	Coal-fired "annex construction" of plant capacity	Coal-fired[c] "replacement"			
Investment (million $/GW)	5550	6175	5135	5555	6180	5140	3980	3700	5540
Capacity utilization (full load)	8760h/a (solar 70%, backup 30%)			8760h/a (solar 80%, backup 20%)			6130h/a	7010h/a	7010h/a
	¢/kWh$_{el}$								¢/kWh$_{H_2(LHV)}$
Capital costs (4% real interest, 45 a)[a]	3.1	3.4	2.8	3.1	3.4	2.8	3.2	2.6	3.8
(Capital costs at 2% real interest)[a]	(2.2)	(2.4)	(2.0)	(2.2)	(2.4)	(2.0)	(2.2)	(1.8)	(2.7)
Operation and maintenance:									
Solar plants	0.7	0.7	0.7	0.7	0.7	0.7	0.9	0.7	0.9
Backup plants	0.1	0.4	0.4	0.1	0.4	0.4	–	–	–
Gas (2.5 ¢/kWh$_{gas}$)[b]	1.3	–	–	0.8	–	–	–	–	–
Coal price[b]	–	EU 90 $/t	EU USA 90 $/t 45 $/t	–	EU USA 90 $/t 45 $/t	EU USA 90 $/t 45 $/t			

Table 4.2 Continued.

2002-US$	Solar power plants									
	Solar power system (solar plus backup power plants)							Power at the plant		
	Spain			Morocco/USA				Spain	Morocco/USA	Solar H$_2$[d)] Morocco/USA
Backup Plants	CCGT	Coal-fired "annex construction"	Coal-fired[c)] "replacement"	CCGT	Coal-fired "annex construction" of plant capacity	Coal-fired[c)] "replacement"				
Coal	–	0.8	0.8	–	0.5	0.5	0.25	–	–	–
Energy cost at 4% real interest	5.2	5.3	4.8	4.7	5.0	4.4	4.1	4.1	3.3	4.7
(Energy cost at 2% real interest)	(4.3)	(4.3)	(4.0)	(3.8)	(4.0)	(3.6)	(3.3)	(3.1)	(2.5)	(3.6)

a) Annuity factor for 45a and 4% real interest: 4.8%; at 2% real interest: 3.4%.
b) Fuel costs:
- Price of natural gas assumed (in 2002-$): 2.5 US-¢/kWh$_{gas}$ (LHV) = 40 $/barrel oil = 6.6 $/MMBTU (HHV). (Compare: 2.5 US-¢/kWh corresponds in Europe – calculated with a purchasing power parity of $1 = 0.96 € – to 2.40 €-cent/kWh).
  Corresponds in 2007-US$ (×1.152): 2.88 ¢/kWh$_{gas}$ (LHV) = 46 $/barrel oil = 7.6 $/MMBTU (HHV) (in Europe (×1.084): 2.60 €-cents/kWh).
  Note: The gas price assumed corresponds at the current monetary value (2007-US$) rounded off to 3 US-¢/kWh (LHV) or 45 $/barrel oil.
  Note also: In the case of increasing natural gas prices, the backup power plants for the solar power system can be supplied with gas from coal gasification. For large-scale users with a separate gas pipeline, this gas, including transport costs, can be obtained for roughly the same price assumed here for natural gas, 2.5 ¢/kWh (2002-US$); see Table 4.3: Production cost of syngas from German lignite ca. 1.8 US-¢/kWh; from US coal ca. 1.9 ¢/kWh (for more information cf. Section 11.2).

4.1 Detailed Treatment of the Costs of the Solar Power System | 77

Comparison: The gas price in the year 2007 on delivery to power plants in the USA (in 2006-US$): 6.9 $/MMBtu (HHV); (Latin: MM = Mille × Mille = 10^6) this corresponds in 2002-US$ to 6.16 $/MMBtu (HHV) = 2.34 ¢/kWh$_{gas}$ (LHV); in Germany 2007 on delivery to power plants (in 2002-US$): 2.48 US-¢/kWh$_{gas}$ (LHV). Gas import price in Germany 2007: 5.5 €/GJ = 2.0 €-cents/kWh (HHV) (2007), corresponding in 2002-US$ to 2.13 US ¢/kWh (LHV) (purchasing power parity $1 = €0.96).

(LHV: lower heating value or net calorific value; HHV: higher heating value or gross calorific value; the HHV includes the heat (enthalpy) of condensation of the water vapor formed on combustion, which as a rule cannot be used.)

- Coal price assumed in 2002-US$: – Europe (imported coal): 90 $/tce (= 1.10 ¢/kWh$_{coal}$); German lignite: 35 $/tce (= 0.43 ¢/kWh$_{coal}$). USA: 45 $/tce (= 0.55 ¢/kWh$_{coal}$)

  Corresponds in 2007-$: – Europe (×1.084): 97.6 $/tce (= 1.20 US ¢/kWh$_{coal}$); German lignite: 38.0 $/tce (= 0.47 ¢/kWh). USA (×1.152): 51.8 $/tce (= 0.64 ¢/kWh$_{coal}$) (1 tce = 8140 kWh (LHV).

  Note: The assumed coal price corresponds at the current monetary value (2007-US$) in Europe rounded off to 100 $/tce and in the USA to 50 $/tce. The reader can thus readily estimate from the tables the effects of other assumed coal prices on the resulting electric power cost.

  Compare USA: The average price for bituminous coal in 2007 was 1.73 $/MMBTU (2006-US$) = in 2002-$ 44 $/tce. (The production costs were about 35 $/tce.) Europe, imported coal: The price rose from January to December 2007 from 78 to 149 $/tce; in 2002-$ from 68 to 129 $/tce. In contrast to the *price*, the *cost of imported coal* at the power plant (costs of coal production and transport) were roughly 50 $/tce (2002-$).) – cf. Chapter 10.

c) Replacement of operating, relatively new coal-fired plants with the goal of using the coal they consume for gasification. The coal-fired plants are then available without extra cost to be used as backup plants – see the text.

d) Solar hydrogen:
- High-temperature vapor-phase electrolysis (as currently planned for nuclear H$_2$ production). Assumptions:
  - Efficiencies: electrolysis 92% (LHV); reduction in power production due to use of steam: 6% (efficiency 94%); overall efficiency 86%.
  - Investment costs: 500 million $/GW$_{H_2}$ (2002-US$)
- H$_2$ transport: Pipeline diameter 1.7 m, pressure 100 bar, transport capacity 25 GW$_{H_2}$; distance (e.g., Sahara-Europe via Sicily): Land distance 3100 km, sea distance 200 km.
  - Losses (power for the H$_2$ pumps for initial pressurization and intermediate pumping stations along the pipeline): Assumed electrical operation of the pumps using power from a transmission line or from the local grid fed by solar power. Under this assumption, the transport efficiency would be 95%.
  - Investment costs for the transport pipeline (million $ (2002)/GW$_{H_2}$): land line (3100 km) 350, sea line (200 km) 110, pumping stations 57; total: 520.
- Overall efficiency for electrolysis and H$_2$ transport: 81.7% (0.86 × 0.95). Investment costs: 3695 million for the solar power plant capacity 1 GW$_{el}$ = $4520 million for a solar plant capacity of 1.244 GW$_{el}$ (compensation of the electrolysis and transport losses, i.e., per GW$_{H_2}$ at the end of the pipeline) + 500 million $/GW$_{H_2}$ (electrolysis) + 520 million $/GW$_{H_2}$ (transport pipeline) = 5540 million $/GW$_{H_2}$
- Without the costs of H$_2$ storage.

Table 4.3 Energy cost from conventional power plants, cost of nuclear hydrogen, of gas from coal gasification and of crude oil at 100 $/barrel.

2002-$	Fossil-fuel base-load plants				Nuclear power plants			
	CCGT	Coal-fired steam plants	Coal-fired IGCC advanced technology[c]		Nuclear (EPR)	Nuclear Pools[f]		Nuclear H_2[g]
				With CO_2-capture	Today[d]	Large scale scenario (US nuclear plants)[e]		
	Electricity							H_2
Investment (million $/GW)	615	1200	2120		2000	1100	1890	1950
Efficiency (LHV)	60%	45%	43%					
Capacity utilization	8000 h/a							
	¢/kWh$_{el}$							¢/kWh$_{H_2}$
Capital cost (4% real interest, 45a)[a]	0.3	0.7	1.3		1.2	0.7	1.2	1.2
(Capital cost at 2% real interest)[a]	(0.2)	(0.5)	(0.9)		(0.85)	(0.5)	(0.8)	(0.85)
Operating and maintenance costs	0.3	0.7	0.8		1.2	1.0	1.0	1.2
Gas (2.5 ¢/kWh$_{gas}$)[b]	4.1							
Coal price[b]		EU 90 $/t	USA 45 $/t	EU 90 $/t	USA 45 $/t			
Coal		2.5	1.3	2.6	1.3			
Fuel cycle[b]					0.4	0.5	0.5	0.6
Natural uranium (130 $/kg)[b]					0.35	0.2	0.2	0.2
Electricity or gas cost (4% real interest)	4.8	3.9	2.7	4.7	3.4	3.1	2.4	2.9	3.2
(Electricity or gas cost at 2% real interest)	(4.7)	(3.7)	(2.5)	(4.3)	(3.0)	(2.8)	(2.2)	(2.55)	(2.85)
plus cost of storing the CO_2 (10 $/t CO_2)[k]				0.8	0.8				
including CO_2 storage (at 4% real interest)				5.5	4.2				

4.1 Detailed Treatment of the Costs of the Solar Power System

Gasification of coal[h]								Oil[i]
Syngas (conventional technology)		H$_2$ (1) (conventional technology)		H$_2$ (2) (conventional technology)		H$_2$ (3) (advanced technology)		at 100 $/b (2008-$) = 84.4 $/b 2002-$
without CO$_2$-capture		without CO$_2$-capture (option 50%)		with CO$_2$-capture		with CO$_2$-capture		
Syngas		H$_2$						Oil
		920		1150		880		
ca 60–65%		54%		50%		64%		
¢/kWh$_{gas(LHV)}$		¢/kWh$_{H_2(LHV)}$						¢/kWh$_{oil}$
		0.75		0.90		0.70		
		(0.60)		(0.75)		(0.60)		
		0.60		0.70		0.45		
EU 90 $/t	USA 45 $/t	EU 90 $/t	USA 45 $/t	EU 90 $/t	USA 45 $/t	EU 90 $/t	USA 45 $/t	
		2.05	1.0	2.20	1.10	1.75	0.85	
ca.2.8 (Tign[j]) ca.1.8	ca.1.9	3.4 (Llgn[j]) 2.1	2.4	3.8	2.7	2.9	2.0	6.3 (2008) 5.3 (2002)[i]
		(3.2)	(2.2)	(3.7)	(2.6)	(2.8)	(1.9)	
–	–	–	–	0.7	0.7	0.5	0.5	
				4.5	3.4	3.4	2.5	

Table 4.3 Continued.

a) Annuity factor for 45a and 4% real interest: 4.8%; at 2% real interest: 3.4%. For the coal gasification plants, an operating lifetime of only 25 years is assumed; the annuity factors then lie at 6.4% or 5.1%. For IGCC plants (Integrated Gasification Combined Cycle), one should actually use the shorter expected operating life of the coal gasification plant (in addition, gas turbines also have an expected operating lifetime of ca. 25 a); here, however, we have assumed 45 a as for the other types of power plant. For further remarks on the operating lifetime, see Section 4.4.1.

b) Fuel costs:
- Price of natural gas see Table 4.2 footnote b).
- Coal price see Table 4.2 footnote b).
- Fuel cycle costs without the cost of natural uranium; spent-fuel recycling costs typical of the USA are included. Assumed here for a large-scale nuclear power scenario: A cost increase from 0.36 to 0.5 ¢/kWh as a result of increasing depletion of natural uranium, cf. Chapter 12.
- Cost of natural uranium – Assumptions:
  - Uranium price 130 $/kg U (= 50 $/pound U_3O_8). This corresponds to the upper cost category in the statistics on uranium reserves. Comparison: the price of uranium attained its highest level up to now in mid-2007, at 350 $/kg U; in mid-2008, it was 143 $/kg U, in August 2008, 170 $/kg U.
  - For the "large-scale scenario," a *future lower fuel consumption of natural uranium* of 14.5 kg U/GWh$_{el}$ is assumed. Currently, the consumption on the world average is 25.5 kg U/GWh$_{el}$; at this consumption level, we obtain natural uranium costs for current power plants of 0.33 ¢/kWh (world average).

c) Coal-fired power plants with integrated coal gasification (IGCC – Integrated Gasification Combined Cycle) with CO_2 sequestration:
- Cost data from (EIA AEO, 2007): Advanced technology (so called *n*th of a kind)
- Including costs for CO_2 separation but without storage and transport costs for the separated CO_2.
- CO_2 storage costs: see footnote k).
- The operating and maintenance costs of the IGCC plants are given in (EIA AEO, 2007) for this complex technology: fixed 39 million $/(a GW), variable 0.39 ¢/kWh, yielding all together 0.88 ¢/kWh. According to STE (2006), they would be: fixed 88 million $/(a GW), variable 0.54 ¢/kWh, yielding all together 1.64 ¢/kWh.
- The lifetime is assumed here as for the other types of power plant to be 45 a. In fact, for IGCC plants, a shorter lifetime should be assumed; cf. footnote *a*, for example, with a lifetime of 30 a instead of 45 a, the capital costs would increase by 0.25 ¢/kWh.

d) Current nuclear plants (in 2002-US$): for example, EPR (European Pressurized Water Reactor); this reactor type corresponds in the Chicago Study (2004) to the upper cost category (see there pp. 3-2 and 9-5): 1800 $/kW without interest during construction = 2020 $/kW with interest. Compare: the first nuclear power plants of the current American type, constructed in Japan and Korea: 2300/2400 $/kW; cf. also EIA AEO (2007): from the fifth plant constructed (i.e., after initial difficulties have been overcome): 1880 $/kW. – Operating and maintenance (O&M) costs as for current plants (after WNA (2005) for the year 2003- see Table 12.3; for the fuel cycle and natural uranium costs, see footnote b).

e) Large-scale scenario for nuclear power plants: assumptions as in the Chicago study (2004); there corresponding to the lower category of investment costs (American power-plant types; cf. there p. 9-5: "the cost range also allows for uncertainty in cost estimates for reasons other than reactor type."); O&M as well as fuel cycle costs as in WNA (2005) – compare Chapter 12:
- Without expected price increases due to the existence of a *de facto* manufacturer cartel (relative to the construction costs as given above).
- Without uranium price increases (for a worldwide large-scale nuclear scenario).

f) Nuclear power-plant pools: distance 1000 km, power transmission with ±800 kV overhead power lines, gas-turbine backup power plants.

g) Nuclear hydrogen: assumptions concerning the electrolysis and H_2 transport: same as for solar hydrogen (see Table 4.2 footnote d); however at a distance of 1000 km (only over land); likewise without costs of H_2 storage.

h) Coal gasification:
Synthesis gas (Syngas), also known as medium BTU gas, refers to raw gas from gasification of coal (H_2/CO mixture) without conversion of CO → H_2; sulfur-free combustion gas, heating value similar to H_2; ca. 20% cheaper than the alternative "H_2 (1)" in Table 4.3.

H_2 production alternative (1) with conventional technology: 50% of the CO_2 can likewise be separated with a certain increase in energy consumption. Separation of this partial output is an integral component of the process. The CO_2 must however be compressed to the required pressure of 200 bar for long-term storage, which also

Table 4.3 *Continued.*

causes a certain decrease in efficiency (not quoted separately in the literature). (In the case of the alternatives "H_2 (2)" and "H_2 (3)", compression of the CO_2 is taken into account.)

The cost of the gas refers to the lower heating value (LHV). For hydrogen, the higher heating value (HHV) is 18% higher than the LHV. This additional amount of energy can be utilized only by low-temperature heating systems with so-called condensing boilers (not completely, but to a large extent). In this case, the energy cost is reduced by more than 10%. Compare the difference between HHV and LHV for natural gas: 10%, and for syngas (depending upon its exact composition): ca. 6%.

i) Crude oil: 1 barrel = 159 l; heating value: 10.0 kWh/l (LHV). Thus: 100 \$/b = 63 ¢/l = 6.3 ¢/kWh (LHV). The oil price in the table corresponds to a fictitious *current* (2008) price of 100 \$/barrel. US inflation (consumer price index) 2002–2008: 1.185 (the inflation value for 2008, which was not yet available, was taken to be the same as in 2007 (2.8%)). \$100 (2008) corresponds to \$84.4 (2002).

j) Lignite in Germany: 35 \$/tce; see Table 4.2 footnote *b*).

k) The storage costs for separated CO_2 were assumed to be 10 \$/t of CO_2 = 36.6 \$/t of C = 27 \$/tce. Depending on the efficiency, this yields for IGCC power plants 0.8 ¢/kWh$_{el}$, and for coal gasification with conventional technology 0.7 ¢/kWh$_{H_2}$; with presumed future technology, 0.5 ¢/kWh$_{H_2}$. Comparisons: the range of current estimates for CO_2 storage costs stretches from 2 to 25 \$/t of CO_2, corresponding to between 5 and 68 \$/t of coal; see also Section 10.3.

the corresponding investment costs (615 \$/kW) were left off, the power cost would be reduced by 0.3 ¢/kWh; in Spain to 4.9 ¢/kWh, in the USA to 4.4 ¢/kWh.

Table 4.3 shows, along with the power costs from fossil-fuel and nuclear base-load plants, also the costs of coal gasification, and for comparison the energy cost of crude-oil at a notional price of 100 \$/barrel.

Some important results from Tables 4.2 and 4.3 are as follows:

- Both for the case of Spain as well as for the USA, the following holds: for the solar power system, either gas CC backup power plants or coal-fired plants lead to the same overall power costs (insofar as the latter are considered to be "annex construction"); in the case of "replacement," solar power with coal-fired backup plants is however noticeably less expensive than with CC backup plants.

- Comparison of the solar power system with CC base-load plants. USA: the power from the solar power system costs the same (4.7 ¢/kWh) as that from natural-gas CC base-load plants (4.8 ¢/kWh at the gas price assumed here); this also holds with coal-fired backup plants in the case of *annex construction* (4.7 ¢/kWh). For the case of *replacement*, solar power is cheaper. Europe (sites in Spain): solar power at 5.2 ¢/kWh (gas-fired backup plants) or 5.3 ¢/kWh (coal-fired, *annex construction*) is only slightly more expensive than natural-gas CC base-load power; with *replacement* coal plants, the cost is the same.

- Syngas from coal gasification can very probably be produced both in the USA as well as in Germany (there using lignite) for less than 2 ¢/kWh. It can thus be provided to CC gas power plants at a price of around 2.5 ¢/kWh; this corresponds to the price of natural gas assumed here. If the price of natural gas rises, the backup power plants could thus be operated with syngas. In the long term, the costs of coal gasification (H_2 production) *including* sequestration of CO_2 will not be much higher, since presuming the successful further development of gasification technology and in particular of gas conditioning ("advanced technology"); the *production* costs of H_2 would be no higher (USA: 2.0 ¢/kWh).

However, the cost of storing the CO_2 (including transport) must be added to this price. At the assumed storage cost of 10 \$/t of CO_2, the H_2 price would be increased by 0.5 ¢/kWh, that is, in the USA to 2.5 ¢/kWh (in Germany, using lignite, a similar value would be expected). In spite of CO_2 sequestration, the cost of H_2 gas would thus not be much higher than for syngas with today's technology.

- This reasonably priced gas from coal gasification would thus provide a decisive alternative for the general energy supply to increasingly expensive natural gas and petroleum. The quoted price of 2.5 ¢/kWh$_{gas}$ (2002-US\$) corresponds to 40 \$/barrel of oil, that is, 46 \$/barrel in 2008-\$. The price for this gas would thus correspond to one-third of the maximum oil price in 2008, 140 \$/barrel. (More information on coal gasification is given in Section 11.2.)

In Tables 4.2 and 4.3, the power costs are also set out for the case of *low interest rates*, namely for 2% real interest, in addition to the basic assumption of 4% real interest. This makes it clear just how strongly a low interest rate shifts the economic advantage toward solar energy. Thus the cost of solar power in the USA (for the case of *replacement* of coal-fired plants) would decrease from 4.1 to 3.3 ¢/kWh at the lower interest rate, while for nuclear power, it would be reduced only from 2.4 to 2.2 ¢/kWh. More comments on the interest rate and on the estimation of capital costs can be found in Section 4.4.

In the following, the costs of power from various competing types of power plants will be considered in terms of their possible future application on a very large scale, complementing the discussion already given in Chapter 2.

4.1.3
Coal-Fired Base-Load Power Plants with CO_2 Sequestration

A massive increase in base-load electric power generation by conventional coal-fired plants would be irresponsible in view of their CO_2 emissions and the problem of climate change. By the year 2030, especially owing to economic and population growth in Asia, an increase in worldwide electric power consumption by ca. 80% can be expected (Appendix C). The proportion of power from coal-fired plants is currently only 40%. If the whole increase were supplied as power from coal-fired plants (only up to the year 2030; consumption will probably continue to increase after that), the amount of power produced by coal-fired plants would triple compared to current values (even with higher overall efficiencies for the more modern power plants). In addition, coal can be expected to be used for gasification as a replacement for petroleum and natural gas.

While solar energy can be developed on a large scale without causing environmental problems, coal can play a comparably major and long-term role for future energy supplies only if power plants and gasification installations are provided with CO_2 sequestration[4] – at least those required by the *increased* use

---

4) In the English-language literature, the concept of separation and sequestration of CO_2 is also often termed "carbon capture and storage" (CCS).

of coal so that today's CO_2 emission levels would at least not increase further; but in the medium term, the CO_2 emissions must in fact be lowered to well below today's levels. In *IGCC power plants*, plants with integrated coal gasification (integrated gasification combined cycle), the CO_2 can be separated from the exhaust gases.

If coal is to be used on a large scale, in addition to the need for separation and secure storage (sequestration) of CO_2, the disadvantage of a correspondingly high consumption rate of the valuable *resource coal* remains. Although coal at the current levels of consumption will be available for human use for many centuries to come, this resource would be consumed within one to two centuries in the case of a long-term increase in consumption by a factor of five or more – cf. Section 11.3; the economically favorable deposits would be exhausted even more quickly. Thus, in the case of a fivefold increase in coal production (for electric power generation plus a complete substitution of *today's* oil and gas consumption through coal gasification), the so-called reserves, that is, those coal deposits which are known at present and which could be tapped economically under current conditions, would be exhausted after only 35 years. Thereafter, one would have to make use of the so-called resources, that is, the coal deposits which have yet to be discovered or which can be exploited only at high cost so that their use is at present not economically feasible. (An example of the latter is the expensive German bituminous coal.) Which contribution will be made in the future by resources that are still to be discovered and can be exploited economically, is uncertain. However, it can be presumed that most of the readily exploitable deposits (thick coal layers at relatively shallow depths in regions of the world which permit favorable transport) have already been discovered.

This fundamental problem already argues against making coal the principal source for electrical energy, even if the resulting CO_2 can be separated and stored. Furthermore, the question arises as to whether suitable storage reservoirs for such enormous amounts of CO_2 can be made available, in particular under *economically favorable* conditions (cf. Section 11.3). This problem will probably represent a stronger limitation on future coal usage than the supplies of coal themselves.

The cost data in Table 4.3 for the IGCC power plants with CO_2 sequestration are those of EIA AEO (2007). As one can see, power from IGCC plants is not cheaper than solar power – in spite of the presumed future ("advanced") technology and at the assumed costs for coal and CO_2 sequestration (or, in the USA, only minimally cheaper).[5] If the costs of solar power are confirmed by a development program such as we advocate here, the future use of coal-fired plants would thus not be reasonable even from an *economic* point of view.

The development of such IGCC power plants is still in its early stages and is *technologically* very demanding. The same holds for coal gasification with CO_2

---

5) This statement refers to a rapid replacement of current coal-fired power plants by IGCC plants; that is, to the case that the not-needed older coal plants could still serve as backup plants for the solar power system. The costs of IGCC are then to be compared with the column "coal replacement" in Table 4.2. (But even for the "annex construction" scenario, the cost difference is marginal.)

sequestration ("advanced technology"), that is, the technology upon which the IGCC power plants would be based. (The situation is, however, different for the production of syngas (medium BTU gas); it is economically feasible already with current technologies.) In contrast to solar power plants, the IGCC plants (with "advanced technology") involve complex chemical processing techniques. These can be made available only by way of several intermediate steps and intensive development programs for the technologies involved in the various process stages (for which often several alternatives exist). Furthermore, there are several fundamentally different technologies for gasification (Texaco process, Shell process, Winkler gasification for lignite), which must be developed simultaneously in parallel.

Only through such a massive development program can the concept of coal gasification be implemented in such a way that it can be economically applied to power plants with the necessary high operational availability. Even if the development efforts are considerably increased, such a program would require considerably more time than the development of technically much simpler solar power plants (compare Section 11.2).

4.1.4
Coal-Fired Power Plants without CO_2 Sequestration

When cheap coal is available, as is currently the case in the USA, conventional coal-fired plants can generate electric power at comparatively low cost. In Europe, power from coal-fired plants at the quoted price of imported coal, leading to 3.9 ¢/kWh, is however not *much* cheaper than power from the solar power system in the case of "replacement." Using expensive German bituminous coal (150 $/t), base-load power from conventional coal-fired plants would cost 5.5 ¢/kWh, and is thus more expensive than solar power system power ("replacement": 4.8 ¢/kWh).

Even countries such as China and India – the major future consumers of electric power – currently have access to low-cost coal (relative to the world market prices). This is, however, due to the low labor costs there, not to favorable geological conditions as in the USA. Reckoned in national currencies, coal production in these countries is considerably more expensive than is suggested by the dollar prices (e.g., for Indian export coal). Given the poor level of modernization in Chinese and Indian coal mines, the coal produced there is even comparatively expensive in the national currencies. In comparing the price of power from coal-fired plants with that of solar power, the low labor costs must, therefore, also be taken into account in estimating the costs of constructing the solar plants so that these costs – insofar as the construction is carried out within the country, which is quite feasible in the case of technically straightforward solar plants – are proportionally lower.

Individual countries with access to low-cost coal could be tempted to expand their electric power generation from coal-fired plants without regard to the worldwide climate effects (as has been the case in the USA). Since, however, worldwide climatic protection does not permit an increase in CO_2 emissions, but instead requires their drastic reduction, it can be assumed that in the *long term* in international treaties and contracts, CO_2 emission limits will be fixed for all countries and the penalties will be so high that base-load electric power generation from coal-fired plants without CO_2 sequestration will no longer be economically advantageous. In the long term, there is no alternative to such a policy – if coal-fired power plants are not to be allowed to become the major cause of a worldwide ecological catastrophe. In Chapter 2, we have already mentioned the fact that the *monetary* effects of climate change have been estimated in the most important studies on this subject to be on average 70 \$/t of CO_2, which translates into an increased cost of electric energy from coal-fired plants of 5 ¢/kWh. In view of these middle- or long-term perspectives (the high external costs of penalties for corresponding CO_2 emissions), it is not reasonable even for countries with low-cost coal supplies to continue along the path of conventional coal-fired plants as sources of electrical energy. For most countries, dependent on imported coal, such a path makes no sense in any case due to the minimal cost advantage over solar power. Furthermore, in addition to the increased costs for CO_2 emission penalties, the problem of uncertain future coal prices must be considered.

The fact that the CO_2 problem affects the solar power system as well, when coal-fired plants are used for backup power generation, was already mentioned in Chapter 2. In the solar power system, the CO_2 emissions are small, however, in view of the relatively low proportion of backup power within the total power output. For this reason, future penalties for CO_2 emissions would not have a major effect on overall power costs. And for CC power plants, which can be operated with H_2 from coal in the future when the "advanced technology" for coal gasification has matured, the CO_2 emissions can be minimized.

4.1.5
Nuclear Power Plants

Nuclear plants generate power at a relatively low cost, insofar as one considers only the operating costs and not those potential costs which would accrue in the case of a nuclear catastrophe; these would be borne by the state or society as a whole.

In considering the numbers in Table 4.3, however, it must be kept in mind that the lowest predicted sums have been assumed for the investment costs of nuclear plants, namely the lower limits of those estimates given in the Chicago Study (2004) for nuclear plants in mass production. In particular, no *cost increases* resulting from the high demand for nuclear plants were taken into account, although these would seem inevitable if a massive increase in nuclear plant construction occurs. The real investment costs (and also the uranium costs) could thus be considerably higher than those shown in the table.

In using nuclear energy, its general disadvantages must be considered: the risk of a serious nuclear accident (especially in an age of potential terrorist attacks on power plants), the proliferation question, the great unsolved problem of long-term waste storage, and the environmental damage resulting from uranium mining. In considering the question of whether nuclear power plants represent a reasonable alternative to current energy supplies based on fossil fuels, an important role is played by the uranium reserves. Their limitations mean that nuclear energy is hardly in a position both to supply future power requirements *and* to serve in the future as a replacement for petroleum and natural gas. Nuclear power plants currently supply only 16% of worldwide electric power. If the total present and future worldwide base-load power consumption were to be generated by nuclear plants (including an increase of 80% by 2030), this would require a 10-fold increase in plant capacity. If in addition, petroleum and natural gas at their present consumption rates were to be completely replaced by nuclear energy (as electrical energy or hydrogen), this would require a further 27-fold increase in plant capacity; all together then nearly a 40-fold increase would be necessary.[6] Even at the lowest predicted future consumption rate of natural uranium, 14.5 kg U/GWh$_{el}$ (current value: 25.5 kg U/GWh$_{el}$), both the currently known reserves as well as the so-called speculative reserves of uranium (all together *14.7 Mtons of natural uranium* with mining and refining costs of up to 130 \$/kg U) would be exhausted within only *10 years* – cf. Chapter 12. (The reserves which can be extracted only at higher cost are discussed later in this section.)

By *reprocessing* the spent fuel elements, the consumption of natural uranium could be decreased by ca. 30%. In effect, this corresponds to an increase in the reserves. It would, however, be accompanied by a cost increase of about 0.6 ¢/kWh$_{el}$.

> For the large German reprocessing plant with high environmental and safety standards planned in the 1980s, according to contemporary publications a price increase for nuclear power of 0.015 DM/kWh (1983) would result (cf. e.g., Deutsches Atomforum, cited by Vosen (1989)); in the year 2002, this corresponds to *1.2 US-¢/kWh$_{el}$* (cf. also Chapter 12).[7] The authors, however, do not state which interest rate for capital investments was

---

6) Current worldwide electric power generation is 1900 GW$_{el}$; expected by 2030: 3400 GW$_{el}$. Of this, 350 GW$_{el}$ is supplied by hydroelectric and other renewable energy sources so that – without these sources – in the year 2030, the power requirements would be ca. 3000 GW$_{el}$. (Compare the current power supplied by nuclear energy: 300 GW$_{el}$). Petroleum consumption (2004): 5000 GW; gas consumption (2004): 3200 GW. The future electrical energy and the current petroleum and natural gas consumption thus all together total 11 200 GW (cf. Appendix C). For the rough quantitative estimates given here, we have made the simplifying assumption that oil and natural gas could be replaced directly by nuclear electric power or nuclear hydrogen. Later, we give more precise estimates: for methanol production using nuclear power *and* coal – in analogy to the production of "sun methanol" – in a quantity which would correspond to the present total consumption of petroleum, 4500 GW of nuclear energy would be required annually (0.9 kWh$_{el}$ per kWh of methanol); cf. Section 11.4.3. Together with the future required electric-power generation, this would necessitate a 25-fold increase in nuclear power output capacity compared to today's value.

7) DM = Deutsche Mark (earlier German currency unit); 0.015 DM (1983) = 0.765 Eurocents (1983) or 1.12 Eurocents/kWh (2002; consumer price index Germany 1983 to 2002: ×1.47). Taking purchasing power parity according to the OECD (1\$ = 0.96€), this gives 1.17 US-¢/kWh (2002).

assumed. Furthermore, it can be presumed that the costs of reprocessing plants would decrease if they were constructed in a large series, that is, if several plants were built at the same time. Possibly, the costs could be halved by applying the interest rates assumed here and for mass production; we therefore use the value of $0.6\,¢/kWh_{el}$. This agrees with the results of a recent American source (UCS, 2008): "The Energy Department recently released an industry estimate that a reprocessing plant with an annual capacity of 2000 metric tons of spent fuel would cost up to \$20 billion to build – and the U.S. would need two of these to reprocess all its spent fuel. An Argonne National Laboratory scientist recently estimated that the cost premium for reprocessing spent fuel would range from 0.4 to 0.6 ¢/kWh."

Nuclear energy could thus become a really "major" energy source only either if uranium can be economically extracted from seawater on a large scale, or else if *fast breeder reactors* are used. The latter are in the general opinion not only considerably more expensive, but also less safe than current reactors so that their large-scale use can hardly be seriously contemplated, and it would not represent a major economic advantage in comparison to solar energy.

No one can give an exact figure for the cost of fast breeder reactors. One can realistically assume that they would be considerably more expensive than current nuclear power plants. As a rough estimate, we assume as an example (and somewhat arbitrarily) for mass production a cost twice as high as for new conventional reactors (lower limit 1100 \$/kW for large-scale construction), that is, 2200 \$/kW. This is less than the present cost of individual reactors of the newest American type (2400 \$/kW), and not much more than the EIA/AEO prognosis of 1880 \$/kW for conventional reactors (cf. footnote *f* of Table 4.3). The use of fast breeders, which is generally regarded with great skepticism for security reasons, would be feasible only with "power-plant pools." The investment costs would then be 3050 \$/kW, and the resulting power cost (not listed in Table 4.3) would be 4.0 ¢/kWh. (The individual contributions in ¢/kWh are 1.9 (capital), 1.0 (operation and maintenance), 0.5 (fuel cycle), 0.6 (reprocessing, see below), 0.0 (natural uranium).) The fast breeder thus probably offers no substantial cost advantage over solar power. Aside from the cost aspect, the question also arises as to whether such a concept would be politically enforceable in the regions where these many fast breeder reactors would have to be built, given the accompanying risks.

The cost of natural uranium is assumed in Table 4.3 to be 130 \$/kg U. An increase in this cost for the case of extraction of uranium from seawater to, for example, 1000 \$/kg U (i.e., nearly eightfold) would lead to a natural uranium cost of 1.6 ¢/kWh$_{el}$ and thus would increase the cost of electrical energy from these nuclear plants by 1.4 ¢/kWh; for mass production, an increase from 2.4 to 3.8 ¢/kWh, for the case of power plant pools from 2.9 to 4.3 ¢/kWh could be expected.

With mass production, only power plant pools located far from population centers would be feasible. At a cost of 4.3¢/kWh, only a marginal cost advantage over solar power would then result so that—taking all the principle disadvantages of nuclear power into account—not even a substantial *microeconomical* advantage would remain.

The uranium reserves mentioned above refer, as stated, to a price of 130 $/kg U. Taking into account reserves with higher extraction costs, the conventionally accessible uranium reserves would be greater. Uranium extraction from such sources would, however, not only be more expensive—whereby the increased risk of price dictation by those who control the sources must be considered—but also, it would be accompanied by increased environmental risks. Table 12.5 shows the *present-day* uranium reserves as a function of extraction costs, with the cost categories 40, 80, and 130 $/kg. The distribution among the cost categories indicates that the expected additional amount of uranium which can be extracted at higher cost is less than proportional to the extraction costs. Thus, the amount of uranium RAR and EAR I up to 40 $/kg U is 2.5 Mt; it increases by only 1 Mt up to 80 $/kg U and by a further 1.1 Mt up to 130 $/kg U, giving a total of 4.6 Mt. A tripling of the cost thus yields less than twice the total amount of reserves. If we assume a doubling of the extractable amount for every tripling of the price, a cost increase from 130 to 400 $/kg U would double the reserves estimated above. The timeline of 10 years (referring to a 40-fold increased power output) would then be increased to 20 years with extraction costs of up to 400 $/kg U (and to roughly 30 years at 800 $/kg U[8]).

If we limit our considerations to electric power generation (worldwide, and referring to power consumption up to the year 2030—that is, a 10-fold increase in plant capacity compared to the current output)—that is if we forget the substitution of petroleum and natural gas via hydrogen production, then the known uranium reserves in the price category up to 130 $/kg would indeed suffice for *one* power plant generation (40 a), and the additional reserves (up to 400 $/kg)—insofar as the above simple estimate can be trusted—would supply a further generation. The exploitation of these additional uranium reserves (extraction *costs* up to 400 $/kg) would, however, have three disadvantages:

1) Increased power costs, which to be sure would still be supportable (0.4¢/kWh, corresponding to a tripling of the uranium cost contribution from 0.2 to 0.6¢/kWh).

2) A further increase in the danger of a price cartel by the uranium suppliers. This danger is present today for the currently used favorably priced reserves up to 130 $/kg U (see below). If the *price* was then increased, for example, from 400 to 800 $/kg, this would result in uranium costs of 1.2¢/kWh$_{el}$ and, if power plant pools were constructed, in a power cost of nearly 4¢/kWh.

---

[8] Here, we presume that a tripling of the extraction costs would lead to a doubling of the available quantity of uranium. A *doubling* of the extraction costs would then correspond to an increase in the reserves by a factor of about 1.5. A cost doubling from 400 to 800 $/kg U would then imply that the reserves would last for 30 years instead of 20 years.

And, in particular:

3) Uranium extraction is associated with very serious environmental dangers, especially in the case of the expensive reserves. In the past, ores with a uranium content between 1% and 10% were available ("high concentration"), and today some ore with a uranium content down to 0.1% is being extracted ("moderate concentration"). With the high-cost reserves, even lower uranium concentrations would have to be accepted. The result would be correspondingly larger quantities of ore which would need to be mined, and to the same extent larger amounts of contaminated residues. Compared to an ore which today is termed "good" with 0.5% uranium, an ore with, for example, 0.05% U would produce the *10-fold* amount of tailings, and with 0.02% U a *25-fold* amount (Section 12.6). The environmental risks from uranium extraction are thus even today considerably greater than in the past, and would increase massively if the additional "expensive" reserves are tapped.

Concerning the danger of a *price cartel*, at present, the major uranium source countries (USA, Canada, Australia) possess only 40% of the currently known or expected low-cost reserves (up to 130 $/kg; cf. Section 12.6). In the case of a massive increase in nuclear plant construction (10-fold increase in nuclear capacity) and the associated great rise in demand for uranium, the limits of currently known reserves would be rapidly approached. The formation of a cartel by the remaining supplier countries would be very probable. This would in turn determine the price very soon; its influence on world prices would increase not just when the reserves of competing countries were exhausted, but already when they were no longer able to compensate a major reduction in supply by the cartel over a longer period of time (increase in extraction rates and use of more expensive reserves over a period of several years). The point at which the previous supplier countries could no longer react to increased demand with low-cost supply increases would be quickly reached in the case of a large-scale nuclear power scenario. The prices could possibly rise rapidly to, for example, 400 or 800 $/kg. With a price cartel, the real price is completely uncertain; it could also be still higher. Note that the price of uranium already reached a maximum of 350 $/kg U in the summer of 2007 (see Table 4.3, footnote b).

Our conclusion concerning uranium reserves is as follows: the low-cost reserves with extraction costs up to 130 $/kg would indeed *theoretically* suffice for one power plant generation for supplying most of the worldwide electric power demands. But considering the distribution of the reserves, one can fear that a similar situation would arise as for petroleum so that the price would not remain long at the level of 130 $/kg if consumption increased strongly.

4.1.6
Weighing Cost Differences

We emphasize again the most important factor in terms of cost differences: in considering the different advantages and disadvantages of various power-generating technologies, it is not so important which of them offers the lowest power cost, but rather whether one can afford an existing cost difference. In Chapter 2, we have already dealt with the economic burden imposed by the additional costs of solar energy for the examples of the USA and Europe (solar plant locations in Spain), making two assumptions:

1) Maximum cost differences, that is, for nuclear energy on a large scale without power plant pools, with a power cost of 2.4 ¢/kWh (lower limit of the conceivable cost range) and for the solar power system (with gas backup plants), a power cost of 4.7 ¢/kWh (USA) or 5.2 ¢/kWh (EU)

2) Massive substitution, that is, replacement of the entire base-load power generating system (80% of the overall electric power) plus substitution of ca. 35% of the current oil and gas consumption by electric energy or hydrogen from electrolysis.[9]

This would yield in the USA and in Europe an increased economic burden of less than 2% of the gross national product (GNP). The previous considerations demonstrate that the cost differences would in reality be smaller. Here, we always assume that solar energy will attain the quoted cost level in the future, and under this assumption, we are dealing with an upper limit for the cost burden (both in terms of the cost difference and also the scale of the substitution), not with the most probable value.

But even if these additional costs occur in full (2% of the GNP), this would be no more than what is attained in 1 to 2 years by the national economy due to the increase in production (economic growth). Making use of the increase in GNP for restructuring energy supplies would require only that the increase in 1 or 2 years be "reserved" over the long term for this task.

The question of whether or not this increased economic burden can be afforded is in fact already answered by these small percentages of the GNP. Two percent of the total economic production would have no noticeable effect on public wealth. Crises, such as can be caused by unpredictable and uncontrollable fluctuations in the oil price, have a much greater influence, as do changes in consumer behavior such as acceptance of vehicles with a high fuel consumption, or of housing with poor insulation. The greatest influence on the prosperity of the majority of citizens has, by the way, the question of the distribution of the national product within a society. Alongside such factors, which have a lasting and major effect on the standard of living, an economic burden of 1–2% of the GNP for a crisis-resistant, secure, environmentally friendly and climate-compatible energy supply is insignificant.

---

9) Together with the *cost-neutral* substitution through coal gasification using coal freed up from electric power production, not considered here – in USA an additional 20%, in Europe an additional 12% of the oil and gas consumption – we are dealing here with a maximum-substitution scenario for petroleum and natural gas, in which roughly half of the overall oil and gas currently consumed would be substituted.

Table 4.4 The cost of solar power from Spain (including power transmission) and the cost of backup power from gas or coal-fired plants.

Spain		Solar	Backup	
			CC	Coal[b]
Investment costs (M$/GW)		4835[a]	715	1340
Capacity factor (%)		70%	30%	30%
Full-load operation (h/a)		6130	2630	2630
		¢/kWh (2002)		
Capital		3.8	1.3	2.5
Operation and maintenance	solar	1.0	–	–
	fossil	–	0.6	1.3
Gas[c]		–	4.3	–
Coal[c]		–	–	2.6
Power cost		4.8	6.2	6.4

a) Investment costs (M$/GW).
　　Spain: Solar plant 3985 (including transmission losses 4335); transmission lines 500; all together 4835.
　　Transmission losses for the case of Spain (2000 km) 8.1%.
b) "Annex construction" case: 50% new and 50% used coal plants plus fast start-up backup plants (see Table 4.1).
c) For fuel prices, see Table 4.2 (Captions); Efficiencies: Natural-gas CC as backup plants: 58% Coal-fired plants new: 45%; used: 40%.

4.1.7
Separate Considerations of Solar and Backup Power Supplies

While in Table 4.2, the costs of solar and backup power plants are shown together (solar power system), Tables 4.4 and 4.5 provide a listing of these costs separately. The cost of solar power given here includes the cost of power transmission. As can be seen, solar power is cheaper than the backup power. The necessity of providing backup power plants to guarantee continuous availability of electric power thus somewhat increases the overall cost of power from the solar power system (depending on the price of the fossil energy source used). However, this increase in power cost due to the small proportion of backup power is only moderate.

Table 4.5 demonstrates that the cost of solar power in the USA, at 4.2 ¢/kWh, is practically no higher than the pure *fuel* costs for *base-load* CCGT power plants (Table 4.3; 4.1 ¢/kWh$_{el}$). The solar power plants can thus be regarded as economic "fuel savers" in connection with gas power plants.[10] (In Spain, the cost of solar power, 4.8 ¢/kWh, is only slightly higher than the cost of gas.)

10) This is only another way of looking at a fact already mentioned; namely that a solar power system in the USA would operate at the same power cost as CCGT base-load power plants.

Table 4.5 The cost of solar power from Morocco or in the USA (including power transmission) and the cost of backup power from gas or coal-fired plants.

Morocco/USA		Solar	Backup CC	Backup Coal[b]	
Investment costs (M$/GW)		4840[a]	715	1340	
Capacity factor (%)		80%	80%	20%	
Full-load operation (h/a)		7010	1750	1750	
		¢/kWh (2002)			
Capital		3.3	2.0	3.7	
Operation and maintenance	Solar	0.9	–	–	
	Fossil	–	0.7	1.8	
Gas[c]		–	4.3	–	
Coal[c]		–	–	EU 2.6	USA 1.3
Power cost		4.2	7.0	8.1	6.8

a) Investment costs (M$/GW).
   Morocco/USA: Solar plant 3695 (including transmission losses 4175); transmission lines: 665; all together 4840.
   Transmission losses for the case of Morocco/USA (3000 km) 11.5%.
b) "Annex construction" case: 50% new and 50% used coal plants plus fast start-up backup plants (see Table 4.1).
c) For fuel prices, see Table 4.2 (captions); efficiencies: natural-gas CC as backup plants: 58% coal-fired plants new: 45%; used: 40%.

4.1.8
Solar Power at the Plant Site

In Table 4.2, the power cost at the solar plant site is also given. It is relevant for hydrogen production (see below) and – with some reservations and as mentioned in Chapter 2 – for power-consuming regions in the immediate neighborhood of the solar plants, in particular in Spain and California. Without the costs of power transmission (investment costs for the transmission system and transmission losses) and the backup plants, solar power is considerably cheaper than that from the complete solar power system. The difference compared to the cost of nuclear power (on a large scale) is then no longer great; in the southwestern USA, it is only about 0.9 ¢/kWh.

As also mentioned in Chapter 2, backup capacity is in fact also required even for power-consuming regions near the plant sites. But in the case of "replacement" of already operating coal-fired plants, the associated capital costs need not be included (the backup plants are "cost-free"). The cost of backup power from the coal-fired plants (Tables 4.4 and 4.5) is then no higher than the cost of solar power at the plant site (Table 4.2) so that there is no increase in overall costs. In the case of "annex construction," no coal-fired plants would be built near the solar plants, but rather the solar and the coal-fired plants would be combined into "hybrid power plants," which together provide all the required base-load power (for details see Section 11.1.3).[11] The power costs would in this case be somewhat higher than for pure solar power, but not much higher.

4.1.9
Hydrogen Production

For solar hydrogen production, the costs of the solar power plants, electrolysis plants, and transport pipelines need to be considered, in addition to the energy losses (during transport, the energy required to compress the hydrogen). To cover the losses, the capacity of the solar power plant must be increased correspondingly. The assumptions regarding investment costs and losses are given in the caption of Table 4.2. The preferred electrolysis process is supposed to be a high-temperature steam electrolysis – which must still be developed! – with an efficiency of 86%, a concept which is currently being pursued also for hydrogen production using nuclear power. The investment costs are adjusted to those usually mentioned for a current conventional electrolysis installation (500 $/kW$_{H_2}$, cf. Section 11.2.9). For the hydrogen transport, electric-powered compressors have been assumed, including the intermediate pumping stations. As can be seen from Table 4.2, the cost of solar hydrogen including transport from Morocco[12] (or at other sites in the Sahara) or in the USA under these conditions is 4.7 ¢/kWh. If the new electrolysis technology can be successfully developed, then the cost of hydrogen would be only 1.4 ¢/kWh higher than that of solar energy at the power plant site.

Even though solar hydrogen production is not a topic of great current interest as a gas supply, these considerations on the one hand show that it could certainly provide in the *long term* a pure renewable-energy alternative to the current use of petroleum and natural gas. The price of energy from this gas, at 4.7 ¢/kWh, would in any case be lower than the energy from oil with a petroleum price of 100 $/barrel (in 2008-$); 4.7 ¢/kWh (in 2002-$) corresponds to 88 $/barrel (in 2008-$), a price with which one has learned to live. It is thus by no means the case, as is sometimes stated, that when the petroleum reserves are exhausted, the end of our

---

11) This is possible only when no long-distance transmission line is present, which could fail (otherwise, the backup power plants would have to be at the end of the transmission line). Through the combination, the power block of the solar plant can also be used for the coal-fired plant (saving investment and operating costs), and one would require only a coal-fired pressureless heater for the molten salt heat-transfer medium, instead of a coal-fired high-pressure steam generator.

12) For solar hydrogen production on a *large scale*, from the European viewpoint only sites in the Sahara need be considered. In Spain, the available land area is too small for this purpose. In computing the transport costs, we have included the more costly sea route from Tunisia to Sicily (200 km undersea distance plus 3100 km over land). This estimate is thus valid not only for Morocco (with a short undersea distance to Gibraltar), but also for other sites in the Sahara.

technological society will arrive, or that energy prices would then be insupportably high.

On the other hand, as already discussed in the "Preliminary Remarks and Summary" at the beginning of this book, solar hydrogen is already of interest today for the production of "sun methanol" (together with coal). Here, again, we would require very large quantities of hydrogen. Due to the shorter transport distance of only ca. 1500 km to the coal fields in the USA, the cost of hydrogen for this purpose would be slightly lower (4.4 ¢/kWh). With very large-scale deployment, thus if H_2 production for methanol were to be provided in addition to electric power for general consumption, the cost might be even lower; a cost of around 4 ¢/kWh would then seem feasible (compare Section 11.4.3.1).

Hydrogen production on a large scale using nuclear power plants would in fact be somewhat cheaper. But nuclear hydrogen production on a genuinely large scale (i.e., on a scale still larger than the future nuclear electric power generation presumed above) is, as stated, already hardly relevant, owing to the limited supplies of natural uranium.

We also wish to mention a further potential possibility (yet to be developed) for producing solar hydrogen. This would be a hybrid process in which the energy input for the decomposition of water into hydrogen and oxygen would be supplied in part as high-temperature heat (from a high-temperature central receiver) for a thermo-chemical process and in part in the form of electrical energy. Here, also, according to a recent publication from Sandia (Kolb *et al.*, 2007b), the possibility exists of achieving costs which are considerably lower than those of conventional electrolysis. However, one must keep in mind that for this process, aside from the production technology for the H_2 itself, a new type of receiver (particle receiver) would also have to be developed.

4.2
Comparison with the Study of Sargent and Lundy

The cost data for solar tower plants used in this book correspond, as mentioned above, to the data which were used in an earlier study by the present authors (1998), that is, the numbers from Sandia (Kolb, 1996a) as well as additional assumptions made by the authors, which appear quite plausible, regarding large-scale production scenarios. In Table 4.6, the values from the Sandia study are given in the left-hand column. These are quoted in more detail in Appendix A. Concerning the indirect costs, we must keep in mind that they were described by Kolb using only a single factor ("indirect multiplier"). The itemization (interest during construction; owner's costs; planning and contracting) as given in Table 4.6 was carried out on the basis of a previous study (Utility Studies: Hillesland, 1988); cf. Section 4.3.1.

In Chapter 2, we have already pointed out the important fact that these numbers agree rather well with those of an American study published in 2003, which was carried out for the US Department of Energy (DOE) by the engineering firm Sargent & Lundy (S&L, 2003). Its title was "Assessment of Parabolic-Trough and

Table 4.6 Comparison of various cost studies – investment costs per 1000 MW of installed solar power generating capacity for locations in Morocco or in the USA (a more detailed comparison is given in Appendix A).

Solar tower power plant – investment costs per 1000 MW

Morocco/USA 2002-$		Kolb (SANDIA) SM 3.7	Kalb/V. Mass production SM 3.7	SunLab 200 MW SM 3.7	S&L 200 MW SM 3.7	SunLab 220 MW SM 3.7	S&L, other heliostat costs + indirect costs
Output power	MW	1000	1000	1000	1000	1000	1000
Solar multiple (SM)		3.7	3.7	3.7	3.7	3.7	3.7
Design insolation (24 h full output)	kWh/m² d	8.0	8.0	8.0	8.0	8.0	8.0
Mirror area	km²	16.97	16.97	16.59	17.01	15.32	17.01
Land area	km²	94.6	94.6	87.4	87.4	80.6	87.4
Mirror area coverage	%	17.9	17.9	19.0	19.5	19.0	19.5
Investment costs	Million $						
Heliostats (per m²)	$/m²	138	83	96	117	76	96
Heliostat field	M$	2343	1402	1592	1991	1164	1633
Tower and receiver (total)	M$	404	404	389	447	342	447
Receiver alone	M$			236	293	197	293
Tower + vertical piping	M$			153	153	145	153
Heat storage capacity	h	16	16	16	16	16	16
Heat storage	M$	436	436	345	345	259	345
Power block (wet cooling)	M$	590	590	425	455	473	455
Land preparation	M$	81	81	46	46	42	46
		–	–	–	–	–	–
Subtotal	M$	3854	2913	2797	3283	2280	2925
plus (Kalb/Vogel):							
Horizontal piping	M$		150	139	139	126	139
Land price	$/m²			0.5	0.5	0.5	0.5
Land costs	M$			43.7	43.7	40.3	43.7
		–	–	–	–	–	–

Table 4.6 Continued.

Solar tower power plant – investment costs per 1000 MW

Morocco/USA 2002-$		Kolb (SANDIA) SM 3.7	Kalb/V. Mass production SM 3.7	SunLab 200 MW SM 3.7	S&L 200 MW SM 3.7	SunLab 220 MW SM 3.7	S&L, other heliostat costs + indirect costs
Total (direct costs)	M$	3854	3062	2980	3465	2447	3108
Indirect costs:							
Construction time (as small vol. prod.)	a	4	2	1		1	
Full interest period (50% of construction time)	a	2	1	(0.5)		(0.5)	
Interest rate	%/a	4	4	(4%/a)		(4%/a)	
Interest during construction	M$	308	122	(4%/a:69)		(4%/a:62)	
Owner's costs (% of investment)	%	6	3				
Owner's costs	M$	231	92				
Planning and contracting (% of investment)	%	9	4				
Planning and contracting	M$	347	122				
Engineering, management, and development (% of investment)	%			7.8	15.0	7.8	7.8
Engineering, management, and development	M$			232	520	191	242
Contingency cost margin (% of investment)	%	7	0	7.4	14.3	8.1	7.4
Contingency cost margin	M$	270		220	496	198	230
		–	–	–	–	–	–
Total (indirect costs)	M$	1156	337	453	1015	389	472
		–	–	–	–	–	–

Table 4.6 *Continued.*

Solar tower power plant – investment costs per 1000 MW

Morocco/USA 2002-$		Kolb (SAN-DIA) SM 3.7	Kalb/V. Mass production SM 3.7	SunLab 200 MW SM 3.7	S&L 200 MW SM 3.7	SunLab 220 MW SM 3.7	S&L, other heliostat costs + indirect costs
Total (overall) (wet cooling)	M$	5010	3399	3433	4481	2836	3581
% less power (dry/wet cooling)	%	8	8	8	8	8	8
Investment costs with dry cooling	M$	5446	3695	3731	4870	3083	3892

(includes higher investment costs for dry cooling towers)

Power-Tower Solar Technology Cost and Performance Forecasts." This study is a high-quality and respected work which is often described as "due diligence" or "due diligence-like." In any case, according to the DOE mandate, it was intended to be a work of this category, and is referred to in these terms by S&L themselves (cf. however the comments below on the NRC statement).

The S&L study offers two things:

Firstly, it describes in detail the expected developments from the authoritative American solar research institution SunLab[13] in terms of the technology and the costs for solar thermal power plants in the time period up to the year 2020.

Secondly – and this was more or less explicitly its mandate – it provides a critical analysis of the SunLab expectations and complements them with its own know-how and with additional research.

The S&L study thus contains two cost estimates:

a) the basic SunLab estimate, which presupposes a rather "aggressive" and essentially successful development program, and

b) a more cautious S&L estimate, which is based mainly on well-established technology. The study is essentially an intensive review of the cost perspectives as seen by the "experts" in the area of solar thermal power generation and/or by its proponents.

Although there can naturally be no absolute certainty in judging the future costs for mass-production series of power plants, and although the S&L study was

---

13) "SunLab comprises researchers from Sandia National Laboratories and the National Renewable Energy Laboratory (NREL) working together on "Concentrating Solar Power" for the Department of Energy." (DOE, 2007)

limited in its time and financial scope, one can still consider the S&L numbers and the SunLab estimates given in the study to represent the best prognoses currently available.

Along with the S&L study – although it also considers heliostats in some detail – a newer Sandia report on the projected costs of heliostats in mass production is also relevant for estimating the cost data which were employed in this book[14] (Kolb et al., 2007a). We consider this point in more detail in Chapter 6.

4.2.1
Costs from Various Studies

4.2.1.1 Investment Costs

Table 4.6 compares the results from various studies, giving the investment costs for an installed solar power capacity of 1000 MW in Morocco and in the USA, with a solar multiple (SM) of 3.7. The *first column* on the right side of the table shows those of Kolb (Sandia) (1996a), which we have used here as a basis (as in 1998, also). We have simply recalculated these values for a mass-production situation (revised heliostat costs and other indirect costs). Also, we extended them – in view of large-scale power plants with power blocks of ca. 700 MW instead of 200 MW – by including the item "horizontal molten salt circuit" and – in view of different site conditions – by the item "dry cooling."[15] The *second* column lists the values from "Kalb/Vogel." The *third* to *fifth* columns show the values obtained from the Sargent & Lundy study[16]: the *third column* shows the expected values for SunLab, the *fourth* those of Sargent & Lundy themselves. (The *fifth* column, which refers to solar power plants with supercritical steam circuits, can be skipped over here.[17])

---

[14] In this report, heliostat costs which lie well below those in the S&L prognosis and rather close to the "SunLab case" of the S&L study are regarded as feasible.

[15] The boundary conditions related to the plant location include also the costs of acquiring the land. For Morocco, in "Kalb/Vogel" no real-estate costs were assumed, since the desert land there is not arable. In Morocco, therefore, the land prices are political – see the remarks to this point later in this chapter. In Spain, a land price of 1.25 $/m² was assumed.

[16] In these columns, our assumptions relating to dry cooling and to a horizontal thermal circuit were also adopted (the latter adjusted to the appropriate land area). We recall that for sites in Spain, wet cooling has been assumed (in analogy to Kolb, 1996a), while for desert sites in Morocco and sites in the USA, dry cooling was assumed. With the assumption that dry cooling causes a decrease in overall power generation by 8% (Section 4.3.7), the net specific investment costs would be increased by 8.7%; in Table 4.6, column "Kalb/Vogel," this corresponds to $295 million.

[17] The *fifth* column likewise shows expected costs for SunLab, however, under the assumption that supercritical steam circuits will be used in the solar power plant, such as are currently being introduced in coal-burning power plants. In the S&L study, this corresponds to the version "SunLab 220 MW." Furthermore, we have assumed somewhat lower heliostat costs here, namely 76 $/m² instead of 96 $/m². This version was briefly referred to in Chapter 2. As already shown there, steam circuits of this type in a solar power plant would require increasing the molten-salt temperature by nearly 100 K. Whether this will prove to be possible and economically reasonable remains to be seen (in

From the *sixth* column, it is clear that the differences between the S&L and the SunLab-predicted values are due essentially to their differing assumptions as to the heliostat costs and the indirect costs. This column shows the S&L predictions again, however, under the assumption of the same heliostat costs and indirect costs (percentage) as for SunLab. At 3892 million $ per 1000 MW, the result differs only marginally from the SunLab estimate of $3731 million.

Note: The indirect costs are itemized differently by "Kalb/Vogel" (cf. Section 4.3.1) from those of SunLab and of S&L; the breakdown in columns 3–6 in Table 4.6 has been adopted from S&L (2003), where the SunLab numbers are also given. The investment costs (excepting the right-hand column in Table 4.6) are quoted in more detail in Appendix A, and there also for sites in Spain. The resulting electrical energy costs are set out in Table 4.8.

As one can see from Tables 4.6 and 4.8, the results of "Kalb/Vogel" and "SunLab" differ only marginally. To be sure, S&L – on the assumption of considerably lower plant deployment rates – arrive at somewhat higher costs. This is, however, almost exclusively due to the differing assumptions with respect to the heliostat costs (cf. Section 6.5.1) and the indirect costs, both of which are strongly dependent on production volume. As can be seen, the overall investment costs from "Kalb/Vogel" (second column) and from S&L (sixth column) are practically identical.

Conclusions: The S&L study confirms the results of "Kalb/Vogel," taking into account the assumption by the latter authors of a much greater scale of mass production. (Equally important as the agreement with the results of the S&L study is the agreement of the predicted heliostat costs with the newer study mentioned above.)

4.2.1.2 Operating and Maintenance Costs

Table 4.7 gives the *operating and maintenance* costs from various studies. They are reproduced here as quoted in the studies, that is, referring to a 200-MW power plant with an SM of 2.7 (Kolb und Kalb/Vogel) or 2.9 (SunLab 200, S&L 200 and SunLab 220). Starting from the total amount of the operating and maintenance costs (Table 4.7, line "Total O&M Cost"), the "O&M cost per kWh" shown in Tables 2.3 and 4.2 was computed.[18]

particular because one then enters the range of decomposition temperatures of the salts thus far being considered). This variant was merely mentioned in the SunLab report, apparently as a "perspective," and was not commented upon in the S&L study. It will, therefore, also not be treated further in this book, although it naturally represents "in principle" an interesting option. Research will first have to demonstrate whether this route can in fact be followed. (The temperature increase to ca. 650 °C is possible with nitrate molten-salt circuits only by employing an O_2 blanket as protective gas.) This column is thus not of very great interest for a comparison of the SunLab prognosis with that of "Kalb/Vogel."

18) *Note*: The costs per kWh listed in Tables 2.3 and 4.2 refer not to kWh of solar energy at the site of the power plant, but rather to kWh delivered by the solar power system, that is, the energy from solar *and* backup plants (utilization: 8760 h/a). Per kWh of solar energy at the plant, they are correspondingly higher (in Spain 0.94 ¢/kWh, in Morocco/USA 0.75 ¢/kWh).

Table 4.7 Operation and maintenance (O&M) costs of solar tower power plants.

O&M Costs (2002-$)		Kolb Sandia, 200 MW SM 2.7	Kalb/Vog. Mass prod. 200 MW SM 2.7	SunLab 200 MW SM 2.9	S&L 200 MW SM 2.9	SunLab 220 MW SM 2.9
References		Kolb, 1996a	(as in Kolb)	S&L, 2003 pp. G1–G6	S&L, 2003 pp. G1–G6	S&L, 2003 pp. G1–G6
Staffing						
Administrative	Persons				7	
Power plant operation	Persons				11	
Power plant maintenance	Persons				7	
					–	
Intermediate balance[a]	Persons				25	
Solar field maintenance and wash crew[b]	Persons				42	
					–	
Totals[c]	Persons			67	67	67
Wages per person	$/a			42.000	50.000	42.000
Staff cost	M$/a			2.81	3.35	2.81
Material and services costs						
Among others[d]						
Service contracts	M$/a			?	0.36	
(Of this weed Control)	M$/a			?	(0.17)	
Cooling water (including mirror washing)[e]	M$/a			ca. 0.1	1.30	
Parts and material (spares)	$/m² a			?	0.3 $/m² a	
Of this: (mirrors)	$/m² a			?	(0.13 $/m² a)	
(Drives)	$/m² a			?	(0.09 $/m² a)	
Miscellaneous	M$/a			?	0.35	
Equipment for O&M[f]	M$/a			?	0.2 ?	
Material and services total	M$/a			1.90	4.28	1.90

Table 4.7 Continued.

O&M Costs (2002-$)		Kolb Sandia, 200 MW SM 2.7	Kalb/Vog. Mass prod. 200 MW SM 2.7	SunLab 200 MW SM 2.9	S&L 200 MW SM 2.9	SunLab 220 MW SM 2.9
Contingency costs	M$/a			–	10%: 0.83?	
Balance					0.67[h]	
Total O&M cost	M$/a	6.0 (1995) 7.1 (2002)[g]	6.0 (1995) 7.1 (2002)[g]	4.71	9.13[h]	4.71

a) S&L p. G-2: "The power block staffing (25) is comparable to the industry average for a 120-MWe combined-cycle power plant."
b) Solar field maintenance including wash crew: S&L p. G-2: "For Kramer Junction, approximately 0.03 maintenance staff is required per 1000 m² of solar field aperture area." Exact value (67 persons divided by mirror area): 26 persons/million m² aperture area.
   (Aperture area: for the curved mirrors of parabolic troughs, not the actual mirror area, but rather the effective area of the mirrors as seen by the sun is quoted. Here, it can be set equal to the mirror area for simplicity.)
c) S&L on the assumptions of the SunLab study regarding personnel (p. G-2): "... is a reasonable estimate. The staffing compares with SEGS power-generating facilities and the recent O&M Cost Reduction Study performed at Kramer Junction (KJCOC, 1999)." (For the reference KJCOC, 1999, see the corresponding footnote in Section 4.2.1.2. "Mirror Cleaning.")
d) In the S&L study, only the assumed costs for particular items are listed under Material and Services, and not the annual sums which result. The latter can therefore be quoted only in part here.
e) S&L *cooling water* for wet cooling (pp. G-3 and G-4):
   Cooling-water costs: S&L: $0.001 22/gallon = 0.32 $/m³. ("... is based on actual cost reports at SEGS ...")
   Cf. SunLab: 0.021 $/m³; only one-fifteenth of the S&L value.
   Cooling-water consumption: cooling tower make-up (2.90) + condensate make-up (0.17) = 3.07 m³/MWh; compare p. G-3: "Water and chemical usage for the power plant thermal part is consistent with industry averages for power plants." With 1292 GWh energy produced by the 200-MW power plant (p. G-1), this yields a cooling-water requirement of 4 M m³/a and cooling-water costs of 1.3 M$/a.
   S&L Washing water:
   For cleaning the heliostats, only 1–2% of the quantity of water is required as for wet cooling. Cf. p. G-4: annually 22 l water per m² mirror area; and p. G-3: "The additional water usage for solar (e.g., mirror wash) is based on the O&M Cost Reduction Study (KJCOC, 1999)." The overall water requirements of the 200-MW power plant are thus found to be 0.022 m³/m² × 2.6 M m² = 57 000 m³, corresponding to 1.4% of the quantity of water given above for cooling, 4 M m³. (However, see p. F-3, for parabolic troughs and for "Mirror Wash Demineralizer Make-up": 0.4% of the overall water consumption with wet cooling.)
f) Equipment for O&M (including mirror wash rig–twister, mirror wash rig–deluge, etc.), all together 0.96 million $ (0.37 $/m² × 2.6 M m²); 5-year equipment life assumed; yields 0.2 M$/a.
g) In Kolb (1996a), only the total is quoted (no itemization of the individual costs). It was adopted by Kalb/Vogel.
h) The list of partial items is clearly not complete in S&L (2003). The intermediate sums given there – boldface in the table – give a different overall total from that quoted by S&L (9.13 M$/a).

The difference between SunLab 200 and S&L results from the *personnel costs* and is due only to the differing assumptions about the average salaries. For the *material and services costs*, one-half the difference of around 2.4 million $/a is due to the differing assumptions on the *cost of cooling water* (cf. p. ES-11, G-6). Furthermore, S&L includes a contingency of 10%.

Table 4.8 Energy costs for the solar power system with natural-gas backup power plants (cf. Appendix A).

	Kolb (Sandia)	Kalb/Vogel (mass production)	SunLab 200 MW	S&L 200 MW	SunLab 220 MW	S&L different heliostat & indirect costs[a]
	¢/kWh (2002-US$)					
Spain	6.3	5.2	5.0	6.1	4.5	5.5
Morocco/USA	5.8	4.7	4.5	5.6	4.1	5.0

a) The right-hand column is not given in Appendix A.

The price of cooling water is irrelevant for sites in Morocco and the USA, where dry cooling is necessary, but it influences the results for Spain, where we have assumed wet cooling. The SunLab costs are evidently based on a site where sufficient cooling water is available, while S&L quotes the cooling-water costs which apply to the California parabolic-trough plants. (In Kolb, 1996a, 1996b, no costs were listed for cooling water.)

For the power plant "SunLab 220," incidentally, the same annual costs are given in Table 4.7 as for "SunLab 200" (in both cases 4.71 million $/a); both refer practically to the same solar plant, the only difference being that for "SunLab 220," less expensive heliostats were assumed. The difference in the O&M costs *per kWh* is the result of a difference in the steam circuits: for "SunLab 220," supercritical steam is planned, which leads to an increase in the electric power output from 200 to 220 MW for the same solar input. The same annual O&M costs are then related to a correspondingly higher annual power output.

Mirror Cleaning The costs for washing the heliostats (personnel, water requirements, equipment) are based on current mirror-washing costs at the Californian parabolic-trough power plants (SEGS); those improvements identified in the "O&M Cost Reduction Study performed at Kramer Junction" *(KJCOC 1999)*[19] are, however, already included.

The personnel costs for "Solar field maintenance & wash crew" are considerable; requiring 42 of 67 persons, they make up two-thirds of the overall personnel (Table 4.7). (The total personnel requirements are taken to be the same in the SunLab and S&L studies.) What fraction of this is for the wash crew is not quoted, but it

---

19) The reference quoted is referred to in the S&L study "KJCOC 1999: O&M Cost Reduction Report, June." It is probably the same as the final report of the O&M cost reduction program cited in Chapter 8 of this book (Cohen et al. 1999). KJCOC (Kramer Junction Company Operating Company) operates five of the all together nine Californian parabolic-trough power plants at the Kramer Junction site.

can be assumed that it is not minor. In contrast, the costs for washing water are practically negligible as a result of the required quantity – only 1–2% of the cooling water consumed; they thus make up only 1–2% of the water costs listed in Table 4.7 (cf. table footnote *e*).

The costs quoted here for mirror cleaning are thus based on current personnel-intensive cleaning methods, not on automated cleaning, as of course would be preferable for a large-scale system, and which will be necessary in the long term in view of the very large heliostat fields. Automatic tracking and cleaning systems (including facilities for recycling the water used for cleaning),[20] which are still to be developed, should lower the operating and maintenance costs of the solar components in the future. Robotic vehicles are already in use today, for example, in container terminals.[21] Along with the development of control mechanisms for automatic cleaning systems, the development of robotic drive systems can also be envisaged as an element of the development of solar power plants. These vehicles can also be used for material transport within the mirror field during the construction phase.

Results of the Comparison of Various Studies Appendix A shows the studies in detail. There, all the values quoted in the studies for a 200-MW power plant are listed, and it can be seen how the assumed investment costs together with the operating and maintenance costs – with similar assumptions concerning the site (SM, capacity factor), power transmission, natural-gas CCGT backup power plants, and interest rates – finally yield the predicted *overall power cost*. Here, we give only the results (Table 4.8).

4.2.2
Response of the NRC to the S&L Study

The S&L study was "guided" and reviewed (review of the review!) by a commission of the National Research Council (NRC),[22] which was constituted for that purpose. The final appraisal of the study by this commission, "Critique of the Sargent & Lundy Assessment of Cost and Performance Forecasts for Concentrating Solar Power" (NRC, 2002) is somewhat ambiguous, not to say contradictory, and it is in sum a bit "unfriendly"– but perhaps this last statement is only an impression resulting from the insufficient understanding of the present (German) authors for the fine points of the English language[23]:

---

20) The water used for mirror washing will be recovered and reused.
21) The robotic vehicles in container terminals orient their movements relative to a guide strip on the ground. For solar power plants, this is not a suitable method; other automatic guide systems must be developed for them.

22) Chair: G. Kulcinski (National Academy of Engineering (NAE), University of Wisconsin.)
23) There is furthermore a revealing response by S&L to the "review" of the S&L study by the NRC commission; indeed, somewhat anomalously, it appears as Appendix I of their "Final Report."

On the one hand, the commission criticizes the fact that S&L concentrated too heavily on the SunLab data and initially used them as the basis for their own study, and furthermore that there was not a genuine "due diligence" study in the strict sense, that is, it did not delve deeply enough into the details of the subject at hand. (The commission, however, explicitly recognized that this would have been impossible all along, owing to the short time available – and this must therefore be understood as a clear-cut criticism of the organization which commissioned the study, the DOE.) "DOE contracted with S&L for a 'due diligence-like' study. However, what S&L did was a capital cost reduction analysis based on an assumed deployment rate. The S&L study assumed a deployment strategy and then calculated a plant cost estimate. Although S&L removed reference to 'due diligence-like analysis' from its report, the committee still must make this fact clear since it is part of the original S&L scope of work. In the committee's opinion neither the time nor resources allotted for the analysis was adequate for a true 'due diligence-like' study, and it is clear that S&L did not do a 'due diligence-like' analysis" (NRC, 2002, p. 11). Previously (on p. 8), they had stated: "To go beyond this very preliminary cost estimate would require a bottom-up, design-sized equipment list, materials breakout and cost analysis at a specific site, and this effort was not within S&L's work scope." Further, they state under the subtitle "Weaknesses of the S&L Assessment" (p. 16): "While the original charge to S&L required a 'due diligence-like' analysis, it is clear that the present report does not represent such an investigation. A 'due diligence' type of analysis has not yet been performed for CSP technology and would be necessary before private investors would fund a CSP plant and before a market assessment based on deployment rate could be developed."

On the other hand – and this is a decisive point here – the S&L study is credited with (1) "objectivity" and (2) "credibility."

Concerning point (1): "The committee found that S&L took any potential conflict of interest very seriously and made a concerted effort to address and avoid it. No obvious example of bias was apparent in S&L's interpretation of the available data nor was there any deliberate omission of pertinent facts. If anything, the S&L analysis was more conservative than SunLab's estimates in assessing areas like time to develop new materials or power conversion technologies."

Point (2): "The committee found that S&L attempted to maintain a credible process by filling in the gaps in its knowledge base with the advice of world-recognized experts. If any fault could be found with the S&L report, it would be in the lack of critics of CSP technologies on the S&L team" (p. 18).

The concluding appraisal reads as follows:

> "The committee finds that within the time and resources available for this study, S&L did a reasonable job in digesting the information provided to it by DOE, S&L expert consultants, and members of the CSP industry. For example, S&L's selection of component costs and economic parameters and assumptions regarding performance is well documented. Nonetheless, the committee also notes that because the CSP community is small,

particularly in that it lacks a large number of commercial companies openly competing in supplying CSP, it is difficult for analysts, including S&L, to obtain cost data representing a variety of perspectives, or to obtain statistically significant samples of data from which to draw inferences" (p. 20).

The review then again becomes very critical: "… S&L gives insufficient attention to factors that could lower costs … does not assess the compound risks associated with the advanced technical developments[24] … In light of these deficiencies, the committee is unable to ascertain whether S&L's projected capital costs and LECs are more accurate than those of SunLab and others" (p. 21);
and at the very end of the review:

"The clear theme of the committee's findings and recommendations is that the limited charge to the S&L team, as well as the inadequate time and resources provided, resulted in an analysis and a report that do not fully answer the questions that DOE seems to be asking – Do CSP plants have the potential to be competitive by 2020? Under those constraints the S&L team did not do a bottom-up cost analysis of the possibilities (or probabilities) of reducing the cost of CSP plants. Rather, it relied on a SunLab model and put in some of its own judgment. A true 'due diligence' study would require four to five times more time and resources …" (p. 22).

In the face of this clear-cut criticism, it is especially interesting that the NRC commission notes at another point (in the section on parabolic-trough power plants) that: "S&L also points out that if a robust, aggressive R&D [research and development] program is supported and proves successful and if policy measures are in place to facilitate deployment then still lower costs of 4 cents/kWh may even be possible." (*Note*: under the boundary conditions on which the S&L report was based, S&L in fact calculated 6.3 ¢/kWh, while SunLab estimated 4.3 ¢/kWh for 2020.) The NRC then writes concerning the S&L estimate for these particular conditions (under which 4 ¢/kWh is held to be possible): "But the committee saw in S&L's report no convincing evidence to that conclusion …, a more plausible estimate would lie somewhere between the two projections (S&L's (6.3 ¢/kWh) and SunLab's (4.3 ¢/kWh) in 2020" (p. 6). The NCR thus at least expresses the expectation that under the conditions given ("aggressive R&D program … proves successful"), the costs (of parabolic-trough power plants) could be lower than the (actual) result of the S&L report.

The criticisms of the NRC can be summarized in two points:

1) The predicted improvement in costs presupposes construction of power plants on a mass scale, as presumed by S&L (*authors' note*: the deployment rate assumed by S&L is still *very* small on the scale of the overall energy economy).

---

24) This was then, however, supplemented in the final version (original "Final Report") by S&L; the NRC review referred according to the DOE (2007, Reference 8), however, to the *draft* report 3.

It was seen, however, as by no means probable that this deployment rate will be reached so that the cost prognoses will hardly be attainable by 2020 (unrealistic owing to a lack of a market and "a sort of chicken and egg situation", p. 15).

This is the main point of criticism, which is stated in detail on nearly five pages (pp. 10–14), but remains groundless, since the investigation mandate for Sargent & Lundy makes sense only in connection with the assumption of a substantial degree of mass production of solar power plants – even if the scale considered by S&L is not as large as would be appropriate from the point of view of the overall energy economy.

2) It cannot be generally excluded that the cost estimates of S&L can prove to be attainable – since no one is currently in a position to predict the costs with certainty (including the NRC). S&L were, however, too optimistic regarding the technical challenges and they pay too little attention to the development risks of this new technology.

These risks were then, however, discussed in the final report so that in the end, only the global criticism remains, that S&L were too optimistic.

It must be kept in mind that the S&L study was in fact not a genuine "cost analysis," which would determine the costs of solar thermal power plants in a very detailed manner; that is, it was not a so-called bottom-up study. Given the brief time available and the limited financial framework, this would not have been possible – as the NRC also ascertained – and was also not expected by the DOE, since in the end the question was whether a more or less intensive support of the development of parabolic-trough and solar tower power plants would be a promising approach for the DOE. In terms of this question, the S&L study is also then represented to the US Congress (indirectly) in the DOE Report (2007, p. iv)) as a due-diligence study:

> "Sargent and Lundy was selected to conduct this analysis on the basis, among other factors, of its independence from the CSP industry and its recognized performance in conducting due diligence studies for the fossil power industry. This approach let engineers experienced in due diligence perform the detailed analysis."

The different appraisals of S&L and the NRC thus concern only future developments (market expectations, expected scale of mass production, development risks), which no one today can predict with absolute certainty. Here, we thus have a case of "opinion *versus* opinion" (see the conclusions below).

4.2.2.1 The Research-Political Context of the S&L Study and the Criticism of the NRC

For a better understanding of the situation, a remark on the "historical" context seems appropriate, in which the S&L study and the resulting NRC response can be seen:

In the year 2000, a very controversial discussion was set in motion by a memorable NRC position paper, in which it was recommended that the US government should completely eliminate all financial support for research and development in the area of solar thermal power plants, owing to their lack of development potential (NRC, 2000):

> "The likelihood of major breakthroughs that will affect cost and performance is small and/or not commensurate with the potential payoff." (NRC, 2000, cited from DOE, 2007, p. 3)

> "The Office of Power Technologies [now the Solar Energy Technologies Program] should limit or halt its research and development on power-tower and power-trough technologies because further refinements would not lead to deployment". (NRC, 2000, cited from DOE, 2007, p. 1)

This downright shocking recommendation was made, by the way, just a year after the 3-year test phase of the SOLAR TWO project was completed – in spite of all of its problems and setbacks, one must say *successfully* – and at a time when the follow-up project was being planned, when a systematic R&D program for parabolic-trough power plants was being worked out, and when furthermore in Europe once again new initiatives for solar thermal power plants were underway on the scientific and the political levels.

It was inevitable that the companies concerned would react strongly and rigorously, among other things with their famous "rebuttal" (CSP Industry, 2000). The DOE was obviously inclined to follow the NRC recommendation, but on the other hand felt duty-bound – in particular with regard to the Congress – to provide a clarification. This was the reason for its mandate to Sargent & Lundy, who were expected to produce a detailed, objective expertise. It, when completed, showed in a rather clear-cut fashion that the standpoint of the NRC was not tenable, and that, on the contrary, a strong case could be made that an appropriate development effort would lead to economically very interesting solar thermal power plants.

That the DOE then planned as a "reaction" to the S&L study to discontinue support for solar thermal R&D, namely in the budget year 2004, was already almost absurd; it was in fact a political scandal, which was however not seriously noticed by the American public.[25] (Already in the year 2001 – promptly following the NRC recommendation – the DOE budget request for the budget year 2002 was reduced from $15 million (in the previous year) to $2 million, likewise for 2003. Only Congress prevented a complete stoppage of research support by appropriating $5 million each year for 2003–2005, before the support was again successively increased in the following years (DOE, 2007)). Figure 4.1 shows the evolution of

---

[25] Equally incomprehensible was the decision taken a few years earlier (ca. 1997/1998) by the *German* Federal Government at the time to reduce drastically its research support for solar thermal power plants, and in particular to withdraw its financial support for the Plataforma Solar in Almeria (Spain).

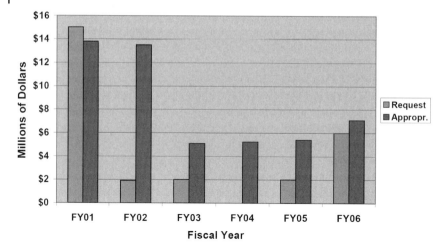

Figure 4.1 Evolution of the budget of the US DOE for the area of "Concentrating Solar Power" (CSP) (DOE, 2007).

the US CSP budget in those years. Furthermore, these budget cuts were not an appropriate response to the "peer review" carried out by MIT and commissioned by the DOE itself, in which we find the statement: "With proper funding, the DOE CSP Program can play an important role in catalyzing further CSP technology advances, which will further improve CSP economics and market penetration. ... The panel noted that support for the CSP program is significantly below the level needed to contribute to the goals of the National Energy Program. Many panel members believe the program is under-funded by about a factor of two to four times" (Tester et al., 2001,[26] cited in (NRC, 2002, p. 10, footnote g)).

4.2.2.2 Conclusions Based on the Current Preliminary State of Knowledge

It is in the end idle speculation to debate the precision with which estimates of the future costs of mass-produced power plants–based on our currently insufficient state of knowledge–could be made by SunLab or S&L (or also by the most recent Sandia study of heliostat costs). At this point, it is decisive to make a consistent effort to improve this state of knowledge. Since certain assumptions and expectations concerning future developments (in terms of plant technology, scale of production, and production methods) entered into each of these predictions, the results can never yield a "proof" of the correctness of particular predictions. Under these circumstances, there will always be differing evaluations–which inevitably are more or less strongly dependent on the basic attitude of those making the predictions with regard to the problem under discussion, and on their domain of special interests.

We could have a much more exact knowledge base within about 2 years, if we began immediately on a serious and extensive program of development, when the

---

26) J. Tester was also a member of the NRC (2002) commission.

results of the suggested large-scale, detailed investigation into heliostat costs under mass production (initially for a standard heliostat) would become available. In this program, the production methods would be simulated "step by step" and information about the mass-production costs for each step would be obtained from that branch of industry which is most competent to carry out the particular step (depending on the component, e.g., automobile or truck producers). It should also be a part of such an investigation into the production of a standard heliostat ("step by step") to develop *in detail* those individual production steps for which there are no comparable conventional models from other industrial areas, and to test them practically through the construction of prototypes. The required development (for determining the costs under mass production) is thus *much* more comprehensive and complex than considered in R&D plans up to now[27]; in these, it was obviously believed that this topic could be dealt with by superficial, all-inclusive studies.

Thus, the NRC held that with "5 times" more money and time than were available to S&L, reliable cost data could be obtained: a naive point of view. It also shows that the NRC had not recognized the *essence* of the task which must be undertaken, and that it has neither a correct idea of the nature of solar-energy research (namely "cost research" or cost appraisal as an essential but new part of research activity), nor a realistic estimation of the question of how solar energy can be effectively promoted in practice.

As we have already mentioned, for each new heliostat design, a comparably detailed investigation must be carried out. Only in this way can it be determined which type of heliostat would be the most cost-effective in mass production. The overall development program would thus include a large number of such detailed analyses of production and fabrication methods. It would, however, represent an *extraordinarily important* gain if initially such a reliable cost estimate for the mass production of a *single* type of heliostat (the standard heliostat) were available. We would then finally have a *secure* starting point.

Together with comparably profound investigations concerning the other components of the solar power plant, one would then have after ca. 2 years the knowledge base necessary to know with "certainty" (i.e., with a small error margin) how much a solar power plant using the technology described here would cost under mass-production conditions.[28]

---

27) The fact that these complex development goals can be attained *essentially* within ca. 2 years was already discussed in Section 2.4. This is related to the fact that these development goals or the cost studies which would have to be undertaken by companies with mass-production experience for each production step can be broken down into a variety of different areas. The investigation would therefore be composed of a number of individual studies which would be carried out by different researchers and could be simultaneously initiated and performed in parallel.

28) To be sure, the situation regarding *tower heat-transfer circuits* cannot be completely clarified within 2 years; for the most important goals, a development period of ca. 4 years is to be expected, and for some special (but not critical) points, perhaps even twice as long will be required (cf. Chapter 7). But after the initiation of a large-scale research program, within 2 years the most important results could be obtained. We could then put much sharper limits on the cost estimates than is possible today.

4.2.2.3 Conclusions Regarding the NRC Report

The data from S&L, or the SunLab numbers quoted there, must be considered to be the best currently available cost estimates, in spite of NRC's criticisms (at least as long as one is contemplating the truly large-scale construction of solar power plants), for the following reasons:

- The cost estimates were – with regard to S&L – developed by a neutral and in terms of conventional power plant construction competent enterprise which had no special interest in giving a *positive* representation of the chances for success of solar energy.

- It should be taken into account that in this book, we are considering very large-scale production of solar power plants, relevant to global energy policies (total substitution of conventional plants), and thus *not* an introductory scenario such as was discussed by S&L and which, in the opinion of the NRC, was still too optimistic.

- We recall that the recent study on heliostat costs under mass production (Sandia heliostat study) reaches similar conclusions to those presented in this book, and thus in this sense also contradicts the NRC estimates.

- The NRC contributed no technical competence of its own to the subject of solar power plants so that its conclusions have only the character of an "opinion"; its report does not represent a "rebuttal study."

- The NRC commission did not consider the problems of solar energy research with a proper view to the global energy-policy relevance of solar power plants, as can be seen from its report (which places too much weight on *problems of introducing the technology*; a scenario involving a large-scale governmental R&D program is not even seriously considered). With this background, the commission's skeptical assessment of the S&L predictions is understandable as far as it concerns attainable solar power plant capacity and cost reductions by the year 2020 within the hitherto existing political framework (business as usual). This report, however, completely misses the point of the challenges considered in this book.

From *today's* point of view, a development scenario as envisioned in the S&L study would appear quite probable – and it corresponds to a great extent to the conclusions of the authors of this book (presuming of course large-scale production of power plants). Naturally, uncertainties remain; but more precise knowledge would be relatively quickly available after initiating a global development program as proposed here.

4.3
Some Special Points Concerning Cost Estimates

We first recall that our estimates differ from those of the SOLAR TWO basic assumptions of 1996 ("Advanced Technology"; Kolb, 1996a) in only two points which relate to small production series, namely the heliostat costs and the indirect costs. In contrast, all the other costs (with the exception of the dry cooling which is necessary at locations in Morocco, and the horizontal molten-salt circuit[29]) are the same as in the SOLAR TWO data. In the case of the heliostat costs, instead of 138 \$/m^2 (for small production series), we have assumed 83 \$/m^2 (2002-\$; for mass production); installation in the field is included in these estimates. We will consider the heliostats in more detail in Chapter 6. In the case of the indirect costs, the global data of Kolb were complemented by data from a preliminary study (cf. the following sections).

4.3.1
The Effect of Mass Production on the Indirect Costs

This cost category includes the interest which accrues during the construction phase, owner's costs, costs for engineering and construction management, and unforeseen (contingency) costs. The values given in Table 2.1 for the construction of a very large solar park will be justified here (cf. also Table 4.6 and Appendix A):

Kolb (1996a) gives the indirect costs only as a total value, not as an itemized list. For this book (Table 2.1), a list was therefore prepared by adopting the data from the preliminary project, which was documented in detail and which essentially formed the basis for the estimates made at Sandia (Kolb, 1996a): the UTILITY studies (Hillesland, 1988). Since the total value quoted there is not exactly the same as that given by Kolb, we adopted only the relative values from this study and adjusted the absolute amounts so that their sum yielded the value quoted by Kolb.[30] Considering the similarities of the two projects (molten-salt tower power plants with a similar scenario), the overall errors introduced by this procedure should not be serious. We will, therefore, not mention this cost breakdown on the basis of the UTILITY studies at each one of the many places within this book where we refer to the data of Kolb.

The time required for construction of a plant depends largely on the scale of mass production. For very large-scale production, the average *construction time* will be taken to be *2 years*. Under the simplifying assumption that—taking into account also the interest on *indirect* costs—the *effective* duration of interest charges on the "direct investment costs" ("full interest period") is equal to half the construction time, the total investment capital (on which the annuity is based) must include interest for *1 year* until the beginning of plant operation.

The fabrication of the components (mainly the heliostats) is the factor which limits the rate of construction. For large-area solar plants, the installation of the mirror fields at the construction site can be carried out in different sectors of the field at the same time without causing problems so that no serious delays should

---

29) We refer here to a horizontal molten-salt circuit for thermal connection of a number of tower units into a larger power plant with a single large steam turbine.

30) The relative amounts were adopted with the exception of the values for "interest during the construction phase." These were *computed*, starting from the assumptions used throughout this book.

arise in the field installation. Assuming overall annual construction rates of one or several gigawatts installed of output capacity, the components of a 700-MW power plant[31] could thus be fabricated within some months and installed immediately so that the assumption of a construction time of 2 years would appear conservative. For comparison, in the S&L report (2003, p. ES-9), the construction time for solar tower plants is taken to be 1 year. For the parabolic-trough plants in California, construction was accomplished in less than 1 year. The new parabolic-trough plants Andasol 1 (Spain) and Nevada Solar One were constructed in 2 years and in 15 months, respectively (although today's deployment rates are still quite small).

The term *owner's cost* includes in particular the measures necessary to prepare the construction site and setting up the necessary infrastructure (especially the construction of railway lines and/or roads). For setting up a large solar park, and in contrast to decentralized single power plants of 100–200 MW output power, the specific costs (i.e., cost per installed gigawatt) would be considerably lower. Roads, rail lines, and possibly harbors can be designed to serve the whole large project region and will be used over many years in the course of the progressive expansion of the plant.

The same is true of the costs for *engineering and construction management*. In the case of the continuous extension of an installation over a large region, these costs are – due to the effects of repetition and the central construction management – proportionately much lower than for the construction of a single solar tower plant.

Contingencies must be planned for, especially during the construction of the first installations. In the case of routine extensions at the same location, there are in contrast seldom major surprises.

4.3.2
Solar Multiple/"24-h Design Insolation"

In a solar power plant *without a heat-storage system*, the incident solar energy must be converted immediately into electrical energy. The power output of the plant, therefore, varies during the course of each day, and its full power (nominal power output) is attained only when the Sun is at its highest in the sky, that is, around noon. In a plant *with a heat-storage system*, the situation is different: if the plant is designed to provide the same output power in the morning and evening, or even 24 h per day, the mirror field (and the receiver) must be enlarged so that they can collect the additional heat energy during the daytime hours, which is needed for power generation in the evening, at night, and in the early morning. SM is the measure of this enlargement factor; it defines the factor by which the field and the receiver must be enlarged compared to a plant of the same nominal output power but without a heat-storage system.

The 200-MW Solar TWO installation designed by Kolb (Advanced Technology, water cooled) had a mirror area of 2 480 000 m^2 (Kolb, 1996b). The solar multiple

---

31) This power output corresponds to the usual type of turbine employed today in large fossil-fuel steam power plants in Europe (in the USA ca. 600 MW).

was given as 2.7 and is based on a maximum irradiation intensity of $0.9\,\text{kW/m}^2$ at noon on the 21st of March (vernal equinox). The value $SM = 1$ (i.e., without heat storage) thus corresponds to a mirror area of $920\,000\,\text{m}^2$ ($2.48\,\text{million m}^2/2.7$). (An installation of this size *without* heat storage would thus also deliver an output power of 200 MW at noon on the 21st of March.)

4.3.2.1 Recalculation for a "Base-Load" Power Plant

The daily average efficiency (annual average) of SOLAR TWO (Advanced Technology, wet cooling) would be 17.6%, according to Kolb (1996b).[32] From this, we find for 24-h full-power operation (at 200 MW) per day a requirement of 27.3 GWh ($0.2\,\text{GW} \times 24\,\text{h}/0.176$) of solar energy.

In order to attain a sufficiently high capacity factor for a *base-load* power plant, the mirror field (and the receiver) must be overdimensioned in relation to the day of the year with the strongest insolation. One then obtains on such "good days" more solar energy than can be converted into electrical energy in 24 h; a certain fraction of the energy must be "dumped."[33] Since in such overdimensioned power plants, not all of the solar energy which is collected by the mirror field over a year can be used for producing electrical energy, a higher SM no longer results in a *proportionally* higher annual power production. SM is thus only a measure of the enlargement of the mirror area in such overdimensioned plants, and no longer implies a corresponding increase in the annual power output.

For such 24-h base-load power plants, one must determine with respect to the specific intended annual power output just what daily insolation (in $\text{kWh/m}^2\text{d}$) is sufficient for 24-h full-power operation. If only a lower daily insolation is available, one requires a correspondingly larger mirror field.

How much energy such a plant can produce per year is then found from the daily values of the insolation at the given location. On days when more solar energy is available than can be converted into electrical energy, a certain portion must be forgone, as mentioned. As one can compute from the statistics for insolation in southern Spain (daily values for several years at the Almeria site), a power plant which is expected to attain a capacity factor of, for example, 72% at this site must be designed so that 24-h full-power operation can be attained at an insolation (DNI) of $6.7\,\text{kWh}/(\text{m}^2\text{d})$. This is thus the daily insolation which is relevant for the dimensioning of the mirror field (cf. the topic "Design Insolation").

The required mirror area is then found from the heat requirements of the power plant per day (i.e., from its efficiency). As already explained above, a 200-

---

32) Since we are speaking here of *one* day, we refer to the "daily" average efficiency, whose annual average is naturally the annual average efficiency.

33) Here, we are dealing with an intentional overdimensioning. A small amount of overdimensioning is also present when the reference day is the 21st of March and not the day with an optimum solar altitude (the 21st of June in the Northern Hemisphere). At noon on the 21st of June, owing to the higher altitude of the Sun and the accompanying higher field efficiency, somewhat more solar energy could be collected than can be accepted by the steam turbine of a power plant designed with reference to the 21st of March. Nevertheless, the solar multiple usually refers to the vernal equinox (21st of March), and then to a certain predefined radiation intensity at noon, namely $0.9\,\text{kW/m}^2$; for example, in Kolb (1996a,b).

MW 24-h power plant (under the assumption of the average efficiency of 17.6% as mentioned) would require 27.3 GWh of solar energy per day. In order to collect this energy at an insolation of 6.7 kWh/(m²d), a mirror area of all together 4 070 000 m² is necessary (27.3 × 10⁶ kWh/(6.7 kWh/m²)). This is, therefore, the required mirror area for a 200-MW plant, in order that it can attain a capacity factor of 72% in southern Spain. Division of this area by 920 000 m² ($SM = 1$) yields an SM of 4.4.

In particularly sunny regions such as Morocco or the southwestern USA, one could, as indicated by the corresponding insolation statistics, obtain a capacity factor of nearly 80% if the plant attains 24 h full-power operation at an insolation of 8 kWh/(m²d). The heliostat field would be less overdimensioned than in Spain (where nevertheless only a capacity factor of 70–75% is possible). The mirror area would be 3.4 million m², corresponding to an SM of 3.7.

> Note: There is as yet no well-defined terminology for the rating factor of base-load solar power plants. For 24-h solar plants with a surplus of solar energy on good days, that is, for base-load plants, SM is no longer sufficient to characterize the attainable annual power production (relative to a plant with an SM of 1). (Only for plants which can convert all the incident solar radiation into electrical energy at the given efficiency does the annual power production increase in direct proportion to SM.) The determining quantity for the mirror field (with respect to the intended capacity factor for a given insolation at the particular site) is then the required daily insolation per m² for the 24-h operation. Up to now, no quantity has been defined to denote the corresponding value. One could call it the "Design Insolation" (DI) or, more precisely, the "24-h Design Insolation" (24-h DI) (this refers to the insolation assumed for the dimensioning ("design") of the mirror field for 24-h operation), or, briefly, the "base-load insolation" (BI). The term "Nominal Insolation" would also appear evident; but this expression is similar to the "Normal Insolation" (i.e., the solar irradiation incident each day normal to the mirror surface), and the similarity could lead to confusion among nonexperts so that it is less suitable. The term "24-h Design Insolation" would seem to be the most obvious choice.

4.3.3
Land Prices in Spain

For prairie or semiarid regions in Spain, where the land is not used for agricultural (cultivation) purposes, or only for (extensive) grazing, a price of *1.25 $/m²* (2002) was assumed. At purchasing power parity for the year 2002 (according to the OECD: $1 = €0.96), this corresponds to 1.2 €/m², comparable to the price of farming land in Germany, which is very fertile in comparison to the Spanish prairie regions. Our price estimate is, therefore, relatively high. Sanchez *et al.* (1996) took a price of 0.8 $/m² (1996-$) for southern Spain. (For comparison, SunLab and S&L in 2002 assumed a price of 0.5 $/m² for land in the USA (S&L, 2003, p. E-4/5).)

From a mirror area of 20.2 km² per 1000 MW and an area-use fraction of 18% (cf. Section 4.3.5), we estimate a required land area of 112 km². At a price of 1.25 \$/m², we then find out the cost of acquiring the land to be a value of *140 million \$/1000 MW*.

4.3.4
Political Costs – North African Solar Energy as a "Relative" Alternative for Europe

For the cost estimates in the case of North Africa (i.e., for energy export to Europe), no land costs are included in Tables 2.1 and 4.6. Here, we are dealing with a "political price," which however cannot be explicitly classified in terms of land costs, but rather must be assigned to the project as a whole.

The political price, that is, the price which can be obtained as the result of negotiations, will in general depend upon the *alternatives* to the North African locations which are available, thus in particular on the cost of generating solar power in southern Spain. Other major alternatives for Europe are power generation from wind energy in the North Sea, and potentially of course power from coal-fired plants using imported coal. The political premium which can be obtained by the North African negotiating parties is limited by the fact that a power-plant site in North Africa must be on the whole less costly than one in Spain, and also less costly than wind energy from the North Sea or power from coal-fired plants (including the sequestration of CO_2). If a North African country demanded a land price which would lead to cost parity with the other potential sources, there would be no reason for a power plant operator to build plants there; a barely marginal advantage would also not represent a sufficient motivation. It must, therefore, be assumed that the economic advantages of the North African locations would in the end be divided more or less equally between the country where the sites are located and the European power-plant operators. It is furthermore clear that the enormous economic, industrial, and social advantages, which would accrue to the country where the sites were located, must also be considered in the overall cost assessment. These general advantages would be obtained not only as a result of the "political premium" (which, as mentioned, could not be all that high), but rather especially from an active participation of these countries in the construction and operation of the solar power plants.

In Spain, an additional political premium is not to be expected owing to Spain's membership in EU. Foreign enterprises have the right to buy real property just as do domestic companies. The same is true of the construction of power lines, for example, passing through France; the EU internal market for electrical energy in principle guarantees the right to free construction of transmission lines.

4.3.4.1 European Alternatives in Negotiations with North African Countries for Potential Power Plant Sites

Offshore wind power plants are, with the present state of knowledge, likewise an economically favorable and large-scale source of renewable energy. (In Chapter 11, this point is discussed in more detail.) In Europe, they are of particular importance, as Europe (in contrast to the USA) does not possess *practically* unlimited land areas for solar power generation on its own territory. In principle, all of today's electrical power consumption could indeed be supplied from sites in Spain; there, sufficient unused land, or land used only for grazing, is available in the southern solar regions, and we can safely assume that the land owners would show an interest in selling their up to now nearly "worthless" real property. It is, however, evident that it would be extremely advantageous to use this potential to only a small extent, limiting the use to the best sites. Insofar as solar power – taking continuity of power consumption into account – is the most economically favorable solution for renewable energy, while wind energy lies in a roughly comparable cost range, one could imagine a division between renewable energy sources in Europe as follows: two-thirds solar energy (one-third each in Spain and Morocco) and one-third from offshore wind power.

If wind energy proved to have an economic advantage, the distribution could be reversed: two-thirds wind energy and one-third solar energy (from Spain and Morocco), whereby the solar power plants in this case would take over the task of making the overall supply more reliable.

Concerning substitution of oil and gas by hydrogen, water electrolysis using wind-generated power would be an alternative to solar power (another alternative would of course be gasification of cheap brown coal or imported coal). For the solar production of hydrogen on a large scale, there are not sufficient land areas available in Europe. Insofar as the potential of wind energy from the North Sea is not limited by the depth of the water (one could envisage floating wind power plants), it would represent a nearly unlimited alternative for this purpose. Particularly for hydrogen generation, which can be interrupted without problems, the lower capacity utilization of wind power plants would not represent a serious disadvantage. In the case of a major cost advantage for wind energy as compared to solar energy, this disadvantage would in fact be compensated. The main alternative to solar-produced hydrogen is, by the way, coal gasification.

Regarding the production of liquid fuels, Europe is not limited to sites in North Africa. As already mentioned in the "Preliminary Remarks and Summary," methanol can be synthesized from solar hydrogen and coal, and this production could be carried out in sunny coal-producing countries such as the USA, Australia, or South Africa. Methanol would then be transported to the consuming regions by tankers, just like crude oil at present.

Regarding the fundamental problem (for solar power), we however take note of the following: if Europe could use *only* the solar energy from North Africa, its potential would be utilized only to a limited extent, since the countries where power plant sites were located could dictate their conditions. Only through the internal European alternatives does North African solar power become politically and practically applicable from the European point of view. These European alternatives define the *upper limit* to the possible demands from the African countries. Imports of solar power from North Africa are in this sense only a "relative" alternative, whose practical value is determined by the spectrum of possible power-generating sources for Europe. Thus the object of negotiation is the difference between the cost of North African power and the costs of the inner-European alternatives, and these costs will most likely be divided roughly equally. Since solar power from Spain represents an immediate alternative to power from North Africa, and there also appears to be a great potential from wind energy, Europe has strong alternatives in terms of renewable energy sources.

Thus, two things are clear: firstly, Europe needs the *European* alternatives! Secondly, Europe has two advantages in terms of importing solar power from North Africa: it provides a definite gain (half the cost difference to the inner-European alternatives), and these inner-European energy sources need be drawn upon to a lesser extent so that efforts can be concentrated on the best wind- and solar-energy sites.

The overall energy-economical potential of solar thermal technology can be realized only on a worldwide scale – this is also true from the European viewpoint. If solar energy becomes an economically attractive alternative to coal-fired power generation, large amounts of coal will become available on the world markets for the purposes of substitution (in particular for gasification of coal). The use of solar energy in the USA and – this will probably also be possible – in Asia (Inner Mongolia, India, and possibly Tibet) would have a long-term price-lowering effect on the world market for the major fossil energy source for the future (coal). The development of solar power plants up to an operation-ready level would then provide *enormous* advantages for the worldwide energy economy, even though solar power plants might not be operable in all countries owing to a lack of suitable sites (or at least only on a limited basis). This also holds for the European countries, which in the past have sometimes tried to steal out of accepting responsibility for a rapid development of this energy technology by pointing out these political problems.

In terms of the European research and development strategy, we must also draw the conclusion that not just solar energy alone, but also its alternatives within the European borders need to be developed – even if North African solar power turns out to be the most economically favorable solution in the end. The research program must therefore be *more comprehensive*, if only for these political reasons. Its goal must be to make all the possible alternative energy sources available as soon as possible; only this will be able to satisfy future needs.

4.3.5
Specific Land-Area Requirements

The ratio of mirror area to land area, which is a measure of the deployment density of the heliostats, is termed the *area utilization factor*. It was not considered in the SOLAR TWO study (Kolb, 1996a) for solar tower power plants. In our study in 1998, we therefore assumed the relatively low value of 18%, in order to be "on the safe side." We adopt this value in the present book for the *estimated costs* of the land. For comparison, SunLab assumes an area utilization factor of 19.0%, while Sargent & Lundy assumes 19.5% (cf. S&L, 2003), and the "Utility Studies" take a value of 22.6% (Hillesland, 1988; Hillesland and De Laquil, 1988). In the case of the latter value, the mirrors thus require one-fifth less deployment area than for the value of 18% assumed here in estimating the land costs. The land costs are, therefore, most likely overestimated in the present study.

The actual area utilization factor is important for estimating the land-area requirements of solar power plants (this is especially important, e.g., in Spain). The 200 MW installation of the SOLAR TWO type (Kolb, 1996a), which is used here as a reference, requires a mirror area of 2.477 million m^2 at an SM of 2.7. Recalculated to an output power of 1000 MW, this corresponds to the following:

In Spain – at an SM of 4.4 – a *mirror area* of 20.2 million m^2 and for the *land area* for the area utilization factor of

- SunLab   (19.0%):   106.3 million m^2
- S&L   (19.5%):   103.6 million m^2
- Utility Study (22.6%):   89.4 million m^2

In Morocco or the USA – at an SM of 3.7 – a *mirror area* of 17.0 million m^2 and for the *land area* for the area utilization factor of

- SunLab   (19.0%):   89.5 million m^2
- S&L   (19.5%):   87.2 million m^2
- Utility Study (22.6%):   75.2 million m^2

Based on 1000 MW output power, for power delivery to, for example, Germany, one has to take the transmission losses into account (from Spain, they are 8.1%, and from Morocco 11.5%).

The deployment density of the heliostats is not a fixed quantity for a given solar power plant, but rather the result of an optimization process. If one has sufficient space and low land prices, a cost minimum for the investment costs of the installation can be aimed at. The geometry of the mirror field depends on the one hand on the height of the tower, and on the other on a suitable spacing of the heliostats. Higher towers are not only more expensive, they are also accompanied by higher pumping losses, since the molten salt must be pumped to a greater height. The heliostats should cast minimal shadows on each other ("shading"), and not interfere with the radiation from other heliostats to the receiver ("blocking"). If they are placed too close together, these optical losses are necessarily greater; if they are further apart, then the distance to the receiver is greater. The latter has as consequences that (1) the light path is longer (stronger absorption by the atmosphere along the path) and (2) the image of the Sun which is projected onto the receiver is larger and, therefore, the losses from radiation which bypasses the receiver ("spilling") increase. As can be seen from the above examples, one can arrive at different area utilization factors as a result of cost optimization depending on the assumptions made about the parameters (costs and optical losses of the mirrors vs. costs and losses at the tower or the receiver).

If the available land area is limited, as is the case in Spain, and as many solar power plants as possible are to be installed on the available terrain, one will *in addition* include the size of the plot of land in the optimization procedure. Then, the height of the towers will be somewhat increased relative to the value obtained from the straightforward cost optimization, and in particular the heliostats will be deployed closer together. One can thus save on land area by accepting somewhat higher costs for the power plants. Just how much the cost will increase for a given savings on land area can be readily computed by using available optimization programs for the mirror field. Unfortunately, no typical values have yet been published. It can, however, be assumed that firstly, the deployment density given in the "Utility Studies" cited above can be achieved, and secondly, with a view to limited land availability, the land requirements can furthermore be reduced by more than 10% without increasing costs to an unacceptable degree. If land is very scarce, one can also choose SM to be somewhat smaller (e.g., only 4.2 instead of 4.4 in the case of Spain). The required land area in Spain could thus be even lower than the lowest value given in the above tabular representation (89.4 million m^2), with only a small accompanying increase in overall costs.

Finally, in this connection we also have to consider the fact that the mirror field of a single solar tower with a molten-salt receiver has a nearly circular (slightly elliptical) shape. When several tower installations are connected together to form a larger power plant, the required land area also depends on how these circular areas are combined. If the mirror field is hexagonal instead of circular, the mirror fields of the individual towers, or several rows of fields, can be fitted together with no waste area (honeycomb pattern). A hexagonal shape instead of a circular one for the mirror field, however, implies a small deviation from the (in terms of the optical efficiency) optimal shape. This must also be taken into account in the simulation calculations, which have yet to be performed.

Furthermore, the optimization process for a solar park with a number of towers is somewhat different from that for single-tower installations, since in the former case a heliostat in general is not associated with a *single* tower, but–depending on the altitude of the Sun–*with several* towers (the same principle that was already mentioned in Chapter 3 with respect to the linear Fresnel system (CLFR)).

It can, therefore, be safely assumed that for solar tower plants in Spain, a land requirement of 90 million m^2 per 1000 MW (at the power plant) will not be exceeded.

For *parabolic-trough power plants*, the deployment density is higher, since the rows of troughs lie only beside each other (and not also in front and behind each other, as do heliostats), and thus can be more closely spaced; their density is ca. 30% (S&L, 2003, p. 4-3). The specific area requirement depends on the overall efficiency of the power plant (apart from the dimensioning of the mirror field for a high capacity factor at a particular site). The efficiency increases with increasing output power and will also increase in the future owing to technical improvements. For the example considered in Chapter 8, the "Long-Term Trough 400 MW" (S&L, 2003, pp. 4-3 and D-3 to D-5), the mirror area (aperture), adjusted for sites in Spain, is 20.8 km^2, and in Morocco/USA it is 17.4 km^2. From these values, we find an area requirement in Spain of 70 km^2 and in Morocco or the USA of 59 km^2, in both cases per GW at the power plant site.

4.3.6
Horizontal Salt Circuits

In large solar power plants, several tower installations can be connected via molten-salt piping. Such large-scale power plants can again be interconnected with neighboring plants. This interconnection of towers and plants offers several advantages. It permits the use of larger turbines, with the advantage of higher efficiencies and lower specific costs. Thus, for example, six large tower installations of the SOLAR TWO type (each equivalent to 200 MW$_{el}$ at an SM of 2.7, corresponding to 123 MW$_{el}$ at an SM of 4.4) can be *thermally* connected – that is, via their molten-salt circuits, to give a base-load plant of ca. 700 MW$_{el}$ (SM 4.4) output power. In this case, the proven and tested steam turbines as currently used in coal-fired power plants could be used. The further interconnection of large power plants can reduce the amount of part-load operation (avoiding reduction of the overall efficiency) and also decrease down time.

For the interconnection of tower installations, the heat-storage reservoirs can remain localized at each tower, with the advantage that the molten salt can be pumped continuously to the central steam plant (power block). One can then employ smaller molten-salt pipes, since if the heat-storage reservoirs were located centrally at the power block, the entire charge of molten salt would have to be pumped there during the daylight hours. Furthermore, daily temperature variations in the piping system are avoided.

On the days with little sunlight, a stand-alone solar plant would have to operate at reduced capacity. If connecting lines for the molten salt are present between neighboring power plant parks, one or more of the turbines in the park can be shut down, allowing the others to operate at full capacity. Likewise, in the winter half of the year, when the full insolation is attained on only a few days, and the plants are as a rule operating only at part load, shutdowns for maintenance of individual turbines can be carried out without losing the solar heat of the corresponding mirror field.

4.3.6.1 **Costs**

As early as 1978, a 300-MW solar power plant with a nitrate-salt circuit was proposed; it consisted of nine interconnected smaller towers with a 12-h heat-storage system (Martin Matietta 1978 and 1979). Due to the small tower installations, a relatively high specific length for the interconnection piping (pipe length per km^2 mirror field) was required. SM was smaller than what we have assumed for a base-load plant. Recalculation to an SM of 4.4 (Spain) shows that the planned Martin-Marietta installation with nine towers would correspond to an overall output power of ca. 240 MW. A large receiver of the SOLAR TWO type (equivalent to 200 MW$_{el}$ at an SM of 2.7 or 125 MW$_{el}$ at an SM of 4.4) thus has the same thermal power as nearly five Martin-Marietta receivers. The interconnection of a few large receivers yields a lower specific piping length than interconnection of many smaller receivers. In the following cost estimate, it will be *assumed* that the piping length, and thereby the piping costs, can be reduced by 25% as compared to the Martin-Marietta design.

The costs of thermal piping for the Martin-Marietta design are given as 7.8% of the investment costs of the solar portion of the plant, that is, $24 million (1978-$). Recalculated to 1000 MW and an SM of 4.4 (Spain) or 3.7 (Morocco/USA), and taking the 25% lower piping costs into consideration, this gives in 2002-$ a cost of $180 million (Spain) and $150 million (Morocco/USA).[34] In such a comparison with the small towers considered in 1978 (likewise only a design study), we are naturally making only a very rough estimate of the costs. The exact costs could be relatively quickly determined within the framework of the investigation suggested in this book, through detailed planning of the interconnection systems.

4.3.7
Dry Cooling

The few currently existing solar power plants are wet cooled, as it was possible to build them near water-supply facilities. For the large-area plants planned in the future, this will often not be the case.

At the solar-energy sites in southern Spain, we can plan for wet cooling owing to the proximity of the seacoast. Even at a very high installed output power, with correspondingly large water requirements that could not be met from the sources on land, it should be possible to construct piping for cooling water to the seashore. It can safely be assumed that fresh water (from rivers) can be supplied by ship or barge and fed into the pipes at a reasonable cost.

At the Moroccan desert sites beyond the Atlas mountains, such a scheme is not practicable, and in the Southwest of the USA, also, the distance to an inexpensive supply of cooling water is too great so that at these sites, dry cooling is necessary. Similar conclusions hold for solar power plants in Tibet[35] – under consideration for a solar power supply to China or India – as well as for solar power plants in Inner Mongolia and in India itself.

Dry cooling has three disadvantages:

1) The most significant is the higher condensation temperature, resulting in a decreased efficiency of the steam turbines. The solar power plant thus generates less power overall.

2) In the case that forced draft cooling is used instead of natural draft cooling towers, the power consumed by the fans must also be subtracted from the net

---

34) $24 million (1978-$) corresponds to $59.3 million in the year 2002, referring to 250 MW and an SM of 4.4. Recalculated for 200 MW and for an SM of 4.4 (Spain), this is $47.4 million, or at an SM of 3.7 (Morocco or USA), $39.8 million. With 25% lower piping costs, we find for Spain (SM 4.4) $35.5 million and for Morocco/USA (SM 3.7) $29.8 million. At 1000 MW, this corresponds to $180 million (Spain) or $150 million (Morocco/USA).

35) Tibet has not only a high direct insolation but also very low air temperatures so that dry cooling there represents little or no economic disadvantage.

power output of the plant. This effect can lie between 0.5% und 1.5% (in certain cases even up to 2%) of the installed output power of the plant.

3) The investment costs for the dry cooling system are higher.

While the first two effects result in *reduced power output* of the plant, the third increases its absolute construction cost. This third effect can, however, be expressed in terms of an equivalent reduction in power output (at constant investment costs). This makes it possible to give a *unified summary* of the three effects in the form "x% less output power."

If a plant produces less power, this means not only an increase in its specific investment cost, but also a corresponding increase in the *specific operating and maintenance (O&M) costs (per kWh_{el})*.[36] This increase in the capital costs *and* the O&M costs effects a proportional increase in the cost of power generation.[37] In the literature, the increased expense due to dry cooling is, therefore, usually expressed in terms of an equivalent "increased power cost" (of the levelized energy cost (LEC)).[38]

As will be shown in the following, the disadvantages of dry cooling correspond to a decrease in power output of *about* 8% for base-load power plants in Morocco or the southwestern USA. (This implies that the electric power produced would be about 8.7% more expensive.) This rough value will be assumed in this book for tower and parabolic-trough power plants. A more precise number is not available at present.

Dry cooling is already in widespread, large-scale use, in particular in three large coal-fired power-plant complexes in South Africa (together ca. 10 GW). It thus represents a mature technology for large power plants so that its costs are in fact

---

36) In estimating the O&M costs, only the *actual* decrease in output power (lowered efficiency and power consumed by fans) is taken into account, not the *computed* decrease including the higher construction costs for the cooling systems. (The latter affects only the capital investment costs.)
37) For fossil-fuel plants with dry cooling, the reduced efficiency must be taken into account also in the fuel consumption rate. This is of course not necessary in the case of solar plants.
38) Here, however, the term "x% less power" is preferred, since then the increase in the specific investment costs (and the resulting increase in capital-investment costs per kWh) are considered separately from the increased operating and maintenance costs, with the advantage that modifications in the assumptions underlying the estimate of capital costs (regarding the interest rate and plant lifetime) can be more readily taken into account. In terms of the computation of "increased power costs," the increase in the specific investment costs and the operating and maintenance costs are not considered separately, but rather only a global quantity: price increase for electric power. This method thus represents a mixed calculation that must be completely repeated if modified assumptions for the capital-investment cost computation need to be considered. In this book, therefore, we always give the capital costs and the operating and maintenance costs separately, and the effects of dry cooling are then taken into account in terms of "x% lower power output." A recomputation involving a different interest rate or plant lifetime (which have no influence on operating and maintenance costs) then gives directly the correct result for dry cooling. In the end, of course, the two representations give the same overall results.

4.3 Some Special Points Concerning Cost Estimates

well known. In addition, there have been a number of studies into dry cooling in connection with solar power plants. Nevertheless, we can give only a *rough estimate* of the costs for the case considered here of large solar parks in Morocco and the USA, specifically for base-load power plants. The precise costs for particular sites must be determined in detail by competent industrial partners.

In the main, data published for solar power plants thus far relate to parabolic-trough plants and not to future large solar tower plants (e.g., in the 300 or 700 MW class with thermal interconnection of several towers); that is, mostly to smaller power blocks and often to plants without (or with relatively small) heat-storage reservoirs. For base-load power plants (with *large* heat-storage systems), the expensive cooling systems must be in operation for nearly twice as long each day (24 h), and in particular nearly half of their operating time falls in the "cooler" night hours, when the reduction of efficiency due to dry cooling is less important.[39] Also, owing to the higher steam pressures and temperatures used in the solar tower plants, the results for parabolic-trough plants can be applied to only a limited extent. Furthermore, the results depend strongly on the air temperature (and on its variations) at the particular site. Finally, in making comparisons with conventional power plants, it must be considered that in their case, fuel costs play an important role in the plant design, in particular regarding the temperature differences within the condenser and in the cooling towers.

In general, various studies exhibit large differences in their cost estimates. It can be assumed that they made use of differing cost data so that the results are to be regarded with caution. Therefore, as already mentioned, it is indispensable that future precise investigations, which must be applied to a whole series of possible sites, be carried out by organizations which themselves construct or plan comparable installations. Such organizations have access to reliable cost data from projects already completed, and these can be used with reference to particular sites for solar power plants.

4.3.7.1 Literature References to Dry Cooling for Solar Power Plants
DOE, 1997, p. 5–21:

> "Levelized energy cost raise by at least 10%" (probably referring to parabolic-trough power plants of <100 MW output power and without heat-storage reservoirs).

> Here, we must take into account that the amount of heat per kWh$_{el}$, which has to be removed from the condenser, is lower by ca. 20% for tower power plants than for parabolic-trough plants: "tower power plants: water usage ... should be about 20% less than SEGS VI." Therefore, the cost increases

---

39) The choice of cooling concept also depends upon the plant capacity factor: for base-load power plants, an indirect cooling system using natural updraft cooling towers would probably be preferable, while for intermediate capacity factors, direct air cooling of the condenser (direct cooling) using fans offers clear-cut advantages. Thus, of the three South African coal-fired plants mentioned above, two are directly cooled and one (with longer operating periods) is indirectly cooled.

due to dry cooling for tower plants should likewise be ca. 20% less than those of trough plants. The power costs would thus increase by only ca. 8%, rather than by 10% (as quoted above).

San Diego Renewable Energy Group, 2005, p. 167:

"Levelized energy cost: +14% (relative to Baseline Wet Cooling)" (parabolic-trough power plant, 100 MW, without heat storage).

Richter, 2007 (cf. also Dersch and Richter, 2007):

"..increasing the levelized energy cost (LEC) by 5 to 10%" (55 MW parabolic-trough power plant with an oil cooling circuit, probably without heat storage, clearly intended for sites in Spain and California).

Kelly, 2006 (cf. also Kelly, 2005):

"Dry heat rejection imposes a 7 to 9% penalty on the levelized energy cost (LEC)" (88 MW, parabolic-trough plant, without heat storage, Barstow climate data).

BMBF, 1996:

Investment costs: +6.1% (Site: Quarzazate (Morocco); parabolic-trough power plant, 80 MW without heat storage).

NREL, 2006:

At an air temperature (daytime) of 29 °C (87 °F): LEC: +10% (rel. to wet cooling).

At an air temperature (daytime) of 23 °C (75 °F): LEC: +6% (rel. to wet cooling).

At an air temperature (daytime) of 18 °C (65 °F): LEC: +5% (rel. to wet cooling).

These amounts (at 29 °C) include:

Capital cost: +8%
Operating cost: −2%

Power consumption for the fans of a 100-MW plant: 1.9 MW$_{el}$ (this is the same as for the fans of a forced-draft wet cooling tower).

Power output (annual performance): −3%

(Parabolic-trough plant, 100 MW, apparently without heat storage, site: Kramer Junction, USA.)

4.3.7.2 Literature References to Dry Cooling for Conventional Power Plants

Maulbetsch, 2006 – relative to a 500 MW steam plant (probably a coal-fired plant):

> Capital cost: +12.5% (of the investment costs for the conventional plant). (Note: relative to the considerably higher investment costs for a solar power plant, the *relative* increase is correspondingly smaller.)
>
> Cooling system power: 3 MW (for a 500-MW power plant: 0.6%)
>
> Plant heat rates: +8%; this corresponds to a decrease in efficiency from the assumed value of 40% to 37%, and thus to 7.5% less power. Considering the low cost of fuel for coal-fired power plants in the USA, the cooling system could be optimized for lowering investment costs in terms of the temperature gradients, thus accepting a greater decrease in efficiency. The estimated decrease in power output can, therefore, be only approximately applied to solar plants. Regarding the question of optimization with respect to investment costs and efficiency (depending on the design temperature), see, for example, NREL (2006).

California Energy Commission, 2002:

> This very extensive and often-cited study referred to combined-cycle power plants (500 MW). The cooling applies only to the last-stage steam process with a power output of ca. 170 MW and with low steam temperatures and pressures. The results are, therefore, not readily transferable to solar plants. (For the combined-cycle process mentioned, additional investment costs of ca. $26 million per 170 MW (i.e., ca. $150 million per GW) were mentioned for a dry cooling system at a desert site as compared to wet cooling.)

GEA Prospectuses (not dated):

> GEA constructed the dry cooling systems for the two power plants in South Africa, Matimba (6 × 665 MW$_{el}$) and Majuba (6 × 660 MW$_{el}$), in both cases using a *direct dry-cooling system*. The description unfortunately does not include a comparison with wet cooling. The power consumption of the fans corresponds to Matimba 1.7%, Majuba 0.9% of the plant's output power.
>
> The cooling systems of the power plant complex Kendal were constructed by the German systems manufacturer Balke/Dürr, using *indirect dry cooling* (Trage and Hintzen, 1989). The description likewise does not include a comparison with wet cooling.

One can note that for relatively small parabolic-trough power plants (max. 100 MW and without heat storage), the estimates vary between a 5% and a 14% increase in power cost. With the exception of the San Diego Energy Group, they all lie at or below 10%, referring to cooling only during the hot daytime hours. For parabolic-trough plants with heat storage, we therefore use the preliminary value of 8%. We also adopt this value for solar tower plants, although in their case, due to their higher overall efficiencies, the disadvantage of dry cooling should be about 20% less, as mentioned above (DOE, 1997).

The required investigations include taking account of the possibilities for cost savings in the dry cooling systems.[40] For the sites where wet cooling cannot be excluded in principle, for example, for some sites in the USA, these considerations must include realistic planning for the large water pipes and the development of water-saving wet cooling processes, among others hybrid systems of various types.[41] The required investigations must of course include planning of the water supplies for the large-area sites in southern Spain, including sea transport of river water to the nearby coast.[42] Since these questions regarding cooling systems involve only planning, it should be possible to answer them within a reasonably short time.

4.3.8
Technical Reliability

Service interruptions of solar power plants for technical reasons would increase the required operating time of the backup plants and would thus result in increased consumption of fossil fuels. As compared to the introductory scenario considered by Kolb (1996a), we assume for the large-scale system that all the components would represent mature technologies, especially in terms of their reliability. The

---

40) An example is a so-called night cooling system. Here, the cooling water exiting from the condensers during the day is stored in special large containers and then cooled during the night hours in the dry cooling towers; cf. (Hillesland, 1988); for a comparison with wet cooling, cf. (Grasse, 1988). The necessary enormous storage capacity for the water may be an obstacle to the success of this concept. Whether the potential of such developments – for example, using different water storage systems from those envisioned thus far – has been fully recognized in the past is hard to judge. The storage systems considered in the past (basins with dams) are suitable only at very level sites.

41) For hybrid systems, see for example (Kutscher et al., 2006) and (Maulbetsch, 2006); for the novel concept of the so-called evaporative condenser or wet surface air cooler (WSAC), see California Energy Commission (2002, Ch. 8).

42) Supplying river water via sea transport for wet cooling at sites near the coast could also be considered for sites near the coast in the extreme southwest of USA. There, water would have to be pumped from the Mexican coast to an altitude of 1000–2000 m. In order to overcome a height difference of 2000 m, one would require 6.6 kWh of electrical energy for 1 m^3 of water at a pumping efficiency of 83%. With a steam power plant efficiency of 40%, the generation of 420 kWh of electrical energy requires 1 m^3 of cooling water (assuming complete evaporation of the water). Thus, to pump the water to an altitude of 2000 m, only 1.6% of the power generated would need to be consumed by the pumps. This can be compared with all together ca. 8% "less power" if dry cooling were used. The altitude of the site thus does not represent a fundamental obstacle to such a high-volume water supply. The distance from the coast to the Mexico–USA border is ca. 100 km. Attractive solar sites can be found just beyond the border.

down time due to the solar components is assumed to be negligible. For example, heliostats with a down time of 1% due to technical defects can be considered as "not sufficiently developed." Similar conclusions hold for the thermal systems, heat storage reservoirs, heat transfer piping, and steam generators.

In contrast to the steam boilers of a fossil-fuel power plant, in the case of a steam generator in a solar plant operated with molten salt, we are dealing with a convective heat exchanger. The latter is thus more comparable to the heat exchangers between the primary and the secondary cooling circuits in nuclear power plants, whose reliability is very high. The steam generator can also be compared to those used in parabolic-trough power plants, which are operated convectively with hot oil – and whose failure rate is likewise very low. A steam generator heated by molten salt could furthermore be constructed as a modular system, which would provide a further reduction in its risk of complete failure.

The most "critical" component in a solar plant is the receiver. However, as long as the exchange of damaged receiver panels can be carried out – as projected – overnight, or at least within 24 h, receiver failures lead to only minimal down times. Their overall reliability would thus be very high, in spite of the occasionally necessary repairs.

In the case of the turbine sets, the situation is different from that of the modular solar components. Here, the assumed high reliability would result from interconnecting many solar power plants over a large area, as mentioned above. This would permit maintenance shutdowns of individual blocks during the winter months, when the supply of solar energy is reduced, with transport of the solar heat to other blocks which would not be operating at full capacity. Planned shutdowns would thus not lead to a serious reduction in power generation. Unplanned shutdowns can be considered to be negligible; with the technologically mature turbine sets of the present generation, they lie "well below 1%," and have for more than 20 years (Hlubek, 1983).

4.3.9
Power Transmission via Overhead Power Lines

For power transmission, we assume that a high-voltage direct current (HVDC) system will be used, in the form of overhead power lines operating at a voltage of 800 kV.[43]

---

43) As already mentioned, the present authors have supported the concept of solar power import from Spain or North Africa to Europe since 1986. In recent years, this idea has been somewhat expanded upon by the research community to give a comprehensive power supply plan for the overall region Europe/North Africa/Near East ("Desertec"). In this plan, the advantages of installing a complete ring transmission line around the Mediterranean Sea, complemented by several undersea interconnecting lines, were considered. We refer the reader to the publications of the German Aerospace Center (DLR-MED 2005, DLR-TRANS 2006) and also to the report of the "Club of Rome" (TREC Initiative since 2003). In the USA, there have also been suggestions in recent years to install a power grid with HVDC transmission lines, especially for electric power from renewable energy sources, and in particular for solar thermal power.

The distance from southern Spain to Central Europe (Germany) is approximately 2000 km (1250 mi.; compare Almeria–Frankfurt: 1800 km). In the case of Morocco, the overland distance consists of 800 km from the Southwest of Morocco to Gibraltar and 2200 km from Gibraltar to Germany (Gibraltar–Frankfurt: 2000 km). The Straits of Gibraltar, which would have to be crossed using undersea cables, has a breadth of 14 km with a sea depth of less than 500 m. In the case of Spain–Germany (2000 km), the assumed transmission losses of 8.1% are comprised of 6.6% in the overland lines (3.3% per 1000 km) and 1.5% at the converter stations (2 × 0.75%). For transmission from Morocco–Germany (3000 km), the power losses in the overland lines would be 9.9%, in the inverters 1.5%, and in the short undersea cables 0.1% – thus all together 11.5%.

In the years 2007 and 2008, the first two 800 kV HVDC systems worldwide were ordered by China from European manufacturers; they are expected to go into operation in 2010 and 2011. These are a 1400-km long 5 GW transmission line (Luxa (Siemens) 2007), and a 2070-km long 6.4 GW line from the hydroelectric plant Xianjiaba to Shanghai, currently the world's longest electric power transmission line (ABB, 2008) (Figure 4.2).

For these new 800 kV transmission lines, the investment costs have thus far not been made public. (Furthermore, part of their production costs will go to the Chinese partners, which operate under different costing conditions from those usually applying in Europe or the USA.) The investment costs (and the above-

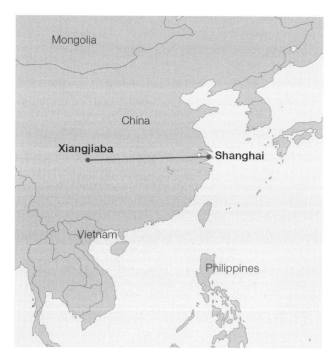

Figure 4.2 800 kV HVDC transmission line from the hydroelectric power plant Xiangjiaba to Shanghai (currently under construction) (ABB).

mentioned losses) were thus estimated by direct extrapolation from those of presently existing 500 or 600 kV lines: for the overhead power lines, $100 million per GW and 1000 km, and for the converter stations, all together $225 million per GW result (see Appendix A). Compare also, for example, DLR-TRANS (2006); there, slightly lower investment costs and losses are assumed.[44]

4.4
Calculating the Power Costs

For the calculation of power costs (in this book always given in units of ¢/kWh, in the literature often also as $/MWh), which are the ultimately decisive quantity for all cost comparisons, essentially three aspects must be taken into account:

- capital costs[45]
- costs for operation and maintenance (O&M)[46]
- fuel costs.[47]

Estimation of the capital costs is described in various ways in the literature:

- nominal versus real costs
- amortization period equal to or shorter than the plant's operating lifetime
- differing interest rates for outside (debt) and equity capital
- different ways of estimating taxes, etc.

Depending on the general goals – a business-economy (microeconomic) or national-economy (macroeconomic) point of view – different estimation proce-

---

[44] The assumptions in DLR-TRANS (2006, pp. 24 and 25) related to a 5-GW direct-current transmission line (800 kV) were:

Specific *Investments:*

Overhead line	70 million €/(1000 km · GW)	= $68 million (2002-$)/(1000 km · GW);
Inverter stations (two)	140 million €/GW	= $136 million (2002-$)/GW;
Undersea cable	500 million €/(1000 km · GW)	= 480 million $ (2002)/(1000 km · GW).

Specific *losses:*

Overhead line	2.5%/1000 km;
Inverter stations (p. 25)	0.9%/station (cf. p. 24: there, 0.6%/station);
Undersea cable:	2.5%/1000 km.

(Inflation Germany 2002–2006: 1.070; purchasing power parity 2002: 1 = €0.96)

[45] The expression "capital costs" is used here as a short form for "the fraction of power cost due to capital costs" ("capital-cost share"), although it is usually used in the English-language literature in the sense of "investment costs." In a given situation, however, there is little danger of confusing the two terms.

[46] One distinguishes between fixed and variable O&M costs: the fixed costs are related to the amount of electrical energy generated (annual fixed costs/annual amount of energy generated → ¢/kWh); the variable costs are given directly in ¢/kWh.

[47] In the case of fossil fuels, the fuel costs (¢/kWh$_{el}$) are obtained from the fuel price (¢/kWh at the lower heating value (LHV) of the fuel) and the efficiency of the plant referred to the LHV.

dures are appropriate. Since the reader will be confronted with different calculation methods in the literature, we treat the topic of the estimation of capital investment costs in some detail here.

4.4.1
Capital Costs, Nominal or Real Interest, Operating Lifetimes

Fundamentally, the following holds: the invested capital is subject to interest and must be repaid by the end of the operating lifetime[48] of the plant. A certain sum must thus be available annually from revenues on the sale of electric power in order to cover the interest payments and repayment of the principal (annual sum of the capital costs). This constant annual sum is the so-called *annuity*. Relative to the capital invested (often expressed as a percentage), one refers to the *annuity factor* (capital recovery (crf), mortgage constant).

The annuity factor (a) is computed by means of the following formula:

$$a = q^n \times (q-1)/(q^n - 1)$$

where

n = amortization time in years
q = 1 + interest rate/100 (e.g., interest rate 4% → q = 1.04)

If, for example, one assumes an interest rate of 4% and an amortization time of 45 years, annually 4.8% of the investment costs must be repaid as debt service. Table 4.9 shows, for several combinations of operating lifetimes and interest rates, the resulting values of the annuity factor.

If we relate the annuity to the annually produced (net) electrical energy, we find the *capital-cost share of the energy costs* (¢/kWh).

The amount of energy produced *annually* by a power plant can be expressed in terms of the number of so-called hours of full-load operation *per year*:

---

48) Here, we use the predicted technical operating lifetime, not the financial amortization time. In industrial practice, the amortization time, which is shorter than the operating lifetime, is often used. One then distinguishes between the period of amortization and the period afterward (i.e., after repayment of the capital costs), the "golden years." With this procedure, the actual power costs during the *total* operating life of the plant must be computed in two steps: initially, during the period of amortization, then in the "golden years." Both must be corrected for interest rates. This method corresponds more closely to operational reality and is standard in many branches of business. Justifications for it are (along with tax advantages) on the one hand, the fact that the real lifetime of a *particular* plant cannot be predicted precisely (in contrast to a power-plant park, which consists of many individual plants) so that one uses the shortest lifetime to be on the safe side. On the other hand, the time horizon for planning in most companies is too short to take real operating lifetimes of power plants into account, for example, here 45 years. These reasons are, however, not important in the present case so that here we calculate using the total lifetime (thus dispensing with the mixed calculation); this simplifies the computation. Taking the expected lifetime into account, we find in this way directly the real costs of power generation (cf. Section 4.4.1.1).

4.4 Calculating the Power Costs

Table 4.9 Annuity factors: constant annual rates (in % of the invested capital) for principal repayment and interest payments on the capital.

Amortization time (years)	Annuity factor at a real interest rate of 4%	Annuity factor at a real interest rate of 2%
20	7.4%	6.1%
30	5.8%	4.5%
40	5.1%	3.7%
45	4.8%	3.4%
50	4.7%	3.2%
60	4.4%	2.9%

Amount of energy = Nominal power output × hours of full-load operation

Example: A 0.5-GW power plant with investment costs of $3 billion, 8000 h of full-load operation, annuity factor 4%.

The annuity is $3 billion × 0.04 = $120 million.

8000 full-load h/a (capacity factor 8000/8760 = 91%) correspond to an annual energy production of 4000 GWh so that a capital-cost share (annuity/annual energy production) of

120 million $/4000 GWh = 3 ¢/kWh

is finally obtained.

In deriving the general formula for the capital cost (capital-cost share), we use the following notation:

I = Investment costs
P = Nominal output power
i = Specific investment costs (I/P)
a = Annuity factor
t_o = Hours of full-load operation per year
C = Capital costs

We then have

$$C = \text{Annuity/energy produced} = I \times a / (P \times t_o) = (I/P) \times a/t_o$$

and thus $C = i \times a/t_o$.

This computational route is in principle the same in all calculation methods for obtaining the capital recovery share in the power cost. There are, however, some differences, for example, in taking inflation into account or in the assumed interest rate.

To take inflation into account, there are two basic computational methods: the *real interest calculation* (which we use here) and the *nominal interest calculation*.

Both of them lead to the same result over the total operating lifetime of the plant under consideration.

As we already mentioned in Chapter 2, a *real interest calculation* is appropriate when making estimates of power costs which are relevant to national-economic considerations and for comparisons between different power-generating systems. The standard real interest rate of 4% used in this book represents, for example, the case of a nominal interest rate of 7% and an inflation rate of 3% (1.07/1.03 = 1.0388 = ca. 1.04; approximation: 7% − 3% = 4%).

For a *nominal calculation*, one employs the nominal interest rate, that is, the interest rate which must be paid to the creditors (without correction for inflation). This leads (together with the repayment of the principal) to a certain annual sum for the capital recovery costs. Here, one thus presumes *nominally* constant annual costs, whose real amount, however, decreases over the years owing to inflation. It is then necessary to compute an average *real* value for all the years of the plant's operating lifetime. This averaging process must be carried out taking into account the interest rates (discounting), since a sum that falls due only in a later year has a lower weight at the beginning of the financing–due to interest–than one that falls due in an earlier year. One must, therefore, determine the averaged, discounted costs (LEC) by means of a relatively complicated calculation. In contrast, with the real interest method, the inflation-corrected interest rate (real interest) is used from the outset. "In reality," both methods lead to the same result when the actual averaged costs during the complete duration of the amortization period are considered (Hansen, 1983).[49]

The real cost computation method is preferable for overall cost comparisons due to its simplicity. The nominal interest method is used in those cases where individual cost items (e.g., the costs of fuel or personnel) with differing inflation rates must be taken into account (Hansen, 1983; Schmitt, 1989). When referring to the real interest rate and the operating lifetime, in this book we will always start from the same assumptions for the solar plants as for fossil-fuel and nuclear power plants. A real interest rate of 4% will be considered to be the standard case, and an operating lifetime of 45 years will be assumed, corresponding to a capital recovery factor of 4.83%. In Tables 4.2 and 4.3, the power costs are also given for the case of 2% real interest rate (crf = 3.4%).

4.4.1.1 Note on the Technical Operating Lifetime

Whether the operating lifetime of a plant will in fact correspond exactly to the lifetime of 45 years which we assume here can of course not be predicted with any certainty. With regard to solar power plants, however, the 23 years of operation of the first parabolic-trough plant in California demonstrates that 45 years is not

---

49) The fact that in a nominal interest calculation for the first years higher capital costs are found than in a real cost calculation (and then lower costs in later years) can result in misleading interpretations: if one, for example, considers only the costs during the initial years (and not the costs averaged over all the years of operation!), then "overall higher costs" can be incorrectly derived.

at all an excessively long estimate. So far, no essential signs of aging have appeared there, which would indicate an impending end to its operating lifetime.

For conventional power plants, also, a lifetime of well over 30 years appears feasible. Future nuclear plants are even being planned for a lifetime of 60 years (e.g., the "European Pressurized Reactor" (EPR)). Thus, the EWI study (2007, p. 15), which was initiated among others by the German Electric Utility Association (Verband der Elektrizitäts-Wirtschaft, VDEW), assumed a unified "technical lifetime" of 45 years in a comparison of bituminous-coal-burning, lignite-burning, and nuclear power plants. We have also chosen 45 years in agreement with this study. (Another example, The Chicago Study (2004, p. 5–17) assumed a unified "economic lifetime" of 40 years for a comparison of nuclear power plants with fossil-fuel plants.)

An operating lifetime of 45 years is not only to be expected for solar power plants themselves, and in particular for their most expensive component, the mirror field, but also for the power transmission system and the backup power plants. To be sure, in the EWI study mentioned above, a lifetime of only 30 years is assumed for combined-cycle natural gas power plants with base-load operation (due to the limited lifetime of the gas turbines); however, as backup plants, these installations have a lower capacity factor in comparison to base-load plants (ca. 25%) so that here, also, a lifetime of 45 years is not unreasonable.

4.4.2
Interest Rates

The question of which real interest rates should be assumed in the future is of considerable importance, especially when systems with very different capital costs are being compared, as here with solar power plants and coal-fired plants. As we shall show, our assumption of 4% interest lies within the usual range for the estimation of costs relevant to national economics. Occasionally, 5% is assumed; on the other hand, for example, in SFOE (2007), a value of only 2.5% is taken.

The long-term real interest rate can furthermore be very different within different major economic regions. For example, over many years, there have been considerable differences between the USA and Japan, and these still persist.

In addition, the interest rate may vary widely in the course of time. It depends among other things on the monetary policies of the central banks. Thus, increases in the real interest rate have often been used as an instrument for combating inflation – a method which has been strongly criticized by many economists, since it does not eliminate the actual causes of the inflation and at the same time makes investments "artificially" less favorable, that is, it curbs economic growth.

For the present cost analysis, not the current but rather the future real interest rate is relevant, and we must keep in mind the following: in principle, the real interest is determined by only two factors: on the one hand by the potential of *economically rewarding* investment objects at a certain interest level, and on the other by the supply of credit. In the long term, at least within the major economic regions of the classical industrial nations, one must expect a reduction in the economic potential of investment objects at the current real interest rate (of ca. 4%) ("the end of growth" or "saturation"). Therefore, many economists presume that a sufficient economic demand (i.e., the avoidance of additional unemployment) will be possible only with lower real interest rates so that in the *middle and long term*, a lower real interest rate than at present, for example, *2% instead of 4%, would appear probable*. For this reason, the power costs estimated in this book are often quoted in addition under the assumption of a 2% real interest rate.

4.4.3
Equity Capital and Outside Capital

In the following, we discuss the frequently used method of a separate interest computation with different interest rates for outside (borrowed) capital and equity capital.

The so-called liberalization (or more precisely "privatization") of the energy economy has led in the past 15 years increasingly to the construction and operation of power plants by private investors. The "independent power producer" (IPP) is just one of the catchwords used in connection with such business models. These investors as a rule expect more than 10% real interest for the equity capital that they bring to the project. The real interest rate of 4%, which we assume here, corresponds on the other hand to exclusive use of outside capital with the (lower) interest rate typical of the capital market. This is only possible for the banks that provide the capital if the investments are made by, for example, publicly owned utilities which can offer sufficient guarantees for the capital borrowed (such public utilities can at least more *readily* provide guarantees), or when the risks which can be due to the markets (e.g., a decreasing oil price) can be covered in some other way, for example, by a governmental suretyship.

With regard to the high interest rates often assumed for equity capital, one must generally consider the following:

This approach for computing the power costs is appropriate to the situation on "entry" into the solar energy market, that is, for financing start-up solar power plant projects. Here, the circumstances are similar to most other investment decisions made by these investors. There are numerous imponderables which are compensated by correspondingly high interest rates for their equity capital. Thus, construction of a power plant entails risks such as delays in construction due to the licensing procedure, accidents, perhaps also the market risks which can affect an individual enterprise, in particular with respect to an insufficient sales volume at the price for electric power dictated by the costs. (The latter holds just as well for coal-fired power plants.) All these potential costs and possible losses must be taken into account in the financing of individual projects by increases to the profit rate, which are based on the past experience of the power plant owners.

The situation in the case of a necessary and rapid conversion of today's energy supply, which is dependent on oil, gas, and coal, is completely different, in particular for replacement of coal-fired power plants by solar plants and alternative use of coal to generate gas. In this case, the goals and the transition plan are laid out by the state.

In the middle term, a renewed strong rise in the price of petroleum is to be expected, and it would seem rather improbable that it would then again decrease (since oil will become increasingly scarce). If, however, the price of oil and other fossil fuels did again decrease – this might occur if "sun methanol" indeed functioned as an effective price brake – with the result that investments in solar energy would no longer be economically competitive, the government – as initiator of the transition – would of course have to assume the guarantees. Apart from the destruction of the solar power plants by natural catastrophes or war, a drastic decrease in energy prices is the only apparent *major* risk. The government must cover this risk, just as it today covers the risk of major damage in the case of a nuclear catastrophe. If solar power plants are financed by private investors, the government would naturally also have to cover the risk due to natural catastrophes or military interventions. For state-operated power plants, this is already the case *per se*.

Finance planning for power plant construction is not a new topic; it was often discussed in the past. Thus, Schmitt (1989, p. 1094) writes on the topic of interest rates and equity/outside capital:

> "The first imperative for estimating the cost factor due to costing-based interest must ... be that it should not be oriented on fictitious minimum and maximum values which are defined once and for all, but rather on the capital market and the predicted inflation rates as well as the particular risk and tax aspects of the electric-power sector. .... (He then initially considers interest rates for outside capital.) .... The greatest problem by far is however that of the determination of the costing-based equity-capital interest rate. For this, the interest rates for long-term outside capital still represent the primary criterion; however, when the net maintenance of assets method is employed, only the current inflation-corrected value of ca. 4.5% is relevant. This real interest rate must be increased by ca. 2% in compensation for the higher risks and especially for the higher tax burden of the equity capital, so that a costing-based equity-capital interest rate of 6.5% would result. In any case, this value – supplemented for costing-based venture risks of 1 to 2%, which will be rather higher in future – must be related to the current material value of the equity-financed assets. This costing-based equity-capital interest rate indeed lies well below the usual orders of magnitude assumed in other industrial sectors, but it would seem to be sufficient, in view of the very different risk factors in the electric-power sector."

The interest rates, whether they be for outside capital or for equity capital, must be oriented toward the capital-market interest rates and toward possible risks (which for solar energy, with the assumed governmental guarantees, are practically nonexistent), and toward tax aspects. These latter can likewise be dispensed with here, where we are trying to estimate the costs to the *national economy*. Outside and equity capital from the utilities should, therefore, be allocated with the same interest rates.[50]

If private investors demand higher interest rates because they can obtain them in other business sectors, one can and must dispense with their participation in setting up a solar energy system. The projects will then have to be carried out by publicly owned utilities, which can finance the necessary investments at a lower cost. The responsibility of electric-power suppliers is in the end not to satisfy the profit expectations of individual financing groups (and this practically without risks), but rather to make electric power available at the lowest reasonable price and with high reliability. Here, there is also a financial competition (public vs. private investors); one will have to choose the most advantageous provider.

The fixation toward private investors and the acceptance of their high profit expectations, which are oriented to other (high-risk) business sectors, can be seen in particular in the guidelines of the US DOE, with the result that DOE studies, or also those carried out by the national laboratories, in general assume high fractions of equity capital and high average interest rates. The DOE quite properly defined a standard to permit different costs estimates to be reliably compared. Whether this standard, however, represents the *correct* method for estimating the costs under the boundary conditions outlined above is another question altogether.

The issue is often treated using methods which are quite different from the DOE standard, especially in Europe.

Thus, the Enquete Commission of the German Parliament (1990, Vol. 3, section on solar thermal power generation, p. 682) assumes a real interest rate of 4%.

A similar rate is assumed by the OECD: in the more recent study IEA/OECD-NEA (2005, pp. 26 and 173/174), as in the corresponding previous OECD studies, a real interest rate ("constant money") is used. In order to cover the bandwidth in the different OECD countries (p. 26), this real interest rate is taken to lie between 5% und 10%. Corresponding to the situation in Germany, 5% is taken for computing the costs for nuclear power plants and other power plants. On page 122, we find: "Electricity generation costs are calculated with an interest rate of 5%. It is assumed that the depreciation time is equal to the technical lifetime of a plant...." (However, later for Germany as well as for the other countries the costs are also estimated for the case of a 10% real interest rate.)[51]

In the section on "Methodologies Incorporating Risk into Generating Cost Estimates" (p. 177), the justification for higher interest rates is discussed in detail

---

50) Admittedly, in the past even publicly owned utilities have tried to justify unreasonably high prices with cost calculations on the basis of higher equity capital interest rates, higher in any case than those corresponding to the actual risks. This has also contributed to the discrepancies between the cost estimates which are to be found in the literature.

51) Except for the United Kingdom, where the power costs are estimated for a 10% interest rate, which is recommended as suitable for this country, but are also discussed as an alternative for 5% interest (p. 150): "Table 2.2 shows the cost projections using a 5% real discount rate. The United Kingdom regards this as unrealistically low in a liberalized electricity market."

(the interested reader is referred to this source!). Thus we find among other statements:

> "Power generation investment risks in the liberalized market...
>
> - Economy-wide factors that affect the demand for electricity and availability of labor and capital.
> - Factors under the control of the policy makers, such as regulatory (economic and non-economic) and political risks, with possible implications for costs, financing conditions and on earnings.
> - Factors under the control of the company, such as the size and diversity of its investment program, the choice and diversity of generation technologies, and control of costs during construction and operation.
> - The price and volume risks in the electricity market.
> - Fuel price and, to a lesser extent, availability risks.
> - Financial risks arise from the financing of investment. They can to some extent be mitigated by the capital structure of the company."

The overall subject matter considered here is also described in very clear fashion in a publication of the Swiss Federal Office of Energy, in "Energy perspectives 2035" (SFOE, 2007), an estimate of the costs of nuclear energy and other methods of power generation. By the way, the study is an example of the use of a *low* real interest rate. In the summary *Vol. 1* (p. 70), it is briefly stated that

> "The real interest rate for all types of plants is 2.5% and the amortization period is taken to be the same as the plant's technical operating lifetime. This corresponds to a macroeconomic cost calculation which differs from that of the individual investors. For nuclear power plants, upgrade, shut-down, liability and waste-disposal costs are taken into account, but not however subjective risk supplements."

In *Vol. 4* (p. 117), this point is considered in more detail (Excursus 9, "Methods of cost estimation (electric power supply)"):

> "The .... required new installations .... will be evaluated in terms of their direct total economic costs. These comprise all those costs to the national economy of electric-power generation by the plants. They include the investment costs for the plants and their financing and operating costs, and the energy-source costs (insofar as they apply). The national-economical viewpoint on the financing costs presumes that the plant construction costs are to be spread over the operating lifetime of the plants as annuities with the long-term real bond interest rate (central bank). This point of view excludes by definition secondary cyclic and allocation effects, such as those which arise from a microeconomic viewpoint via shorter amortization periods,

higher interest rates and internal interest-yield expectations. As an illustration: Amortization periods which are shorter than the operational lifetime lead to lower operating costs after the end of the amortization phase (and therefore for constant electric-power prices to higher profits). This must however be paid for through higher capital costs ... during the financing phase. This 'production at the golden end' of the operation of paid-up plants is not reflected in the macroeconomic point of view; there, the actual plant and operating costs are spread over the entire operating life of the plants.

Likewise, allocation effects through distribution of yields and profits are excluded. All the costs and prices are calculated without considering taxes and duties as well as subventions, since these from the macro-economical viewpoint only give rise to redistributions between consumers and the state. Thus the pure macro-economical resource availment for the national economy due to the investment in and operation of the plants is considered. ......

This calculation applies to the costs of new plants. The existing power-plant park is not evaluated, likewise the power transmission system. Under the assumption that the investments in infrastructure are not all too different for the various scenarios, this method of cost estimation is suitable for comparisons (differential examination) of the macro-economical costs among the various schemes. .... It is not suitable for the estimation of the costs or prices from the viewpoint of individual players (for example utilities or power consumers)."

4.4.3.1 Conclusions

The widely used method of cost estimation assuming high equity interest rates, which is in the end adapted to a high-risk situation for private investors in the energy sector (including solar energy, but only during the risky entry phase), has caused much confusion. The bottom line is that it falsifies cost estimates for the various future energy options. The risks mentioned are in fact nonexistent for the industrial construction of major solar parks on a national energy-economic scale, which are embedded in a national energy plan (guarantees or public utility operators). The decisive question relating to energy policy is the burden on the *national* economy. The cost-estimation methods used to date, especially in the USA, lead for capital-intensive energy sources such as solar energy to considerable deviations of the estimated costs from the actual costs to the national economy, and in the comparison of various energy technologies with different capital-cost rates, incorrect relations are obtained.[52] The computational methods used thus far, in

---

52) The Sargent & Lundy study of solar thermal power plants (S&L, 2003, pp. B-7/8) assumes nominal interest rates for equity capital ("return on equity") of 14%, and for outside capital ("debt interest rate") of 8.5%, at an inflation rate of 2.5% and with an amortization time ("debt repayment period") of only 30 years (i.e., less than the technical operating lifetime), and including taxes. This then leads to completely different estimated power costs from those given in this book (costs to the national economy).

particular the guidelines employed by the US DOE, are thus at the very least in need of amendment, and the discussion concerning standards for arriving at cost comparisons should be reconsidered internationally – and especially in the USA.

Readers who wish to grapple with this complex topic in more detail are referred to El-Sawy *et al.* (1979) (the discussion which led to the current standards is to be found there), and to the extensive article of Short *et al.* (NREL) (1995).

5
The Potential of Solar Thermal Power Plants for the Energy Supply: Capacity Factor, Availability of Solar Energy, and Land Availability

In this chapter, we deal with the potential of solar thermal power plants in terms of the availability of solar energy and land disposability in individual countries. Emphasis is placed on solar sites in Spain, Morocco, and the USA; however, we also briefly discuss other sites in North Africa as well as in Asia.

5.1
Overview

As will be shown in the following, in southern Spain with a "24 h Design Insolation" (see Section 4.3.2) of 6.7 kWh/m² d (corresponding to a solar multiple (SM) of 4.4), an annual capacity factor of about 72% can be expected. In the USA, in contrast, owing to the considerably higher insolation, an annual capacity factor of 80% and higher would already be reached for a 24-h design insolation of 8.0 kWh/m² d (i.e., at an SM of 3.7). The same is true of good solar sites in the Sahara. In Morocco, this applies only to the very best sites, of which there are only a few. The available solar energy is less at most of the potential sites there so that (at an SM of 3.7) a capacity factor of possibly only ca. 75% is attainable. In comparison to Spain, however, note that at sites in the USA, in the Sahara, and in Morocco, in spite of a smaller mirror field (SM of 3.7 instead of 4.4) – and the resulting reduced investment costs for the mirror field – the annual energy yield is higher. Furthermore, a certain effect on the uniformity of power production due to dry cooling should be taken into account, as it is necessary at almost all these sites.[1]

---

1) The average daily efficiency of 17.6% quoted by Kolb (1996a) refers to water cooling. In Morocco and other dry regions, however, we must assume that dry cooling will be used. In this case, at the air temperatures which are found in Morocco or the southwestern USA, power production would be about 8% less (cf. Section 4.3.7). An overall lower power output would have no influence on the capacity factor. The reduction of efficiency due to dry cooling depends, however, strongly on the air temperature. In the cooler winter half of the year, which is also the period with lower insolation, the efficiency loss due to dry cooling is not as great as in the summer. In the summer half of the year, this efficiency reduction is greater than the annual average; in this period, however, there is an oversupply of solar energy for many days. Dry cooling

---

Large-Scale Solar Thermal Power. Werner Vogel and Henry Kalb
© 2010 WILEY-VCH Verlag GmbH & Co. KGaA, Weinheim
ISBN: 978-3-527-40515-2

(In southern Morocco, however, it might even be possible to use wet cooling.) This effect is not included in the above values of 80% or 75% capacity factor so that the actual annual capacity factor might be somewhat higher. In Morocco, as an average over potential sites (and with dry cooling), it would thus lie somewhere between 75% and 80%. In the cost tables in this book, for simplicity we have assumed a rough value of 80% for Morocco, also. This numerical example is, therefore, typical of the good sites in the Sahara as well as in the USA (particularly since the distances over which the power must be transported are similar in these two cases).

While in the USA, practically unlimited land areas are available with the insolation cited above, and the topography is well documented, similar data for Spain are very uncertain; to a lesser extent, this holds also for Morocco. In Spain, the land supply depends decisively on the slope of the ground which can be tolerated for solar tower plants, on whether in addition to grazing lands also pasture and arable lands will be used for the construction of power plants, and on whether the present insolation data correctly reflect the long-term situation. Depending on these factors, the available land with a "good" insolation (5.5–6 kWh/m² d = 2100 kWh/m² a) varies between 6000 and 60 000 km², corresponding to a solar tower plant capacity (for SM = 4.4) of 60–600 GW. The upper limit shows that "if needed" – this means accepting certain disadvantages such as somewhat poorer insolation, increased costs due to a slope of the land at the site, or higher land prices (farmland) – there would be sufficient land in Spain alone for potentially supplying the electric power requirements of all of Europe. In Morocco, a *minimum area* can be specified with more certainty using the available data. The area usable for solar tower plants should be ca. 40 000 km² (referring to a daily insolation of 6–6.5 kWh/m² d (2280 kWh/m² a) – this yields the capacity factor of 75% cited above; of this area, ca. 20 000 km² lies within the territory of the Spanish Sahara, which now belongs to Morocco. This total of ca. 40 000 km² corresponds to a solar power plant capacity of about 400 GW. (An area estimate by the DLR, which however applies only to the particularly favorable year 2002, arrives at a total area with good insolation ca. four times larger. It would seem to strongly exaggerate the area available on a long-term basis.)

For comparison, the total power consumption of Europe (EU 25) in 2004 was 360 GWa$_{el}$. In Section 11.1.4, on the combination of solar energy with wind energy from the North Sea, however, we will discuss how this power-generating potential might be split up into three equal parts as solar energy from Spain, solar energy

---

thus has an *evening-out* effect on power production. The oversupply of solar energy in the summer can only be made use of if the turbine section of the power plant is designed to accept a higher heat input. For this purpose, only the steam generator and the turbines (including the waste-heat cooling system) need to be designed to be somewhat larger (e.g., by 10%). The heat storage reservoir is not affected by this consideration. In the standard case, it will be designed to store heat for 16 h of operation per day, in order to be able to store sufficient heat for the long nights in winter. For the shorter summer nights, it will thus have sufficient capacity to allow a certain increase in total heat input per day.

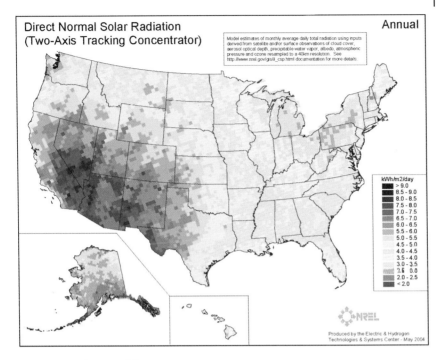

Figure 5.1 Solar map of the USA (NREL, 2004).

from Morocco, and wind energy from the North Sea. If we assume a base load of 300 GWa$_{el}$, which is to be substituted by renewable energy, this would correspond to ca. 100 GW from each source. For a scenario of this type, the available land areas for solar sites in Spain (and even more so in Morocco) would be more than sufficient.

The most important source of information for the following estimate of the available land areas are the maps that show the insolation in each region (solar maps); cf. Figures 5.1–5.8. These maps were prepared using satellite data. A satellite, however, cannot *measure* the solar radiation at the surface of the Earth. The insolation is *computed* for the respective altitude of the site making use of satellite data for the average cloud cover, the aerosol content of the atmosphere, the relative humidity, the air and surface temperatures, and complementary data from ground stations on humidity and other parameters. In particular, in the case of partial cloud cover, a precise quantification of the radiation which reaches ground level is difficult. (To a lesser extent, the aerosol content of the atmosphere also leads to inaccuracies.) These effects are the main sources of error. In the considerations that follow, we must keep in mind certain reservations concerning the accuracy of these information sources (solar maps). Some doubts concerning the informational value of these maps are in order, particularly for certain regions, and especially since the results obtained by various institutions (NREL and DLR)

5 The Potential of Solar Thermal Power Plants for the Energy Supply

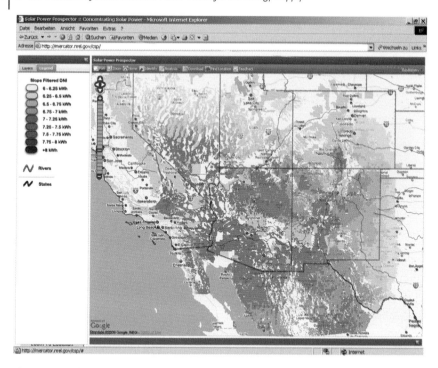

Figure 5.2 Solar Power Prospector Tool: DNI (1998–2005), filtered with "less than 4 percent slope" (NREL, 2008).

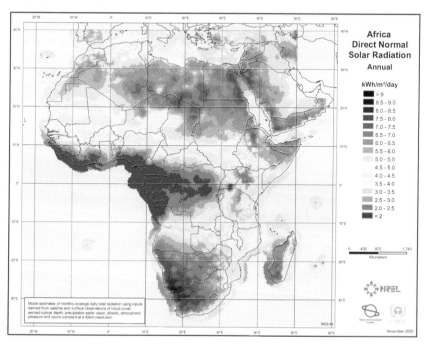

Figure 5.3 NREL map of Africa, showing also southern Spain (NREL, 2005).

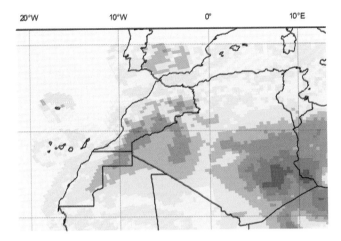

Figure 5.4 Enlarged detail of the NREL map of Africa (2005).

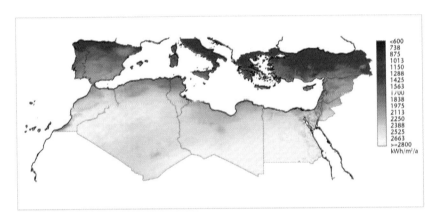

Figure 5.5 DLR map: Direct normal insolation (DNI) in the year 2002 (Hoyer-Klick et al., 2006). (cf. map in DLR-MED-CSP 2005 Summary p.10; there including Saudi Arabia).

are to some extent contradictory. This is true in particular for the high-lying Tibetan region in Asia. In Spain and most parts of North Africa, a considerably better agreement is found among the different maps. The uncertainties will be discussed in detail at the conclusion of the chapter. With an effort of roughly $100 million for a worldwide measurement program, these questions could be answered with certainty. After only 1–2 years of operation of the stations (i.e., ca. 3 to 5 years after beginning the program), we would have much more reliable data sets for insolation over time than are at present available (for all the years of the past).

146 | 5 The Potential of Solar Thermal Power Plants for the Energy Supply

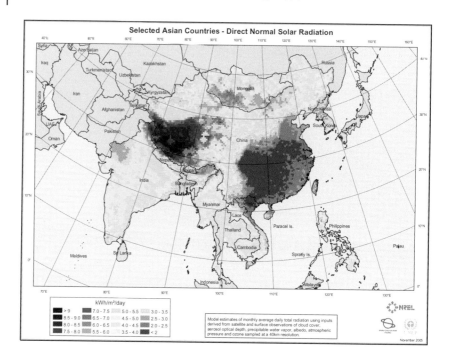

Figure 5.6 NREL map of South and East Asia (NREL, 2005).

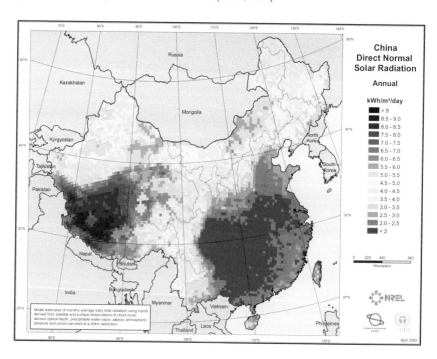

Figure 5.7 NREL map of China (NREL, 2005).

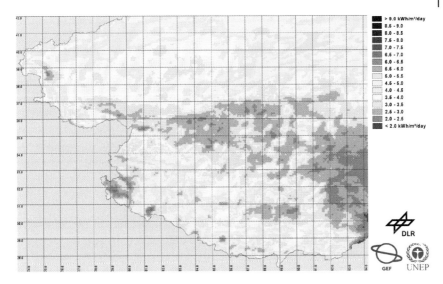

Figure 5.8 DLR map of western China (for the year 2003 only) (Schillings, 2004a).

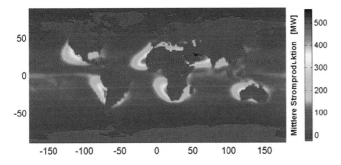

Figure 5.9 Coastal sites for downdraft power plants – the potential annual power output of a plant with a tower 1200 m high and 400 m in diameter (color scale: annual mean power from 0 to 500 MW) (Czisch, 2005).

5.2
Spain: Capacity Utilization and Insolation

At the PSA[2] (in the present book referred briefly as "Almeria"), near the town of Almeria in southern Spain, since the mid-1980s a measurement station has been in operation which determines the direct normal irradiance (DNI).[3]

---

2) The Spanish solar energy research center: Plataforma Solar de Almeria.
3) The solar irradiation is composed of direct and diffuse radiation; their sum is the global

irradiation. A portion of the solar radiation is scattered by the atmosphere, that is, its direction of propagation is changed; it reaches the earth's surface as diffuse

Unfortunately, the wish of the present authors to provide readers with these measurement data—namely the daily insolation values over the entire period of the measurements since the 1980s—could not be fulfilled. Only the values from a few recent years can be made available. For our earlier study, we obtained the daily values for the years 1990–1994 (PSA, 1997), and as a result of the present renewed request, only the data for the years 2001–2007 (PSA, 2008). Whether these 12 years represent in total better or poorer years for insolation could be judged only by comparing with the annual sums for the complete period of the measurements (since the 1980s). Unfortunately, the annual totals were also not made available. In an earlier data release (Ruiz, 1989), however, the annual sums for 1985 through 1988 were quoted[4]; their average value was 1947 kWh/m² a.

At present, it is thus not possible, even for scientific purposes, to present the data for Spain in complete form for at least *one* site.

Figures 5.10 and 5.11 show the examples of annual variation of the daily solar irradiation for a nearly average year[5] (1990), and for a less sunny year (1993) (PSA, 1997). Within the 12-year period mentioned, these daily values are known. In computing the resulting annual capacity factor, it must be kept in mind that in Spain, with the chosen SM of 4.4, only a maximum irradiation of 6.7 kWh/m² d can be used (the power plant is then running 24 h per day at full output power); the supply of solar energy that exceeds this value for many days during the year must be rejected. Table 5.1 gives the annual sums obtained from the daily values for the 12 years listed; in the left column, the total solar irradiation is shown and in the right column the irradiation taking the 6.7 kWh/m² d limit into account. As an average of these 12 years, an annual sum of *usable* solar energy of 1767 kWh/m² a is found.

When the supply of solar energy is insufficient, the backup power plants must fill the gap. On many days during a year, solar energy is not completely lacking but is only reduced. On such days, the backup plants need supply only some fraction of the required power production.

---

radiation. Using mirrors, only the directed or direct radiation can be focused and deflected. For solar thermal power plants, therefore, only this portion of the solar radiation is of interest (with the exception of chimney power plants). In measuring the solar radiation (or more precisely the radiation flux density), one distinguishes further between irradiation onto a horizontal surface area and the irradiation of a surface which is perpendicular (normal) to the direction of the sunlight (normal radiation). The irradiation onto a horizontal surface depends upon the angle of incidence (and thus on the geographic latitude of the site and the time of day). Solar power plants make use of movable mirrors, which can be guided to follow the Sun (even though the mirrors do not always point exactly perpendicular to the direction of the sunlight). As a measure of the supply of solar energy to a solar power plant, one uses the *direct normal irradiance (DNI)*.

4) 1985–2063 kWh/m²; 1986–1926 kWh/m²; 1987–1828 kWh/m²; 1988–1970 kWh/m².
5) The year 1990, at 2076 kWh/m² a, was only 4% above the long-standing average (2008 kWh/m² a).

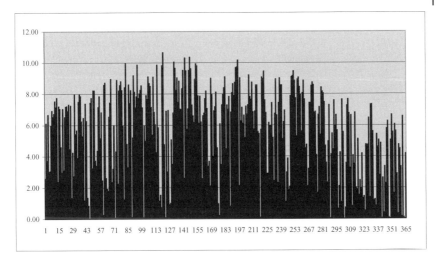

Figure 5.10 Daily solar irradiation in Almeria (Spain) during a nearly average year 1990–2076 kWh/m² a (PSA, 1997). DNI is given in kWh/m² d (abscissa: days).

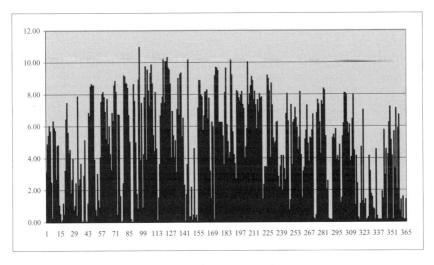

Figure 5.11 The daily solar irradiation in Almeria (Spain) during a poor year, 1993–1771 kWh/m² a (PSA, 1997). DNI is given in kWh/m² d.

A comparison of these 12 years with all the annual values is, as mentioned, not possible. If one includes at least the available values for the years 1985–1988, also, then the annual sum is reduced (from 2028 kWh/m² – Table 5.1) by 20 kWh/m² a to an average value of 2008 kWh/m². The "annual sum of the daily values for utilization of *at most* 6.7 kWh/m² d" would then decrease by this same amount:

Table 5.1 Solar irradiation in Almeria (Spain): The direct normal irradiation (PSA 1997, 2008).

Year	Annual sum of the solar irradiation kWh/m² a	Annual sum for utilization of at most 6.7 kWh/ m² d
1990	2076	1843
1991	1584	1489
1992	1941	1731
1993	1771	1587
1994	2035	1801
2001	2170	1807
2002	2230	1919
2003	2099	1786
2004	2041	1777
2005	2300	1943
2006	1952	1702
2007	2141	1821
Average values	2028	1767

it would then be 1747 kWh/m² a (instead of 1767 kWh/m² a). (The limitation to 6.7 kWh/m² d means that 13% of the available solar energy is rejected (1747/2008 = 87%)).

This corresponds to a usable average daily irradiation of 4.79 kWh/m² d. From this value, the average capacity factor of a solar plant designed for a maximum of 6.7 kWh/m² d at Almeria would amount to 71.4% in the years 1985–1988, 1990–1994, and 2001–2007. Over the long term, one can thus expect an annual capacity factor at the Almeria site of 70–75%.

The insolation situation at Almeria (PSA, a site near the coast) is probably not really representative of those Spanish sites which are located *more toward the interior*. As an example, we consider *Guadix*. (It lies ca. 50 km east of Granada or 75 km northwest of Almeria and is ca. 60 km from the coast.) On the one hand, the sites in the interior lack to some extent the advantage of the southeastern coastal region, namely a certain amount of screening by the Sierra Nevada mountain range against the clouds, which come mainly from the west. On the other hand, these sites are considerably higher (Guadix is at an altitude of 900 m, Almeria (PSA) at 400 m) so that the supply of solar energy on cloudless days is generally greater owing to the lower absorption losses. For Guadix, the annual values of 1990–2000 are available (PSA, 2008). They are clearly higher, at an average of 2136 kWh/m² a, than the value quoted for Almeria (2008 kWh/m² a – whereby the latter value refers to the 16 years mentioned above, which are not identical with the measurement period for Guadix). This value of the solar irradiation for Guadix (and its altitude) should be typical of the solar sites which are located more toward the interior.

According to the solar map of NREL (2005-Africa – Figure 5.3; see also the larger scale detail section in Figure 5.4), the solar irradiation in the major portion of the southern half of Spain lies between 5.5 and 6 kWh/m² d, corresponding to ca. 2100 kWh/m² a (5.75 × 365). At the southwest corner (Almeria), it is equal to only 5–5.5 kWh/m² d, corresponding to 1920 kWh/m² a. Only when reliable daily values for a whole series of sites in Spain become available (see below) can the question of the insolation situation be answered precisely.

The distribution of the supply of solar radiation over the year would, however, be somewhat less homogeneous at these interior sites than at Almeria, since at those sites, the screening of the clouds by the mountains would probably be less effective than in Almeria. The overall higher solar energy supply should compensate for this reduced homogeneity, at least in part. Furthermore, a less homogeneous supply of solar energy would lead to a different optimization of the size of the mirror field.[6] It can be assumed that all together, the same or at least very similar economic effectiveness would be obtained as in Almeria (presuming that also at these sites, which are somewhat further from the coast, wet cooling would still be feasible). But we must wait for reliable data. This is even more relevant to the question of land disposability in Spain, which we will discuss below.

What we have said above depends, as mentioned, on whether the NREL map of Africa (Figure 5.1) is correct.[7] According to NREL, the accuracy of this map lies in the range of ±15%, whereby the error limits depend on microclimatic influences and increase with increasing distance from the meteorological measurement stations. These stations deliver additional data on the humidity of the air, its aerosol content near the ground, etc., from which – together with the information from satellites on cloud cover (and likewise on aerosol content and humidity) – the solar irradiation is obtained computationally. According to NREL, in individual cases errors of over ±25% can occur. (Whether this estimate of their own accuracy by NREL is likely to be correct will be discussed below.) From the NREL specifications, it is unfortunately not clear on which years the data in the NREL maps (2005) are based. At least in terms of the NREL map for the USA (2004), the important

---

6) In this book, the optimization was carried out roughly for the Almeria site (with SM = 4.4). When the irradiation is less homogeneous, the optimization would result in a lower solar multiple (i.e., a correspondingly smaller and cheaper mirror field), and the acceptance of a somewhat higher fraction of backup power. (This last aspect would be particularly worth considering in the case of a combination of solar power and offshore wind power.)

7) The DLR map for North Africa, which also includes Spain, applies only to the good year 2002 (DLR-MED, 2005, Summary, p. 10). Considering the fact that in Spain several solar power plants are in operation or under construction, and that measurement stations are maintained there, one would indeed expect that the NREL data could be checked there, at least for the last few years. However, these data have not been published by the private project operators. Thus, Schillings et al. (DLR) (2004b) refer to "sensitive data." The data were, however, used for a first comparison to the DLR map, but the result is hard to interpret. The comparison of the data with those of a public measurement station in Morocco, however, indicates considerable failings in the methodology of the DLR – see below.

data on cloud cover are probably based on only 8 years (1985–1992) (Figure 5.1). This possibility is, however, contradicted by the fact that the maps (for the USA) appeared in the year 2004 (and were newly issued in 2006 for the USA; this new map edition, however, holds only for the years 1998 through 2005). Thus probably all the years back to 1985 are taken into account (also in the NREL maps from 2005 for Africa, Asia etc.).

The above estimate of the capacity utilization (71.4%) was made for the Almeria site using the sum of 2008 kWh/m² a for annual irradiation. The NREL map (2005), which gives 5.5–6.0 kWh/m² d (2100 kWh/m² a), lies somewhat higher so that here, a certain "reserve" for inaccuracies is included. Even if the annual irradiations were somewhat lower than quoted there, a capacity factor of 70% should be attained, at least "approximately." Even if in some spanish regions only a factor of 65% should be attained (as in Almeria in the poor year 1993 (Figure 5.11)), the cost calculation (here 70%) would not change very much (higher fuel costs for the backup plants); cf. the maps in Suri et al. (2009), Meteotest (2009) and (with lower values, but derived from the Global Insolation and not commented) Cayetano (2009).

The *land disposability* in Spain will be treated next (after first considering the USA, where the situation is particularly clear). Under this topic, we will also discuss the insolation measurements of the DLR (for the "good" year 2002).

5.3
The USA

For the USA, there are unfortunately no data on the annual trends in the daily insolation over *several* years. These trends, which are important for computing the annual capacity utilization, are as a rule not specified in the literature. In the Mediterranean study (Klaiß and Staiß, 1992, Vol. II, pp. V-31 and 33), however, the irradiation data for Barstow (USA) were also given for comparison, and indeed included the annual trends for 2 years: for 1976, as the best year up to that time (measurements up to 1992; annual value 2850 kWh/m² a), and for 1984, as the poorest year up to 1992 (2370 kWh/m² a); see Figure 5.12. The evaluation of these daily values under the condition that at most 8.0 kWh/m² per day could be utilized yields a capacity factor of 86.2% for 1976 and of 75.9% for 1984. The annual trends for these 2 years show that for 2550 kWh/m² a (corresponding to 7 kWh/m² d), a capacity factor of 80% can be attained.[8] (Later years, however, indicated that for Barstow itself, we can assume that the long-term average in fact corresponds to a higher insolation than was suggested by the measurement period up to 1992, namely 7.5 kWh/m² d (2740 kWh/m² a) and higher. For a solar power plant site at Barstow with these values, one can thus presume a capacity factor of more than 80%.)

8) Taking the two reference years 1976 (2850 kWh/m² a = 86.2% capacity factor) and 1984 (2370 kWh/m² a = 75.9% capacity factor) would yield a capacity factor of 81.1% for 2610 kWh/m² a, while for 2550 kWh/m² a it corresponds to 79.2%.

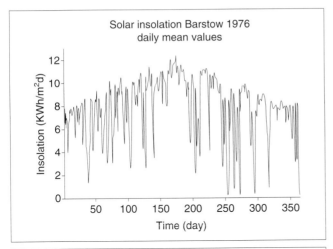

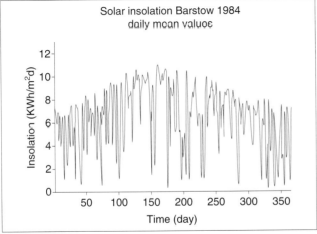

Figure 5.12 The direct normal irradiation (kWh/m² d) for Barstow (Mojave desert, California) in a good year (1976–above) and in a poor year (1984–below) (Klaiß and Staiß, 1992, p. 33).

Figure 5.1 (NREL, 2004) and Figure 5.2 (NREL, 2008) show the insolation in the USA. The new high-resolution solar map in Figure 5.2 is based on the insolation from only 8 years (1998–2005)[9] and it shows only those regions with a slope of the terrain of less than 4%. (A map without slope filter can be seen in NREL, 2007.) As is readily apparent from Figure 5.2, with this maximum slope of the terrain (which will be discussed below), in California, Arizona, New Mexico, and Nevada there are enormous areas that have a solar energy incidence of over 7 kWh/m² d

---

9) Since the NREL map (2004) is based on a different series of measurements (1985–1992), we show it here for comparison.

(2550 kWh/m² a).[10] For these regions (shown in red in the figure), a capacity utilization of "around 80%" is thus applicable.

From Figure 5.2, it can be seen that the region with over 7.0 kWh/m² d should be all together larger than the entire state of Arizona (cf. Arizona – 296 000 km²; New Mexico – 315 000 km²); as a rough value, this area would be 350 000 km². Together with the regions with an insolation between 6.5 and 7 kWh/m² d (average of 6.75 kWh/m² d) – shown in olive and brown colors in the map – the total area should be greater than 500 000 km². (In the prospector tool (NREL, 2008), the reader can see the area with other assumptions for the slope of the terrain. With "5% slope," the area is noticeably greater.) With a land requirement of 89 km²/GW (at an SM of 3.7) without transmission losses, and taking these to be 11.5% (for 3000 km lines), an area of 100 km² corresponds exactly to a power plant capacity of 1 GW$_{el}$ (cf. Section 4.3.5), that is, 500 000 km² correspond to 5000 GW$_{el}$. However, it should be mentioned that the older solar map in Figure 5.1 shows somewhat lower values (in Arizona and New Mexico, in many regions the difference is 0.5 kWh/m² d), and that with respect to the roughness of the terrain (at a given total slope) there could be some further restrictions concerning the usable land (see below).

The above-quoted potential power capacity is to be compared with the electric power consumption in the USA (2004), which totaled 450 GWa$_{el}$ (in 2004, the energy consumed from petroleum was 1250 GWa, and that from natural gas was 770 GWa). So the USA has sufficient land area to provide for every conceivable type of energy substitution.[11]

The figures quoted, however, hold *only for solar tower plants*. In the case of *parabolic-trough plants*, the requirements for flatness of the site are considerably more stringent. In a study of the San Diego Renewable Energy Group (2005) on these power plants, a maximum permissible slope of the ground of only 1% was assumed. In this study, other reasons for excluding sites were also considered (Indian reservations, national parks, nature preserves, etc.). Nevertheless, for the southwestern states in the USA all together, an available area of 130 000 km² was found. Concerning the required insolation, this study makes nearly identical

---

10) Compare: 7.25 kWh/m² d corresponds to an annual total of 2650 kWh/m² a.
11) The total area is probably large enough to provide – theoretically – even the entire world with oil substitutes from solar energy and coal. The world oil consumption in 2004 was 5000 GWa (3.8 billion tons of oil). To produce this amount of liquid energy carriers from solar hydrogen and coal (methanol), roughly 450 000 km² would be required (90 km²/GWa-methanol, cf. Section 11.4.3.3). This includes the oil consumption of USA. For the US power supply (450 GWa$_{el}$), an additional ca. 50 000 km² would be needed, thus 500 000 km² in total. Even with more restrictions than those shown in Figure 5.2 (somewhat lower insolation and unusable land due to roughness), there should be large areas with an insolation of 7 or at least 6.5 kWh/m² d kWh (e.g., 300 000 km²). That would still be enough to permit a great influence to be exerted on the oil price. (And by using additional regions with 6–6.5 kWh/m² d, as Figure 5.1 shows, even in this case there should be a usable area of more than 500 000 km² in total.)

assumptions as we did in the above estimates, namely at least 6.75 kWh/m² d (2460 kWh/m² a). The supply of sites in the USA is thus enormous, for parabolic-trough power plants also. In mountainous Spain, this is not the case, and the same is true – but to a much lesser extent – of Morocco.

5.4
Solar Tower Plants – Permissible Slope of the Terrain

As just mentioned, parabolic-trough power plants have stringent requirements with respect to a horizontal and flat terrain for setting up the reflectors. Solar tower plants, in contrast, do not in principle require a planar ground area. It is intuitively clear that when a large number of mirrors are set up, the array can be more readily adapted to the topological situation of the site than in the installation of parallel troughs, which are ca. 100 m long. This difference plays a particularly important role when land is overall in short supply; therefore, we treat this aspect of the slope of the terrain for solar tower plants briefly here, before taking up the question of the availability of land in Spain in the next section.

A certain *slope of the flat terrain* represents practically no disadvantage for solar tower plants in terms of the geometry of the mirror field (losses by shadowing of the heliostats). To be sure, if (in the Northern Hemisphere) the slope is to the north, the spacing of the heliostats must be increased, while if the slope is to the south, the spacing can be closer so that on a large scale, the two effects average out. For the geometry of the mirror field, however, the light path from the heliostats to the tower also plays a role (losses from "blocking"), in particular with *rough (wavy or hilly) terrain*. Here, the decisive factor is the slope relative to the local direction to the tower. If, for example, the land is rough, but overall horizontal,[12] there are an equal number of sections that slope away from the tower as toward the tower. In the case of the former, the spacing of the heliostats must again be increased; for the latter, it can be reduced. The compensation is, however, only a global trend, but it functions all the better if the terrain has a slope as a whole more toward the south.

Solar tower plants are thus relatively tolerant with respect to the terrain. The recent Mediterranean study of DLR (DLR-MED, 2005) assumed an allowable slope of the mirror field of 4%, referred to the areas actually used, that is, on a small scale. Since the analysis could not take this small-scale tilting of the terrain into account (the grid spacing was 1 km), the simplifying assumption was made that the slope of the grid could be at most 2%, in order to include all the areas whose subareas on the whole had a tilt of less than 4%. (One can see how rough this assumption is from the fact that the value is derived not from a calculation, but is simply an intuitively defined limit. This also becomes apparent when one considers that no distinction was made between an overall slope to the north or to the south.) This reduction to half the permissible slope avoids an overestimation of the usable

---

12) This also holds for a terrain which slopes as a whole toward a certain direction.

land area. A *lower limit* is thus obtained, not the true potentially usable area, which is particularly important in Spain. In the North African Mediterranean countries for which the DLR study was mainly conceived, this factor is less important, since there for the most part an oversupply of land area is available. The study was intended to show, as the authors emphasize, where the attractive regions are to be found in each country. The study is, therefore, less meaningful precisely for Spain. The limitation to half the allowable slope (2% instead of 4%) furthermore ignores the possibility of at least a partial compensation of smaller areas, which slope toward or away from the tower within a rough terrain; the assumption is for this reason alone inconclusive – but it was indeed just a stopgap solution. For the solar maps of the USA (NREL, 2007, 2008), this topic is not discussed. Therefore, it is possible that there could be some restrictions on the areas shown in Figure 5.2.

In any case, we can presume that for areas that on the whole slope toward the east, the south, or the west, the average slope itself is not so much of an obstacle to the construction of solar tower plants; in particular for a southern exposure, a relatively steep slope would in principle be acceptable – but rather the nature of the terrain determines its usefulness, especially the *accessibility* of the area during construction and later operation of the plant.

In the earlier Mediterranean region study (Klaiß and Staiß (DLR) 1992), a slope of up to 5% was even allowed as a criterion (and indeed without any limitation due to the roughness of the terrain). The statements made in that study for the useful land area in Spain are, however, less significant, since the altitude determination was made using a very rough grid. (The slope is determined by comparing the altitude of one cell with that of a neighboring cell in the grid, thus giving an "average" angle of inclination. The size of the cells with the then-available satellite data was, however, relatively large, namely 10 km × 10 km, so that only limited information on the true inclination of the land areas within the cells is available.) At that time, no more precise data were available. For this reason, probably many areas with still steeper slopes (than 5%) were included within the analysis. All the same, a slope of "at most 5%" was considered to be permissible.

Speculation about the allowable angle of inclination in any case does not help us to deal with the problem; we require concrete investigations of various types of terrain, that is, the actual planning of mirror fields on different terrain types which are exemplary of the different classes of possible sites.

The following steps thus need to be taken:

- site classes must be defined in terms of the roughness of the terrain;
- mirror fields should be designed in detail for each such class;
- from the design positions of the heliostats, the geometric losses need to be computed.

Future *terrain maps* must, therefore, show not only the altitude, as was the case up to now, but also the roughness of the terrain, obtained from satellite data that record the altitude profile of each region with sufficiently high resolution. With the TerraSAR-X satellites launched in 2007, "altitude profiles of down to 1 m areal resolution can be obtained" (DLR, 2008); their altitude precision is also less than

1 m. In contrast, in the year 2000, the limit was still 17 m areal resolution and an altitude precision of 8 m (DLR IMF, 2008). To answer definitively the question of which area inclinations and roughness are acceptable for solar tower plants (to the extent that the terrain can be prepared for use with only minor additional costs), it will not be necessary to construct test plants in each area. But the mirror fields of the plants must be planned in advance, using the data base that will apply to the actual construction (satellite data – possibly complemented by an exemplary verification of the satellite-based data through aerial surveys or even terrestrial examination). Building on these data, large-scale maps with the respective terrain classes must be prepared (at least for Spain, Turkey, and other mountainous locations with a limited potential). This includes also the important Asian locations such as Tibet (Chinese: Xizang) and possibly the Qaidam Basin (Chinese: Chaidamupendi – altitude ca. 3000 m) as well as possibly some mountainous regions in Inner Mongolia and in the Indian solar regions. And in the USA, the total amount of usable land (with respect to oil substitution) could be determined using the defined site classes.

An authoritative definition of the "permissible" types of terrain (or more precisely, a determination of the additional costs which can be tolerated in connection with the use of a particular site) is of fundamental importance as a research topic for solar thermal power plants, in particular since the most favorable locations for solar energy are to be found at higher altitudes. (This is in part for meteorological reasons; on the other hand, it is also related to the fact that at higher altitudes, the absorption losses are noticeably smaller in the less dense and usually aerosol-poor atmosphere there.) High altitude locations are as a rule more mountainous, and therefore rougher (with the exception of high plateaus). Together with exact measurements of the insolation as mentioned above, such a precise survey of the terrain with respect to roughness (terrain classes) is thus of great importance.

5.5
Spain: Availability of Sites

The Spanish landscape is considerably more mountainous than, for example, that of Morocco, which we will discuss below. Here, the available area will depend to a large extent on how much of a slope can be tolerated by the solar tower plants. Another important factor is the degree to which agricultural lands must be excluded as potential sites; this applies in particular to arable croplands.

Arriving at a conclusive estimate of the suitable land area in Spain has thus far not been possible. We can only hope – by making use of diverse sources of information – to obtain some indications in this direction. However, one can say that in the case of *serious need* (depending on preferences, either accepting a lower insolation, a greater roughness of the terrain, or possibly higher land costs due to use of arable lands), the entire European Union (EU 25), with an annual overall energy consumption of 360 GWa, could be supplied with electric power from Spain.

For the estimation of available land areas in Spain, the two studies by the DLR already cited above are at our disposal. Both the newer one (DLR-MED, 2005) and the older one (Klaiß and Staiß, 1992) entail estimates of the potential for the entire Mediterranean region. In the DLR-MED study, a "real" slope of 4% in the terrain was taken as the upper limit for solar tower plants, but the analysis of potential was carried out assuming a large-scale slope of less than 2%.

In the older Mediterranean region study, on the other hand, a permissible slope of 5% was assumed, which should be more applicable to the case of Spain. Both studies were furthermore aimed primarily at the North African countries near the Mediterranean coast, for which such rough assumptions are sufficient; for Spain, their significance is, therefore, reduced, as mentioned above.

Both studies also include insolation data, and use them to estimate the potentially available land areas. In the newer study (MED), these data are, however, based upon the radiation values from a single year, namely 2002 (see Figure 5.5), which – especially in Spain – was an excellent year for insolation and, therefore, not typical (cf. Table 5.1). In the older study (1992), the minimum value for the insolation was taken to be very low, at 1700 kWh/m^2 a so that the required areas calculated there do not document the true economic potential. For this reason, in the following estimates we make use only of the maps showing the slope of the terrain or showing all the possible reasons for exclusion of particular sites from these two studies. These data are then combined with the insolation data for a number of years from the NREL map (2005, Africa) (see Figures 5.3 and 5.4). For the exclusion map, the older DLR chart with 5% maximum terrain slope is of particular interest. If one overlays the regions from the NREL map having an insolation of 5.5–6 kWh/m^2 d (2100 kWh/m^2 a) over this exclusion map from the DLR study (Klaiß and Staiß, 1992), then a large region in the southern half of Spain remains: ca. 60 000 km^2, corresponding to a solar tower plant capacity of 600 GW (for an SM of 4.4) if this region were completely utilized.[13] Here 2100 kWh/m^2 a (NREL map) is around 5% more than the average value found above for Almeria (over 16 years), 2008 kWh/m^2 a. We have already pointed out that the distribution of insolation throughout the year is possibly somewhat less favorable here, that the NREL map may overestimate the actual insolation, and that the DLR map (which defines the excluded areas) is in fact more adapted to the situation in North Africa (with its relatively coarse resolution of 10 km × 10 km), so that this latter map describes the terrain in Spain with less precision; furthermore, along with veld-like regions (used mainly for grazing), this area also contains pasture land and arable land. The effectively usable areas will thus be less than this estimate. When arable lands are excluded, they will in any case be smaller.

The newer study (DLR-MED, 2005) has, in contrast, the disadvantage of the lower permissible slopes considered (max. 2%), the complete exclusion of all

---

13) In Spain, the heliostats would be placed closer together owing to the higher land prices than in the USA or in Morocco. Then, in spite of the higher SM of 4.4, one obtains nearly the same land area requirements as in Morocco/USA, with an SM of 3.7 (cf. Section 4.3.5).

"agricultural" lands (arable and pasture lands[14]), and especially of having a very limited database for the insolation, using only the data from a single, atypical year. The terrain slope was, however, obtained with a higher areal resolution of $1\,km^2$. Taking the minimal insolation value to be $2000\,kWh/m^2$ a, one finds as the result a potential of $1278\,TWh/a$ (based upon a 30% area utilization);[15] with the typical area utilization factors for solar tower plants of 20%, this is $850\,TWh/a$ ($98\,GWa/a$). Taking the capacity factor for solar power plants to be typically 70%, this corresponds to a solar plant capacity of $140\,GW$. Of this, 54% is in regions with an insolation range of 2000–2100, and 23% in the range of 2100–2200, 16% in the range of 2200–2300, and 6% in the range of $2300–2400\,kWh/m^2$ a. This, however, applies only for the good year 2002, in which a high annual insolation was registered in many regions. This then indicates that over a long-term average, the potential could be perceptibly lower;[16] possibly only as great as would correspond to the insolation range over $2100\,kWh/m^2$ a in the good year 2002. This would give only $60\,GW$ (instead of $140\,GW$). Nevertheless, this study also indicates a *great overall potential* (with $2000\,kWh/m^2$ a) in Spain: namely by the use of locations with slopes of >2% and including grazing lands, perhaps even arable lands. Omission of these two exclusion criteria would make the final result very different.[17] The possible power plant capacity based on available land areas thus lies – depending on preferences – between $60\,GW$ and theoretically $600\,GW$.

The question of arable land is in the case of Spain rather academic. In the cost estimates, a land price was arbitrarily assumed that corresponds roughly to the price of German arable lands ($1.25\,\$/m^2$ – see Section 4.3.3). Concerning pasture lands or lands used only for grazing in the arid regions of Spain, this price is *by far* overestimated in terms of the agricultural value of the land. It is, in any case, not the goal to cover Spain as far as possible with solar plants, and certainly not to use arable lands for that purpose. While negotiating the terms for construction of solar plants by the European Union countries with the *North African* states, the Spanish sites are, however, the decisive alternative. Therefore, the question of utilization of arable lands in Spain, with this more theoretical background, arises in the sense of a "potential" increase in usable sites. Even if, at a price of $1.25\,\$/$

---

14) Pasture land is probably very difficult to distinguish from unused land or land used only for grazing (e.g., for sheep) in the maps.

15) The value of 1278 TWh quoted in the study refers to an area utilization factor of 30%, typical of parabolic-trough power plants (DLR-MED, 2005-Resources, p. 60). In recalculating for solar tower plants, their lower area utilization factor (of 20%) was taken into account. For simplicity, however, the same efficiency was assumed as in the DLR study on parabolic-trough power plants, namely 15%; compare solar tower plants ca. 17% (with water cooling).

16) The value of 54% of the area for 2000–$2100\,kWh/m^2$ a (i.e., on average $2050\,kWh/m^2$ a) in a good year thus indicates that the NREL map, with its large areas of $2100\,kWh/m^2$ a, overestimates the long-term insolation situation. On the other hand, the regions with still higher insolation are missing on this map (due to its coarser resolution).

17) The regions found in the DLR study are (owing to the stringent requirements on the slope of the ground) for the most part not contiguous areas, but rather a patchwork of smaller fields. (On the main, however, they lie within the zone of the NREL map with $5.5–6\,kWh/m^2$ a.) If more strongly sloping land and pastures could be utilized, one would find larger contiguous areas.

m², arable land is already taken into consideration, the effects of still higher prices are also interesting: if, for example, one assumes prices which are a factor of three higher (3.75 \$/m²), the investment costs for the solar power system would increase by 6% and the power costs (including backup power) by 0.16 ¢/kWh or by 3% relative to the base estimate[18]; the land costs would then make up ca. 8% of the overall investment costs. In the political negotiations, one could thus point out this alternative, if in fact not enough land area turns out to be available in Spain excluding arable croplands. The demands of North African countries would thus find an upper limit (cf. Section 4.3.4). The same is by the way true for potential areas with greater slopes than would be allowable in the normal case (with additional costs caused by this factor), and naturally also for areas with somewhat poorer insolation values. "Potential" areas (which, however, are not quite as good as the base assumptions) are thus available in large quantity in Spain.

We have already pointed out that a division of the future European power-generating capacity (e.g., a base-load capacity of 300 GW)[19] into 100 GW power from wind energy, 100 GW from solar energy in Spain, and 100 GW from solar energy in Morocco is a realistic option. (Without wind energy, one could plan for, e.g., 150 GW each from Spain and Morocco. Other sites close to Europe such as Tunisia or Turkey are not included in this concept.) Per 100 GW (SM 4.4), only 1.7% of the area of the Iberian Peninsula would be required. There are thus no plans to use "all of Spain" for this purpose (although the owners of the practically valueless lands would doubtlessly be interested in selling more of it).

5.6
Morocco/Sahara

As is shown by the NREL map (2005, Africa), in Morocco, the insolation values in the regions relevant to utilization for solar power plants lie in the range of 6–6.5 kWh/m² d; these correspond to an annual sum of 2280 kWh/m² a (6.25 × 365). These regions thus correspond over a long-term average rather closely to the insolation in Barstow (USA) during the poor year 1984 there, with 2360 kWh/m² a. As mentioned above, the year 1984 (at a usable insolation of 8 kWh/m² d) would yield an annual capacity factor of 76%. For Morocco, there are no data on the trends of the daily insolation values in the course of a year. Therefore, we use a simplified model based on the yearly insolation in Barstow in 1984 as a reference for the daily values of the insolation class 2360 kWh/m² a; this method was already applied in the older Mediterranean region study (Klaiß and Staiß, 1992, Vol.2, p. II-54/57). The NREL insolation class 6–6.5 kWh/m² d represents nearly the

---

18) For arable land, an area utilization factor of 22.6% was assumed (as was used in the required area estimates), while for the cost analysis (base computation), an especially low value of 18% was taken (cf. Section 4.3.5).

19) The total electrical energy consumption of Europe, including the Eastern European states (EU 25) in 2004, was 360 GWa.

same annual sum (2280 kWh/m² a); it is only *slightly* smaller. Therefore, the capacity factor should be only a little less than 75%. Including the above-mentioned effect of dry cooling, it should however lie in the range between 75% and 80%. (For more precise values, we must wait for the results of further investigations.)

Which regions in Morocco are relevant as potential sites for solar power plants can be seen by comparing the NREL map with a map from the newer Mediterranean region study (DLR-MED, 2005-Resources, pp. 60 and 62); the latter shows the excluded areas for North Africa (including Morocco).[20] (This map is also reproduced in (May, 2005, p. 65).) It can be seen there which regions are not suitable due to a slope of the terrain of more than 2%, and other exclusion criteria.

A simple, rough comparison of the two maps (NREL, DLR) makes it clear[21] that in Morocco, in particular three regions can be considered to be suitable:

- the northwest corner of the country, with insolation values of 5–6 kWh/m² d;
- a region in central Morocco near the border with Algeria,[22] likewise with values of 5–6 kWh/m² d; and
- an especially attractive region in southern Morocco near the border between the previously Moroccan territory and the former Spanish Sahara. This region stretches out to the Algerian border and to the north up to the Wadi Draa (Qued Draa), and has insolation values of 6 kWh/m² d. (The "Spanish Sahara," which currently belongs to Morocco, is politically not uncontroversial.)

The last-named region alone includes an area of roughly 10 000 km² within the original Moroccan territory; this corresponds to 100 GW. An additional ca. 20 000 km² with a similar insolation lies to the south in the territory of the former Spanish Sahara.[23] In southern Morocco alone (including the "Spanish Sahara"), there is thus a usable area of roughly 30 000 km² with a high insolation ($\geq$ 6 kWh/m² d).

In the regions in the northwest corner and in central Morocco, only a small area lies within the insolation class 6 kWh/m² d (here, we always mean "from 6 to 6.5 kWh/m² d"); these areas cannot be clearly defined with the exclusion map. Nevertheless, an area of an estimated 5000 km² with these high insolation values

---

20) This exclusion map also contains insolation values, so that a comparison with the NREL map should in fact be unnecessary. These values can however not be precisely read off the map and furthermore apply only to a single year (2002). This holds true also for a map of Brösamle *et al.* (2001, p. 5), which reflects only the year 1998.

21) To this end, the maps must be displayed on the computer screen with a high magnification!

22) This region stretches roughly from the 28th to the 30th parallel and is bordered on the west by the cities Enachidia, Goulmima and by a line running from Tinernn toward the south to the national border.

23) According to (DLR-MED, 2005, Resources, p. 60), there are practically no excluded areas. As one can see from current atlases, the land in the south of Morocco and in the relevant part of the former Spanish Sahara is not mountainous, only hilly, with an altitude of apparently somewhat more than 500 m (in any case less than 1000 m).

is probably available. (In addition, these regions each include an area of ca. 10 000 km² in the insolation classes 5.5 and 5 kWh/m² d.) In contrast to the relatively flat southern region, which could be used almost without restrictions, the topography of the terrain plays a more important role in the northern and central regions. Here, the available area might be increased (at 6 kWh/m² d) by allowing greater inclinations of the terrain, for example, by an additional 5000 km². In the central and northern regions, one would then have ca. 10 000 km² with the high insolation of 6 kWh/m² d (along with a further 10 000 km² with >5.5–6 kWh/m² d). Together with the southern region (including the "Spanish Sahara"), one then arrives at a total area of ca. 40 000 km², corresponding to 400 GW, with the high insolation values.

The DLR study (DLR-MED, 2005-Annex, p. A-2) indicates the solar energy potential for only a single year (2002). Whether this year, which in Spain was *especially* sunny, had an insolation above the long-term average also in Morocco is not mentioned there, but this can be assumed in view of the relative nearness of the two countries. The much higher potentials in comparison to the above estimates (many more "good" solar areas in the year 2002) are thus probably exaggerated in relation to an average year. (Thus, in 2002 there were many sunny sites even on the west side of the Atlas mountain range; these are mainly lacking on the NREL map.) At insolations of more than 2300 kWh/m² a, according to this study (recalculated to an area utilization factor of 20% for solar tower plants), there would be possibly 12 400 TWh/a of solar energy available (even without the "Spanish Sahara"!), corresponding to 1410 GWa/a (8.76 TWh = 1 GWa). At a capacity factor of 80% for the solar power plants, this corresponds to nearly 1800 GW. Of this, ca. 85% would be due to areas with an insolation of over 2500 kWh/m² a (in the year 2002). Even though it is not clear just which areas could be expected for the long-term averaged insolation, this study still demonstrates that the potentially usable areas in Morocco are "very large."

The large regions in the south of the country (and also in the former Spanish Sahara) would by the way be suitable in principle for a potential cooling-water supply via transport of *river water* by *sea*. The concept envisages the transport of cooling water from rivers in Central Europe using special, simple transport barges at low cost via the sea to the Moroccan coast. Along with the other advantages of this region, one would have the additional benefit in this case that the reduction in the efficiency of the power plants due to dry cooling (which is in fact typical of desert sites) would be avoided. The region is in principle suitable for this type of water supply, since it is not separated from the coast by mountains and its altitude is not very great (500–1000 m). Insofar as this type of cooling-water transport proves to be feasible in general, it would thus be relevant for this region.

The distance of these southern sites to Germany is ca. 3300 km, rather than the 3000 km assumed in the cost-estimate tables. The transmission losses (without the inverter stations), which would be about 10% for ±800 kV over a distance of 3000 km, would then increase to 11%. Likewise, the cost of the transmission lines would increase by 10%. The investment costs for the whole system would be increased by only 1.8%, and the power cost by only 0.06 ¢/kWh.

Tunisia provides no real alternative to Morocco from the European point of view, but rather at most to Spain. Tunisia is known as the country with the poorest insolation in North Africa (Klaiß and Staiß, 1992). The insolation values in the south of this small country lie in part below those for Spain, and as a rule (according to the NREL map) are only around $5\,kWh/m^2$ d ($1920\,kWh/m^2$ a). There is just one small area with $5.5\,kWh/m^2$ d ($2100\,kWh/m^2$ a). Major portions of the region with this insolation are furthermore sandy desert and thus must be excluded as sites for power plants. The overlap of the NREL map with the exclusion map (DLR-MED, 2005-Annex, p. A-3) indicates that the region with an insolation of $5.5\,kWh/m^2$ d could be around $3000–5000\,km^2$ (corresponding to 30–50 GW); only the region with a lower insolation of $5\,kWh/m^2$ d has an area of possibly $10000\,km^2$ or somewhat more. (The major portion of the remaining territory of Tunisia has, according to NREL, only $4.5\,kWh/m^2$ d.) Concerning the estimation of the potential by the DLR-MED study itself, the same holds as in the case of Morocco: due to the exceptionally good year on which the study is based, the potential was no doubt considerably overestimated.[24]

In the western Sahara, for example, in Algeria, only the best sites are equivalent to the better American sites in terms of insolation. Large areas exhibit insolation values of $7–7.5\,kWh/m^2$ d. These include the northern foothills of the Hoggar mountain range (Ahaggar) in the south of Algeria. As shown by the exclusion map (DLR-MED, 2005-Annex, p. A-4), the usable area here could be $20000\,km^2$ or even somewhat more. To the west and south of this region, there is a large area which still has $6.5–7\,kWh/m^2$ d; of this, at least $40000–50000\,km^2$ should be usable.[25] The DLR-MED study (Annex, p. A-4) finds for the year 2002 once again much higher values: using only the very high insolation value of $2800\,kWh/m^2$ a ($7.5\,kWh/m^2$ d), it arrives at a potential energy production of $30000\,TWh/a$ (30% area utilization!), corresponding to $20000\,TWh/a$ for solar tower plants with a 20% area utilization factor (2280 GWa/a); recalculated to 80% capacity factor, this leads to a capacity of 2800 GW. Allowing a lower insolation ($\geq 2500\,kWh/m^2$ a), the total usable area would be ca. four times larger. For comparison, for the USA, we assumed above a "typical" insolation value of $2500\,kWh/m^2$ a. Thus, in this DLR study, the potential for Algeria was probably considerably overestimated.

The best insolation values are to be found (with the exception of a relatively small area in the south of Libya) in Egypt, east of the Nile. At 7–7.5 and $7.5–8\,kWh/m^2$ d (NREL), large regions are similar to the best American sites. Eastern Egypt

---

24) This study arrives at the following conclusion for Tunisia in 2002: the potential (recalculated for solar tower power plants with an area utilization of 20%) would be 4600 TWh/a (525 GWa/a) for an insolation of over $2300\,kWh/m^2$ a (whereby this value can be up to $2600\,kWh/m^2$ a for certain areas). At a capacity factor of 80% for the solar plants, 525 GWa/a corresponds to a capacity of 650 GW. On a long-term average, the regions with a high insolation would, however, presumably be considerably smaller than would appear on the basis of the year 2002.

25) To the north, according to the NREL map, there is a still much larger region with an insolation of $6–6.5\,kWh/m^2$ d; these are values comparable to those in the south of Morocco. Here, there are practically no limitations for the use of solar plants.

is also a candidate for wet cooling of power plants, using water from the Nile. However, the Nile does not have sufficient surplus water on a year-round basis for the cooling of solar power plants so that wet cooling would in principle be possible during only a part of each year. The best approach would thus be hybrid systems, which could be switched to purely dry cooling when insufficient water is available for wet cooling. This hypothetical possibility needs to be investigated carefully. Whether or not Egypt can be considered for sites to supply electric power to Europe, in particular to Eastern Europe, depends also upon the transmission of power across the Mediterranean Sea via the route Egypt–Crete to the Peloponnesus. Some sections of this route lie at depths of up to 2500 m; the undersea distance is also relatively long (ca. 650 km, depending on the exact route taken). Therefore, economically feasible power transport will be possible only if the cost of the undersea cables at this great depth is not much higher than for the cables in shallower water. The deepest undersea cable yet installed runs from Italy to Greece at a depth of 1000 m. (Comparing the sea depths, Gibraltar 500 m; Morocco–Spain (eastern, direct route) 500–1000 m; Tunisia–Sardinia 1000–1500 m; Tunisia–Sicily 500 m.)

Finally, we wish here to mention Turkey – although it does not lie in North Africa. The available area for solar power plants is difficult to estimate with the present data. There are, to be sure, some regions with high insolations (6.5–7 or 6–6.5 kWh/m^2 d). Since, however, these regions are very mountainous, their useful area is probably rather small. More detailed investigations are required to clarify this question. (The DLR-MED study (2005), with its stringent requirements on the slope of the terrain (≤2%), finds here a vanishingly small potential: with insolations of over 2000 kWh/m^2 a (as in Spain), it would be only 80 TWh/a (9 GWa/a), assuming an area utilization of 20%; this corresponds to a solar tower plant capacity of 13 GW (at a capacity factor of 70%).)

5.7
China, India, and Potential Sites in Tibet – Inaccuracy of the Available Maps

Asia, owing to its large population, the expected broad-scale industrialization, and the demand for air conditioning, will be the great electric-power consuming region in the future. This important region will determine whether solar energy can be *the* major source of electric power, or whether coal and nuclear energy will continue to be required on a world scale for electric power generation. Solar power plants are important for these countries especially because only these would permit a switch from coal burning, freeing the coal which is needed for other uses, rapidly and by their own effort (without imported technology as in the case of nuclear power), to an alternative energy source. Therefore, we consider this region in some detail in the following, even though the data situation is particularly fragmentary here. In addition, there has evidently as yet been no concrete planning for the use of solar energy on a large scale in this region so that much of what we say in the following must necessarily remain speculative.

Particularly in Asia, the reliability of the two solar energy maps breaks down, namely in the important region of western China. Thus, the NREL map (2005, East Asia) or the enlarged version NREL (2005, China) (Figures 5.6 and 5.7) indicates for the sunny Tibet area, which is especially interesting, within the main region between the 80th and the 90th lines of longitude and between the 30th and 35th parallels of latitude, mainly values around $8\,\text{kWh}/\text{m}^2$ d. The DLR map (Schillings *et al.*, 2004a) – cf. Figure 5.8 – gives completely different values there, namely only 3.5–$3\,\text{kWh}/\text{m}^2$ d (at some locations where NREL gives $8\,\text{kWh}/\text{m}^2$ d, it shows only $2.5\,\text{kWh}/\text{m}^2$ d). The discrepancy could hardly be greater. Furthermore, both institutions place the uncertainty of their maps at ca. $\pm 15\%$. Either one of the maps is grossly incorrect, or they are both incorrect to a considerable degree.[26]

The first priority must be to find out which of these conclusions applies. This question could be decided very quickly by immediately installing a few measurement stations in Tibet. Even within the first year of their operation, one would ascertain which of the maps is incorrect or whether both must be revised. In the following, the topic of such measurements will be discussed. (As a preliminary effort, much could be clarified by a more detailed analysis of existing data – see below.)

In the next paragraphs, we will nevertheless consider the situation concerning a possible solar electric energy supply for China, under the assumption that the NREL map is correct – which is in fact quite possible. (If the DLR map were correct, then the Tibetan region would be completely uninteresting as a location for solar power plants!) We find the following situation: Tibet is shown in this map as the only region in East Asia with *very* favorable insolation (cf. Figure 5.6) Given the very large amount of available *land area* (see below), all of East Asia could potentially be supplied with electric power from here (and possibly even all the energy required in the future, by producing hydrogen). This, however, presumes that this very high-altitude region can be developed economically in terms of the other relevant points. These are as follows:

- the ground conditions (the ground on the Tibetan high plateau freezes in winter down to a depth of 3 m (China Physical Atlas, 1999));

- transport facilities (construction time and investment costs);

- the particular characteristics concerning the construction of transmission lines;

- the weather conditions (e.g., hail is much more frequent in high mountain regions than in lowlands; the maximum snow depth, however, appears to be relatively shallow) (China Physical Atlas, 1999).

China, India, and the countries between lie within a circle with a radius of only 3000 km around the Tibetan region (Figure 5.6). Thus, Tibet could in principle be

---

26) Tibet has been repeatedly characterized in the literature as a region with the highest insolation values in the world; so, for example, by Zhou *et al.* (2005, p. 10): "Especially, Central and South Qinghai-Tibet plateau possess the richest solar energy of the world." (This, to be sure, refers to the global irradiation.)

the most significant energy region for this enormous territory of East Asia – at least in terms of the size of its population. It is less than 1000 km from North India, the most populous area in India; however, there are high mountain passes between that would represent barriers to power transmission lines.[27] To a lesser extent, depending on the route chosen, this also holds for transmission lines to eastern and southern China.

If the NREL map is in the main correct, Tibet has not only excellent conditions for solar energy (better than in many locations in the Sahara), but also this advantage is combined with a low air temperature. On the Tibetan high plateau (80th–90th eastern longitude, 33rd–37th north latitude), the average annual temperature is $-4\,°C$;[28] to the south (down to the 30th parallel), it is around $0\,°C$ (China Physical Atlas, 1999, p. 75). With this cool air, dry cooling is no longer a disadvantage, but instead it potentially even has the advantage of a higher efficiency. The latter is, however, true only with special cooling technology, which remains to be developed and optimized for such low temperatures.[29] This advantage could possibly compensate for some of the drawbacks of the region as mentioned earlier.

However, in the case that the solar map of the DLR turns out to be correct, the entire region, with typical insolations of $3–3.5\,kWh/m^2\,d$, would lie well below the values for Spain and thus, in view of its other disadvantages, would be completely useless for solar power generation. The DLR map of Tibet holds, as mentioned, only for the years 2000, 2002, and 2003 (one map for each year; these maps, however, exhibit no noticeable differences). It was not checked against values from a measurement station; that is certainly true of the NREL map (for Tibetan territory) also. We will return to this point again later.

First, we give some data on the availability of land in Tibet in relation to the future energy requirements in Asia:

> As the NREL map shows, the solar irradiation (DNI) in the major portion of western Tibet (roughly bounded by the 80th and the 90th degree of longitude) lies at values of over $7–7.5\,kWh/m^2\,d$, corresponding to

---

27) Especially with power transmission using superconducting cables, mountain passes should not represent a fundamental obstacle.
28) The monthly average for May is 0 °C, in August 6 °C; in the winter it is correspondingly colder.
29) With a conventional steam cycle, lower outside temperatures would not allow a notable reduction in the condensation temperature (owing to the enormous volume increase of the steam with a further decrease). However, since the 1960s (and also in the course of the development of nuclear power) there were plans to combine the steam cycle at its "cold end" with cryogenic circuits. In this case, the waste heat could be extracted at a much lower temperature (well below 0 °C), for example, at night, which would increase the overall efficiency of the process. What advantage this might yield in a concrete case (cost-benefit ratio), and what special conditions would have to be taken into account (e.g., frequent de-icing of the cooling surfaces) remains to be investigated. This would require the conceptual design of such a cooling plant and possibly the construction and testing of individual components (in the form of small test installations which contain all the relevant components as modules and which could be set up and tested in the region of interest).

2650 kWh/m² a. (In more than ca. 85% of this area, the values lie over 7.5 kWh/m² d, some even over 8.5 kWh/m² d.) This region (>7 kWh/m² d) includes in the north the western portion of the Tibetan highlands and in the south (southwest) the high mountain region. A rough indication of the usable portion of this enormous area can be obtained from the newly prepared (1999) edition of the very extensive China Physical Atlas. It shows the different types of terrain. To exclude mountainous regions, the category "Intermountain wide valley or basin" is particularly interesting. At least these "valleys" or "basins" should be suitable for solar tower plants in terms of their topography. A rough evaluation of such regions shows that within the area having high insolation values, a total area of this category of all together 80 000–110 000 km² is available, sufficient for roughly 800–1100 GW of solar base-load power generation. There is, however, no information given about how accurately the atlas reproduces the areas in each case. The greatest portion of this solar region, namely that part belonging to the Tibetan highlands, is denoted in the atlas as "hilly plateau." This region has overall (after excluding the "valleys ...") an area of at least 300 000 km² distributed over a region of roughly 500 km by 700 km (containing also the excluded areas), which would correspond to a generating capacity of ca. 3000 GW if the area were to be fully exploited. Whether or not "hilly plateaus" are generally usable for solar tower plants (or what percentage of them would be usable) could no doubt be rapidly ascertained by exemplary investigations of the terrain on the ground. The *available area* within the Tibetan region would represent a total capacity of over 4000 GW if it could be used unrestrictedly, and with the additional areas at the boundaries of this zone (regions with somewhat less insolation), even somewhat more. This represents, to be sure, an "upper limit." Furthermore, it must first be established whether the entire region can be made accessible to modern transportation. We are thus left with four questions: is the insolation indeed as high as shown on the NREL map? What fraction of the area can in fact be utilized in terms of its topography? To what extent can the region be accessed by transportation? And what limitations will be imposed by ground conditions and other environmental factors?

If it turns out that these points do not impose any serious restrictions, then Tibet would provide a region with the best insolation values, which could supply the electric-power requirements for all of South and East Asia over the long term. By the year 2030, an increase in worldwide electric energy consumption of ca. 1500 GWa$_{el}$ is predicted (starting from the present consumption of 1900 GWa$_{el}$), which will be mainly due to South and East Asia (including India). It is probable that the increases will continue in the ensuing decades. Assuming that India can supply its needs from its own solar power regions, the potential of Tibet should suffice for the rest of East Asia including other energy applications (transportation, heating, etc.). Comparisons: the 450 million inhabitants of the EU (25) in the year 2004 consumed – with the comparatively high energy usage for transportation and the sizeable heating requirements in the cold climate of Europe – 920 GWa of oil and 560 GWa of natural gas, all together around 1500 GWa. And this reflects the present extravagant energy consumption, typical of the low energy prices in past years. For the energy supply of Asia, with a thrifty use of expensive energy, the potential area mentioned earlier should be almost sufficient to cover all the energy needs of the region (excepting India). In particular, in China the area of Inner Mongolia is also available (however with a poorer insolation); in the Northeast alone, counting areas with at least $5\,kWh/m^2$ d according to the NREL map, it includes a region of roughly 300 km by 500 km, that is, 150 000 km², yielding 1500 GW. Furthermore, there are large areas in the West (south of the border with Mongolia and possibly the Zaidam Basin), which however all together are only half as large as those in eastern Inner Mongolia. Furthermore, there are other large areas within the territory of Mongolia. These are, as mentioned, all upper limits, and these conclusions presume the correctness of the NREL maps on which our rough area estimates are based.

Keeping in mind the enormous discrepancy in the data for the insolation in China, we can ask how well the two map systems of NREL and DLR agree for the other regions which are covered by both maps. As stated above: the uncertainty in the solar data is quoted by NREL and DLR as ±15% for their maps. In many regions of North Africa, the maps from these two institutions appear to agree so that the 15% error margins are justified here. In other, often important regions, the differences are however greater, sometimes significantly or even grossly. Here, we must remember that the DLR map for North Africa reflects the data from only one year (2002). In Spain, this map unfortunately cannot be very precisely evaluated owing to the fact that its spatial resolution is too rough and the free-flowing coloration is too heterogeneous. One can at least discern that there, the same regions are denoted as favorable as on the NREL map. In Morocco (precisely in the most relevant regions), the DLR map exaggerates the features relative to the NREL map. In southern Algeria, the eastern edge of Egypt, and in the south of Saudi Arabia, the differences in the two maps are particularly noticeable.

We can, however, expect, or at least *hope*, that the NREL map (in the case that it should in fact be found to be incorrect for Tibet) exhibits large errors only for the *high-altitude* regions. This is indicated by the fact that the other regions where the two maps (NREL, DLR) show gross disagreement are for the most part mountainous; this is particularly clear, for example, in southern Arabia. Furthermore, the agreement of the two otherwise so different maps is very noticeable in individual regions in western China, which are not at high altitudes, for example, at the northwestern corner of the DLR map (along the northern border of the Tarim Basin).

Insofar as at least the regions outside the high mountains are reliably represented on the NREL map, there are still other potential solar regions in China and India, although they are much less favorable in terms of insolation, for example,

Inner Mongolia (within China) and the western arid region of India – cf. Figure 5.6. Here, one would find a situation similar to that in Spain, and the areas with a high insolation are likewise extraordinarily large. Even though large portions of this region in India cannot be used (sandy desert), it would appear that still large areas (with 5–5.5 and 5.5–6 kWh/m^2 d according to NREL) remain, possibly around 100 000 km^2 (this corresponds to, e.g., 300 km by 330 km). Thus, an energy production of 1000 GW appears possible. (The quoted values of 5.5–6 kWh/m^2 d correspond to 2100 kWh/m^2 a. In Ehrenberg (1997, p. 8), for Jodhpur 2290 kWh/m^2 a is quoted, and in DOE (1997, p. 5–37) a value of 2200 kWh/m^2 a is given.) The average distance to the energy-consuming centers in India is only ca. 1000 km. The Chinese province of "Inner Mongolia" has a similar situation; its boundary is only ca. 500 km northwest of Peking. Here, also the potentially usable regions appear to be very large. The presupposition that the insolation data are correct remains to be verified.

If Tibet cannot be used, the energy supply for Asia from a solar region with "very good" insolation values would be possible only from Saudi Arabia. In technical and economic terms, this would appear to be quite feasible, especially if superconducting cables can be used. (In that case, distances of even 6000 or 8000 km would be economically possible, which of course would not be necessary for the energy supply to nearby India.) However, in many regions of Saudi Arabia, there are large differences between the two maps. The Southeast of the Arabian Peninsula, which faces India, would be of primary interest. The good insolation values shown here by the NREL map (2005-Africa) (Figure 5.3) are not found on the DLR map.[30] The power would thus – in the distant future – possibly have to be transmitted from the northwest of the Arabian Peninsula or from Egypt, which has a nearly inexhaustible potential for solar energy. If superconducting transmission technology develops as expected, this would be feasible, from a purely economic standpoint.[31]

5.7.1
Conclusions

The uncertainties are too great to allow a reliable picture of the potentials of either China or of India to be established on the basis of the currently available insolation maps. The maps thus far published claim an accuracy which is clearly unrealistic.

---

30) DLR map (for 2002) with a large section (including Saudi Arabia); not reproduced here – in (DLR-MED, 2005, Summary, p. 10).

31) The future political situation is another question. Iran and Pakistan can, however, potentially be circumvented by means of an undersea transmission line so that India could be connected directly to Saudi Arabia via Oman. The distance from the south of the Arabian Peninsula to the center of India is ca. 3000 km. If the overland route for transmission lines through Iran and Pakistan is not usable for political reasons, an undersea line ca. 850 km long with a maximum depth in the range of 2000–4000 m would be necessary. These are somewhat more stringent requirements than for a sea cable from Egypt to Greece.

Table 5.2 Calculated values (Global Horizontal Insolation (GHI)) from NREL (2005) and DLR (Schillings et al., 2004a) in comparison to individual measured values in Tibet.

Location	Longitude (E)	Latitude (N)	Global insolation Zhou et al. measured value (MJ/m² a)	Global insolation Zhou et al. Measured value kWh/m² d	Global insolation NREL computed kWh/m² d	Global insolation DLR (for 2003) computed kWh/m² d
Shiquanhe	ca. 80°	ca. 32.5°	(7808)	5.9	5.5–6	5–5.5
Ngari	ca. 81.5°	ca. 32.5°	(7925)	6.0	6–6.5	4.5–5
Shigatse	89°	29.3°	(>7500)	>5.7	5–5.5	4.5–5
Lhasa	91.3°	29.8°	(7784)	5.9	5–5.5	4–4.5
Nagqu	92°	31.5°	(6557)	5.0	4.5–5	3.5–4

These uncertainties could, however, be eliminated to a large extent within 1 to 2 years, and completely within only a few years. The questions of possible restrictions due to other criteria (topography, condition of the ground, availability of transportation, special aspects of cooling technology, routing of the transmission lines) must first be more carefully investigated.

Note on the problem of transportation:

In Tibet, railroad lines are an option only in exceptional cases (for a few main lines), owing to the topography, the properties of the ground, and the necessity of rapid construction of transport routes. Here, one must keep in mind that transportation routes must be provided to the often-isolated sites of solar plants. There is a rail connection to Lhasa (from the north). Transport will thus mainly be over roads, which must be constructed or provisionally improved (possibly unpaved roads). With today's transport vehicles, this is in principle feasible.

However, there should be more economically favorable solutions, which are especially optimized for the particular transport problems associated with the construction of solar plants. Considering the high volume of material to be transported in a short time (due to the rapid construction rate of the power plants), special transport vehicles might be used, similar to the "long trucks" in Australia (which were developed especially for the particular transport conditions found there), with several trailers. Since what is required is a direct connection (in contrast to a branched network of roads), these lines could even be electrified, like the bus lines in many cities. The trailers could then be powered on each axle. This would allow convoys with more trailers than are used in Australia, even on roads with steep grades. If the trailers could in addition be steered automatically, for example, guided by the overhead electrical lines or on a dedicated electrical signal line, then relatively long convoys would be possible without problems in negotiating curves. Such convoys would be similar to railroad trains. The development of such powerful vehicles, which would be capable of following curves on a mountain road with a minimum of personnel and at the same time would be energy efficient (through their electrical drives), can be considered to be a branch of the development of solar power plants, for sites in western China also. It should be possible to rapidly estimate the costs of such "trailer trains" roughly, since only certain key components (automatic steering, electric drive, and electrical contacts) would need to be designed. A simple restructuring of existing truck-trailers would allow tests in practice. An important point is that here, in contrast to rail lines, we would not be dealing with a long-term infrastructure item, but rather with a high-capacity transport connection, which would be required only during the relatively short period of power-plant construction. Similar considerations hold for the bridge construction, which would be initially set up as efficient two-lane temporary structures. Later, these could be replaced by permanent structures (e.g., single-lane bridges). Especially in Tibet, where there are extended areas (sufficient for several GW of power generation), widely separated between mountain ranges, the development of road vehicles of this type would be particularly important. If the highlands were to be provided with solar plants according to a fixed plan, one would have sufficient time for constructing the transport lines needed. At the beginning of this construction phase (when the preliminary preparation time would be short), the construction would be concentrated in the nearest areas so that the transport lines would be comparatively short.

5.8
Insufficient Accuracy of the Insolation Data; Measurement Program

Given the discrepancies between the solar maps of NREL and DLR, which is inexplicably large in particular in western China, we consider here some possible reasons for the inaccuracies in more detail.

Satellites yield only data which can be used to calculate the insolation on the ground: the ozone content of the atmosphere, cloud cover, humidity, aerosol content, as well as the ground and air temperatures (Schillings et al., 2004a). (For the NREL maps, these data are then combined with weather data which are obtained from a network of meteorological stations: among others local atmospheric haze, humidity, aerosols.) From these data, taking into account the altitude at each location, the direct solar radiation level on the ground is computed. The main problem is the quantitative consideration of the effects of clouds and, for certain sites, the atmospheric haziness (aerosol content); the remaining values evidently give rise to only small errors.[32]

The *real cause* of the inaccuracies in current insolation maps is, however, the lack of measurement stations at which the computed results could be compared with direct measurements. NREL and DLR both point out that there are too few measurement stations and that every additional station would be welcome (e.g., Stoffel and George, 2007, p. 55; Schillings, 2008). Thus, in the description of the computational techniques, it is emphasized over and over (e.g., in Hoyer-Klick *et al.*, 2006, p. 27): "The information basis for the available solar resources is still relatively poor." (Note: In Hoyer-Klick, this includes also the estimates of the available suitable ground area.) And: "At present, much of the information which determines the availability of solar radiation is not at our disposal with sufficient precision. This is especially true of aerosols, which have clear-cut effects on the direct insolation. Likewise, there is a need for more precise descriptions of the optical properties of clouds and their influence on the insolation."

On the precision of the computed NREL data, Stoffel and George (2007, p. 17) write, evidently with respect to the calculated and measured values in the USA[33]:

Estimated Uncertainties (optimal)

	Global/diffuse	Direct normal
Calculated (SUNY model)	±8%	±15%
For comparison: measured	±6%	±5%

Suboptimal values can range up to 25% based on ASOS cloud observations used in METSTAT, snow cover, or high-latitude locations for "SUNY model."

---

32) According to Schillings *et al.* (2004b, p. 4), the various factors act as follows:

Ozone (absorption), ca. 1%;

Air molecules (scattering, absorption), ca. 15%;

Aerosols (scattering, absorption), on average ca. 15%, maximum up to 100%;

Clouds (reflection, scattering, absorption), maximum 100%,

Water vapor (absorption), ca. 15%.

In addition to cloud cover, the aerosol content in the atmosphere thus also plays an important role. This effect is also present in a solar tower plant along the light path from the heliostats to the receiver. It is, however, in general not included in the insolation data. A turbid layer of air near the ground can have a noticeable effect on the light yield (i.e., the radiation density at the receiver). (For parabolic-trough plants, this effect is unimportant due to the short distance between the mirror surface and the absorber tube.) The measurement program should, therefore, include this ground-level effect, especially at locations which are subject to aerosol exposure. This could possibly be the case in some regions of Tibet (wind-carried dust), although the data for aerosols of Schillings *et al.* (2004a) do not indicate this (the map shown there is, however, only for February). Aerosols thus cannot be the cause of the large discrepancies between the two solar maps of Tibet.

33) On the NREL homepage (NREL, 2004) under the topic "How the maps were made," the error limits of ±10% evidently refer to the global irradiation.

The DLR likewise quotes errors of roughly ±15% (see below), whereby this limit can be noticeably higher, however, within *localized* regions due to microclimatic influences. The discrepancy between the maps from the two institutions is, however, much greater at many locations (on a macroscopic scale) than it should be, given the quoted error limits.

How do the DLR values compare with the results of measurements?

The values in Spain were checked at seven measurement sites. Since the insolation values from these (privately operated) measurement stations are not publicly available, one cannot obtain much information about the comparison. The computed values were also tested at a site in *Morocco* and the result is given in Schillings *et al.* (2004b, p. 20). There, the completely inadequate precision in the presence of *partial cloud cover* becomes clear. With a measured hourly insolation (DNI) of $0.4 \, \text{kWh/m}^2$ h – typical for partial cloud cover – the computed values (for the many hourly measurements given) range from 0 to $1 \, \text{kWh/m}^2$ h, and not even a maximum frequency of values near $0.4 \, \text{kWh/m}^2$ h can be discerned. Precisely with cloud cover – and this is an important situation – there are apparently *very great* uncertainties in the computed values. Only when the sky is clear, do the computed values agree with measurements. Thus, Schillings *et al.* (2004b) make no conclusive statements about the uncertainties (only the accuracy of individual daily values is shown graphically). This is also true for the investigation of western China (Schillings *et al.*, 2004a), which is important for Asia (here, as mentioned, no comparison with measurements was possible), and for the other publications of the DLR.[34] Only in the work of Brösamle *et al.* (DLR, 2001) can one conclude that the uncertainty lies in the range of +5% to −15%.[35]

Nonexistent (or too few) ground stations are *the* problem worldwide, even in the USA. Thus in 2007, there was not a single measurement station in Arizona, and only one each in California, Nevada, and New Mexico, in Utah two (measurement of the DNI (Stoffel and George, 2007, p. 12). In Morocco, there is, to our knowledge, also only one station, as already mentioned, which by the way is not

---

34) Also in the newer Mediterranean study (DLR-MED, 2005, Resources), there are no statements about the uncertainties. There, only the mean square deviation is quoted (p. 59), "which is usually on the order of ±5%."

35) Only Brösamle *et al.* (2001) provide some conclusions about uncertainties: the comparison with measured values from sites in Spain (Almeria) showed in sum good results, "but unsatisfactory for cloudy days." Further but less intensive comparisons were carried out at one measurement station each in Morocco, Jordan, and Egypt: "... compared to our results showing differences from +5% to −15% with respect to the annual totals of DNI." (Here, it is not made clear whether the agreement found at a few arbitrarily chosen stations should be taken as a general statement about the errors in the computational procedure.) The presumed small error for the values from 1998 is, however, in contrast to the later values computed for 2002 and quoted in DLR-MED Resources (2005, p. 59). The distribution of good sites within a country is in part very different from that shown on the map of Brösamle *et al.* (For example, the values for the south of Morocco and in western Algeria were completely different.) Even considering that these values refer to two different years (1998 and 2002), this discrepancy is sizeable and indicates a larger error margin than claimed.

located within the typical solar region. In India also, there is probably only one station for DNI measurements, namely in Jodhpur – in any case, in Ehrenberg (1997, p. 8) for Jodhpur 2290 kWh/m² a (DNI) is quoted; and all together only 14 stations for the global insolation (Maxwell, George, and Wilcox, 1998). A similar situation holds for China. Thus, the DLR map for western China could not be compared with measured values, since there are evidently no measurements of the DNI. (In China, there is only a series of stations for measuring the global insolation, and in the 1990s in Tibet, there were only four (Zhang and Lin, 1992, p. 189); today, there are a few more, see below.) Zhang and Lin (1992, p. 189) also point out the necessity of a separate computation of the direct solar irradiation at high altitudes (such as Tibet), since the conventional computational methods are not applicable here: "... conventional formulas applicable to the plains would not be adaptable.")

There is thus a blatant lack of fundamental information. This also points up serious *structural* defects in the organization of solar research (as can also be seen from many other examples): it becomes clear that research up to now was (almost) exclusively acquisition-oriented. (An acquisition-oriented behavior of the research institutions, however, unfortunately does not lead to research and development which are adequate for solving the current energy and climate problems.)

Concerning the especially striking discrepancies in the two maps with reference to Tibet, we can add the following: NREL and DLR have not only published maps showing DNI but also the global horizontal insolation (GHI). Here, the DLR values are also lower, but the differences are considerably smaller than for the DNI. The global insolation maps can even be compared with occasional measured values, which however became available only after the publication of the DLR maps; they are quoted in (Zhou et al., 2005, p. 11) for Tibet. In Table 5.2, these measured values are compared with the computations of NREL and DLR. One can see that the NREL data show better agreement with the measurements. (There is a surprise in comparing the DNI maps of NREL (Figure 3.9) and DLR (Figure 3.10) on the one hand with the corresponding GHI maps (not shown here) on the other: NREL gives higher DNI than GHI values (as would be expected in the ideal case), while the DNI values are lower in the DLR data, in some cases considerably.)

On the whole, we can make the following statements about this unexplainable discrepancy:

- A scientific discussion between the two institutions NREL and DLR about its causes has not taken place (Schillings, 2008). The topic has indeed been briefly addressed (in more or less incidental conversations), but not in fact analyzed (Schillings: "That is also a question of a budget for this purpose.") Only generalized explanations were offered: differences in the input data and different computational methods[36] (and of course insufficient opportunities for comparison with measurements). As a whole, the participants were more

---

36) It is emphasized over and over in the descriptions that the differing computational procedures yield in part very different results, even with the same input data.

or less perplexed. (Schillings: "In the final analysis, we need ground-based measurements.")

- No knowledge of which (possibly singular) principal factors gave rise to the low values computed by the DLR is available. As we have already mentioned, the aerosol concentration cannot be the cause, as the corresponding map presented by the DLR confirms (Schillings et al., 2004a, p. 5), although this map again refers only to the month of February. Schillings (2008) also did not indicate that this could be an important factor.
- The differing discrepancies in the direct and the global insolations were evidently also not considered.

This debate – in the sense of a systematic cause-and-effect analysis – should be carried out first of all. It could clarify much in a short time, perhaps even the real cause of the discrepancies, and thereby would lend support to the reliability of one or the other of the maps. The relatively minor resources which would be required to carry this out should be made available in a nonbureaucratic manner.

- The DLR points out the higher spatial and temporal resolution of its computational method. A higher resolution with smaller pixels would, however, improve the analysis only within this area, in comparison to a lower resolution with larger pixels, but it could not explain differences in the *average* computed values of the larger pixels; in the NREL maps, an "averaging" is carried out over larger spatial localities and longer time intervals. The reason for the poorer resolution of the NREL computations is the fact that they are not based on data from a geostationary satellite, but only from one in a polar orbit (large pixels and only three passes daily over a particular region). (Note: in the course of international cooperation, it should be possible in the future to organize an exchange of data.)
- Clouds. A satellite measures only a gray-scale value. Two problems make the interpretation of a gray-scale value in terms of cloud cover difficult: for one thing, the background plays a role – in Tibet, in particular, snow has to be taken into account, while even light desert sand presents a certain problem. Secondly, the transmission of solar radiation through clouds is strongly dependent on the cloud type. Different types of clouds can lead to the same gray-scale values, although they may have quite different transmission properties.
- In the Far East there is a new phenomenon, which in future will have to be considered: the "Asian brown cloud," the problem of large-scale air pollution (high aerosol concentration), which has appeared in recent years over large areas of Asia. This Asian brown cloud is not yet included in the data of the DLR, which date from 2003, and evidently also not in those of NREL, which were published in 2005. To what extent this will affect the Tibetan region remains uncertain. However, at least for the solar regions in India and in China ("Inner Mongolia"), it will be of importance.

To clear up the remaining uncertainties, only a few additional measurement stations will not be sufficient; rather, it will be necessary to set up a whole *network of measurement stations.* This network must cover all the relevant solar regions of each country and encompass every smaller region with special microclimatic properties. Such networks must in addition cover all the regions worldwide which are under consideration for the construction of solar power plants. Only such a measure will do justice to the present situation, which demands rapid decisions in all countries about the course to be followed in the coming years. It is in the interest of all nations that they – and the others – must use fossil energy sources to the smallest extent possible for generating electric power in the future.

Concerning the question of timing, the following must be considered: newly constructed measurement stations collect the relevant data only year by year; they thus do not allow immediate long-term data series, as can be obtained from satellite-supported computations. But even the measured values from the first year are very significant; they allow not only a test of the precision of the previous computed values from satellite data, but also a calibration, that is, the correction of future computations (individually for each measurement site).

If a deviation is found in the first year (or in the first 2–3 years), which may show seasonal variations, then the computed values can be modified in the simplest case by a corresponding correction factor for each location (taking the seasonal variations into account). (In this sense, at least a rough calibration could be carried out.) The measured errors would then be corrected globally on a *seasonal basis*, which would already represent a definite improvement.

The values could also be corrected individually with respect to certain reference data. The main cause of the uncertainties is as mentioned the cloud cover. In the interpretation of the satellite-based cloud data starting from the gray-scale values, the result depends decisively on the optical background, in particular with partial cloud cover. This background, however, changes during the course of a year, for example, with changes in the vegetation. (Plant growth affects all regions except for purely desert sites, especially Spain but also many areas in North Africa such as in Morocco.) The landscape (which forms the background for the satellite images) changes not only with the seasons but also depending on weather (rain or snow) and the temperature changes during the year considered at each individual site, in a characteristic manner (again in dependence on the altitude of the site and its basic appearance, e.g., the lightness or darkness of the ground and the shape of the terrain). The first year (or the first 2–3 years) of observations would already yield this characteristic appearance (corresponding to the seasonal status of the *vegetation*); and at the same time would give data on the amount of solar radiation actually reaching the ground for a particular gray-scale value (cloud cover) against the background coloration at the selected site. This would thus allow a calibration of the computations according to the appearance of the *vegetation-dependent* image background (in contrast to a calibration by season alone).[37] In the

---

37) The appearance of the landscape can be directly determined from satellite images on clear days. The state of the vegetation can also be derived at each site from the season (with corrections for the amount of water available and the temperature progression of the particular year), oriented on the known appearance from the previous year.

special case of Tibet, with its low temperatures, the interpretation of the gray-scale values would – in addition to changes in the vegetation – also need to include snow cover and possibly also ground frost. There, the reference values for the vegetation background would have to be combined with information on snow cover and ground temperature (whereby the data for the latter two quantities could also be obtained from satellite observations).

This refined procedure thus makes use of *reference values* (measured values of the light transmission for a particular gray-scale value over a characteristic image background), which are employed for individual corrections. It should thus provide a notable improvement with respect to global (date-dependent) correction. (However, the influence of different cloud types – with the same gray-scale value – could not be included.) Thus, only a few years of measurements should suffice to improve the computations based on satellite data, by means of such a differentiated calibration *in retrospect* over the years of the satellite observations.

Furthermore, one would know exactly just how precise the computed series of values are, since the computed values could be compared year by year with the measured data from each individual site.

Within a total of ca. 4 years – including 2–3 years for setting up the apparatus and 2 years of measurements – one would thus have obtained considerably more precise data. At least then for some individual representative sites in every country, long-term values could be computed retroactively (even if due to the short time and the continuing installation of measurement stations not yet for the whole country). The installation of the stations would be completed soon after in all the regions of interest, and every year thereafter, new measured values would be obtained.

Measurement apparatus for the DNI are comparatively simple devices. The sensor tracks the sun, similarly to a heliostat (cf. Figure 5.13). A measurement station can however include – along with the DNI apparatus – additional measurement devices. It would be reasonable to set up different *classes* of measurement stations with different instrumentation:

1) Simple stations with just one device for the DNI. They could be set up in large numbers.

2) Stations which in addition can measure the global and the diffuse insolation.

3) Stations with all the relevant measurement instrumentation (Schillings et al., 2002).[38] With these stations, in particular the turbidity of the lower atmospheric layers could be determined from the ground (horizontal absorption measurements). Such stations would yield more information, useful for calibrations.

---

38) Schillings et al., (2002): "Each of these cross-validation sites consists of a fully equipped ground platform including global and spectral, direct and diffuse radiometers, all-sky camera and LIDAR for a fully automated, real-time state of the sky monitoring, sunphotometer, nephelometer for ground visibility measurements and standard meteorological measurement."

Figure 5.13 Measurement of the direct normal insolation DNI. The pyroheliometer sensor tracks the path of the sun (NREL).

To improve the calculated values, it would however not be necessary to equip each measurement station with all possible instrumentation. A small proportion of "completely equipped" stations (class three) would yield very valuable information for the correction of computational programs in relation to the local situation within the larger region under consideration. It is evident that a direct measured value (DNI) already contains the most important single piece of information. It can, therefore, be assumed that only a *very* small proportion, for example, 10%, of all the stations would have to be of class three.

Currently available devices for DNI measurements cost roughly US $15 000.[39] (Here, the very small number of such instruments produced to date – more or less single-item production – must be considered.) There is, however, a still simpler device for determining the DNI. It uses a single sensor (employing a rotating shutter to block the sensor at regular intervals) to measure both the global and the diffuse insolation so that from the difference, the direct insolation can be computed (rotating shadow-band radiometer (RSR)) (Figure 5.14). This apparatus is less costly than those for DNI measurements alone, and it measures the global and diffuse insolation at the same time, is less sensitive toward contamination,

---

39) Cost of measurement devices according to Rosenthal and Roberg (1994, p. 4) in 2006-$ (US inflation 1994–2006: 1.36):

Global horizontal:	$2 700
Diffuse horizontal:	$4 800
Direct normal:	$13 800
Compare RSR apparatus: global, diffuse and direct normal:	$7 800

(RSR: rotating shadow-band radiometer).

5.8 Insufficient Accuracy of the Insolation Data; Measurement Program | 179

Figure 5.14 Simple measurement apparatus for global, diffuse, and direct insolation: a rotating shadow-band radiometer (RSR) (NREL).

and has longer technical maintenance intervals (for recalibration).[40] However, it is apparently not very precise for measuring the diffuse radiation. It is thus evidently less precise particularly in the presence of partial cloud cover (the most important source of error in the presently available data) for determining the direct insolation (but this imprecision is possibly not too great and can be further reduced (Wilcox *et al.*, 2005, p. 2)).

However, this instrument also has thus far no automatic cleaning function. Thus, Schillings (2008) remarked in regard to the problem of contamination: "The quality of the measurements must also be kept in mind. Ground-based measurements are reliable only when the measurement stations are well maintained. The great effort thus far required results from the necessity of almost daily cleaning." See also (Stoffel and George, 2007); they quote (on p. 17) the errors of ground-based measurements (DNI) as (optimally) ±5%.

With the generally relatively low cost of the instruments, which would decrease still further if they were produced in large quantities, it would not be necessary to use only the cheapest apparatus mentioned above. A more important factor would

---

40) According to Rosenthal and Roberg (1994), the RSR device would have technical maintenance intervals twice as long as the other instruments ("for recalibration"), namely 2 years instead of 1 year.

be the development of automatically operating stations to reduce the personnel costs, that is, in any case of apparatus with automated self-cleaning.

If one assumes that for large-scale production, the cost of a DNI instrument including automatic cleaning would be around $15 000, worldwide, for example, a total of 3000 measurement stations for the DNI could be set up at an overall cost of probably less than $100 million, the price of a large military aircraft. A total of 3000 stations for $15 000 each would cost $45 million. For the costs of the (comparatively few!) stations of class 2 and 3, which would be more completely equipped, and for computer programs for data processing, ca. $50 million would still be available, without exceeding the arbitrarily defined sum of $100 million.

With 3000 stations worldwide – this number was not justified in detail – it should be possible to cover all the relevant sites on the Earth, including the many areas with distinctive microclimatic features within larger regions. The personnel costs for operating the apparatus were not estimated here; but, assuming that the stations were automated to a large extent, the personnel expenditures should not be all that great. In that case, the operating costs over several years should not exceed in total the acquisition costs for the apparatus ($15 000/station). The low labor costs in many countries (exceptions are, e.g., Spain and the USA) are also to be kept in mind here.

Taking into account the individual information from the separate stations, the computing effort would certainly be much greater than for the current global computations. Corresponding costs for the extension and refinement of the computer programs must, therefore, be expected. Given the computing power presently available, this would be more an organizational problem than a financial one (immediate operational availability of a correspondingly larger personnel capacity, which however would be needed for only a few years, e.g., through cooperation with universities).

With this measurement program, and the resulting more precise computation of insolation values from satellite data, we would finally have a clear picture for all the relevant regions, worldwide. The data from a few reference areas in each region would be evaluated first of all. With a correspondingly large-scale application of computing power, it should then be possible to evaluate the data for all the particular microclimatic areas that would have to be treated individually (i.e., all the 3000 stations assumed here) in a relatively short time, in each case for a long-term series of data points. We would then quickly obtain an optimum amount of information.

One may certainly assume that with the hypothetical 3000 measurement stations, the whole world would be adequately covered (possibly many fewer would suffice). In the case that, contrary to expectations, still more stations were required, these higher costs would have to be borne. Then the price of, for example, *two* "large military aircraft" would have to be expended.

In any case, we need reliable data soon; the energy-political decisions must be made in the near future. To this end, we must know what solar energy can deliver in terms of the given geographical preconditions.

6
Heliostats

In discussing the cost of the most characteristic component of a solar tower power plant, the heliostat field, we employ the same arguments as were used in a previous study (Kalb and Vogel, 1998). We arrive at almost exactly the same cost estimates for heliostats as were obtained in a recent US study; we also compare with the extensive report of Sargent and Lundy (2003) on solar thermal power plants.

With this somewhat retrospective approach, we want to make it clear that the renewed interest in solar thermal power plants that seems to be developing on the current political scene cannot be understood as a praiseworthy reaction to drastic (and unexpected) improvements in the cost perspectives, not even those for heliostats. Rather, the cost-reduction potential for heliostats under mass production and with appropriate research and development has been practically ignored by the relevant political institutions, especially in Europe. This was also the case in latter years in the USA, although there in the 1970s and 1980s, parallel to systematic development of the technology, the first large-scale cost studies were carried out. Subsequent US governments apparently ignored their own studies.

In Europe, throughout a period of 30 years, practically no government-supported or even government-initiated systematic heliostat development (with the possible exception of CIEMAT in Spain) was carried out. In particular, individual initiatives by private companies continued the development to a modest extent (but nevertheless with often noteworthy results). Those responsible in the relevant ministries and parliamentary committees refused to appropriate the necessary research funds. This was typically accompanied by the remark that no "market" was available for solar power plants. At the same time, however, it was emphasized that a market was not available because the heliostats were too expensive. This is a self-inflicted vicious circle that was invoked by the political decision-makers.

6.1
Estimating the Heliostat Costs

The cost estimate that we present here is based in the first instance on a publication dealing with heliostat production in very small numbers for first applications,

Large-Scale Solar Thermal Power. Werner Vogel and Henry Kalb
© 2010 WILEY-VCH Verlag GmbH & Co. KGaA, Weinheim
ISBN: 978-3-527-40515-2

Table 6.1 Early heliostat cost prognoses (Sandia) for a production rate of 50 000 heliostats/a.

	1986-$	1995-$
Stretched-membrane[a] (Solar Kinetics, Inc.) ($/m^2)	42	58
Stretched-membrane[a] (SAIC[b]) ($/m^2)	47	65
Conventional heliostats ($/m^2)	56	78

a) As a result of later investigations, these cost estimates for stretched-membrane heliostats are generally considered to be too optimistic. Also, different methods of cost estimation led to different results.
b) Science Applications International Corp.

namely, 2500 heliostats per year (Kolb, 1996a). The small-batch price for glass-mirror heliostats would be 138 $/m^2 (2002-$), according to this article.

For a large-scale scenario, however, the production rates would be considerably greater. To permit the annual installation of 1000 MW of solar base-load generating capacity (with a solar multiple of 4.4), the production of ca. 130 000 heliostats per year (each with 150 m^2 of mirror area) would be required. By "mass production," we thus refer here to production rates of the order of well more than 100 000 items per year. For such production scenarios, no really *detailed* cost analyses have yet been published. We therefore take the same approach to arrive at a first approximation as in 1998, as mentioned above. At that time, several different publications were available that gave long-term (more precisely, referring to a mature state of development) cost perspectives (these were, however, often just termed "cost goals").

6.1.1
Examples

Winter (DLR, Germany) gave a cost estimate in 1991 for conventional heliostats of roughly 60–80 $/m^2, and for the stretched-membrane heliostats, discussed below, of 40–60 $/m^2 (Winter et al., 1991).

Mavis (1989), in a Sandia Report (cited by Kolb et al., 2007a), quoted costs (for production series of 50 000 heliostats/a) as given in Table 6.1.

In a *US-DOE* Report, in 1997 (for 50 000 heliostats/a) as a "rough estimate," ca. 70 $/m^2 (1995-$) was quoted (DOE, 1997).

Independently of this latter report, we assumed heliostat costs of 70 $/m^2 in 1995-$ as the result of a qualitative argumentation. Recalculated to 2002-$, this gives the value that we use in this book, that is, 83 $/m^2. (For comparison: the production cost of mirror glass in mass production would lie in the range of 10 $/m^2 (S&L, 2003, p. E-34, and Kolb et al., 2007a, p. 49).

A number obtained in such a manner can, however, only serve as a rough indicator. The production costs of heliostats must be determined precisely by detailed investigations within the broad-based development program advocated here. For the well-founded economic evaluation of a large-scale future solar technology, the use of so-called learning curves (as is done to some extent in the prognoses of SunLab and S&L) can provide only a provisional method for obtaining rough estimates; these results will always be subject to controversial discussions.

In the following text, we adopt our earlier general argumentation for justifying the deviation from the original Sandia estimates (Kolb, 1996a). The assumed price reduction for conventional heliostats with a mature technology and mass production (from $138\,\$/m^2$ for 2500 heliostats annually (initial price) to $83\,\$/m^2$ for production of ca. 100 000 heliostats annually, mature technology) would appear plausible based on the following considerations:

The previous cost estimates – for single items or small production series – are derived from current prices for rapid delivery. For large-scale series, prices that are nearer to the production costs are relevant. With continuous and predictable installation of large solar parks over a longer period, at a constant and high rate of construction and with long-term contracts, the production companies can set up new facilities dedicated to these production series. The usual price surcharges for variations in capacity utilization will not apply. Likewise, surcharges for development, marketing, and sales costs, which can increase costs for small production series, will be inapplicable.

For the fabrication of the mechanical components – for example, welding and assembly of the reflector framework, fabrication of the gear boxes and positioning motors – in mass production, a high degree of automation can be presumed. Such a nearly personnel-free production is already common today in automobile factories. Not only the fabrication of the individual parts but also the materials flow, assembly, and painting are largely automated. Compared to previous cost analyses, especially for assembly technology, considerable progress in the use of robots has been made, and this development will continue in future. In particular, for the nonmaterial-intensive, but rather fabrication- and assembly-intensive components such as power trains or electric motors, with mass production in the future a further cost reduction will be possible.

Cumbersome parts such as the reflector framework could be produced in factories built nearby to large solar parks concentrated in a particular region, to eliminate transport problems; for small solar power plants at remote sites, of course considerably higher transport costs will have to be borne.

In the past cost analyses, the price of control and regulation (electronics) was a major item, which today is already much cheaper. In order to specifically reduce this cost item (as well as that of the drive units), the heliostats can be made larger in comparison to earlier constructions. In general, owing to the continuing dynamic development of microelectronics, we can presume a continuous price decrease. Compared to the 1980s, the situation has already changed drastically. This holds also for signal transmission and energy supplies, for which one has today fundamentally different technical possibilities from those of 20 years ago.[1]

---

1) In Spain, in recent years, the concept of an "autonomous heliostat" has been developed, which is self-sufficient in terms of its control and especially its power supply (Garcia *et al.*, 2003, 2004). Whether this concept can in fact be realized is still uncertain and depends among other things on the reliability of the batteries.

Assembly in the field could also be highly automated in the case of a large-scale plant construction scenario. For this purpose, transportable rigs for partially automated assembly can be envisioned.

On the question of *heliostat size*, we note that the tendency to larger and larger heliostats is not completely unequivocal. It would appear that a size of the order of 150 m^2 is generally seen as an optimum, but other sizes are often mentioned, some considerably smaller, for example, around 90 m^2, but also some substantially larger, for example, 300 m^2 ("mega heliostat"). In some of the newer projects, extremely small heliostats are planned, for example, with an area of only 7 m^2 (LUZ II, 2008). This shows, on the one hand, that the development process is still not at an end. On the other hand, this broad size range is a result of the fact that the optimization of the heliostats depends on many boundary conditions, such as the overall capacity of the power plant, the total number of heliostats being delivered, their production rate, and the transport situation, to mention only a few.

Kolb *et al.* (2007a) in their most recent heliostat study discuss the reasonability and necessity of very large heliostats, whereby they emphasize in particular that the main point is to spread the relatively high cost of the drive units over as large a mirror area as possible. They point out, however, that smaller heliostats have thus far simply not been tested intensively. S. Jones (like G. Kolb also at Sandia) discusses the topic of heliostat size extensively in an article that is attached as an appendix to the recent study by Kolb mentioned above (Jones, 2000).

6.1.2
Preliminary Conclusions

Even if a price of 83 US $/m^2 cannot be attained for conventional heliostats, there remains the possibility that this price can be reached for other types, which we describe below. The probability is thus high from today's viewpoint that at least with *one* system, this cost level can be attained. As an upper limit for mass production with present knowledge, a price of 100 $/m^2 (2002-$) seems plausible. With this value, 17 $/m^2 higher than our basic assumption for the cost of the mirror field, the investment costs of the solar power system (with respect to, e.g., 1000 MW in Central Europe after power transmission) would increase by $400 million, and the power cost by 0.25 ¢/kWh. The cost level would thus not be fundamentally changed.

In the last 10 years, a significant development effort was evidently made only for conventional heliostats (particularly in Spain). The high-risk development of light-weight heliostats could not be pursued to anywhere near the necessary degree due to the lack of governmental support.

From today's point of view (and even from the viewpoint of the 1990s), one can or could consider it quite possible that the costs of glass-mirror heliostats with future mass production rates will be in the range of 80–90 $/m^2 (2002-$). With a high probability, in our view, at least one of the two systems (conventional or membrane) will be available within this price range. If it turns out that the development of membrane heliostats is truly successful, this price could even be markedly reduced.

SunLab (USA) considers a price of around 75 $/m^2$ to be possible in the long term at the production rates considered here (see Section 6.5.1). Uncertainties will, however, remain until detailed investigations are carried out.

6.2
Necessary Measures for the Precise Determination of Costs in Mass Production

To clarify the uncertainties described above, extensive investigations are required; these must be carried out mainly by companies that have experience in cost analysis for mass production, for example, by automobile producers. As already discussed in Section 2.4, for each design variant of a particular type of heliostat, the costs under mass production must be obtained, in an interplay with parallel operational testing (see below), both for the fabrication of individual parts and for the assembly procedure. Such a cost determination is typical of the automobile industry and in certain cases even necessitates the conception and construction of prototype fabrication facilities. In the case of heliostat development, this would include, for example, the laser welding equipment for metal-foil heliostats (cf. the following section) or splicing rigs for assembly.

Since there are various types of heliostats, each with numerous possible variants for construction methods, this results in an extensive program of investigation. These cost analyses must therefore be considered to be an autonomous field of investigation, as is usual for the analogous tasks in the design sections of automobile producing firms. The costs of this important field of research would probably lie, depending on the particular type of heliostat (including individual developments in fabrication technology), in the range of a few million dollars up to as much as several tens of millions of dollars per heliostat type. The overall package (including all the heliostat versions) would thus probably cost well over $100 million (possibly as much as $200 or 300 million); it is, however, conceivable that it would be under $100 million.

It is not to be expected that automobile producers or other institutions will carry out such elaborate investigations on their own initiative, although the sum involved, compared with the usual standards for energy research, is not large. Without a radical change in energy policy, there would be no apparent incentive for them, in particular since the results of such investigations cannot be protected by patents or copyright. They would provide no competitive advantage to the company in terms of later awarding of contracts. This is therefore a typical governmental research task. It must, however, be carried out by qualified industrial firms or at least with their participation. If such investigations were taken on independently by companies alone, the results would in general not be made public and would thus not be available for public discussion. (This problem is also pointed out in (S&L, 2003, p. 5-1).)

The estimated costs in the range of one (or two or three) hundred million dollars are, in view of the importance of the research and in comparison to other expenditures for energy research, not excessive (especially in comparison with today's costs for importing oil). Only with the results of such studies will it be possible to recognize and evaluate the option of solar thermal energy correctly! These investigations fulfill a key function. This remains true in the face of the fact that considerably higher expenditures will be required, for example, for the fabrication of prototype thermal installations (receivers, heat-storage reservoirs). The currently beginning worldwide construction of so-called commercial power plants, "commercial" under consideration of special power-input reimbursements and other incentives, also does not invalidate the importance of such investigations.

Of course, cost analyses must also be carried out for the thermal plant components for the case that solar thermal power plants are to be constructed on a large scale. For example, in the molten-salt receiver, these include the costs of the receiver in mass production, the cost of the pumps, the insulated piping for the molten salt circuits, the costs of the salt heat-transport medium, and the costs of the insulated containers for heat storage, down to the costs of the concrete towers and electronic control systems. All of these plant components must be fabricated in mass production, even though the numbers will be small (e.g., 10 towers annually, or up to 100 molten-salt pumps).

6.3
Stretched-Membrane Heliostats

Great hopes have been placed in this type of heliostat since the 1980s, but its development has nevertheless been neglected in recent years and has thus stagnated; we will therefore discuss it in some detail in the following sections. Owing to the materials savings of this construction principle, it is one of the most important development options for heliostats.

6.3.1
Technology

A considerable portion of the cost of *conventional* glass-mirror heliostats is due to the reflector. Along with the actual mirrors, it consists of a flat supporting frame, a welded structure on which the mirrors are supported. This frame must be so stable (bending stiffness) statically and dynamically that it is deformed only within the allowed tolerances by the normal wind pressure in its operating position and can resist maximum wind forces in its horizontal safety position during storms. The welded frame must be adjusted to permit the exact alignment of the mirrors. They are not held over the whole area, but only at individual support points. To prevent them from sagging, glass mirrors with a usual thickness of 3–4 mm are used (ultrathin glass mirrors cannot be employed).

At this point, the principle of the membrane reflector comes into play. Here, a foil (membrane) is stretched onto a circular steel ring and forms a flat surface without further support structures; it carries the actual mirror. On the back side is in general a second membrane, so that a hermetically sealed space is formed between the two membranes. A slight vacuum can be produced within this space, so that a parabolic, weakly focusing mirror surface results (which can be simply adjusted to the given distance to the receiver).

The stretched-membrane heliostat in its original conception consisted of a steel ring,[2] which was stiffened by several radial struts. A thin aluminized or silvered *plastic foil* was stretched over this ring; the metal layer formed the reflecting surface and was protected by a thin, transparent film. The ring must be dimensioned in such a way that it can support the radial tensile force of the foil without being deformed. Since this force depends on the thickness of the foil, a very thin foil of 0.15 mm thickness was used. While the measured optical properties were good, there were problems with the mechanical stability of the plastic membrane (Weinrebe, 2000).

For this reason, the term "stretched-membrane heliostat" now usually refers to *metal-membrane* heliostats. In this type, a steel ring with a diameter of several meters (depending on the particular construction, 3–14 m) is covered on both sides by a *stainless steel* foil, which is welded to the ring under tension. A uniform radial force thus acts on the ring, so that deformations are largely eliminated and no additional stabilizing elements are necessary ("self-stabilized," the tension-spoke/rim principle).[3] The result is a stable and material-saving construction that is light in comparison to a conventional heliostat. On the front side, the carrier foil is covered either with a *silvered plastic film* or with a *thin glass mirror*.

In the case of the plastic film, there is, however, a problem which has yet to be satisfactorily solved: The film degenerates under the influence of UV radiation. Polymer material with a greater UV resistance is still too expensive, likewise repeated replacements of a low-cost material. The development of UV-resistant polymers would require a considerable investment, so that the – thus far private – developers could not take this route.

This materials problem is eliminated if the flat or very slightly curved metal foil carries a glued-on thin glass mirror with a thickness of less than 1 mm. This small thickness in addition to saving glass has the advantage that low-cost (iron-containing) green glass can be used. In a conventional heliostat with thicker glass, the iron content reduces the reflectivity of the mirror significantly.[4]

The focal length of these mirrors can be adjusted via the air pressure in the inner space or also by mechanical deformation of the back-side foil; all together, the optical quality of this type of reflector is very good. A typical initial problem is, however, the current availability of steel foil and mirrors with an optimal thickness (recently also the generally higher stainless-steel price).

---

2) Various different designs were investigated: a hollow toroid with a rectangular cross section, or a rectangular externally open cross section, or an H profile. For the optimization of the ring height, on the one hand, the material costs play a role; on the other, the improved stiffness with increasing height.

3) In this connection, it is interesting that precisely due to the *double* membrane, the stabilizing effect is particularly strong in this design (Murphy et al., 1986).

4) The strong absorption is due to Fe^{2+} ions.

Figure 6.1 A second-generation stretched membrane heliostat prototype (USA) – pedestal type (SANDIA).

Beginning with the fundamental principle of the stretched-membrane heliostat, several variations are possible. The basic structure consists first of all – as in a conventional heliostat – of a foundation, a mount, and the drive unit (motor, gear box, control system). The reflector sits on this base with its center of gravity above the mount, connected to the drive unit above the center of the main support (see Figures 6.1–6.3).

The rectangular mirror surface of a conventional heliostat is – as mentioned above – made of individual facets of, for example, $3\,m^2$ each. The reflector of a stretched-membrane heliostat is composed likewise of a number (e.g., 10–20) of small facets (see Figures 2.9 and 2.10), or else possibly of only two larger circular facets; the latter construction originally proved to be especially favorable after intensive tests at Sandia in the early 1990s (Strachan and Van Der Geest, 1993) (Figure 6.4). In 1999, the manufacturer (SAIC), however, did not mention this dual-element stretched-membrane heliostat in answer to a query, but rather described one with 22 individual facets, which had in total a mirror area of not just $70\,m^2$, like the design with two mirrors, but rather $170\,m^2$. A single large reflector (with one stand and a central support) is in any case not feasible with the stretched-membrane type when a glass mirror is not employed as reflecting surface, but instead a silvered plastic foil. With a single-mirror design, the face-down protective position would not be possible; but it is necessary for a foil reflector because the reflecting foil is especially susceptible to damage by hailstorms.

Figure 6.2 A stretched membrane heliostat prototype, seen from the rear (SANDIA).

In principle, then, a membrane heliostat is distinguished from a conventional heliostat only by the construction of its reflecting surface. The requirements for the tracking system and the substructure (mount and foundation) are the same for both types. The mechanical stress is due (as mentioned), on the one hand, to wind forces in the operating position (precision of the tracking mechanism and stiffness of the mounting structure), with a view to the related optical quality; and on the other hand, to the extremely strong forces during a storm when the heliostat is in its protective position. The weight of the reflectors therefore plays only a minor role for the support and tracking mechanism.

In the 1990s, the German company Schlaich, Bergermann, and Partners (SBP) (together with the Steinmüller Company) developed a different mechanical support system for the reflector,[5] in connection and in parallel with their development program for metal-membrane heliostats (using thin glass mirrors). The reflector is mounted within a "rotation mount" or carrousel mount constructed from steel tubing. This framework can be rotated as a whole around the vertical axis (on

---

5) This construction is also employed in the SBP dish systems, as well as in a similar form already for the Australian "Big Dish."

Figure 6.3 A stretched membrane heliostat (SBP) with a single large support tube (Photo SBP).

Figure 6.4 A dual-element stretched membrane heliostat (SAIC) (Photo SANDIA).

wheels), so that the reflector itself need be rotated only around a horizontal axis, that is, it is gimbal-mounted (see Figures 6.5–6.7). The tracking drive is accomplished by means of rack and pinion gearing and electric motors, which are mounted away from the rotational axes. The rotation mount suggests itself here, since the reflector must be suspended from its outer steel ring in any case. A *central* suspension (with a few radial struts) is in principle also possible for a metal-membrane heliostat, but the rotation mount will probably become relatively less costly with increasing reflector size.

6.3.2
Development Aspects

The glass-mirror-membrane type is not yet ready for mass production at its current state of development and has yet to be optimized for this purpose; there is still a clearcut potential for cost reduction. This and a few other typical requirements for further development can only be sketched here, mainly following (Schiel (SBP), 1997).

The choice of material for the prototypes thus far was dependent on the availability of particular foil thicknesses. Thus, for the SBP heliostat, a relatively thick steel foil (0.4 mm) was used, since this thickness was readily available on the market. The mechanical stress on the pressure ring of strip steel, however, depends on

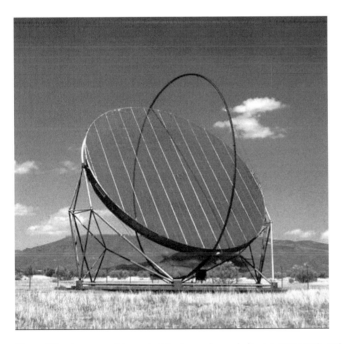

Figure 6.5 A carrousel type stretched membrane heliostat (ASM-150, 150 m^2, built by Schlaich, Bergermann, and Partner (SBP)/Steinmüller, Germany).

Figure 6.6 Aerial view of the stretched membrane heliostat ASM-150 (SBP).

Figure 6.7 Mounting of the metal foil on a prototype of the SBP heliostat (SBP).

the foil thickness. Since the tension on the foil cannot be reduced, in order to avoid sagging, the stress on the ring can be diminished only by reducing the foil thickness. With a view to the static properties of the reflector, a foil thickness of 0.1–0.2 mm would suffice. Both the amount of stainless steel required for the foil as well as the material needed for the pressure ring would be decreased.

Furthermore, thus far only glass mirrors of 2 mm thickness have been used. However, thinner glass mirrors of around 0.6 mm could be employed. So far, these have been manufactured only for exotic applications, such as for small cosmetic mirrors, and are not comparable to the standard mirror thicknesses in terms of price. With higher production rates, they would, however, become cheaper than the thicker mirrors. Their main advantage, though, lies in their lower weight, which reduces the necessary pretensioning of the steel foil, with the consequences described above for the overall costs.

Precise cost estimates have thus far been difficult to make, in part due to the material prices for small production series. Thus, the price of the steel foil varied strongly in the past, depending on the market situation.

An important point here is that the thinner steel foils which could be used for stretched-membrane heliostats must be rolled out to the desired thickness in an additional milling step. There are no cost estimates for this procedure; thus far, there has evidently been no demand for this thickness within the industry. The technical difficulties increase with decreasing foil thickness, so that the typical costs cannot simply be extrapolated. This question must be answered by a special appraisal by the rolling mills.

The demands on rolling technology depend on the width of the foils to be manufactured. Difficulties may be encountered, in particular, for large widths. If economic limits should become apparent due to this factor, the width could be reduced. This would have the following consequence: Since the complete foils for the heliostats are produced by welding together individual strips, narrower strips would require more welding seams. This is, however, only a negative cost factor if the welding is done in single-weld facilities, in which one strip after another is welded on; the strips must then be unrolled and fed into the machine individually. For mass production, one would use multiple-welding machines and weld all the strips in a single pass to give the required width.

A different problem results from matching of the width of the mirror layers. Since these cannot be laid over the welded joints due to the roughness of the latter, they must be fabricated and glued on with the same width as the steel carrier strips. This problem could possibly be solved by applying an intermediate layer of plastic, of the same width as the foil strips, to compensate for the thickness of the weld joints.

For the optimization, cooperation with companies specializing in the various fields required (welding, rolling, production of thin glass mirrors) is necessary. The cost of the requisite analyses and planning would presumably lie in a range below $10 million. The work would initially consist of feasibility studies, later of preliminary planning of the required facilities (e.g., welding machines or process modifications for the production of thin glass layers). The subsequent construction of the plants, which would be carried out only if the corresponding heliostat type were found to be optimal within the overall planning, is of course not included within these costs.

Aside from this optimization in terms of materials, for the general development of heliostats, a number of other particular developments are necessary, for example, more precise aerodynamic calculations and measurements in a wind tunnel. The computations for the SBP heliostat up to now are based on wind-tunnel experiments carried out with similar, but much more strongly curved dish reflectors. The results can be transferred to heliostats only with considerable reservations. This notable uncertainty with respect to the final form of the heliostats must be taken into account by "surcharges" on the estimated costs. The financial expenditure for wind-tunnel experiments with models is on the order of only a few

$100 000. Such amounts are insubstantial compared to the overall costs, but it cannot be expected that they will be borne by private developers. Such expenses make no sense in terms of a private investment as long as a serious political interest in this technology is not apparent and therefore no really large-scale market can be foreseen, for which an optimization of this type would be profitable. The companies doing development work carried it out over a number of years with practically no public support. The politicians responsible expected them to finance developments for a nonexistent market. As a result, important development steps were not taken for the lack of relatively small amounts of financial support. Reliable cost prognoses for mass production are therefore at present not possible.

Possibilities for cost reduction are also present in the rotating mount, which, as mentioned, is not the only option for stretched-membrane reflectors. Thus far, it is conceived as a structure made up of rods connected by flexible corner screw joints ("nodes"). The statics of the structure have not been optimized. This holds also for the connecting points of the rods at the vertices using screws. In mass production, such stands would be manufactured by automatic welding equipment with automatic part feeding. Both the statics calculations and also the planning of such welding production lines (and thus the cost estimates) are goals that must be met in the course of the system development.

The control systems are thus far typical of prototype plants or small production series. At the rotational axis, the angular positioning of the reflector is detected by a measurement device and passed on to the control electronics. For this purpose, so far, equipment was used that was originally developed for a very different application in tool-making machines. The electronics associated with the angle measuring device (at the current price for small-volume deliveries) makes a significant contribution to the costs. In the case of a specially developed system for heliostats, the costs could be drastically reduced. This would, however, entail development costs of several $100 000, which again from the viewpoint of the developers was so far not worthwhile. Such potential for cost reduction exists, as mentioned, throughout the entire control electronics and also for the signal transmission system.

This type of heliostat illustrates the whole plight of solar energy research in the past. It was tested intensively in the 1990s with very positive results. Its further development (in Germany) has, however, not been supported by any governmental funding, and it has remained up to now only a single prototype heliostat.

6.4
Installations for Operational Testing of the Heliostats

For the further development of heliostats, it is essential that new (possibly less expensive) construction types be quickly tested with respect to their stability under various kinds of environmental exposure. The method of simply setting up the heliostats and waiting until storms, sandstorms, ice and snow, hail or even earthquakes occur is not feasible within a reasonable period. We therefore require a test park with all the necessary test facilities for simulating every stress factor to which a heliostat might be exposed. Most test facilities would not have to be newly developed, since they already exist in a similar form for other testing applications. An exception to this is a test facility for resistance to storms.

Heliostats must be designed to withstand well-defined maximum wind velocities. For very large heliostats with diameters of up to 15 m, tests cannot be carried out in a wind tunnel. Therefore, up to now, one has used models with diameters of only a few meters and extrapolated the results to larger heliostats. Although such extrapolations are usual in other technical areas and in general yield good results, it would be an advantage with respect to the broad acceptance of solar technology to be able to demonstrate the stability of the heliostats "directly." The maximum wind velocities to be tested lie in the range of 150 km/h.

If the construction of extralarge wind tunnels is not an option due to its high cost, it would be useful to investigate whether tests could also be carried out on movable platforms, for example, on rails with two parallel tracks. Instead of moving the air around the stationary heliostat, the heliostat would be moved through the air at a corresponding velocity. Manufacturers of rail vehicles would have to design suitable platforms and supply cost estimates. Possibly, a single track could be used with a parallel rail at a distance of some meters to support an outrigger that would prevent the tipping of the platform. Since an overhead power line would not be suitable, diesel-electric locomotives would have to be used to provide motive power. A simple solution without rails might be to perform the tests on an airport runway or a suitable section of roadway. The length of the test segment is not decisive; more important is the velocity attained. In this case, suitable towing vehicles to provide the necessary acceleration and maximum velocity would have to be built. The overall cost of such mobile test platforms should certainly not exceed $20 million (a rough estimate!).

The remaining facilities of the test park are nothing new. Large earthquake test beds already exist, on which even whole buildings can be tested. A platform carrying the test object is "shaken" by hydraulic cylinders at the frequency and strength of an earthquake. The effects of ice formation and frozen-on snow can be tested in a large refrigerated warehouse using snow machines. During snowfall, the heliostats would be placed in a vertical position, so that no snow would pile up on the glass surface; the tests relate to drifting snow and to its slumping down on melting. The sensitivity of the gear boxes and other moving parts to sand could be tested in sand chambers using blowers or sandblasting equipment. If necessary, time-lapse tests to investigate scratching of the mirror surfaces could also be performed. The mirrors in the parabolic trough power plants in California exhibit no damage due to sand. Such tests could, however, become necessary if a stronger sandstorm activity was expected at other locations, or if more sensitive reflector materials were used.

In addition to the simulation facilities, different test sites should be employed, in particular for tests of the wind resistance of the heliostat designs. In mountainous regions or coastal areas with frequent high winds, much higher wind velocities occur than at most potential sites for solar plants. In such locations, the maximum wind velocity for which a heliostat was designed would occur, depending on the particular site, not only after decades but possibly every year, so that one could test the real behavior of the heliostats relatively quickly. Several specimens of each newly designed type of heliostat under consideration should be set up at such extreme locations for testing purposes.

During hailstorms, the heliostats would be placed in their vertical protective position (possibly with a slight inclination to the vertical); a heliostat, however, shows its greatest resistance to storm winds in the horizontal position. Hail would thus represent a major problem if it was accompanied by such strong winds that the horizontal position would have to be used. At the California sites, so far neither windstorms nor hail have proved to be a problem. The question of whether hail and maximum wind velocities would occur simultaneously of course depends on the particular site; whether the experience gained in California can be applied to other locations is not clear. This should be investigated as precisely as possible by meteorologists for all of the potential sites for solar power plants worldwide. If hail accompanied by very high wind velocities is to be expected, only those types of heliostats that could be placed in a horizontal protective position with their mirror surfaces downward ("stow position") could be used at that particular site. The other, simpler types for which this "face-down" position is not possible would not be adequately protected. Thus, the question of the probability of the combined occurrence of hailstorms and very high wind velocities should be clarified in advance for *each site considered*. At problematic sites, heliostats permitting the "face-down" position would have to be used. These investigations should also clarify the question as to whether so-called "hail abatement," as is used in some large cities, could make a contribution toward the protection of the heliostats. In this technology, storm clouds are seeded with silver iodide crystals from aircraft or rockets in order to prevent the formation of large grains of hail. For this question, precise meteorological conclusions are more important than testing facilities. Such testing facilities for hail damage exist, and they have already been used during the early development stages for heliostats (cf. (Boeing, 1978, Table 2.1.-1)).

In the case of heliostats that employ a plastic reflecting foil, the question of what material to use, especially with reference to its resistance to UV radiation, plays an essential role. New materials must be developed especially for heliostats, and new testing facilities for these materials are likewise required, in particular accelerated-time tests for aging of plastics under UV exposure. Facilities with UV radiation sources, and the corresponding test procedures, must therefore be developed.

6.5
Comparison of the Cost Assumptions with Those of Other Studies

6.5.1
Heliostat Costs in the S&L Study

In this extensive study of parabolic-trough and solar tower power plants, already discussed in Section 4.2 (S&L, 2003), the cost perspectives for the heliostats were investigated in detail. The same method of cost analysis was used for the heliostats as for other aspects in the study. The basis for its cost estimates were the detailed data of SunLab. They were (in keeping with the commissioning of the S&L study) critically checked for plausibility[6] that was also complemented by expert knowledge and direct information from industry.

At the end of a 16-year development period considered in the study (to begin in the year 2004), that is, after the year 2020, with an assumed total installed solar tower plant capacity of 2.6 GW, S&L predicted that heliostat costs of 117 $/m^2 (2002-$) would be realistic (S&L, 2003, p. E-64), without assuming any major technological strides forward and (according to S&L) without undue optimism. Parallel to this, the S&L study shows how SunLab estimated the development potential for heliostats. This was based on a more optimistic overall appraisal, on the assumption of a more "aggressive" technological development, and with emphasis on the development potential as a whole, as well as the assumption of a cumulative installed capacity of 8.7 GW by 2020, instead of only 2.6 GW as in the S&L basic scenario. SunLab, on this basis, maintains that heliostat costs of 75 $/m^2 (2002-$) are realistic. According to S&L (2003, p. 5–13), this value, however, does not refer to a conventional 148-m^2 heliostat, but rather to an "advanced" heliostat (p. 5-1) in connection with the concept "SunLab 220." In contrast to this, SunLab estimates a heliostat price of 96 $/m^2 (S&L, p. E-64) for the case of the solar power plant scenario "SunLab 200"–which assumes a smaller "cumulative deployment" of only 3.9 GW installed capacity, and presumes a conventional type of heliostat; see also Section 4.2 (Table 4.6) and Appendix A.

6.5.2
The Sandia Heliostat Study

Given the fact that the heliostat field is the single largest individual cost factor for a solar tower power plant (ca. 40% of the total cost), it was decided in the USA that in addition to the S&L study, a special "heliostat cost-reduction study" should be carried out. It was published in 2007 (Kolb (Sandia) *et al.*, 2007a) and is notable for two reasons:

1) This Sandia study presents, after nearly 20 years, once again an extensive, both broad and in-depth investigation. On the one hand, the entire previous development (more than 30 years) up to the present in terms of technological approaches and various cost analyses is included, on the other hand, the current state-of-the-art and the presently foreseeable spectrum of development potential are treated.

2) This was not just a "normal" Sandia study, but rather it incorporated the worldwide definitive knowledge about heliostats. The Sandia study group

---

6) Verification of "plausibility" was carried out in particular with the help of so-called "learning curves," which were adopted from other comparable technological development histories.

discussed the topic intensely during two workshops with ca. 30 heliostat and manufacturing experts from Europe, Australia, and the USA, considering far-reaching technological options and ideas (including "brainstorming" sessions). The results can therefore be deemed to be the most well-founded and current among those published to date (2008) on heliostats.[7]

The principal conclusion from this study is that heliostat costs of 90 $/m^2 (2006-$) should be feasible, to be sure with a production rate of 50 000 heliostats/a. Converted to 2002-$, which we use as the general pricing basis in this book, this cost corresponds to 80 $/m^2. We can consider this value to be the best currently available estimate. In comparison to it, one is on the safe side in using a price of 83 $/m^2 as in this book – and as explained at the beginning of this chapter – resulting from an earlier study by the authors.

A second conclusion of the Sandia study is also very interesting: The price of 90 $/m^2 (2006-$) can, in the opinion of Kolb *et al.* – starting from today's price of 126 $/m^2 – be achieved through the usual learning effects and through an R&D investment of "probably" only around $5 million (2006-$). Here, it is assumed that the learning effects and the results of the systematic R&D program would each contribute roughly half of the overall reduction of 36 $/m^2.

The sum of $5 million includes only the most important development goals (taking into account risks, benefits, and costs – six projects); it corresponds to the case of a "limited budget." For these six research projects, only very low costs are listed, each in the range of $1–2 million, that is, altogether ca. $8 million. Even with only $5 million, one could – according to this study – clear up some of the most important questions. Even with this modest R&D budget, there would be a certain hope of being able to reduce the costs still further, for example, by 24 $/m^2 instead of only 18 $/m^2. This, together with the "normal" learning effects (in the sense of increases in productivity), would lead to a price of 82 $/m^2 (2006-$). This furthermore implies that with a considerably higher R&D budget, the heliostat costs could probably be reduced even further. In that case, the additional research projects that were envisaged in the Sandia study could also be carried out. If one presumes that an appropriate R&D program will be funded, it would appear (taking the Sandia study into consideration) that in the medium term (2015–2020), the heliostat costs could be reduced not only to 80 $/m^2, but even to 70–75 $/m^2 (2002-$).

In the first (and more extensive) part of the Sandia study, estimates of the costs of conventional and stretched-membrane heliostats corresponding to the current state of the art are obtained; we give the results here. Only the most important numbers are listed, and in general only those that hold for a truly relevant production rate, that is, 50 000 heliostats/a. This corresponds to a good approximation to

---

7) One must, however, emphasize that it is not an actual (bottom-up) cost analysis. In particular, from the point of view of methodology and breadth, it cannot be compared with considerably more comprehensive investigations into the production costs of heliostats, such as are proposed in this book. Instead, it represents a "preliminary study" – to be sure, a very valuable one.

a production rate of the order of 100 000 heliostats/a, which we have thus far considered to be relevant.[8]

The numbers were adopted without change as 2006 $, since here only the relative values are important. To recalculate in terms of 2002 $, one would have to divide all the costs listed by the corresponding inflation factor (using the US Consumer Price Index, this factor is 1.08.)

First of all, in Table 6.2, we give the total costs for completely installed heliostats with current technology. Here, one should note that the "conventional heliostat" referred to is the concept of the small US company Advanced Thermal Systems (ATS). This 148 m² heliostat was already constructed several years ago and was tested intensively. The stretched-membrane heliostat listed was built a few years ago with an area of 90 m² and – like the conventional heliostat – it was constructed and tested as a pedestal-type heliostat. In the following tables, that is, in the basic tables from the Sandia study, a 150 m² stretched-membrane heliostat, which has not yet been constructed, is assumed.

In Table 6.3, the rough cost structure for current production of the ATS heliostat is listed.

Table 6.2 Heliostat costs for different production rates (with today's technology, prices in 2006-$).[a]

	5000 Heliostats/a	50 000 Heliostats/a
Conventional heliostat ($/m²)	164	126
Stretched-membrane heliostat ($/m²)	170	133

a) Note: In these values, a blanket contribution of ca. 15% for "overhead/profit" is included (more precisely: 20% relative to the "direct production costs"; cf. also Table 6.3).

Table 6.3 Cost components for a conventional ATS heliostat (compare Kolb et al., 2007a, Table 3-14).

	$/m²
Drive	31
Mirror module	23
Support structure	21
Pedestal	17
Fabrication direct cost	92
Overhead/profit (20%)	18
Total fabrication cost	110
Field assembly/wiring	16
Total installed cost	126

8) Sandia give throughout their discussion the costs for both the case of a production rate of 5000 heliostats/a as well as for the case of 50 000 heliostats/a.

Table 6.4 Costs of the motion drive components for a heliostat (compare Kolb et al., 2007a, Table 3-13).

	Cost/Heliostat	Cost/m^2
Azimuth subassembly	$3000	20 $/m^2
Elevation subassembly	$1000	7 $/m^2
Gear drive	$4000	27 $/m^2
Electrical components	$550	4 $/m^2
Overall motion control	$4550	31 $/m^2

Table 6.5 Cost reduction as a result of mass production – the importance of the costs of the mechanical drive system (compare Kolb *et al.*, 2007a, Tables 3-14 and 3-15).

	5000 Heliostats/a ($/m^2)	50000 Heliostats/a ($/m^2)	Reduction ($/m^2)
Gear drive (including 17% overhead/profit)	58 (35%)	32 (26%)	26
Other parts (electrical components, mirror support structure, mirrors, pedestal, field assembly, field wiring)	105	94	11
Total installed cost	163 (100%)	126 (100%)	37

In Table 6.4, the special cost structure for the drive system is given. One can see that the electrical components (including the drive motor and the control electronics) are practically negligible. Concerning the mechanical drive components, in particular the gear box, we note the following: The elevation drive rotates the mirror during the course of a day around a horizontal axis. It consists essentially of a ball screw, which is technically standard. The azimuth drive rotates the mirror in the course of a day around the vertical axis and is technologically and in terms of its fabrication the most demanding component. The Sandia study considered a special construction (expressly denoted as "clever") of the US company Peerless Winsmith, a geared transmission with an extremely large gear ratio (33 000 : 1) and very little backlash. At a production rate of 5000 units/a, the elevation drive is presumed to cost $1500, whereas the azimuth drive would cost $5700. It is now decisive that in going to the higher production rate of 50 000 heliostats/a, the cost of the elevation drive would decrease to $1000 (i.e., by $500), whereas the cost of the azimuth drive would be reduced to $3000, corresponding to a reduction by $2700. This is based on a statement ("expectation") by Winsmith, which, however, is considered in the Sandia study to be "plausible."

In Table 6.5, the significance of the mechanical drive for the cost regression of the heliostats under mass production is shown. On going from 5000 to 50 000 heliostats/a, the cost would be reduced by 37 $/m^2, of which 26 $/m^2 corresponds to cost reductions due to the mechanical drive system.

Table 6.6 shows the cost structure for a stretched-membrane heliostat in comparison to a conventional heliostat at a production rate of 50 000 heliostats/a. As mentioned above, the stretched-membrane heliostat would be more expensive by 17 $/m^2 according to this estimate; this is essentially due to the higher cost of the mirror module, including the nonnegligible cost of the vacuum control system for the stretched-membrane heliostat (cf. Tables 6.7 and 6.8).

In comparing conventional and stretched-membrane heliostats, one must keep a particular fact in mind: The effective cost difference is less serious than it may seem, since the stretched-membrane heliostat earns a bonus of 10 $/m^2 due to its better optical quality (among other things, more precise focusing), so that a realistic comparison (as shown in Table 6.2) would yield:

- Conventional heliostat   126 $/m^2
- Stretched-membrane heliostat 133 $/m^2

The second part of the Sandia study deals with various options for development of heliostats. In the following section, we give an impression of the results and the methodology of this part (cf. Kolb et al., 2007a, pp. 81–90).

First of all, we list the "opening questions" (slightly abbreviated) used in the three brainstorming sessions that dealt with the areas (1) conventional metal/glass heliostats, (2) stretched-membrane heliostats, and (3) innovative concepts.

Glass/metal brainstorming session

- What are the biggest technological problems that need to be solved to make glass/metal heliostats economically viable?
- Which components of the heliostat offer the greatest potential for cost reduction?
- What is the most economically viable way to manufacture the glass/metal array?
- Which manufacturing technologies offer the greatest potential for cost reduction (given a suitable volume)?

Stretched-membrane brainstorming session

- What are the biggest technological problems that need to be solved to make stretched membrane heliostats economically viable?
- What are the alternative technologies available to fabricate the stretched membrane? Which is the most economically viable?
- What are the pros and cons of stretched membrane versus glass/metal heliostats?
- Is the stretched membrane heliostat fundamentally less costly than the glass/metal heliostat?

Innovative concepts brainstorming session

- Is the pedestal mount the best approach?
- Are ball-jack screw elevation and planocentric azimuth the best drives?
- Does closed-loop control offer advantages over the current open-loop approach?

Table 6.6 Cost comparison of conventional and stretched-membrane heliostats at a production rate of 50 000 heliostats/a (cf. Kolb et al., 2007a, Table 3-22).

	Conventional ($/m²)	Stretched-membrane ($/m²)
Drive	31	30
Mirror module	23	43
Support structure	21	19
Pedestal	17	17
Total direct cost	92	109
Overhead/profit (20%)	18	22
Total fabricated price	110	131
Field installation	16	12
Total installed price	126	143

Table 6.7 The cost of the mirror modules for stretched-membrane heliostats (cf. Kolb et al., 2007a, Table 3-22).

	$/m²
Ring	5
Membranes	12
Mirror	9
Focus system	11
Mirror module tooling	1
Mirror module labor	5
Total module cost	43

Table 6.8 The cost of the mirror modules for conventional heliostats (cf. Kolb et al., 2007a, Table 3-11).

	$/m²
Glass-mirror facets	10
Hat sections	10
Cross members, adhesive, fasteners, assembly	3
Total module cost	23

- Which materials are the "best" reflectors? Possibility of cost reduction with a new material?
- What new technologies are on the horizon for heliostats?

The most important approaches in the view of the participating heliostat experts are as given below:

Glass/Metal

- Explore whether drive specifications are too conservative
- Heliostat designer and drive designer work together to minimize drive cost

- Use brake to loosen drive backlash specifications
- Study "pipe-in-pipe" drive concept

Stretched membrane

- compare pedestal-drive system to alternative drive concepts;
- for carrousel-type drive, explore precast, truckable concrete bases;
- evaluate three-point ground-mounted drive concept;
- find source or develop capability to use wider stainless steel strips to make stretched membrane.
- study use of an impregnated fabric as the membrane instead of stainless steel.
- study use of a polymer sheet (covered with glass) as the membrane instead of stainless steel.

Innovative concepts

- new Mexico Tech water-ballasted heliostat
- mega-heliostat systems study (>300 m^2)
- hydraulic drive study
- study latest closed-loop control options including signal mirror technology
- coat mirrors with SuNyx[9] to eliminate mirror washing.

The suggestions and ideas from the brainstorming sessions were evaluated in the subsequent months by the Sandia team (in particular, from the viewpoint of a cost/benefit analysis). In this process, six overriding development projects were identified that – taking account of the weighting by the participants of the brainstorming sessions – comprise about 75% of the originally suggested ideas.

1) Large (150-m^2) single metal-based stretched-membrane facets
2) Less-conservative, high-volume, pedestal-mounted azimuth drive
3) Pipe-in-pipe azimuth drive
4) Large (150-m^2) carrousel-type stretched-membrane heliostat
5) Large (150-m^2) single-fabric-based stretched-membrane facet
6) Transform large (>300 m^2) APS photovoltaic tracker to a heliostat
7) New Mexico Tech water-ballasted heliostat with closed-loop control

The following description of these projects was adopted (nearly without change) directly from the Sandia study. We also cite the first project, later evaluated as being unpromising, namely, the scaling-up of the stretched-membrane reflector (with a classical pedestal construction) to 150 m^2. From the justification for excluding this project from further consideration, it is not clear whether it was believed to be generally hopeless, or only in view of the limited R&D budget of only a few million dollars.

Project 1: Large stretched-membrane facet
In this project a large (150 m^2), stretched-membrane-facet is developed that can be integrated into a pedestal-type heliostat. In the USA, only 50-m^2 facets have been

---

9) Generally, self-cleansing coatings on the basis of "nanotechnology" are referred to here (SuNyx is a German company).

built. Scale-up to 150 m² was proposed to reduce cost on a dollars-per-square-meter basis. In effect, the ATS glass/metal structure and mirror modules above the drive would be replaced with a single stretched-membrane mirror module. Early evaluations in the 1980s suggested that this would result in a heliostat that cost 20% less than the glass/metal heliostat. However, analyses presented in Section 3.6 [of the Sandia studing] indicate that this type actually results in a higher cost. The project is therefore eliminated from further consideration.

Project 2: Less-conservative azimuth drive
As described, the azimuth drive is the most significant heliostat cost contributor, especially at low production volumes (5000/year). During the brainstorming, Winsmith stated that the design of their "gear-type" azimuth drive may be too conservative and that a less-conservative, less-costly drive might be developed if Winsmith could get a better understanding of the wind loads and torques on the heliostat drive. Significant cost reduction can also be achieved through highly automated production-line manufacturing techniques. A production line does not currently exist. The R&D project would provide a detailed price estimate for a less-conservative gear-type azimuth drive given differing amounts of manufacturing automation. A 33% price reduction is targeted. If detailed analysis indicates that this target can be achieved, a new prototype drive would be built and tested.

Project 3: Pipe in pipe azimuth drive
The brainstorming group explored different approaches to the conventional gear-type drive historically built by Winsmith and Flender. At the White Cliffs plant in Australia, a pipe-in-pipe approach was successfully used to position relatively small (7-m²) solar dishes. In this concept, azimuth motion is achieved by rotating a pipe within a fixed pedestal. The driving motor is located at the bottom of the pedestal and the wind loads on the drive are distributed along the length of the pipes, as opposed to a single point within the Winsmith. Cost reductions relative to a gear-type drive appear feasible because manufacturing of the pipe-in-pipe could be simpler. This R&D project would provide a detailed price estimate for pipe-in-pipe drive that is suitable for a 150-m² heliostat. A 33% price reduction relative to the current Winsmith azimuth drive is targeted. If detailed analysis indicates that this target can be achieved, a new prototype drive would be built and tested.

Project 4: Large carrousel-type stretched-membrane heliostat
A large (150-m²) heliostat like this has been operating at Plataforma Solar de Almeria, Spain, for 10 years (called ASM-150). The optical performance of this heliostat is significantly better than the ATS glass/metal type. According to the DELSOL (software) analysis, this optical advantage is worth ca. $10/m². In addition, analysis conducted in the 1990s indicates the cost of this heliostat should be significantly lower than a glass/metal heliostat built by a Spanish company. However, a few in the brainstorming group suggested that the concrete foundation for the ASM-150 is too costly. The group then explored ideas on how to drastically reduce the cost of the foundation. Precast concrete foundations that "roll off a

truck" were thought to be a possible low-cost solution. This R&D project would provide a detailed price estimate for a large carrousel-type stretched-membrane heliostat with a low-cost foundation. More than 10% of the capital cost reduction relative to the ATS (glass/metal) is targeted. This appears feasible because it weighs ca. 50% less than the ATS. Combining this with the performance improvement of ca. $10/m^2 should result in an overall cost reduction of ca. 20%. If detailed analysis indicates that this target can be achieved, a new prototype drive would be built and tested.

Project 5: Large single-fabric-based stretched-membrane facet
As described, today's stretched-membrane facets are created by welding multiple strips of stainless steel across a ring. The welding process is complex and cumbersome. The brainstorming group thought that significant cost reduction for the facet could be achieved if the stainless steel strips were replaced with a single large piece of fabric. Besides eliminating expensive stainless steel, connection to the outer ring could be greatly simplified by using an "embroidery-hoop" method, that is, two concentric hoops are press-fit together to form the connection between the material and the ring. The fabric must not leak air to maintain the vacuum within the facet plenum. Thus, the fabric would need to be impregnated with a sealer. This R&D project would provide a detailed price estimate for a large fabric facet. Rough calculations suggest this facet could lead to an additional cost reduction of ca. $7/m^2 relative to the carrousel heliostat described in Project 4. If detailed analysis indicates that this target can be achieved, a new fabric-based facet would be built and tested.

Project 6: Mega heliostat
Arizona Public Service currently operates several large-area two-axis PV concentrators. The largest is about 320 m^2. This device could be converted to a heliostat by replacing the Fresnel-PV modules with mirrors. At this size, the use of hydraulic type azimuth and elevation drives appears to be justified. The brainstorming group generally concluded that hydraulic drive systems are more complex and require more maintenance than mechanical drive systems. However, they are very strong and could be the preferred low-cost approach for mega heliostats. This R&D project would provide a detailed price estimate for a mega heliostat of more than 300 m^2 in size. Engineering scaling laws indicate that the cost of this heliostat could be $21/m^2 less than 148-m^2 ATS heliostat. However, the optical quality of the mega heliostat will be worse than the ATS because the reflected beam will be larger. Thus, the net cost reduction is ca. $18/m^2.

Project 7: Water-ballasted heliostat
Students at New Mexico Tech are exploring innovative "water-ballasted" heliostats. Heliostat tracking is achieved by pumping water between chambers located on the back of a mirror. This eliminates the use of costly gear drives. Two different approaches are being investigated. In the rolling ball concept, flexure of the ball structure and ground-surface irregularities will result in pointing errors that will require correction by using a closed-loop control system. A few in the brain-

storming group suggested that signal-mirror technology can be used to close the loop. After the initial heliostat workshop, New Mexico Tech began to investigate a nonball approach. Water is still pumped between chambers, but the mirror does not move until electric brakes are released at the pivots. New Mexico Tech has their own funding from the Environment Protection Agency to explore these concepts. SAIC has given New Mexico Tech several 8-m^2 facets and Sandia is part of the review committee. We will keep abreast of their progress. No DOE funding is requested at this time.

To conclude, two remarks on the results of the Sandia study seem particularly important to us (cf. Kolb *et al.*, 2007a, p. 91 ff):

1) The estimation of the cost of development (order of magnitude: $10 million) that would be required to attain the goals defined is interesting and very useful in terms of future requirements toward initiating a systematic development program. In particular, a (rationally justified) indication is given for the order of magnitude to be considered here. Still, it must be taken into account that the argumentation in this study is based on a very simple stochastic model. Its presuppositions (assumed probability distributions referring to individual development results, probabilities of success of the projects, roughly estimated development costs, and subsequent production costs) are rather uncertain. The lower and upper limits for the benefits held probable by the experts in each case lie far apart. Briefly stated, this means that a development program funded at the suggested level of $5–8 million harbors a barely-calculable risk of failure.

    But even if we accept the expectations of a developmental result (aside from the uncertain presuppositions), this value cannot be taken recklessly as the basis for decision-making.[10] It is at most an *aid to decision making* that naturally is especially useful when decisions must be taken about thousands of projects each year in a ministry and the goal is to employ the funds available in the most effective manner possible. Heliostat development is, however, so important that its development program should not be financially limited to correspond to the (apparently very reasonable) expected value from this study, but instead must be so generously conceived that the risk of its failure is minimized. Furthermore, a sum of the order of $5 million for heliostat development also appears very low, because the expenditures in the past have already exceeded $20 million. The results were indeed quite positive, but by no means sufficient. For the "last 20 $" (per square meter), one should expect even a still larger expenditure. The principal risk is

---

10) In principle, the practical significance of the concept "expected value" and of stochastic calculations depends on *many* (in practice at least a number) of statistical experiments, so that one must be very cautious here in discussing the one-time implementation of a development program using stochastic arguments and calculations. Furthermore, a development program is essentially not a stochastic process, but rather is dependent on objective problems, so that stochastic laws can be applied to only a very limited extent.

certainly that the political and bureaucratic decision makers will stop the development on the basis of "lack of feasibility" if the originally foreseen funds should prove to be insufficient.

2) Finally, we should emphasize once more that the Sandia study clearly verifies what can be termed one of the fundamental proposals of this book: Our knowledge of the real cost-reduction potential for heliostats is still too limited.

Concerning the most important cost item, the azimuth drive, the Sandia study states:

> "Is the price drop for the azimuth drive from $5700 to 3000 as quantity increases from 5000 to 50 000 drives per year believable? Here the answer is a qualified yes. To estimate the true savings, an in-depth study of the drive needs to be done … along with economic analysis of the specific capital equipment required for a dedicated line." (Kolb et al., 2007a, p. 79.)

But we still do not even know the precise demands that must be placed on the drive due to wind loading:

> "It would be valuable for Sandia to instrument a couple of heliostats to really understand the relationship with wind, since forces and torques generated by wind are critically important to the design. … It will become really clear after a year or two how wind really affects the heliostat field.
>
> … The current 20.5-inch Winsmith drive is probably conservative for the 150-m^2 heliostat. … It might be useful to work with Winsmith and others to determine the largest heliostat size they would consider possible. … If Sandia were to put a couple of larger heliostats in the field and study them using the instrumentation project just described, there could be a significant benefit with relatively little cost." (p. 78).

In the Sandia study, the following general statement is made concerning the uncertainties in the cost estimates:

> "Multiple factors make it difficult to generate precise cost estimates for a study like this one. In some cases (e.g. mirrors), the production quantities involved are beyond many suppliers' capacities, requiring them to make extrapolations of their present costs to new facilities that would be required. Also, it is difficult to get suppliers to apply large efforts to make accurate cost estimates for a nascent and unfamiliar market. In some cases, suppliers feel that they are under no competitive pressure, and they therefore tend to be more conservative in their estimates. A production pricing study that includes representative manufacturing companies could possibly provide a higher degree of certainty." (p. 66).

The funding for studies of this type has over a period of many years not been forthcoming in the necessary amounts, not from the US government, nor from the EU Commission, and also not from the German government. In this connection, we must also point out the failure of the responsible members of parliament in exercising their control function over the appropriate institutions (e.g., in the US Congress, in the EU Parliament, and in the German Parliament); aside from the fact that in a parliamentary democracy, in the end no one other than the parliament itself is responsible for not appropriating urgently needed development funding.

7
Receivers

7.1
SOLAR TWO: Development Requirements for the "Advanced Receiver"

7.1.1
Costs and Basic Technology

7.1.1.1 Costs

In the cost estimates for future large-scale installations given by Kolb (1996a; "Advanced Technology," 200 MW), which were used in the cost comparisons in the preceding chapters, it was assumed that the receivers for solar radiation will be further developed. This is also presumed in the study by Sargent and Lundy (2003) cited earlier. In the "SOLAR TWO" power plant in Barstow, USA, tested from 1996–1999, the receiver technology was still preliminary.

For a 200-MW solar tower plant, Kolb gave the costs for the receiver and the tower[1] as $50 million (1995). (Corresponding to the solar multiple (SM) of 2.7 which was assumed by Kolb, 200 MW$_{el}$ require a receiver with 1400 MW$_{th}$ power capacity (mounted on a tower of 200–250 m in height).) Such a receiver would thus deliver a thermal power of 1400 MW$_{th}$ to the molten-salt circuit at the nominal insolation, a certain outside temperature, and a given wind speed. Using the inflation factor of 1.18 (US consumer price index 1995–2002), we find $59 million (2002-$), and for an electrical output power of 1000 MW$_{el}$, then $295 million. If – in order to achieve a larger annual capacity factor – the mirror field and the receiver are overdimensioned to a still greater extent (i.e., a larger SM is used), we find for the example sites in Spain and Morocco/USA-Southwest the following Receiver/Tower costs, as shown in Table 2.1:

- **Spain (SM 4.4):** 480 million $/1000 MW$_{el}$
- **Morocco/USA (SM 3.7):** 405 million $/1000 MW$_{el}$

---

1) The tower cost includes costs for the vertical molten-salt piping in the tower that connects the salt tanks on the ground to the receiver at its top.

Large-Scale Solar Thermal Power. Werner Vogel and Henry Kalb
© 2010 WILEY-VCH Verlag GmbH & Co. KGaA, Weinheim
ISBN: 978-3-527-40515-2

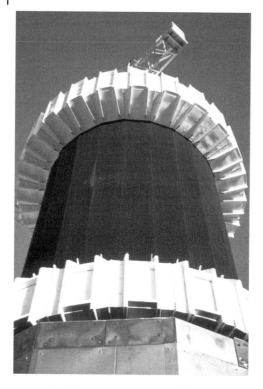

Figure 7.1 The molten-salt receiver of SOLAR TWO (Barstow, USA) (SANDIA).

7.1.1.2 Design and Function

First of all, we demonstrate the basic design of a salt receiver and the function of the molten-salt circuit, using the example of the SOLAR TWO receiver.

This receiver consisted of 6-m-long stainless-steel tubes of 2 cm diameter, of which 32 were connected to form a flat "panel." Twenty four of these 6-m-high panels were then arranged to form a large, hollow cylinder, whose diameter was about 5 m (Figure 7.1). Roughly, the same ratio of diameter to height is also used for larger receivers. The geometric "size" of the receiver is defined by the surface area of the mantle of the cylinder; in the case of SOLAR TWO, it was 100 m². Its nominal power was 42 MW_{th}. In the case of SOLAR TRES (Spain), a power capacity of 120 MW_{th} is planned, whereas the large-scale receiver of a 200 MW_{el} plant (with an SM of 2.9) would have, according to S&L (2003), a nominal power of 1400 MW_{th} (diameter, e.g., 22 m, height 28 m, surface area 1900 m²).

The molten salt flows in parallel through all the tubes of a panel, but in the opposite sense through neighboring panels. If, for example, cool molten salt – at a temperature of 290 °C, coming up to the vertical pipe ("riser") from the cool salt storage tank – flows from the top to the bottom of the tubes in the first panel, it will then be collected in a manifold and passed on to the next panel. There, it flows from the bottom to the top and is further heated, etc. The salt flow is thus along a serpentine through a series circuit of 12 panels (in the case of SOLAR TWO) that form one leg; the receiver consists of two such legs. The two legs, which are in parallel, start at neighboring panels on one side of the cylinder, run in opposite directions, and end at two neighboring panels on the opposite side. From there, the hot molten salt (565 °C) passes down a vertical pipe ("downcomer") to the hot storage tank, where a portion is diverted directly to the steam generator in the conventional part of the power plant (steam turbine).

The fundamental problem now consists of employing measuring devices, pumps and valves to fine-tune the flow of molten salt and the heliostats so that the exit temperature from the receiver is always 565 °C. This must in particular be assured even when clouds are passing over and during start-up in the mornings and shut-down in the evenings. It is especially important that no sudden disturbance of the molten-salt circuit be allowed to occur, which could cause the receiver tubes to overheat.

Each evening, the salt is drained out of the receiver and the entire piping system of the receiver circuit. Before it can be pumped back into the circuit the next morning, the piping system must be preheated by a complex electrical heating system, the so-called heat tracing system (heating tapes around the piping and valves, referred to for short as the "heat trace"). The receiver is also preheated by focusing a suitable radiation density onto its surface from a limited number of heliostats. We note here that the heat-trace system – after causing some initial problems at SOLAR TWO – later functioned well, and should not represent a decisive problem in future power plants.

7.1.1.3 Developmental Requirements

Molten salts have notably lower heat conductivities than liquid metals (e.g., sodium). For this reason, the permissible power density of the concentrated solar radiation onto the receiver surface (i.e., the radiation flux density) had to be kept rather low in the absorber tubes used at Barstow. SOLAR TWO worked with radiation flux densities of up to ca. $0.85\,\text{MW}/\text{m}^2$ (compare the steam receiver of SOLAR ONE: maximum only $0.3\,\text{MW}/\text{m}^2$). The short-term development goal (for SOLAR TRES) is 1.2 or at least $1.0\,\text{MW}/\text{m}^2$ (Zavoico, 2001, Lata et al., 2006); the longer-term goal (for 2018) is $1.6\,\text{MW}/\text{m}^2$ (S&L, 2003). This would permit the receiver to be made smaller and saves not only on weight and investment costs, but, in particular, it would reduce the energy losses (which are proportional to the receiver's surface area).

The permissible wall stresses in the tubes of the receiver are a limiting factor for the radiation density, and these stresses, which arise from the temperature gradient in the tube walls, depend on the wall thickness. For the SOLAR TWO receiver, the wall thickness was 1.25 mm. In the "advanced receiver," the use of thin-walled tubing is planned, with a wall thickness of approximately 1 mm or less (precise values have not been published, but according to Zavoico (2001, p. 57), the use of certain nickel alloys should permit an increase of the radiation density from $0.85\,\text{MW}/\text{m}^2$ at SOLAR TWO to $1.0\,\text{MW}/\text{m}^2$). Such thin walls have not been used so far, although they would be feasible in terms of mechanical technology, since the receiver is essentially a pressure-free system. Suitable materials and corresponding fabrication techniques must be developed and tested, for example, for welding the thin-walled tubing. This will thus represent a special development program for the particular application conditions at hand.

The Sandia estimate (Kolb) started with the assumption that among other things, the use of thin-walled tubing (and correspondingly smaller receivers), together with new coating materials for reducing the losses due to infrared reradiation, would permit the efficiency of the receiver to be increased from 78% to 87%. In a study published in 1997 for the US Department of Energy and the Electric Power Research Institute, they assert that (by using selective coatings) in the long term (2020), "around 90%" would be possible (DOE, 1997). Sargent and Lundy cite a possible increase from 76% (SOLAR TWO) to 83.5% (SOLAR 200) by the year 2012 and likewise mention a "decrease of receiver emissivity from selective coatings" (S&L, 2003, p. E-50).[2]

In addition to the fabrication of thin-walled tubing, this presumes the development of suitable coating materials. The latter is a strongly "basic research"-oriented task. To accomplish it, university research institutes and the laboratories of appropriate chemical firms could be engaged. This development involves a mixture of basic research and applied process technology. For the latter part, only the chemical industry can be considered. The fundamental investigations, in contrast, could be the subject of an invitation for proposals worldwide. Suggestions and concepts could be collected and further developed through research contracts. Organizationally, the basic research program should form a separate part of the overall development effort.

The search for an optimal tubing material is of a similar nature (materials research). The material that was used for the SOLAR TWO receiver, that is, stainless steel, is not sufficiently corrosion resistant under the influence of the hot nitrate salts at 570 °C.[3] In particular, the cause of the corrosion is chloride impurities in the salt mixture, as well as moisture from the air that enters the piping whenever the system is drained of molten salt. On the one hand, this is "uniform" corrosion; on the other, it also involves intergranular corrosion.[4] Corrosion is promoted by thermal loading (mechanical stresses), which can also cause damage to the protective oxide scale. Generally, stresses due to the high temperatures (thermal fatigue) are very important, especially the cyclic thermal loading of the material due to the frequent rapid temperature variations (low cycle fatigue). In

---

2) In the publications on SOLAR TWO and SOLAR TRES mentioned below, these selective coatings are not treated. Evidently, this topic is still at the stage of basic research, in any case for the relevant temperature range of around 700 °C, so that it played no role in the short-term planning of solar tower plants. For the absorber tubes of parabolic-trough plants, however, recent developments are showing signs of success. (An excellent overview of this very complex area of research is offered by Kennedy (2002).)

3) For the "cool" components of the molten-salt circuit (including the cool salt storage tank), normal carbon steel is adequate.

4) Corrosion along the grain boundaries, often also referred to as "intergranular attack."

particular, the welded joints are subject to corrosion (at the transition zones between the receiver tubes and the upper and lower valves, the so-called headers). The greatest thermal stresses occur in the transition regions from the thick-walled headers to the thin-walled header nozzles. However, this problem may possibly be solved by using the novel (patented) construction of a "thin-walled header" (Marko, 2004).

After the SOLAR TWO receiver exhibited leakage, among other measures, a single panel made up of a high-nickel alloy was tested with a positive result (which, however, must be considered preliminary due to the brevity of the test phase). CIEMAT and SENER are carrying out systematic investigations in connection with SOLAR TRES using various high-grade alloys based on nickel (superalloys with ca. 60% Ni, ca. 20% Cr, and varying fractions of Mo, Co, Nb, Ta, and W), as well as with an Austenitic chrome–nickel stainless steel (30% Ni). The results have not yet been published (Lata et al., 2006).

As mentioned, these investigations have as their initial goal the discovery of a material that would be suitable for a radiation density of $1–1.2\,MW/m^2$. It is clear that considerable effort will still be required to finally attain the longer-term goal of $1.6\,MW/m^2$ quoted in the Sargent & Lundy study.

Along with materials and fabrication questions, geometry plays an important role. Reduction of the tubing wall thickness has already been discussed; in addition, the tubing diameter is also relevant (for SOLAR TWO, ca. 20 mm). On the one hand, it should be as small as possible because then the flow velocity of the molten salt is high, favoring heat transfer from the walls to the salt. On the other hand, a small diameter increases the pressure gradient along the tubes, causing a corresponding power loss (pumping power). In the optimization procedure, it must also be taken into account that fabrication costs increase when thinner tubes are employed (Lata et al., 2006).

The developmental work that is still required in view of these problems was on the whole presaged by experience with SOLAR TWO, and has to some extent already been carried out in the USA (insofar as this was possible given the restrictive research support policies of the Federal Government). Especially in connection with the SOLAR TRES project in Spain (CIEMAT, SENER), intensive development of the molten-salt receiver is being continued. The well-documented and large pool of experience in the USA (Sandia, Rocketdyne[5]) serves as a basis for this development, as well as their own experience from the 1980s (CIEMAT, molten-salt receivers in Almeria).

After the most important industrial know-how bearer, namely, Rocketdyne (Boeing) opted out of the SOLAR TRES project, it was necessary to continue developing the molten-salt receiver in Spain. To some extent, this development was begun again from the beginning; even the basic concept of the tube receiver was reconsidered (and reaffirmed) in comparison with other concepts, among

---

5) Rocketdyne belonged to Boeing until 2005 and was then taken over by the United Technologies Corp.

others with the plate receiver and also with the film receiver, in which an internal salt film (or even directly an external salt film) is heated.

Although there have been no publications, it can be assumed that Rocketdyne has continued with further development of the molten-salt receiver subsequent to the SOLAR TWO project.[6] At least there have been plans in South Africa since around the year 2000 to build a 100-MW molten-salt tower plant, as well as (since about 2007) newer, concrete planning of solar tower plants in the southwestern USA, in particular with the direct participation of United Technologies Corp./Rocketdyne via the firms Hamilton Sundstrand and SolarReserve.

Regarding the technology and the developmental requirements and perspectives for the molten-salt receiver, we refer the reader to the extensive reports on the overall SOLAR TWO project (PIER, 1999, Zavoico, 2001, Reilly and Kolb, 2001), to the specific report of Litwin (2002) on the SOLAR TWO receiver, and to the Sargent & Lundy study (S&L, 2003), as well as to a pair of likewise very interesting short articles on current developments at the Spanish SOLAR TRES project (Lata et al., 2006, Ortega et al., 2008).

7.1.2
System Development: Molten-Salt Circuits and Receivers

The systems development of modular-structured systems is characterized among other things by the fact that the individual components can be developed separately, and this avoids time delays due to the mutual coordination of the partial developments. In the case of solar tower power plants, these would be: the *heliostat*, the *receiver*, and (here) the *molten-salt circuit*. These can be designed to a large extent independently of each other, and independently tested and investigated in terms of their costs under mass-production conditions.

For complex systems such as nuclear power plants, in contrast, the "scaling-up" method is employed (however, not exclusively; see the note below). Through the construction of small- and medium-sized pilot plants, the interactions of the individual plant components in practical operation are tested. The experience gained is then transferred to the next-larger plant. In the case of solar power plants, a pure scaling-up would mean that the receiver and the molten-salt circuit (i.e., the heat-transfer piping, pumps, heat-storage tanks, materials development etc.) could only be tested when large heliostat fields are available. This, in turn, would for economic reasons require a precondition that the heliostat development process be essentially completed before plant testing could begin – with the result of a considerable and unnecessary delay in the overall development. Naturally, even for a solar power plant, the interactions of all the subsystems play a certain role, but this is more or less secondary in the framework of the overall development process.

---

6) In the Sargent & Lundy study, it was stated in this connection: "Boeing is presently spending significant money (not disclosed due to confidentiality) on industry research and development." (S&L, 2003, p. E-49).

The development of nuclear-power technology was indeed carried out in the main by the scaling-up method; however, there were also elements of a modular development. Where expedient, individual components (modules) were developed independently of the complex structure of the entire power plant. Thus, for the German high-temperature reactor, a large helium-gas turbine was planned. Since there was no existing experience with this type of turbines, a helium turbine was first constructed on a small scale (50 MW$_{el}$) and tested using natural-gas combustion as the heat source (helium gas turbine power plant, Oberhausen, Germany). This test installation had all of the important technical characteristics of the planned large facility (Bammert and Deuster, 1974). According to the existing development plan, by the time the large plant was to be put into operation, there would have been more than 10 years of operational experience with the smaller turbine. This efficient method should be an absolute matter of course for the development of solar energy. And just as in the case of nuclear energy, where a fossil-fuel power plant was constructed to test nuclear components, a similar path could be followed for developing solar energy plants (see the following section).

7.1.2.1 Molten-Salt Circuits

With modular systems, the individual modules can be tested directly at full size. For cost reasons, however, one could start with an intermediate-sized version that could be scaled up, in particular, for the heat-storage tanks and the components of the molten-salt circuit. As with the development of nuclear power plants, these could be initially constructed on a semitechnical scale and tested practically, using fossil fuels to heat the salt. The complete salt circuit could be tested in this way and developed to maturity. Such a test installation would comprise the complete thermal part of a solar power plant (e.g., with 30 MW$_{el}$ output power) using a molten-salt thermal circuit, in which, however, the receiver would be replaced by a salt heater. The heat-storage tanks would indeed be smaller than those planned for full-scale use, but of similar design. The same holds for the molten-salt piping, pumps, steam generator, and so on. Possibly, one could dispense with the power block. The steam produced in the steam generator could instead be fed into an existing fossil-fuel power plant or heating plant, which would be energetically more favorable and less costly. (The steam turbines for the planned future large solar power plants are already standard today, and would require no special testing.) A long-term test of the thermal part of the plants would also not cause additional energy consumption, since the steam could be used for other purposes.

All of the individual technologies are fundamentally known and would need only to be optimized and tested for the present application. Thus, the proposed molten-salt test bed with fossil-fuel heating (but without its own turbines) could be planned within one year and start its operation within 2–3 years. With a power output as mentioned above of 30 MW$_{el}$ – in base-load mode – such a semitechnical plant would be ca. 10 times larger, that is, would produce daily 10 times as much energy as the test installation SOLAR TWO (10 MW, SM 1.1).[7] Heat-storage tanks, steam generator, piping cross sections, pumps, the electrical heating of the molten-salt piping,[8] and others could all be scaled up.

The riser piping to the receiver could be tested by using a loop which could perhaps be mounted in the chimney of the fossil-fuel power plant (where the steam from the test plant would be input). For simulating the horizontal piping (i.e., the thermal connection between individual solar towers) including insulation, expansion joints, monitoring equipment, etc., for example, a loop at ground level could be employed. The test installation would not have to be located on the site of the fossil-fuel plant, but instead could be built at some distance and connected to it via this horizontal piping.

The operation of the molten-salt circuit could be simulated in time-lapse mode in such a test installation. In this way, the materials behavior of the piping, regulation of the pumps with variations of the thermal power level, for example, due to passing clouds, and the start-up and shut-down cycles (day–night cycles) could all be tried out under realistic conditions. One could start the system up and shut it down more frequently than once every 24 hours, for example, 4–6 times per day, and thus within 5 years could simulate the stress due to cycling in 20 years of operation.

Such a semitechnical installation would be systematically oriented toward the planned future large plant. All of the technologies to be used there would be included in the test installation; the test results would otherwise not be completely meaningful. We mention these obvious points once again here precisely because they were often not observed during the development of solar energy thus far. So, for example, the results from a small test installation for air receivers that was put into operation in Spain in 1993 were transferable in only very limited fashion to larger plants (scaling up), and in particular gave essentially no useful information on their costs.

All of the individual novel and possibly cost-saving technical developments that are held to be feasible for future solar power plants would have to be integrated into the test installation (determination of the innovation potential!). It would not be the goal of the test installation to prove the functional capability of a plant of semitechnical size, but rather to demonstrate the "future technologies" of the large-scale plant. It would therefore have to be designed to be flexible, that is, also capable of extension, so that newer developments could be immediately tested in an intensive and realistic manner. It should be emphasized here that the presumed implementation of the plans for SOLAR TRES – which is very much to be hoped for – or of the South African or the new planned American plants – does not affect the necessity for such test installations, since they are important for the maturing and scaling-up of new technologies.

The test installation would be a complete fossil-fuel power plant (but without its own turbine), so that its cost can be roughly estimated by comparison to the usual investment costs for power plants: Taking an assumed upper limit of 4000 \$/kW$_{el}$

---

7) Compared with the Spanish SOLAR TRES project (15 MW base load), such a test installation would be twice as large.

8) Heating is necessary, since the salt solidifies at ca. 240 °C; heating the piping therefore belongs to the necessary procedures for daily start-up of the plant and in the case of longer downtimes.

(for comparison: coal-fired plants including turbine cost ca. 1200 $/kW$_{el}$), a test installation of 30 MW$_{el}$ would cost at most ca. $120 million.

7.1.2.2 The Development of Hybrid Boilers

Solar energy power plants for export to countries within the "sun belt" or in general for use without long-distance power transmission are as a rule equipped with fossil-fuel-heated auxiliary boilers for backup power on days with little or no solar radiation (*hybrid power plants*). Aside from natural gas, coal is also a preferred fuel for reasons of availability. In addition to a heating unit fueled by natural gas, also two different coal-fired heaters for molten salt should be constructed, one of them with a fluidized-bed combustion system (for smaller solar power plants) and one with pulverized coal combustion, as is usual in large plants. The goal of the construction and operation of these components would be, first, a practical test of the molten-salt heat exchanger; and second, in the course of the development, design documentation would be obtained that could serve as a basis for estimating the investment costs of future large-scale plants. This would hold for all of the components of the molten-salt circuit. A molten-salt heat exchanger is in terms of costs quite different from a steam generator, since the salt is at normal pressure and has very different heat-exchange properties, and since the combustion air must be preheated to a relatively high temperature. The costs could thus only be determined reliably by using such a test installation.

7.1.2.3 A Test Installation for Receiver Development

In order to be independent of solar radiation during materials testing for the receiver tubes and the testing of complete receiver panels (containing many parallel tubes), it should be clarified whether the construction of a special fossil-fuel heated facility is feasible. In analogy to the boilers used in fossil-fuel power plants, a radiation chamber could possibly be constructed in which the same radiation flux density would be attained as at the solar tower of a solar power plant. In the boiler of a fossil-fuel power plant, a major portion of the heat is transferred to the steam tubes via radiation. The intensity of the radiation depends on the combustion temperature and the geometric arrangement of the boiler. The situation within the boiler is comparable to that in a receiver (cf. Bammert and Seifert, 1981). Along with the special construction of the combustion chamber, an increase in the temperature of the natural-gas flame by preheating the combustion air would be necessary.[9] The energy produced by the test facility could once again be used for steam generation. The receiver test facility should furthermore be a separate installation, independent of the test plant for the molten-salt circuit, since different cycles would need to be programmed from those used for the testing of the long-term operation of a molten-salt circuit. The facility could, however, be built adjacent to the molten-salt test plant and could feed its steam into the same turbine.

In developing a molten-salt receiver, along with the usual optimization of the overall technical design, two essential problem areas must be distinguished:

---

[9] The resulting high degree of nitrogen oxide formation can be compensated by an NO$_x$ catalytic filter in the exhaust of the combustion chamber.

1) The *reliability* and the *service life of the molten-salt piping system* (thin-wall technology) in view of the high radiation density and the frequent temperature variations. This could be investigated in long-term time-lapse tests. In 24 h, for example, 5–10 start-up and shut-down cycles and a corresponding number of insolation variations due to passage of clouds could be reproduced. As with the molten-salt circuit, within two years of testing, 10 to 20 years of operation could be simulated.

2) The development and testing of the coatings intended to reduce reradiation of heat, that is, the *selective absorber* coating. Its long-term thermal stability could likewise be tested in the radiation chamber, but, however, not its optical properties, since the radiation spectrum within the chamber would not be the same as with solar radiation. The optical properties could be determined by parallel testing at one of the large solar research centers (e.g., Sandia, NREL, PSA).

As in large boilers or steam generators, in the test facility, the combustion region would be separated from the tube walls (tubes carrying molten salt) by a suitable geometrical arrangement. The air flow would be adjusted so that their surface temperature would correspond to that of the tubes in a receiver. This could be accomplished, for example, by using a cool-air curtain (air inflow). The distribution of radiation on the receiver surface in solar operation (the radiation intensity is highest at the center and decreases sharply toward the upper and lower edges) could be simulated using heat-resistant (ceramic) baffles between the combustion chamber and the tube surfaces. The length of the tubes in the test facility could probably be considerably shorter than in the actual receiver. This should have no significant effect on the test results, since the expected radiation densities could be simulated on shorter tubes just as well and the relevant test results are those applying to the zones of the tubes subject to the greatest stress. The test facility could thus be smaller and less expensive. The cost of such a panel test facility – that is, a (most likely) relatively small installation – could be in the range of some $10 million, it should be, however, less than $50 million.

Here, we must emphasize that within the framework of an extensive development program – in view of the central importance of a low-cost and reliable receiver – even higher costs would be justified and manageable.

7.2
Air Receivers

7.2.1
Technology

An alternative to the molten-salt heat transport circuits (SOLAR TWO) is the system developed mainly by European firms (among others Sulzer, Steinmüller, Kraftanlagen München (KAM)) and research institutions (DLR, CIEMAT, SIJ at

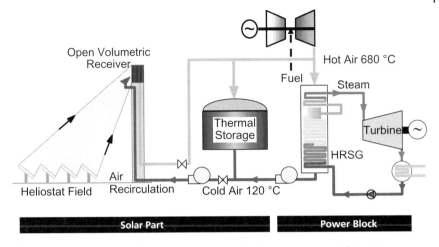

Figure 7.2 A tower power plant with an atmospheric volumetric receiver (Photo SIJ).

the Aachen University of Applied Sciences) that uses air as the heat-transfer medium (*atmospheric volumetric receiver*)[10]; it has been discussed over a number of years under the name "PHOEBUS Concept." This innovative concept exhibits attractive advantages but also some disadvantages. Overall, it appears to us from *today's* viewpoint – as was also true in the 1990s (Becker and Klimas, 1993) – to be marginally inferior to the molten-salt concept. This assessment can of course be only preliminary, since the advantages and disadvantages of the two concepts can be reliably quantified only after the completion of a serious development program. An important advantage of the PHOEBUS concept is that it would permit hybrid operation with a natural-gas-fueled combined-cycle circuit (Hoffschmidt, 2007) (Figure 7.2).

From today's viewpoint (2009), there is certainly a chance that the PHOEBUS concept would prove to be less expensive than the use of molten salts. It thus represents a very interesting line of development. This is all the more the case in view of the possibility that the hoped-for progress in the molten-salt concept may not be achieved.

An important milestone for the air receiver is the construction of a first demonstration and test power plant with an output power of 1.5 MW$_{el}$ in Jülich (Germany); it has gone into test operation at the beginning of 2009. The project, initiated by the DLR and the Solar Institute in Jülich (SIJ, Germany), is being constructed by KAM as general contractor and will be marketed in the future. Here, it is also certainly interesting that the plant, which will be operated by the Jülich Public Services Department, will allow for a continuous further development by the research team. (Whether the actual development will be carried out

---

10) For the pressurized version cf. Section 3.2.

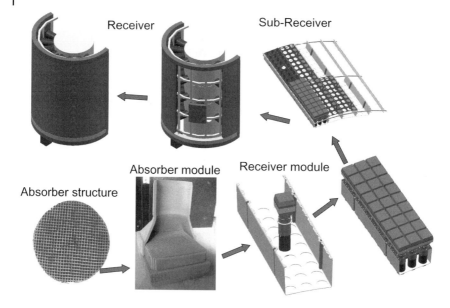

Figure 7.3 The elements of the volumetric receiver (Photo SIJ).

to the necessary extent and with the required breadth is, however, still an open question.)

The special feature of this concept is the "open" air circuit that operates at atmospheric pressure and the use of a "volumetric" absorber. Instead of tubes as in the molten-salt receiver (and in earlier air receivers), the first models for a volumetric absorber consisted of a "wire mesh" that is heated by the solar radiation. In recent years, the development has moved toward the use of porous ceramics, as they have proved to be more stable. The receiver is then assembled out of many individual rectangular or hexagonal ceramic modules through which air from the surroundings is input (see Figures 7.3–7.5). This air is heated by the hot ceramic surfaces to roughly 700 °C and is passed down to the foot of the tower, where it is used to generate steam for the conventional power-plant section.[11]

Air as heat-transfer medium makes it possible – together with the ceramic absorber material – to construct a very simple, light, corrosion-resistant, and inexpensive receiver that can accept a high radiation-flux density. Furthermore, volumetric receivers have a "remarkably low thermal inertia" and "quick sun-following properties" (Marcos *et al.*, 2004).

However, this elegant principle – the intake of atmospheric air – possesses two disadvantages:

11) In current designs, the conventional part of the plant, that is, the steam generator, thermal-storage reservoir, and the turbine-generator, are assembled at the top of the tower, in order to minimize losses due to the flow of hot air through piping. In large-scale systems, this conventional part would be on the ground.

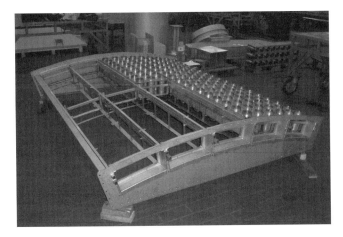

Figure 7.4 Subreceiver (Hoffschmidt, 2007).

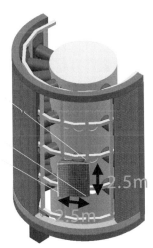

Figure 7.5 Complete receiver, schematic (right), and testing of one subreceiver (left) (Photo SIJ).

In the steam generator, only the temperature difference between the hot air at 700 °C and ca. 170–200 °C can normally be exploited with an optimal thermal efficiency by the power block (Becker and Klimas, 1993). (Of course the air can be cooled to a lower temperature, but the effectiveness of transforming its energy content into electricity is in principle somewhat poorer at the lower end of the cooling process.) The remaining temperature difference (e.g., from 170 °C down to the ambient air temperature) is initially useless. This is compensated by passing the air after it has been cooled to 170 °C in the steam generator back through a second airflow pipe to the top of the tower. There, the still hot air at 170 °C is blown through an air-channel system in front of the intake zone of the receiver (*air-return system*). In the theoretical ideal case, preheated air at 170 °C would thus be input to the receiver. This would represent a closed air circuit with an air-recovery quota of 100%. Because of the mixing with outside air, which naturally depends on the wind velocity, the receiver geometry, and the particular return system,[12] one, however, can achieve an air-return quota of only ca. 60%; the rest of the preheated air is lost to the atmosphere. This return quota is a preliminary design value; it must still be verified for large receivers. The average intake temperature would thus be only 110 °C.[13] This means that ca. 10% of the energy that was initially contained in the hot air after its passage through the absorber would be lost.[14] In the new design, recently realized in a prototype plant in Germany, the air is cooled to 120 °C (Figure 7.2). Cooling the air in the steam generator to this temperature (instead of to 170 °C), with a "small" reduction in thermal efficiency, is possible only by reducing the temperature difference (between air and steam) in the steam generator, and this requires a somewhat more expensive generator. Here, not only are the problems of the air-return system reduced (at the same air-recovery rate, the losses are less by one-third), but also air at 120 °C is more favorable for cooling the metal structure of the receiver. Thus, this concept is currently preferred.

Regarding the problem of air return, we refer to a publication by Marcos *et al.* (2004), in which this topic is treated in connection with a report on the corresponding computer simulations. The authors emphasize that an air-return quota of over 80% must be the goal, in contrast to the range of 45–70% thus far achieved. Using open volumetric receivers in more efficient thermodynamic circuits with air-return temperatures of more than 400 °C would (according to Marcos *et al.*) require a return quota of even up to 90%. They give their justification for the development of the simulation tool as: "The ignorance currently existing as to what relates to the phenomenology of air return and its complexity, as well as the large number of possible geometries. ..." It is clear that – on the basis of these simulations – the corresponding development at the hardware level must be expedited by the appropriation of sufficient support funding.

The second disadvantage of the air receiver may turn out to be the question of *heat storage*.

One of the storage concepts is based on a recipient with a packed bed of small *concrete* or *ceramic* balls or especially shaped so-called saddles (of ca. 2 cm in size). To charge up the reservoir, the hot air arriving from the receiver is passed through this bed and warms the material in it. Discharging is the reverse process: cool air is passed through the bed and then into the steam generator.

---

12) In the currently favored air-recovery system, the returning air is fed through the gaps between the absorber modules in the front of the absorber, whereby the "cold" air is at the same time used to cool the mounting structure of the absorber.

13) 170 °C − 20 °C (ambient air) = 150 °C. 150 °C × 0.6 = 90 °C. 90 °C + 20 °C (ambient air) = 110 °C.

14) 700 °C − 110 °C = 590 °C. Losses: 170 °C − 110 °C = 60 °C.

Concerning the costs of such storage reservoirs on a large scale, we can at present make no hard and fast estimates. In principle, they could be less costly than the salt tanks for a molten-salt circuit; in particular, concrete as heat-storage medium would be considerably cheaper than salts. Whether this still holds for the overall concept, including the preparation of the packed beds, airflow channels, air blowers, insulation, etc. cannot be stated with any certainty at this point.

The packed beds would have to consist of several layers, since the layer thickness cannot be too great due to pressure gradients. A large system has to our knowledge not yet been designed, but rather only a small test plant, whose design cannot be directly applied to the construction of a large-scale facility. Most likely, a multistoried structure of individual heat-storage layers with separate air channels for each layer will be required. For such a multistoried design, there are as yet no cost estimates, just as for the overall large-scale plant. In particular, the fabrication costs of the grains for the solid bed in mass production are unknown, as are those associated with the special transport requirements. In general, it can be stated that the concept development for solid-bed storage reservoirs is tending toward ceramic materials, owing to their superior high-temperature stability as compared with concrete.

An especially interesting and relatively new storage concept (which was already mentioned in Chapter 3) is the *sand-based heat-storage reservoir*. Here, the hot air arriving from the receiver is passed through quartz sand, which is thereby heated and then falls directly into the storage reservoir. Making use of a fluidized-bed heat exchanger for steam generation, the reservoir can be discharged directly. The cooled sand is then stored in a holding tank.

The sand reservoir has two main advantages relative to concrete or ceramic solid-bed reservoirs:

a) The materials costs are extremely low.

b) The pressure gradient within the air circuit is smaller and also does not depend (as for solid-bed heat storage) on the size of the reservoir, thus yielding lower energy losses or a higher overall efficiency.

The goal is now to develop a sufficiently inexpensive and reliable technology within a short time. The first steps have been taken in Germany in the last few years. On the basis of computer simulations, in which parameters such as the air velocity, grain size of the sand, or external dimensions of the reservoir play a role, the first experiments with a prototype have been carried out on a laboratory scale. These have shown that "a heat transport system based on the working principle described, which can meet the requirements for heat transfer, is feasible" (Warerkar et al., 2007). On the contrary, the experiments showed – as was to be expected – that the technological development will not be trivial and that a large number of further investigations are still necessary.

For an example, the air intake and outflow take place through walls that are permeable to the air. These could be made of a fine steel mesh (with openings smaller than the sand grains), which, however, would have to be followed up by dust filters. Another possibility is the use of porous walls made of silicon carbide.

However, there are mechanical problems (crack formation), and furthermore, the pressure gradient is too high, in spite of the limited thickness of the walls (2.4 mm) and their high porosity (45%).

The fundamental problem of the sand-based heat-storage system lies in its stringent requirements for the thermal stability of the construction materials – as in all high-temperature systems – and at the same time, their extreme abrasion resistance due to the hard quartz sand used.

A disadvantage of the air receiver at first view is also the fact that with air, in contrast to molten salts, the interconnection of a number of towers into a large-scale power plant (with a single turbine) is not feasible; such interconnections would, however, probably be very expedient for solar parks. Enlargement by adding more towers, and the resulting reduction in the specific cost of the steam-turbine plant, as well as the possibility of supplying heat to neighboring plants during solar operation at reduced power output (to minimize the partial-output losses in the steam cycle) would then not be an option. However, using a sand-based thermal storage system, several towers could be interconnected since the hot sand can be transported (here, conventional solutions are available).

If the receiver temperature can be increased, the hot sand (at >800 °C) could also be used in a pressure-loaded fluidized-bed heat exchanger. In this case, one would obtain a 100% solar-heated, base-load CC power plant (since it would have a large heat-storage reservoir). Thus far, there is no other design that would make a 100% solar CC power plant feasible at all. This process has been patented by the DLR. Work on the sand-based storage reservoir represents a first step in this direction (Hoffschmidt, 2007).

Conclusions: The high-temperature sand-based heat-storage reservoir possibly represents a decisive option for solar power plants using an air receiver. It in any case offers an important potential alternative to the concrete or ceramic solid-bed heat storage systems, in case the development of the latter should be delayed or even fail. Since it in addition may prove to be more economical, its real potential should be clarified as soon as possible through a dedicated research program.

Two general disadvantages of heat storage for an air receiver are the energy required for pumping the air (especially through the solid-bed reservoir) and the temperature gradient between charging and discharging. The latter can, however, possibly be partially compensated by optimization toward the direction of higher input temperatures, that is, higher receiver temperatures,[15] so that this disadvantage may not be too serious. The operating steam parameters would then be the same for night operation as for daytime operation.

A still not precisely known cost item for the air receiver is the fabrication of the long *airflow piping* from the top of the tower to the ground and back. In the originally planned (in the 1990s), practically construction-ready PHOEBUS project

---

15) A temperature increase can be accomplished for a volumetric receiver with a lower accompanying increase in loss rate than for a salt receiver, since due to the absorption within its volume, the corresponding increase in the surface temperature, which determines losses through reradiation, is not as great.

(which was then, however, not built – mainly for financial reasons within the company), no thermal storage system was planned, since the facility was designed as an "export power plant" for the entry market. Without a heat-storage reservoir, there was no need to pass the air from the receiver to the ground. Instead, the steam generator was to be mounted on the tower just below the receiver. The problem of long air channels was thus circumvented in this type of plant. Using natural-gas heating on the tower between the receiver and the steam generator, the nighttime hours and cloudy days were to be bridged over (this was not a baseload plant!). The costs of airflow piping were not further investigated in the following years.[16]

In this connection, another (potential) advantage of the air receiver should be mentioned: With the air receiver, higher towers, larger receivers (in the sense of a higher thermal power capacity), and larger heliostat fields tend to be more readily attainable than with a molten-salt receiver. With molten salt, the hydrostatic pressure of the column of fluid increases with increasing tower height, which results in an increase in the required pumping power; and also the lower weight of an air receiver relative to its power capacity favors larger installations. On the contrary, in the cost optimization, the costs of the longer airflow piping come into play – a typical example of the multiple aspects of the optimization problem for solar power plants.

7.2.2
Development

Within the framework of the system development, in addition to the further improvement of the receiver itself (including materials research), which we shall not discuss further here, it is important to treat the three problem areas for the air circuit, each with its own subprogram: the air-return quota, the thermal storage system, and the airflow piping. We begin the description with the last-mentioned, simplest task, but we should also point out that a sand-based thermal-storage system would render the airflow pipes to a large extent obsolete.[17]

7.2.2.1 Airflow Piping
Determining the costs of the airflow piping requires a detailed design study and cost analysis. If a practical test and review should prove necessary, hot-air piping of the required dimensions could be laid on the ground. The air could be heated using natural gas and fed into a steam generator (with further use of the steam generated, as in the case of a molten-salt circuit, in a power plant or heating plant). Various techniques for dealing with thermal expansion, different insulating mate-

---

16) Subsidizing this version of the power plant with public funds, as was suggested at the time, would thus have yielded no information at all about the costs of the thermal storage system or those of the airflow piping.

17) Sand would thus be used not only as *storage* medium but also as *transport* medium for heat energy.

rials, the air blowers, regulation of the blowers to compensate for variations in insolation, etc. could in this manner be tested, and realistic operating conditions could be simulated.

7.2.2.2 Heat Storage Systems

Development and testing of an inexpensive large-scale heat storage system (*concrete or ceramic solid-bed*):

A test installation for the thermal storage system could readily be integrated into the natural-gas-heated test facility described above. The thermal reservoirs to be tested – of a size which would allow scaling-up – should exhibit all of the design characteristics of the planned large-scale storage system and should therefore be planned as a modular system, with all of the components of the large-scale reservoir. Possible modifications could be built into the test facility on demand. The facility would therefore have to be designed to be readily modified and to allow a rapid retrofitting of desired revisions. Every variant that would possibly be less expensive in mass production, for example, balls or saddle-shaped bodies of different sizes or different materials, variations in airflow-channel design, or a different static solution for a multistoried bed, would have to be able to be tested in modular fashion. In the final version, such a test facility might be of a size corresponding to a power output of $10\,MW_{el}$ with a storage capacity of 15 h. Its costs including airflow piping should not exceed $30–40 million (at most $4000 $/kW$_{el}$).

Parallel to the design and testing, cost analyses for mass production would need to be carried out: For example, regarding the concrete balls, the prefabricated parts for the multistoried structure of the reservoir, the insulating material, or the air blowers of appropriate power, this always assumes very high production rates. To determine the cost of the fabrication of concrete parts, it would possibly be necessary to carry out large-scale preliminary tests with the machinery to produce the shaped particles (balls or "saddles").

A development program for the *sand-based heat storage system* – as suggested above – would need to be set up and carried out in a similar manner.

7.2.2.3 Air-Recovery System

A major goal is the improvement of the design in order to increase the air-return quota. A reduction of air losses in the intake zone of the receiver is extremely important for the economic performance of the concept. If the air-return quota could be increased from 60% to 80%, the initial principal disadvantage of this concept would be rebutted. For the development and testing of a "sophisticated" layout for the airflow, operation within a real solar power plant would not be necessary. This operation can most probably be simulated using cool air (with the advantage of greatly reduced energy consumption) and using towers of a modest height. If the air for testing were heated by a gas burner to, for example, 10 K above the temperature of the ambient air, and if the temperature of the intake air were 6 K above the ambient, then one could read off an air-return factor of about 60%. (However, a correction factor for the stronger thermal motion of the hot air in realistic receiver operation could, if necessary, be included in the evaluation of the test results.) The energy consumption of such a test facility would be negligible owing to the minimal temperature differences, especially since in this facility, no long-term tests need be carried out. For the latter reason, these tests could, however, also be performed at the actual air-return temperature, in case they proved not to be feasible with cooler air.

The test receiver could be mounted on a very low tower (of height, e.g., 20–40 m). It could have the actual dimensions of a large-scale receiver, but would not necessarily be constructed of the same materials. The air-return factor depends to a considerable extent on the wind velocity, and this in turn on the height of the tower[18]; similar velocities would be present at the low test tower, but not with the same frequency of occurrence.

To improve the air-return quota, an active regulation of the air-recovery system will be necessary. This regulation – in various versions – could be tried out in the test facility and continuously improved with the aid of the measurement data obtained. (As mentioned, these would not be long-term tests, but rather a series of measurements under different wind conditions.) The air-recovery system (a purely mechanical system) could also be developed and optimized within a short time given a suitable commitment of personnel, at least so far that one would have considerably more reliable data than at present. The costs of such a facility and of the continuing development of the air-recovery system can at present not be estimated. After completion of the preliminary studies, we would have more precise figures.

7.2.2.4 Test Installation for Receiver Development

For the development and testing of the air-receiver modules, a radiation chamber similar to that described for the molten-salt receiver would be required. Absorber design, regulation of the air intake (depending on the radiation density), and also different absorber materials could be subjected to long-term tests in this facility. The radiation chamber would be different, in terms of its geometrical configuration, from the one described for molten-salt receivers – at least in part. For the air receiver, the strong airflow due to the intake would have to be taken into account. The input air would have to be brought to the receiver, possibly using compressed-air jets, and the distance to the flame wall producing the radiation would have to be greater.

---

18) The wind velocities at the corresponding height at planned sites could be measured, for example, in Spain from the chimneys of fossil-fuel power plants

8
Parabolic-Trough Power Plants

8.1
Basic Facts

The Californian parabolic-trough power plants (Figures 8.1 and 8.2) are – apart from a 64-MW plant in Nevada and the Spanish 50-MW plants (Figures 8.3) world-wide the only solar power plants that are commercially operated in the sense of "utility-scale" operations, and this has been true for more than 20 years. They have thus far generated power only for *peak load* use, as they lack a thermal-storage system. Their total installed capacity is 350 MW (at a solar multiple (SM) of 1). Converted to the corresponding *base-load* power plant capacity, with its much larger mirror field (SM ca. 4), this corresponds to only ca. 90 MW. These plants began operation in the years 1985–1990, so that their construction rate was equivalent to only around 20 MW of *base-load* capacity per year. For a large-scale deployment scenario, in contrast, well more than 1000 MW *base-load* per year would be necessary. Probably, in the great power-consuming regions (e.g., the USA or Europe), the construction rate would even be more than 5000 MW per year, and this over a period of 20–40 years.[1] One should always be aware of this relation in order to understand that solar thermal power plants, including parabolic-trough plants, still require an immense amount of development (on the one hand, with respect to their technology, but especially also with respect to preparation for their mass production), which is ignored by many of the supporters of the solar power. This is anything but a "mature technology," as it is often referred to, and in particular, today's construction costs have nothing in common with real mass-production costs. A decisive point here is, however: "The order of magnitude of the required investments in the coming 50 years will lie in the range of trillions of dollars." In comparison to this sum, a development cost of even – improbably – some billions of dollars is vanishingly small and completely justified.

For the generation of peak-load power, the parabolic-trough plants were more or less competitive at a time of high oil and gas prices,[2] but even then only with the aid of generous subsidies. They were in competition with peak-load gas-turbine

---

1) 5 GW/a corresponds in 20 years to only 100 GW.
2) The gas price in 1985 was higher than today (2007), taking inflation into account.

Large-Scale Solar Thermal Power. Werner Vogel and Henry Kalb
© 2010 WILEY-VCH Verlag GmbH & Co. KGaA, Weinheim
ISBN: 978-3-527-40515-2

230 | 8 Parabolic-Trough Power Plants

Figure 8.1 The parabolic troughs of the power plant at Kramer Junction, CA (USA) (SANDIA).

Figure 8.2 Aerial view of the parabolic-trough plant at Kramer Junction (5 × 30 MW) (SANDIA).

Figure 8.3 The parabolic-trough plant *Andasol* (Spain) (Photo DLR).

plants that were operated with petroleum or natural gas fuels. The price of natural gas decreased during the 1990s to ca. one-third of its earlier maximum. This brought an end to construction of the solar plants as early as 1990, and new plants were for many years no longer economically competitive even for peak-load power generation.

In a parabolic-trough mirror, the solar radiation is concentrated onto the focus line and impinges on an absorber tube there, which in currently operating plants is cooled with a synthetic thermal oil (cf. Figures 2.15–2.18). The long rows of collectors are rotated in the course of a day around their axes, in general arranged in a north–south direction, in order to track the sun. The hot oil is used to produce steam for the turbines in a steam generator.

For 24-h operation, a heat-storage reservoir must be integrated into the oil circuit. In the first instance, liquid or solid storage media are used to store the heat energy ("sensible heat") corresponding to their heat capacities and the temperature difference employed. As a material for the solid-state heat-storage medium, concrete is readily used; a common liquid storage medium is the molten nitrate salt used in thermal circuits. The latter is in fact employed in the 50-MW power plants Andasol I and II, currently under construction or in operation in Spain.

A remark about tracking of the reflectors: The north–south orientation is generally an advantage, especially when a power plant is to be optimized for lowest power costs, since more power is generated annually than with an east–west orientation. However, the seasonal variations are less pronounced with the east–west orientation, which thus yields a higher annual capacity factor and tends to favor a base-load scenario (with a capacity factor of ca. 70%).[3] Whether the somewhat lower annual power yield compared with the north–south orientation is compensated by the higher capacity factor needs to be investigated and furthermore depends among other things on the geographic location (latitude) (ENEA, 2001).

8.2
Costs

8.2.1
Preliminary Remarks

Our estimates for parabolic-trough power plants in future large-scale applications are based on the numbers from SunLab, which are quoted in the above-mentioned cost study of Sargent & Lundy for the US Department of Energy, carried out in 2002 (S&L, 2003; see also the discussion of this study in Section 4.2). The SunLab data are based on a state of development that the parabolic-trough plants could attain in the year 2020, if their currently known potential is realized to the full, and presuming that by then, an installed capacity of about 4 GW has been deployed using the not completely mature (i.e., more expensive) available technology.

The goal set for S&L by the US-DOE – as already mentioned in connection with solar tower power plants – was to review these SunLab results. S&L come in the end to the conclusion that SunLab obtained their data in principle correctly, but they, however, also emphasize that these data correspond to an "aggressive" and more or less successful development program. This is termed in part a "high-risk" development, and S&L present their own numbers for comparison, which are based on verified data or tested components. Thus, they arrive at power costs that are roughly 50% higher than those of SunLab. The general conclusion of S&L (and also of the NRC, see Section 4.2) is that the future power cost from parabolic-trough plants will probably lie somewhere between the two estimates. We make use of the SunLab data, since our goal is to point out the chances offered by a massive development program. Furthermore, the SunLab and the S&L data (like those for solar tower plants) are not directly comparable, since S&L presume a slower-paced deployment scenario.

SunLab as well as S&L make the assumption that in the course of this 4-GW construction program, a large portion of the necessary development work will be accomplished. That this will require 15–20 years is, however, not a law of nature, and a state-supported systematic R&D program could surely shorten this development period considerably. Also, it is by no means certain that the presumed research would in fact be carried out to the evidently necessary extent and with the required breadth without a governmental initiative.

---

3) It is, however, noteworthy that Mills *et al.* (2004) state for *linear Fresnel systems* on the basis of simulation calculations that one would obtain a more uniform power yield throughout the year with a north–south orientation.

Table 8.1 Technical data on which the "SunLab long-term" case is based (S&L, 2003, p. 4–3).

Net power	400 MW
Capacity factor	57%
Annual efficiency	17.2%
Solar field	3.9 km²
Land area	13.2 km²
Land usage	30%
Heat-transfer fluid	"advanced"[a]
Operating temperature	500 °C
Thermal storage	Direct thermocline (12 h)

a) According to S&L, p. 4–36: HitecXL.

Table 8.2 Elements that contribute to the annual efficiency in the "SunLab long-term" case (S&L, 2003, p. 4–5).

Solar field optical efficiency	60%
Receiver	85%
Piping thermal efficiency	97%
Storage	99%
Electric power generation system	40%
Electric parasitic load	93%
Power plant availability	94%
Annual solar-to-electric efficiency (total)	17.2%

In the S&L study, we are first told: "Deployment provides a means for continued research in technology improvements. ..." Almost immediately thereafter, however, clear-cut skepticism is expressed as to whether this is necessarily the case: "...The actual strategy employed by the plant suppliers can be ... with more emphasis on near-term cost reduction with a minimum of risk. The trough plant suppliers may opt to provide multiple plants in the 50 MW$_e$ to 100 MW$_e$ size with no thermal storage but with a supplemental steam generator, replicating the proven technology of the existing SEGS plants. The suppliers can rely more on initial production volume to reduce costs as opposed to efficiency and technology improvements and scale-up factors." (S&L, 2003, pp. 4–37 and 4–38.)

How SunLab assesses the mature technology in detail is shown in Tables 8.1 and 8.2. Note, in particular, the efficiencies that at 17% are greatly increased in comparison to the 10% obtained from the California power plants. In this connection, it is interesting that the first step, namely, an increase to more than 14%, can practically be considered to be the current state of the technology. This corresponds in the series of developmental steps that were considered in the S&L study and by SunLab to the "near-term" case expected for the year 2004. A further increase of the efficiency to more than 17% has then to be seen in connection with the transition from using thermal oil to molten nitrate salts as heat-transfer medium, which, however, will still require a considerable research effort.

We mention here—without going into detail—a few typical examples of the individual steps that are planned to lead to a cost reduction and, in particular, an increase in efficiency:

- mechanically, optically, and thermally improved *collector system*
- more stable *absorber elements*
- *Connections between the absorber tubes* and between the absorber tubes and the collection piping: The use of ball joints instead of flexible hoses should increase both the reliability and the overall efficiency of the piping system. According to S&L (2003), the pressure drop within the piping system of the collector field, which causes a decrease in overall output power, can be reduced by 50%. (The pumping losses will already be strongly reduced due to the transition from thermal oil to molten nitrate salts with their higher density, larger heat capacity, and greater operational temperature difference.)

8.2.2
Investment Costs

In the first column of Table 8.3, the SunLab cost data used (S&L, 2003, p. D-5) are listed. They correspond to a 400-MW power plant, but have been recalculated to 1000 MW. The mirror area (aperture area) of around $9.8\,km^2$ (1000 MW) and the heat-storage reservoir with a thermal capacity of 32 700 MWh_{th}, which suffices for 12 h of operation at full power output, would yield an annual capacity factor of 57% at a very favorable site according to SunLab (site: Kramer Junction, California; average annual insolation in 1999: $8.0\,kWh/m^2\,d$, that is, around $2900\,kWh/m^2\,a$). The costs of the collector field and the heat-storage reservoir correspond to specific

Table 8.3 Investment costs (rounded values) for parabolic-trough power plants[a].

	SunLab	Spain	Morocco/USA
	Million $ (2002)/1000 MW		
Collector field (120$/m²)	1175	2500	2100
Heat storage (11.7 $/kWh$_{th}$)	385 (12 h)	510 (16 h)	510 (16 h)
Conventional parts and the rest of the plant	360	360	360
Land (1.25 $/m²)	–	85	–
Direct investment costs	1920	3455	2970
Indirect costs (including interest during construction)	290	505	445
Overall investment costs (wet cooling)	2210	3960	3415
Dry cooling (+8.7%)	–	–	295
Total investment costs	2210	3960	3710

a) The costing is based on SunLab data cited in the Sargent & Lundy study (2003, p. D-5) for mature-technology 400-MW plants with 12-h thermal storage and a capacity factor of 0.57.

costs of 120 $/m² (including 4 $/m² for "structures and improvements") or ca. 12 $/kWh$_{th}$ (storage capacity).

Beginning with these values, we recompute the cost for two sites with less favorable insolation conditions, the first in the range of 2000 kWh/m² a (5.5 kWh/m² d) and the second in the range of 2300 kWh/m² a (6.3 kWh/m² d). The first case corresponds, for example, to a site in southern Spain, and the second to a site in the Sahara Desert (e.g., Morocco) or in the southwestern USA. We emphasize once again that there are – on the one hand – large areas in North Africa and in the US Southwest that have considerably higher insolations, but on the other hand, the limitations on the maximum slope and roughness of the ground are more stringent for parabolic-trough power plants than for solar tower plants (see Chapter 5).

We make the following assumptions, similar to those made for solar tower plants, in order to recalculate the SunLab cost data – using rough estimates of the annual variation of the insolation as already discussed (cf. Section 5.1):

1) In order to achieve a capacity factor of 70% at the first site (Spain – 2000 kWh/m² a), the plant must be designed so that on a day with an insolation of 6.7 kWh, the energy collected is sufficient to operate the turbines for 24 h at full power.

2) At the second (more favorable) site (Morocco or the USA – 2300 kWh/m² a), to allow a capacity factor of ca. 80%, the plant must be designed so that on a day with an insolation of 8.0 kWh, the energy collected is sufficient to operate the turbines at full power for 24 h.

This is not quite correct since a precise calculation would have to take the time dependence of the *efficiency* into account in addition to that of the insolation. (A parabolic-trough power plant with a SM of 1 (i.e., without a heat-storage reservoir), whose troughs are set up in a north–south direction, generates per day a somewhat higher amount of power than a solar tower plant, due to its higher efficiency during the morning and evening hours; with an east-west orientation, the reverse is true.) Which annual capacity factor should be aimed at for a solar power plant is, however, in the end a question of the overall optimization for the large-scale system discussed here, so that this error should play no role in our fundamental considerations. It is in any case clear that aiming for a high capacity factor, that is, a high proportion of solar energy in the annual overall power output of the complete solar power system, will cause an increase in the power price relative to the optimum case considered by SunLab with the collector area assumed there and a capacity factor of only 57%. The economic significance of a more or less high capacity factor in the end depends on the cost of the fossil fuel that is used in the backup power plants, apart from the question of climate protection, which is of very fundamental importance.

Calculating for simplicity with a daily average efficiency[4] of 17.2%, we find for the site with an insolation of around 2000 kWh/m² a a required collector area of

---

4) The daily average efficiency varies in the course of a year; with north–south trough orientation, it is noticeably higher in the summer than in winter. Here, we use the annual average efficiency quoted by SunLab. In a more precise calculation, one would have to take the annual variations into account and also to distinguish between north–south and east–west orientations.

8 Parabolic-Trough Power Plants

Table 8.4 Investment costs for the complete solar power system[a].

Investment costs, parabolic-trough plants	Spain	Morocco/USA
	Million $ (2002)/1000 MW	
Solar plant per 1000 MW at the site	3960	3710
Solar plant per 1000 MW after transmission	4310	4190
Transmission lines	500	665
Backup power plants	715	715
Total investment costs	5525	5570

a) A total of 1000 MW at the end of the transmission line means for the case of Spain (8.1% losses) 1090 MW at the solar plant site; for the case of Morocco (11.5% losses), it requires 1130 MW output power at the plant site.

20.8 km^2/1000 MW.[5] For the site with an insolation of ca. 2300 kWh/m^2 a, the result is 17.4 km^2.

The cost of 120 $/m^2 quoted above gives collector-field costs of 2500 and 2100 million $/1000 MW for the two sites, respectively. In both cases, a 16-h heat-storage reservoir is taken into account, at a cost of $510 million (SunLab: 12-h storage, proportionally extrapolated). If the conventional parts and the rest of the plant are presumed to cost the same at both sites, we find all together $3500 or 3000 million as direct investment costs. As does SunLab, we add 15% indirect costs to the direct costs (for the site in Spain, 15% of the direct costs without the cost of the land). This finally yields $3960 million for Spain and $3415 million for Morocco. As for a solar tower plant (compare Table 2.1), at the desert site in Morocco, we assume dry cooling, with a corresponding decrease in efficiency. We again presume an increase in the specific investment costs of 8.7% (cf. Section 4.3.7), so that a final cost of $3710 million results for Morocco.

Table 8.4 shows how the investment costs per 1000 MW for the overall solar power system are obtained from this: Corresponding to the assumed transmission losses of 8.1% or 11.5%, respectively, the solar capacity is multiplied by a factor of 1.088 or 1.13 from 1000 MW to give 1088 MW (Spain) or 1130 MW (Morocco). To this, we must add the cost of the backup plants (full capacity) and the transmission lines, finding total investment costs of around $5550 million for 1000 MW of delivered power capacity from the complete base-load system.

5) In computing the first value (20.8 km^2), it was assumed that a daily insolation of 6.7 kWh/m^2 is sufficient for operation at full-power output (see footnote 1). Then per kilowatt of output capacity, in 24 h, 24 kWh$_{el}$ would be generated. At an efficiency of 17.2%, this requires 139.5 kWh of solar energy (daily insolation, DNI). Under the conditions assumed, this gives: 139.5 kW/(6.7 kWh/m^2) = 20.8 m^2. For 1 GW, one thus requires a mirror area (aperture) of 20.8 km^2. Under the assumption that 24-h full output is achieved with a daily insolation of 8 kWh/m^2 as in the second case, a collector area of 17.4 km^2 is obtained.

8.2.3
Operating and Maintenance Costs

At the small Californian plants with the maintenance methods which were common in the early 1990s, the maintenance costs were around 4 ¢/kWh (2002-$). These high specific costs are due, in particular, to the turbines with their small size (up to now 50 or 80 MW) and their short operating times (thus far, only ca. 12 h/d – without heat storage but with some hours of additional gas-burning operation). With base-load operation and larger turbine units, the specific costs would be lower. A second reason for future cost decreases can be found in the optimization of previous strategies for operation and maintenance. Already in the past years at the Californian plants, a systematic program to lower the maintenance costs has been put in place (Cohen et al., 1999). On the basis of this, one can expect for *new* power plants a reduction "by a factor of 1.5 or more in future plants that incorporate the findings of the program." This refers, for example, to "the solar field control system, data acquisition and handling for performance and maintenance needs, solar field performance data, and plant maintenance planning methodologies." (Pilkington, 1996, p. 5-10). Price and Kearney (2003) quote a reduction from 4.6 to 2.8 ¢/kWh (in the "near term"). It will be very interesting, especially for the newly commissioned plants, to observe how these improvements affect the costs.

Up to now, the mirrors have been manually cleaned. For large-scale plants, only automatic cleaning systems are conceivable, for example, using spray-jet robots similar to the spray-painting robots in the automobile industry, combined with fully or partially automated drive systems. These systems must be developed to maturity with a high priority.

In addition to the size increase of the individual power blocks, one can also envisage the central monitoring and control of several blocks within a large solar field from a single control center, thus reducing the personnel requirements.

A point which is very important for further development (or has been for over 10 years) is the replacement by ball joints of the flexible hoses used thus far, which – as mentioned – connect the rotating absorbers with one another and with the piping that collects the heat-transfer medium. This will not only save on maintenance costs, but will also avoid loss of thermal oil through leakage, which was not a rare occurrence in the past. The development of these ball joints for thermal oil seems to have been successful. These connections are, however, one of the most important points in a development program toward "advanced technology," that is, toward higher operating temperatures with cooling and heat transfer using molten nitrate salts[6] or direct steam generation. In Germany, a corresponding research project is currently underway for direct steam generation.

---

6) The seals in the ball joints have thus far been made of graphite. Above about 400 °C, the problem arises that graphite is oxidized by nitrates. As replacement material for the graphite, boron nitride is under discussion; another possible solution would be to maintain the temperature of the graphite seals below 300 °C by means of internal insulation. It may prove necessary to return to the use of flexible metal tubing for molten-salt circuits.

Table 8.5 Energy cost from the solar power system using parabolic-trough plants.

Cost of energy from the solar power system		Solar power system (¢/kWh (2002-$))	
		Spain	Morocco/USA
Capital costs[a]		3.1	3.1
Operation and maintenance	Solar	0.7	0.6
	Backup	0.1	0.1
Gas[b]		1.3	0.8
Energy cost		5.2	4.7

a) 4% real interest rate, 45 years operating lifetime.
b) Gas price: 2.5 ¢/kWh$_{Gas}$ (LHV) = 6.6 $/MMBTU (HHV); efficiency: 58% – see also the caption of Table 4.2.

One can thus assume that for large-scale deployment, including carrying out the measures mentioned, and with further development, including a high degree of automation and larger generating blocks, the costs for operation and maintenance would be considerably reduced. Thus, the major cost reduction included in the tables for the final development stage and for the case of large solar plant parks seems plausible.

The costs listed in Table 8.5 for operation and maintenance (solar) are based on the SunLab value of 0.5 ¢/kWh (S&L, 2003, p. F-1), which was recomputed as follows:

For the "SunLab power plant" (recalculated output-power capacity: 1000 MW; capacity factor: 57% (5000 h)), with an annual electrical energy output of 5000 GWh, 0.5 ¢/kWh corresponds to an annual expenditure of $25 million. To a rough approximation – which is sufficient for our purposes here – extrapolating linearly to the larger collector fields, this corresponds to $53 million (Spain) or $44 million (Morocco) annually. Correcting for the transmission losses – again with the factors 1.088 or 1.13 – and then dividing by the annual electrical energy output per gigawatt of the overall solar power system (8760 GWh), we obtain 0.7 or 0.6 ¢/kWh, respectively.

8.2.4
Power Costs

Table 8.5 shows the power costs from the solar power system, that is, the cost of power from the combination of solar power plants and fossil-fuel backup plants. For comparison with the power costs from coal-fired and nuclear power plants, we refer the reader to Chapters 2 and Section 4.1, there in particular to Tables 2.3, 4.2 and 4.3. From the investment costs, with an annuity factor of 0.0483 (45 years operating life, 4% interest rate) and 8760 operational hours per year, we find the capital cost per kilowatt-hour. The maintenance costs for the overall solar power system consist of a portion for the solar components (see above) and a portion for the backup plants (see Chapter 10).

The cost of power from a solar power system based on parabolic troughs is exactly the same as from solar tower plants, given the assumptions made here. However, one must take into account that for the trough power plants, we have considered the most favorable case according to the SunLab report (as quoted in the S&L study). For solar tower plants (with additional assumptions regarding mass production), the somewhat older and more conservative data of Kolb (1996a) were used. If one takes the most favorable SunLab data for the investment costs for solar tower plants as given in the S&L study ("Solar 220," see Section 4.2), one arrives at a power cost that is around 0.4 ¢/kWh lower. With additional consideration of the lower operating and maintenance costs, the power price in the most favorable case considered by SunLab would be 0.7 ¢/kWh lower than from parabolic-trough plants. This again corresponds to the remark made at the outset that for large-scale systems, the tower principle can be considered in general to be the more economically favorable technology in the long term. The relative difference between trough and tower plants is, however, more clear-cut for the *pure solar power costs* (*at the power plant*) than for the whole solar power system, where also the costs of the backup plants must be included. According to the S&L study, the future expected cost of power from trough plants in the more conservative S&L scenario is around 15% higher than from tower plants; in the more optimistic SunLab scenario, it would be around 25% higher (S&L, 2003).

Remark: Above, we pointed out that the change in the sun's position in the course of a year has a stronger effect on parabolic-trough plants than on solar tower plants, and that with the present assumed size of the collector field (owing to the simple estimates made), the capacity factor is overestimated if the values 70% or 80% are used. If we assume in Spain and in Morocco/USA in each case a lower capacity factor (by 5%), that is, 65% and 75%, respectively, this results in an increase in the estimated power price by 0.2 ¢/kWh. From an economic point of view, this difference would have relatively little significance, but it corresponds to a 17% or 25% higher consumption of fossil energy (backup power). This could of course be avoided by using a sufficiently large collector field, which would, however, be more expensive, since at a higher degree of utilization, the cost of an additional percentage rapidly increases.

From the above numbers and prognoses, one cannot yet *conclusively* deduce that trough power plants in large-scale use would be economically inferior to solar tower plants, and certainly not in the case of small plants. A conclusive balance will be possible only when more exact investigations into the costs under mass production and into the potential for further development have been carried out for both types of plants. A major uncertainty is furthermore due to the assumptions about operating and maintenance costs.

Of course, current cost prognoses for parabolic-trough plants are being improved by the construction of a whole series of new plants in the range of 50-MW output power. The need for detailed cost analyses is, however, still present due to the many technological options for trough plants; the currently deployed plants represent "classical" technology, on the whole. Along with certain improvements in the mechanical components and in the absorbers, the decisive innovation is, in particular, the introduction of a 7-h molten-salt heat-storage system (in Spain).

8.3
Development Program and Cost Estimates for Mass Production

What was stated in the chapters on heliostats and receivers in connection with future production costs, which depend both on the development of improved construction designs and also on improvements in fabrication technology, holds in analogous form for parabolic-trough power plants as well. Within the framework of a large development program, in the course of the technical development of the collector components and the oil circuit, and especially the development of cooling circuits using other heat-transfer media,[7] the economic feasibility (for mass production) of the present suggestions must be verified. (Also, the improved technology for operation and maintenance needs to be implemented on a trial basis in the existing power plants and tested there.) This has of course been carried out already on a small scale in the course of the development to date. A good example was the test of a European innovation for the collector mounting structure[8] (known as "SKAL-ET") in California (ca. 2003–2005), which is now being used in Spain; this is furthermore also a very good example of reasonable international cooperation in spite of economic and scientific competition. A similar example is the development of a new absorber in Germany (by the Schott Corporation) and its subsequent testing in a power plant in California.

We can not go into more detail here.[9] The further development in any case must include all of the options that are currently under discussion, of course making use of preparatory work which has been carried out by manufacturers and operators of solar plants up to now. We mention a few of the important aspects requiring research:

a) continuing improvements in mechanical design;
b) development of processes for fabrication and assembly, in particular, keeping in mind mass production and large-scale systems;
c) quality control for absorbers and mirrors;
d) the application of thin-glass mirrors or advanced reflector materials;
e) development of selective absorber coatings for temperatures above 500 °C;

---

7) This of course includes, along with a molten nitrate salt circuit, also direct steam generation – with the reservations regarding heat storage already mentioned in Section 3.3.

8) An increase in collector length from 100 to 150 m was achieved, accompanied by a reduction of the specific mass (relative to the length) and a simultaneous improvement in stiffness. These measures allowed savings in particular for the drive units and the absorber connections (ball joints), as well as in the required pumping power.

9) The entire development potential is described and discussed in great detail in the S&L study (2003). For additional in-depth reading on development strategies, we refer readers to the European ECOSTAR study (ECOSTAR, 2004).

f) intensive research and development toward using molten nitrate salts as heat-transfer medium, including thermal storage.

8.3.1
Test Plants

The various improved components and technologies, after completion of the actual development work (including individual testing), must be demonstrated in a small prototype parabolic-trough power plant. Here, a thermal-storage reservoir should also be integrated into the plant. A test plant of ca. 5–10 MW output power would be sufficient.[10] If it were connected to an existing trough plant in California (or recently also in Spain), it would not even require its own steam turbine. The size of a test plant is not of critical importance, in particular, for a parabolic-trough installation, since larger plants have already been deployed. Adding, for example, an additional 30 MW instead of 10 MW would not be particularly helpful. The installation could be connected, for example, to the first parabolic-trough power plant constructed in California, with 14 MW output power. Since its oil temperature (mineral oil) is lower than that in the plants built later, the thermal energy from the new components could be transferred to the process there using an oil/oil heat exchanger and could thus be completely utilized. The existing plant could thereby be upgraded to allow 24-h operation.

In such a prototype plant, all of the new developments (mirrors, cleaning techniques, etc.) including all of their variants would be tested. It would thus be a demonstration plant for individual technologies and, at the same time, for base-load operation. As a very rough estimate, we can assume that the specific cost of such a small plant, which would be rather different from those currently operating, could be in the range of 10 000–20 000 $/kW. Then, a 5-MW base-load installation would cost at most about $100 million.

8.4
Heat-Storage Systems for Parabolic-Trough Power Plants

8.4.1
Preliminary Remarks

The parabolic-trough power plants built thus far in the USA (using a thermal oil as heat-transfer medium) operate – with the exception of the first one constructed – without a heat-storage reservoir (peak-load operation only). In the case of the first

---

10) The fact that even small test installations can make a contribution to rapid development is illustrated by the integration of a SKAL-ET test loop in a California solar power plant. It was, however, also typical that the German government could not (or would not) finance the necessary few million dollars from its regular research budget, but instead required a so-called Future Investment Program that was financed from the proceeds of an earlier sale of telecommunications licenses.

plant, an inexpensive mineral oil was used as heat-transfer medium. Because of its low materials price, a heat-storage reservoir could be implemented. Large amounts (3000 m^3) of the hot oil were simply stored in insulated tanks, analogous to the molten-salt storage system of SOLAR TWO, and for the cool oil there was an equally large tank. The storage capacity of 110 MWh$_{th}$ was sufficient for at least 3 h of full-power operation of the 15-MW power block.

In the plants built later, in order to attain a higher efficiency, the oil temperature was successively increased from 300 to 390 °C. This was possible only by using synthetic oil. Therminol[11] VP-1, the oil used, is a eutectic mixture that remains sufficiently stable up to 400 °C, but is, however, considerably more expensive, so that its use for *direct* thermal storage in oil tanks was not feasible.

Molten salt (solar salt), as is used in solar tower plants, is also being considered as a heat-transfer medium for parabolic-trough plants. Its melting point lies above 200 °C and it thus requires ancillary electrical heating for the piping to keep the salt molten and for starting up the plant. For the widely branched piping network of a parabolic-trough power plant, this was previously held to be impracticable and absolutely not economically feasible. Interestingly, today – ca. 20 years later – the nitrate salts as heat-transfer medium for trough plants are held to be one of the most important cost reduction options. A reason for this rethinking is certainly the technical experience that has been gained in the development and testing of the SOLAR-TWO molten-salt circuit. Why this rethinking occurred so late is due to a problem that is quite typical of the development of solar energy thus far: The needed financial support for systematic research and development of all – or at least the most important – options was simply not made available by those politically responsible. With the extremely sparse support at hand, it was necessary to concentrate almost exclusively on areas that promised short-term success, and that meant in this case a thermal circuit using oil.

A fundamental problem of heat storage with parabolic troughs and an oil circuit is the limited temperature range between the oil input temperature (290 °C) and the oil output temperature (390 °C). For heat storage, this range of at most 100 K is available. With solar tower plants using a molten-salt receiver, the temperature range in contrast is nearly 300 K (290 °C input, 565 °C output temperature). For a molten-salt storage reservoir in a trough power plant, one would therefore need nearly three times the amount of salt, so that here a particular interest in less expensive storage media is understandable.

In the following text, we shall limit our discussion to the two options of molten-salt and concrete storage media, that is, to storage in the form of sensible heat.

---

11) Therminol VP-1® from Solutia (previously Monsanto) is identical to Dowtherm A® (Dow); it is a eutectic mixture of 74% diphenyl oxide and 26% biphenyl.

Diphenyloxide   Biphenyl

The latent heat-storage systems already mentioned in Chapter 3 are at present still in the early stages of development, and there are hardly any reliable estimates of their future cost for large-scale applications, so that we dispense with a more detailed treatment here.

Very briefly, we give a few words on the essentials of latent heat-storage systems:

In general, it should be stressed that research work in this area has been intensified just in the last few years, especially in connection with the development of new types of compound materials. These combine the actual storage medium, that is, the phase-change material (PCM), with other materials that have a high thermal conductivity. PCMs for the temperature range of interest here (around 300 °C) are among others alkali nitrates. The material with a high conductivity currently being used is so-called expanded graphite, a type of graphite with an extremely low density. The latent-heat-storage systems are being developed especially in connection with direct steam generation in the absorber pipes, since the two systems — phase change for thermal storage and direct steam generation from water — are compatible in that the phase transitions (solid/liquid in the PCM or liquid/gas in steam generation) are both processes occurring at constant temperature. The two transition temperatures must naturally lie very close together.[12]

A strong intensification of this special research area would of course be a part of a large, systematic R&D program.

8.4.2
Molten-Salt Heat-Storage System

The molten-salt storage reservoir (in any case the two-tank system) is technically relatively simple and is thus already in use at the first Spanish 50-MW solar plants (7 h storage capacity; see Figure 8.4).

Hot oil from the absorber tubes transfers thermal energy in a heat exchanger to a molten salt ("solar salt," a eutectic mixture of $NaNO_3/KNO_3$), which comes from the cold tank and is then stored in the hot tank at roughly 390 °C (indirect storage). At night, this hot salt is pumped through the steam generator (or the heat exchanger, respectively) and then, at a temperature of 290 °C, back into the cold tank. Disadvantages of this storage concept are the large amounts of salt required, as mentioned (a factor of 2.75 more than for solar tower plants), and the need for a heat exchanger. The latter not only entails additional investment costs but also leads to a somewhat lower steam temperature, owing to the temperature gradient between the oil and the molten salt, and thus to a reduced thermal efficiency. The cost of the Andasol storage system was unfortunately not published, so that we cannot quote a precise value for the cost of this simple storage technology here. It can, however, be assumed that the overall cost (taking the reduced

---

12) It is possible that in future, latent-heat-storage systems will not be used only for direct steam generation. With an oil circuit, it would then, for example, be reasonable to use several different temperature steps (between the upper and lower temperature limits of the oil circuit), each with a different PCM. Such a system is referred to as a "cascade storage reservoir" (Michels and Pitz-Paal, 2007).

Figure 8.4 Power plant *Andasol* (Spain), with a molten-salt heat-storage system. In the right foreground are the large "hot" and "cold" salt tanks (DLR).

efficiency into account, also) would be ca. four to five times that of the heat-storage system for a solar tower plant; for the latter, the cost is expected to lie in the range of 260–440 million \$/GW$_{el}$, depending on the study quoted (cf. Table 4.6 in Section 4.2). One can therefore safely assume that the cost of heat storage here would be at least more than 1000 million \$/GW$_{el}$ (possibly as high as 1500). According to SunLab, direct costs (without including the reduction in efficiency) of 960 million \$/GW$_{el}$ are to be expected (S&L, 2003, p. 4–18).[13] Therefore, a number of developments for less elaborate heat-storage systems have been suggested.

The required amount of salt can be greatly reduced by adding low-cost solid materials (e.g., a mixture of silica sand and quartzite, a mineral derived from sandstone) that store the major portion of the heat energy. This filler material forms a layer in a single tank. To charge the storage reservoir, hot molten salt is pumped into the tank from above and cold salt is pumped out from below. The narrow transition zone between the regions with the higher and the lower temperature[14] moves from above the filler layer to below it. For discharging, the cold

---

13) For the future "mature technology," as presupposed in Table 8.3, SunLab presumes the use of a molten-salt circuit within the mirror field (at a temperature of 500 °C). Then, not only could the heat exchanger be dispensed with but also the available temperature difference for heat storage would be greater than with an oil circuit. Furthermore, instead of a two-tank system, a thermocline system was presumed.

14) "... With the hot and cold fluid in a single tank, the thermocline storage system relies on thermal buoyancy to maintain thermal stratification." (S&L, 2003, p. D-35). Such layering is also possible within a liquid.

heat-transfer medium is pumped into the lower zone of the tank and the hot medium is pumped out from above; the temperature transition zone again moves upward. This type of storage system is called a thermocline reservoir (or simply a thermocline). The second tank can thus be dispensed with,[15] as well as a major portion of the salt mixture. Incidentally, this type of storage system is not very new; it was, for example, already in use in the solar tower plant SOLAR ONE (rock/mineral oil). Thermocline storage systems using nitrate salts have been investigated in recent years especially in the USA. We refer the reader to an excellent description of the principle and of recent research work at Sandia by Pacheco et al. (2001).

In the case that nitrate salts are also employed as heat-transfer medium in the solar field, that is, with *direct* thermal storage, one would likewise have the two possibilities of a two- or a one-tank storage system. However, the latter, a thermocline reservoir that conserves salt mixture, is particularly important since in future, the heat-transfer medium – if nitrate salts are used at all – will quite possibly not be the binary "solar salt," but instead the more expensive, likewise eutectic ternary mixture of KNO_3, $NaNO_3$, and $Ca(NO_3)_2$, which is called "HitecXL."[16] Its advantage lies in its lower solidification temperature of only 130–150 °C (depending on the mixing ratio) as compared with that of solar salt (220 °C), which greatly facilitates the avoidance of solidification in the piping and reduces the need to thaw frozen pipes.[17] The S&L study (2003) also presumes that in future "advanced" parabolic-trough power plants, a HitecXL–heat-transfer circuit will be used, and this forms the basis for the cost estimates in Table 8.3.

There is thus a trade-off between solar salt, with its advantages of low materials cost and somewhat higher maximum operating temperatures, and HitecXL, with the advantage of solidification at a considerably lower temperature. In the development work in Italy (see Section 3.3), thus far solar salt was given preference.

In recent years, there seems to have been some success in the search for special inorganic salt mixtures that solidify even below 100 °C: "Multicomponent mixtures

(The term "thermocline" is also employed in connection with the very stable thermal layering in the oceans at differing water temperatures). The stability of the boundary layer is, however, improved by the solid filler material, which inhibits convective heat transport.

15) The container used is only slightly larger than one of the tanks of the two-tank system.
16) In practice, and mainly for cost reasons, the exact eutectic concentration ratio is not employed, that is, the mixing ratio with the lowest solidification temperature possible with the given materials. The solidification temperature is, however, not very sensitive to the exact mixing ratio. Pacheco et al. (2001), like Kelly et al. (2007), for example, give for HitecXL a composition of 15% $NaNO_3$, 43% KNO_3, and 42% $Ca(NO_3)_2$, whereas Kearney et al. (2003) quote 7% $NaNO_3$, 45% KNO_3, and 48% $Ca(NO_3)_2$. The eutectic mixture lies at a concentration ratio of 7%/30%/63% (ENEA, 2001).
17) For solar tower receivers, the use of HitecXL is not planned since its thermal stability extends only up to around 500 °C. With tower receivers, it is even being considered to raise the salt temperature from 565 to ca. 650 °C. This would be possible even using the more thermally stable solar salt only if oxygen were used as shielding gas. (Oxygen prevents the thermal decomposition of the nitrates according to the reaction $NO_3^- \rightarrow NO_2^- + 1/2\, O_2$.)

of nitrate salts (Na, K, Li, and Ca) remain liquid at relatively low temperatures compared with binary and ternary mixtures." (Bradshaw (Sandia), 2008). In Germany also, reports of new salt mixtures (nitrate/nitrite) have been published or even filed for patenting (Gladen et al., 2008).

A few years ago, great hopes were placed on the theoretically very interesting option of employing "ionic liquids" ("organic salts"), in particular imidazolium salts, as heat-transfer media (Wu et al., 2001, Moens et al., 2003). These materials have the potential advantage of a solidification temperature that lies near room temperature or even below 0 °C. Unfortunately, the optimistic goal of synthesizing ionic liquids that remain stable up to ca. 450 °C has been somewhat dampened by the results of the careful investigations of Blake et al. (2006).

8.4.3
Heat-Storage Systems Based on Concrete

The concrete thermal-storage reservoir was conceived for parabolic-trough systems with an oil-based thermal circuit. Their construction is (in principle) simple: pipes carrying the thermal oil are laid into a latticework of steel matting, so that their spacing is fixed. This structure is then poured full of concrete.

The heat-transfer properties depend on the number of pipes. In a large-scale storage reservoir for base-load operation, heat is stored slowly over an entire day and is removed even more slowly during the nighttime hours. In contrast to the originally designed and investigated short-term heat-storage reservoirs (buffer reservoirs), which were conceived for input and output storage times in the range of 1 h, for 24-h operation less piping per kilogram of concrete (or fewer pipes per square meter of cross-sectional area) is required, so that the specific costs are lower.

The oil temperature is lower on discharging than on charging (temperature gradients due to the twofold heat exchange processes). This temperature decrease cannot be compensated by an increase in input temperature in the case of a parabolic-trough system owing to the limited thermal stability of the oil.[18] At night, one must therefore operate with lower oil and steam temperatures, with the result that the efficiency of the plant is lower. This effect is naturally also present with a salt-based thermal-storage system (indirect, with a heat exchanger), but is not as serious as with concrete storage systems.

Thus far, there are only two test installations for concrete thermal-storage reservoirs which are larger than a laboratory model: (i) In Almeria (Spain) in the years 2003–2005, four 25-kWh$_{th}$ modules were tested (Figure 8.5), and (ii) since 2006, in Stuttgart (Germany), there is a larger installation in which the DLR has tested a 400-MW$_{th}$ pilot module. Since 2008, tests are being carried out on a second module of the same storage capacity (but not in parallel). For comparison, the storage modules in a solar power plant would have a capacity on the order of 10 MW$_{th}$.

---

18) This is, for example, possible to some extent using a concrete thermal-storage reservoir in an air-cooled solar tower plant with a volumetric receiver.

Figure 8.5 A concrete thermal energy storage system – the test facility (without thermal insulation) in the PSA, Almeria (Spain). Four modules, each for 25 kW, 80 kWh, and test operation 2004–2005 (DLR).

This type of storage reservoir has been investigated with extensive calculations and laboratory experiments in the last 15 years – especially in Germany (by the DLR). The fact that practical development over and above the laboratory scale has started so late is again a typical example of the years-long neglect of the development of storage systems, in particular, in terms of governmental support.

First estimates of the investment costs for thermal-storage systems with a high heat withdrawal rate as they have thus far been designed (discharge of the reservoir within 1 h) lie in the range of 30–60 $ (2002-$)/kWh of storage capacity (Pilkington, 1994). At the same time, for large-scale plants, there have also been estimates in the range of 20 $/kWh. For the case considered here, that is, base-load power plants (with less piping per stored kilowatt-hour) plus mass production, the lower price range is relevant. In the S&L study (2003), it was generally stated (referring also to concrete along with phase-change materials): "Although all of these storage options are in the early stages of development, they provide alternative paths to achieving cost targets in a range similar to HitecXL." R. Tamme at the 2006 Trough Workshop named 20 $/kWh$_{th}$ as a realistic goal: "Feasibility of solid media thermal energy storage concept successfully demonstrated ..., Cost goal 20 €/kWh capacity can be achieved" (Tamme, 2006). For a 16-h storage reservoir (base-load operation), this yields investment costs of ca. $870 million for an output capacity of 1000 MW$_{el}$. The resulting power cost from the solar power system would be around 0.3 ¢ higher than with the use of HitecXL (direct storage, 12 $/kWh$_{th}$[19]).

---

19) Cf. also Kelly et al. (2007): ".. decreasing to 13 $/kWh for binary salt storage (solar salt) at 540 °C."

The option of a concrete storage reservoir is of great significance for the not-impossible case that the development of advanced heat-transfer circuits will not be successful. For an oil circuit, *indirect* storage with nitrate salts is only an intermediate solution; it is too expensive in the long term, at around 30 $/kWh, especially for very large storage reservoirs.

In the course of a major development program, these thus-far rough cost analyses, in particular, for large-scale systems must be placed on a reliable basis. This must be done in parallel with the experimental investigation of thermal designs, and this is also true for solar tower plants.

Here, we must naturally point out that in the last few years the question of thermal storage has again moved into the foreground of solar research. In the meantime, with the new German test installation mentioned above, we have already entered a sort of demonstration phase. However, although the construction industry has already become involved, this area of research is still completely underfinanced.

8.4.4
Test Facilities for Solid and Thermocline Heat-Storage Systems

While for heat storage using molten salts, the cost and the corrosiveness of the salt mixture used, as well as the materials properties of the tanks and the heat exchangers are at the focus of research, for concrete storage modules and combined salt-solid storage reservoirs (thermoclines), along with materials questions – for the materials problems of the thermocline (cf. Brosseau *et al.* (2004)) – also questions of the construction design and the operation of the storage system must be clarified. To this end, suitable test installations are needed.

These need not be connected to parabolic-trough plants, since the heat-transfer medium being tested can be heated in a gas-fired boiler, for example. Since no solar field needs to be planned and constructed, such test installations could be set up within a short time. For oil and molten-salt thermal circuits, the charging and discharging of thermal-storage reservoirs can be tested under the same conditions as in a solar power plant, that is, at the same temperatures, the same charging and discharging rates, etc. A first such test installation was set up at Sandia a few years ago for the testing of a thermocline storage system using the combination solar salt/quartzite. It had a storage capacity of 2.3 MWh$_{th}$ and was operated with a propane salt heater (2.9 MW) (Pacheco *et al.*, 2001).

In the case of *concrete storage reservoirs*, large test installations would allow the important long-term tests to be carried out on the thermal-storage modules. These tests would be required for all possible constructional variants of the concrete/piping system. To test the thermal-cycling stability, individual storage modules could be charged and discharged at short intervals (e.g., five times per day). In four years of such testing, the possible material fatigue during a 20-year operational lifetime could be simulated. From this, one could draw preliminary conclusions about the behavior of the system during a planned operational lifetime of 40 years or more.

8.4 Heat-Storage Systems for Parabolic-Trough Power Plants

Concrete thermal-storage reservoirs are constructed in modular form. They consist of a large number of identical concrete blocks. The design and long-term tests need only be performed on a small number of modules. A relatively small and inexpensive test installation would suffice; it would, however, have to be large enough that different modifications of the blocks could be tested in *parallel*. Possibly other solid-state storage media besides concrete, from among the many concepts for combination with an oil thermal circuit that were previously tested (cf. e.g., Dinter et al., 1991), would enter into the short list of candidates. These have so far not been intensively investigated due to a lack of research support; they would also have to be subjected to module tests.[20] The same holds for the latent heat-storage systems.

With solid-state and thermocline storage systems, it must be kept in mind that the thermal gradient is shifted internally owing to the slow heat exchange in the interior of the reservoir. This must be corrected by heating the individual modules of the reservoir at certain time intervals all the way (or nearly all the way) up to the input temperature of the heat-transfer liquid. Such operational aspects could be optimized in detail in the test installations, above all on a large scale.

After a deployment period of 1 or 2 years, the accompanying detailed analyses of fabrication costs of the modules would also be completed. These are relatively simple to conduct for the concept of the concrete thermal-storage system, since the modules consist of only a few components. In estimating the cost of fabricating the piping system, it can be assumed that automatic welding machines would be used both for welding the piping itself and for the steel framework that holds the piping during casting of the concrete. These production steps would presumably be carried out at least in part in a central factory.

The essential questions in connection with thermal-storage systems – materials problems, thermal behavior, operational strategies, mass-production costs – could thus probably be answered within ca. 3–5 years within the framework of a serious development program. Furthermore, within the limits of the present book, we cannot judge in detail to what extent some parts of the measures suggested here have already been carried out in recent times using the newer test installations, or will be in the near future. The fundamental requirement of a massive development program, however, remains valid.

After the completion of this actual research phase, in which different storage media would be compared, a larger demonstration plant could be constructed to test the materials finally chosen and the optimal system (possibly several systems of similar quality). This plant could again operate continuously using gas heating (possibly as a heating plant to make good use of the thermal energy). The storage system could be of a size that would correspond, for example, to a base-load power plant capacity of ca. 5 MW$_{el}$ with a 16-h thermal-storage reservoir (storage capacity of ca. 200 MWh$_{th}$). Using concrete storage modules, this corresponds to 20

---

20) According to Tamme (2006), in the meantime, the concept of concrete storage reservoirs seems to have prevailed among solid-state storage systems, in particular as compared with ceramic storage media.

modules, each with a storage capacity of 10 MWh$_{th}$, as mentioned above. A thermocline tank for a storage capacity of 200 MW$_{th}$ – linearly extrapolated from the Sandia pilot plant (2.3 MWh$_{th}$, 37 m^3) – would have a volume of ca. 3700 m^3 (cylinder of, e.g., 16 m diameter and ca. 18 m height). This is, however, just a numerical example; the optimal size for individual tanks is still unknown. For comparison: The two molten-salt tanks of the Spanish Andasol plants (50-MW, 7-h storage reservoir, ca. 1000 MWh$_{th}$) each have a diameter of 36 m and a height of 14 m, that is, a volume of ca. 14 000 m^3.

As the next step, the storage systems (of the types then favored) could be integrated into the 5-MW demonstration power plant mentioned above. With two "large" storage reservoirs, the real 24-h operation of a solar power plant could be "demonstrated." This would possibly not yield any further new or decisive technical knowledge, since all the essential points would have already been investigated in the module test installations. However, aside from the confidence gained in the overall concept, the estimated item costs could be precisely verified, which is of course of great importance for the final evaluation of the economic feasibility of the system.

9
Solar Updraft Power Plants

9.1
Introductory Remarks

This type of solar power plant was developed to a great extent in Germany – apart from the very first *ideas* proposed more than 100 years ago: It originated at the perhaps most innovative private German development agency working in the area of solar energy, the engineering firm ("structural consulting engineers") of Schlaich, Bergermann und Partner (SBP) in Stuttgart. This team not only developed the concept of the chimney power plant as a completely new type of solar power source from its inception up to construction-ready plans, but they have also contributed to the development of important aspects of other types of solar power plants. We have already mentioned this in connection with the development of heliostats. Parallel to that, they constructed several variants of the small solar-dish power plant (which is beyond the scope of this book), which in terms of its reflector is closely related to heliostats. SBP was particularly successful recently with their further development of the mechanical design of parabolic-trough collectors. Their recent construction (SKAL-ET) is already in use in the new Spanish solar power plants.

The Stuttgart firm in 1996 employed, for example, six staff members for solar-energy development. They were financed mainly by profits from the core business of this statics and construction agency with 35 employees (construction of bridges, buildings, and so on). The firm belongs within a group of private institutions who have been highly innovative, such as – to name another example from Germany – the BOMIN Solar Research company (Kleinwächter). This latter firm had already developed a novel solar dish installation by the end of the 1970s and made analogous suggestions for the improvement of heliostat systems. The company, however, had no financial basis outside the field of solar energy. Owing to the lack of public support, it was able to carry out long-term developments for the future only to a very limited extent. Its innovative approaches, especially related to the cost reduction of heliostats and solar-dish systems, could therefore not be implemented.

Without a separate economic basis, the same would have been true of the research group associated with Prof. Schlaich. The exceedingly important developments accomplished by this team to date are therefore due to "chance," so to speak,

Large-Scale Solar Thermal Power. Werner Vogel and Henry Kalb
© 2010 WILEY-VCH Verlag GmbH & Co. KGaA, Weinheim
ISBN: 978-3-527-40515-2

Figure 9.1 A prototype updraft power plant in Manzanares, Spain (SBP).

namely the private initiative of unusually responsible individuals, rather than governmental research policy. Thus far, only one project was genuinely supported with public funds: the construction of a 50 kW demonstration chimney plant in Manzanares (Spain) (Figure 9.1). This project, however, was phased out (after several years of operation) already more than 20 years ago, and further developments have been mainly privately supported and more or less limited to theoretical work.

9.2
The Principle

The chimney power plant functions according to a well-known and simple principle: Under a large-area glass roof (many square kilometers), the air is heated by the ground that is in turn heated by global solar radiation.[1] At the center of the glass or plexiglass roof is a concrete tower, the *chimney*, in which the air inside, warmed by ca. 40 K relative to the outside air, flows upward (see Figure 9.2–9.4). This is the *chimney effect*, which, simply stated, is due to the difference in densities of warm and cool air. The air flow is used to generate electrical energy by means of air turbines that are mounted in the inflow area at the base of the chimney. It should be kept in mind that these "wind turbines" are not similar to the turbines used in a normal wind power-generating system, but correspond rather to a water turbine, in terms of the underlying physical principle. The difference in velocities is not the definitive quantity; this is the *pressure difference* between the regions in front of and behind the turbine.

---

1) The sum of direct and diffuse solar radiations.

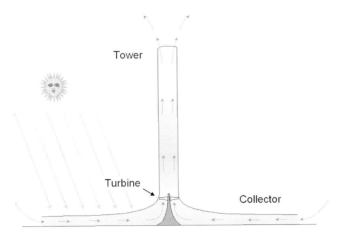

Figure 9.2 Solar updraft power plant (principle) (SBP).

Figure 9.3 Updraft power plant – artist's conception (chimney, e.g., 1000 m high, collector roof diameter 6 km) (SBP).

The dimensions of a typical installation (100 MW, with heat storage) would be:

Tower height: 1000 m
Tower diameter: 110 m
Collector diameter: 6 km
Collector area: 28 km².

First of all, it is interesting that the ground itself represents an inherent *heat-storage reservoir* that evens out small daily fluctuations in the insolation. It is, however, then a decisive advantage that the integration of an additional heat-storage system in a truly "simple" manner is (most likely) possible, permitting 24-h operation (Figure 9.5): Voluminous plastic hoses filled with water (or "water cushions") are laid out under the collector roof. Owing to the high heat capacity of the water, only

Figure 9.4 Updraft power plant – artist's conception: Chimney with spoked stabilizing inserts (SBP).

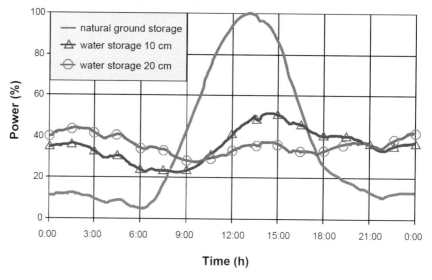

Figure 9.5 Effect of heat storage underneath the collector roof using water-filled black tubes. Simulation results from Schlaich et al. (2005).

a part of the solar energy entering the collector during the day is immediately passed on to the air; the rest heats the water. Conversely, at night, the cooler air entering at the outer rim of the collector roof is heated by the warmer water, keeping the air turbines running.[2] These water cushions need not cover the entire

---

2) The strength of the heat exchange between the water and the air can be regulated. To this end, within the water cushions, a minimal water flow is maintained by simple electrically driven miniature propellers (forced-convective heat transport). The newest concept, however, is based on the assumption that the natural convection would yield a sufficiently effective heat exchange (Weinrebe (SBP), 2008).

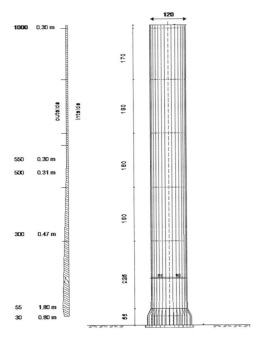

Figure 9.6 Tower cross section (SBP).

Figure 9.7 Glass roof installation in the prototype plant at Manzanares, Spain (SBP).

ground area; the collector area given above and the cost data in the following tables correspond to 25% coverage with water cushions.

The power plant consists essentially, at least in principle, of conventional components such as the concrete tower (Figure 9.6), the glass roof (Figure 9.7), and the air turbines. These differ from those used in other applications merely in their dimensions. Only the water-based thermal-storage system is genuinely new.

The solar-electrical efficiency of a chimney power plant is proportional to the height of its chimney. For a commercial-sized plant, the chimney would be about 1300 m (or up to 1500 m) in height, whereby the optimization of its height depends on the relative construction costs of the tower and the collector roof. For statics reasons, such high towers must have a certain minimum diameter. This in turn requires a large collector roof area; otherwise, the chimney would be overdimensioned. Thus, for economic reasons, large plants in the range of 100–200 MW output power are favored. For this type of power plant, even more so than for the optically concentrating solar thermal plants, it is clear that intermediate-sized plants are less economically feasible.[3]

Chimney power plants require roughly three times the land area as compared with solar tower plants and more than four times that of parabolic trough plants; this is simply the result of their very low efficiencies (for a chimney of about 1000 m in height) of only around 1%. If, for example, a large-scale power supply system is envisaged for Europe, with capacities in the multigigawatt range, only the Sahara would be appropriate for locating chimney plants; Spain is out of the question. In North Africa, the area argument is nearly irrelevant in view of the size of suitable regions, although one should not forget that a greater required land area in general gives rise to increased infrastructure costs. Furthermore, we note that the requirements on the topography of the terrain are less stringent for a chimney power plant than for the optically concentrating plants.

9.3
Investment and Power Costs

The cost of the chimney, the collector roof, and the air turbines can be estimated with relative precision. With the exception of the wind turbines, we are dealing with mass-produced items (cement, steel, glass, or plastic). In constructing the concrete chimney, the technology used for cooling towers can be applied. Owing to the great height of the chimney, new transport technology for the concrete must be developed, but such height differences of 1000 m and more must also be overcome in mining technology, so that rough comparison values for the material transport are available. The steel structure for mounting the glass roof represents more or less conventional technology.

The remaining uncertainties in the cost estimates can be eliminated in the course of further development. For the chimney, the major uncertainties are related to construction at heights of more than 1000 m. The cost progression and the costing risks at such heights were perhaps underestimated in the 1990s. In the year 1996,

---

3) The Australian solar-chimney project of the Enviromission Corporation was reduced to 50 MW during the planning phase, in order to keep the capital sums required of the first investors from becoming too large. A chimney plant planned for Spain because of the favorable power-input allowances there will have an output capacity of only 50 MW (Weinrebe et al., 2006).

a joint working group from SBP and two large German electric power suppliers[4] assumed a chimney of 1500 m height for their cost estimates (Schlaich et al., 1996). (The annual capacity factor was quoted to be 85%!) In the cost tables of more recent SBP publications, tower heights of only up to 1000 m are listed (Weinrebe (SBP, 2008)); however, Weinrebe speaks of an "optimal tower height more in the range of 1300 m." It is still unclear whether the published cost estimates were changed only on the basis of more precise investigations or because efforts were made to arrive at a more *conservative* estimate. A particular source of uncertainty is represented by the water cushions (heat-storage system), for which thus far no experience has been accumulated (except for small-scale experiments).

Air turbines for this application with a corresponding output power likewise will require new development – with some uncertainties regarding their cost. In a 100-MW plant, various turbine configurations are conceivable:

1) One could use, for example, six turbines, each with an output power of 17 MW (oriented along the axis of the chimney: vertical-axis machines).

2) One could use a larger number of small wind turbines with their axes horizontal, arranged around the lower perimeter of the chimney (e.g., 33 turbines, each with 3 MW).

3) Alongside these two "conventional" solutions, there is in principle also the possibility of generating electrical energy using a single 100-MW turbine with its axis oriented vertically that would fill the whole cross-sectional area of the chimney at a certain height above the ground. This concept naturally represents a special technological challenge, but from the initial investigations, it would appear to be feasible, and it is conceivable that it would result in lower costs than the use of many smaller turbines.

The planners of the first chimney-plant projects appear to prefer the second concept, that is, of horizontal-axis turbines (Weinrebe, 2008).

In Table 9.1, the costs are listed as estimated for the first plants (Weinrebe et al., 2006). The conversion to dollars (2002-$) was made in the usual way.[5] For mass production – with an annual deployment of ten 100-MW base-load plants or still more – we must take into account a corresponding potential for cost reduction due to the *learning curve*. According to Schlaich (1997), this reduction (based on then-current low base values) could lie roughly in the range of 30%; Weinrebe (2008) holds (on the basis of newer cost data) a reduction of 15% to be realistic. Here, we assume following Weinrebe that the reduction would amount to 15%. This yields *overall investment costs of 5280 million $/1000 MW*.

The investment costs for the complete solar power system are collected in Table 9.2. Taking the transmission losses into account (e.g., from Morocco to Germany,

---

4) The two utilities "Energieversorgung Schwaben" and "Badenwerk" merged in 1997 to become the "Energie Baden-Württemberg."

5) Exchange rate for Euros (2006) to 2002-$ with an inflation factor (Germany) of 1.047 and purchasing-power parity (2002) of 1.043 $/Euro.

Table 9.1 The specific investment costs for updraft power plants, referred to an output power of 1000 MW, corresponding to 10 plants of 100 MW each.

Investment costs	Million $/1000 MW (2002-$)
Chimney	1560
Collector (roof and thermal storage)	3370
Turbines	750
Other costs	530
Overall investment costs	6210
Total investment ("mature")	5280

Table 9.2 Investment costs for the complete solar power system (for a site in Morocco, transmission distance 3000 km, with transmission losses of 11.5%).

Investment costs	Million $/1000 MW (2002-$)
Investment per 1000 MW at the site	5280
Chimney plants (1130 MW at the site)	5970
CCGT backup plants	715
Transmission system	650
Solar power system	7335

Table 9.3 The energy cost from the solar power system based on chimney power plants and CCGT backup plants.

Energy cost		¢/kWh (2002-$)
Capital		4.0
Operation and maintenance	Solar	0.3
	Backup	0.2
Gas		1.5
Energy cost		6.0

3000 km) at 11.5%, one obtains investment costs for the chimney plants (1130 MW at the plants) of $5970 million per gigawatt of the power delivered at the end of the transmission line. Together with the costs of the transmission lines and of the CCGT backup power plants,[6] we obtain *$7335 million (2002-$) per gigawatt* for the *solar power system*.

In Table 9.3, the *power cost* resulting from these investment costs is given. The individual contributions are as follows:

From the investment costs and the basic assumptions on interest rates (4% real) and operating lifetimes (45 a), we obtain *capital costs of 4.0 ¢/kWh*.

---

6) The investment costs for the backup system are based on Chapter 10.

The *costs for operation and maintenance (O&M) of chimney power plants* were given by Weinrebe *et al.* (2006) as Euro26 million per 1000 MW and year, which for the solar power system per gigawatt (with a factor of 1.13 due to transmission losses, currency conversion as above) corresponds to $29 million (2002-$). Relative to the annual energy output of the solar power system (8760 GWh), we find from this O&M costs of *0.3 ¢/kWh*.

The O&M costs assumed here are based on experience at the demonstration power plant near Manzanares, mentioned earlier. In contrast to the other types of solar power plants, we can expect only minor operating and maintenance costs for a chimney plant – apart from unforeseen repairs and the cost of maintaining the thermal storage system, which is not yet clearly defined. The operating personnel need only monitor the air turbines.[7] Except for their rotors, including those of the generators, the plant has practically no moving parts that would require adjustment.

The O&M costs and also the gas costs for the backup power plants depend on the capacity factor of the chimney plant. Weinrebe *et al.* estimate for an excellent site with an annual *global* insolation of 2300 kWh/m^2 a that the yield of electrical energy would be 6180 GWh per GW and year. If we presume that the "good" sites in the Sahara have an insolation of 2150 kWh/m^2 a and take (following Weinrebe) a roughly linear dependence of the yield on the insolation, we obtain an annual energy production of 5780 GWh (5780 h of full-power operation). The backup plants would thus need to operate (on the average) for 2980 h at full power per year (34% capacity factor).

The variable O&M costs of CCGT plants are 0.17 ¢/kWh (cf. Chapter 10). With a weighting factor of 0.34 (34% fraction of backup power), we obtain variable costs of 0.06 ¢/kWh for the complete system. The annual fixed O&M costs of the CCGT plant (10 million $/GW) are related to the 1 GW of the complete system that produces 8760 GWh, giving 0.11 ¢/kWh. All together, we thus find *O&M costs of around 0.2 ¢/kWh* for the *backup power plants* (solar power system).

The *gas costs* at a gas price of 2.5 ¢/kWh of gas and at an efficiency of 58%, with a 34% capacity factor, amount to *1.5 ¢/kWh$_{el}$* (for the solar power system).

9.4
Development Program

9.4.1
The Development of Components

Since the collector roof is purely a design and construction problem, and not a subject for research, the essential development topics would be as given below:

---

[7] Conventional wind-energy plants as a rule operate automatically, without any long-term manual monitoring.

- air turbines (3, 17, or 100 MW)
- construction of a chimney of the required height
- thermal-storage cushions.

Here, we shall not consider the first topic in any detail. It is a matter of the usual development tasks for optimizing turbine blades, bearings, reduction gearing, and generator, as are well known from the general development of wind and water turbines.

9.4.1.1 The Chimney

The chimney of the 100-MW plant, at a height of 1000 m, would have, as mentioned, a diameter of 110 m. In connection with a large demonstration plant, the construction of a tower of this height and diameter is *in fact* not necessary, since the costs of towers of various sizes can be calculated to relatively good accuracy. There, however, appears to be a *psychological problem*, which in the past has proved to be an inhibiting factor to the introduction of this technology: This is the fact that the construction of a tower with a height of more than 1000 m is hard to imagine.

Such psychological factors are not subject to logical arguments and can frequently represent more formidable obstacles than do real, factual problems. Construction of a tower of somewhat *more than* 1000 m height would eliminate the subjective impression of its being "utopian." This prototype tower should be, for example, 1050 m high. (It is easier for many observers to extrapolate from 1050 to 1300 m than, for example, from 750 to 1000 m. In both cases, however, the statics can be calculated with the same precision.)

In the first step of the development program, the computations presented by the development company must be verified by multiple independent assessments (or the assessment should be repeated, insofar as it has already been carried out[8]). Although the psychological objections have then been refuted scientifically, as a second step a corresponding tower would be constructed (see below).

---

8) Such assessments have in fact already been presented and are mentioned in particular in the final report of the working group of SBP and two German utility companies mentioned above (Schlaich et al., 1996). Here we give just four examples:

a) Prof. Dr. Ing. Göde, Dr. Ing. Ruprecht: "Investigation and Calculation of the Flow Characteristics and Turbine Design for a 200-MW Chimney Power Plant," Institute of Fluid Mechanics and Hydraulic Machinery, University of Stuttgart (Germany), 1996.

b) Construction Firm Dyckerhoff & Widmann AG: "A Suggestion for the Construction of the Chimney," Dresden (Germany), 1996.

c) Prof. Ben C. Gerwick: "Solar Chimney Constructability Report," San Francisco, 1996.

d) "The turbines were designed and specified with the support of Dipl. Ing Kohler (expert for turbines and former leader of the design group at the Voith Hydro company), and the design was refined in extensive consultations with the firms Voith Hydro Kraftwerkstechnik, Heidenheim (Germany), Sulzer Hydro, Ravensburg (Germany), and ABB Kraftwerke AG, Mannheim (Germany), … As a validation, we [SBP] have a quotation from the Voith Hydro company for the suggested price of 32 single turbines of 6.25 MW each."

In designing this tower, depending on its location, various forms of tectonic activity must be taken into account. Their effects can be investigated using computer simulations in which the influence of earthquakes of various strengths and types on the structures can be demonstrated. (Fundamentally, only regions in the lowest class of earthquake risk can be considered to be suitable sites.) Insofar, as different research groups arrive independently at the same conclusions regarding the critical points, the correctness of the analysis would be "demonstrated" even for the nonexperts, in particular for those politically responsible for decision-making. A real verification of earthquake security by means of a demonstration plant is hardly feasible, since the stability of the plant would be proven only by a real and correspondingly strong earthquake (as assumed in the simulations).

In this connection, it would seem important to mention that locations in Morocco, if one is thinking of an electric power supply for Europe on the basis of chimney plants, are problematic. Considering the earthquake risk, among locations with a good insolation, only the more southerly sites there are risk-free. The size of their overall area is probably marginal, given the fact mentioned above that chimney plants require about four times the land area as do optically concentrating solar power plants. (For this reason, locations in Morocco would probably not be sufficient to meet the needs for supplying all the electric power to Europe from chimney plants there, even neglecting the risk of earthquakes). Thus, most likely sites in the Sahara (e.g., in Algeria) would also be needed. Schlaich *et al.* (1996) state in this connection: "… in terms of suitable sites, there could be difficulties which would increase costs, e.g. due to the risk of earthquakes or problems with the terrain. There are, however, sufficient stony desert regions, especially in the Sahara, where there is no earthquake risk and a good substratum is present."

9.4.1.2 Heat Storage

The heat-storage system can be developed and tested within a short time in a small test installation constructed for that purpose. A modest glass roof of a size similar to that of the previous test plant in Manzanares would suffice. It could take the form of, for example, a narrow segment of the complete large roof and could be closed at its sides. A width of 50–100 m with a length of, for example, 500 or up to at most 1000 m would be appropriate. The air would be drawn in at one end, pass under the roof in the long direction, and would enter the chimney at the other end. The chimney could in this case be built without turbines and would fulfill only the task of generating the air flow under the glass roof. To simulate the air flow with turbines, variable baffles could be installed in the chimney. In order to test different sizes and shapes of the water tubes simultaneously, the roof could be subdivided by partitions in the long direction. A much smaller test installation might be adequate, in which the air flow would be produced not by a chimney but by a blower. The length would, however, need to be adequate so that the simulation of the air flow for different roof heights could be carried out.[9] Taking the

---

9) The collector roof of a chimney power plant is low at its outer perimeter and rises toward its center, since there, a larger radial air current per unit of ground area must be accommodated.

higher specific costs for a very small installation into account, the larger test installation described above (with a glass roof of 1000 m × 50 m) would cost ca. $10–15 million, at most $20 million (2002).

Such a small (or indeed very small) test installation could be set up within 2 years. After 2 more years, the relevant operational data would be at hand, for example, on the heat exchange between air and the water cushions for different sizes and shapes of the cushions. The realistic heat-storage operation could thus be simulated, experimentally documented, and optimized.

9.4.2
A Demonstration Plant

After the completion of this detailed development program, a demonstration plant with a chimney tower of the height mentioned (more than 1000 m) would need to be constructed. Here, we mention some important points: A high tower requires a corresponding diameter for statics reasons, in this case 110 m. This tower would be suitable for a 100-MW plant from the point of view of its flow characteristics. Nevertheless, a demonstration plant would not have to be constructed with such a high power-output capacity, since it would not require a full-sized collector roof.

A single sector should be sufficient, for example, one-sixth of the complete circular collector area. In this case (in accord with the concept using horizontal-axis turbines), only about 5 of the 33 turbines (each with 3 MW installed power) would need to be mounted. More important than the installation of 33 turbines would be the demonstration of turbines of a similar size to those that would later be employed in a full-scale plant, or at least close to that size. As in the test installation for the heat-storage system, the collector roof could be subdivided by radial partitions. This would enable variations in its construction to be tested, for example, in its height, and would allow a precise determination of the operational data. In the test plant, all the components and functions of the full-scale plant could be verified: thus, the air flow and the design of the roof near the chimney, the air flow at the chimney inlets, and the air flow into the air turbines. Further, and this would be a decisive point, the feasibility of constructing chimney towers of more than 1000 m in height would be demonstrated.

The remaining sectors of the roof and the additional turbines belonging to them could be completed later, on a commercial basis. The small-scale plant would indeed show the *feasibility* of a complete plant to the whole world and in particular to potential investors and banks. It is more important for the implementation of this technology on the world market to construct a plant immediately (even if it is not full-sized) than, for example, to bring a larger, full-scale power plant on-line only some years down the line. The later completion of the test plant could not be expected to yield essential new insights.

The costs of the *test plant* could be around *$300 million*, whereas a complete, full-scale plant would cost ca. $600 million (2002).

These costs were roughly estimated (as in Table 9.1):

- The *chimney* of a 100-MW plant (without heat storage) with a height of 1000 m would cost ca. $156 million. The required tower of *more than* 1000 m height, for example, 1050 m, should thus cost at most $180 million.

- The *collector roof* of the 100-MW base plant (including heat storage for 24-h operation) would cost $337 million. A segment of one-sixth of the overall area would thus cost ca. $60 million.

- The horizontal-axis *turbines* of the 100-MW plant would cost $75 million. One-sixth of these would thus cost $13 million.

- For "other costs," we estimate an additional $50 million.

- *Total cost*: around $300 million.

This demonstration plant should be constructed in such a manner that different versions of the air turbines could be installed. Alongside the five horizontal-axis machines mentioned (which would correspond to the roof segment of one-sixth of the full diameter), one or more different vertical-axis machines could be installed. There is sufficient room for this since the tower is large enough to accommodate six large vertical-axis machines. This, however, would require additional funds.[10] If the financial backing for the plant were correspondingly extended, all the different designs and types (e.g., different rotor types) could be installed and compared in terms of their output power and reliability. If there were more versions than could be installed at once, they could be tested for several months in sequence.

9.4.3
Detailed Cost Estimates

As with the other types of power plants, we need to specify the costs of chimney plants under mass production: thus, for example, the price increases included by the suppliers for small series (in contrast to large-scale deliveries), or price modifications resulting from market fluctuations for steel, concrete, and glass; and also the advantage of constant and guaranteed annual purchase quantities, that is, the cost markups for capacity utilization risks of the production facilities. In addition, there are the fabrication costs of the water cushions and the cost of their installation at the construction site. These cost estimates should also be compared redundantly by several independent experts.

---

10) In rough analogy to the costs of wind-energy plants, we could expect the following expenses: On the basis of 1000 $/kW for a complete wind-energy plant – and setting the cost of the turbine equal to the entire cost of the wind plant – for turbines with a net output power of 100 MW, we would expect a price of ca. $100 million. Depending on how many different turbine designs needed to be constructed and tested, the total power output should in any case remain well under 100 MW. (The demonstration plant, a *partial-scale plant* with one-sixth of the capacity of a 100-MW plant, would have an output power of only 18 MW.)

Table 9.4 Estimated development costs for the updraft power plant.

Development costs	Million $ (2002)
Chimney tower	
Assessment by several independent experts, computer simulation of earthquakes	ca. 10
Air turbines	
Development analogous to wind power plants (development only, without construction costs)	<30
Heat-storage system	
Development costs	ca. 15
First installation with test collectors	ca. 10
Test roof for heat-storage tests (of similar size to the demonstration plant at Manzanares (see Section 9.1))	ca. 10
Demonstration plant (without a test program for air turbines)	ca. 300
Test program for air turbines in the demonstration chimney	e.g., 100
Additional development requirements	ca. 50
Total	530 (ca. 500–600)

In the course of the construction of the test installation for heat-storage systems, some initial experience would be obtained, for example, in terms of installation costs, charging with water, and, if necessary, power requirements of the circulation pumps. Remaining uncertainties concerning the operating lifetime of water cushions made of different materials would have to be clarified during the technical development, as early as possible.

9.4.4
Development Costs

The chimney-plant technology is an example of how a completely new type of power plant, which generates electric power *without consuming fuel*, could be developed and demonstrated for an expenditure of ca. $500 million (Table 9.4), that is – to simply point up the order of magnitude – for roughly one-fourth of the current cost of a single new nuclear power plant (ca. 2000 million $/GW). It is therefore completely beyond comprehension that this development was not carried out long ago.

10
Fossil-Fuel Power Plants

The cost data for fossil-fuel power plants are summarized in this chapter. On the one hand, this applies to backup plants (gas or coal-fired) that enter into the costs of a complete solar power system; and on the other hand, to the fossil-fuel base-load power plants, to which the solar power system is being compared.

In the framework of this book, it is not possible to include a deeper discussion of the often diverse cost specifications for fossil-fuel plants to be found in the literature. As the basis for our cost estimates, we have used essentially two sources:

1) An often cited and very extensive American study entitled "The Economic Future of Nuclear Power"; it was prepared at the University of Chicago on the request of the US Department of Energy and completed in August 2004. In the following, this study is referred to simply as the "Chicago Study" (2004). This study contains, on the one hand, the whole scale of cost expectations for future gas and coal-fired as well as nuclear power plants; on the other hand, it also lists typical detailed cost data that are useful for the purpose of comparisons in this book.

2) The "Annual Energy Outlook" from the US Energy Information Administration (EIA AEO, 2007).

The spectrum of the investment costs which the Chicago study quotes from other studies is relatively broad, both for the fossil-fuel plants and for the nuclear power plants. This is surprising since one would expect that for "conventional" plants, the costs would be precisely known. This gives rise within the cost comparisons given in this book to a double "uncertainty," that is, uncertainties not only in the costs of solar power plants but also in those of the conventional plants with which we wish to compare them.

Large-Scale Solar Thermal Power. Werner Vogel and Henry Kalb
© 2010 WILEY-VCH Verlag GmbH & Co. KGaA, Weinheim
ISBN: 978-3-527-40515-2

10.1
Natural Gas Plants

10.1.1
Investment Costs

For a natural-gas combined cycle gas turbine (CCGT) plant in the 600-MW class, we assume (for an output power of 1000 MW) that the investment costs will be $580 million (2002-$; cf. Chicago Study, 2004). Taking a construction period of 3 years into account (1.5 years full interest period, 4% real interest), we find for a *base-load* power plant, *615 million $/GW* (in 2002-$).

For the 600-MW installations, which are also planned as backup plants, we include a cost increase for an additional gas turbine, for the following reason: In a CCGT plant, the output power is roughly divided into two-thirds from the gas turbine and one-third from the steam turbine. In case of an unforeseen interruption in the long-distance transmission line from the solar power plants, the gas turbine can be brought up to full power within a few minutes (less than 15 min); it has a "rapid startup capability." It can be routinely started up in 15 min; in emergencies, this time could be even shorter. In contrast, the steam turbine circuit at present requires around 1 h for startup from the preheated state[1]; in newer plants, this time can evidently be reduced to ca. 30 min (E&M, 2007). In order to guarantee a fast startup with the *full output power* of the CCGT plant, an additional gas turbine of the same nominal power as the steam turbine section (i.e., one-third of the full output power of the plant) must be installed. The specific investment costs of the complete backup system are then found as follows: A CCGT plant costs 615 million $/GW (as above), complete peak-load gas-turbine plants (200-MW class) cost about two-thirds as much (AEO, 2007), that is, around 400 million $/GW.[2] Since the fast-startup gas turbine is integrated into a CCGT power plant, it can be assumed that its cost would be less than that for a complete gas-turbine plant. According to the AEO (2007, p. 79), the cost of the turbines constitutes 50% of the overall cost of a gas-turbine plant.[3] For this *additional gas-turbine installation* within the CCGT plant, we therefore estimate 300 million $/GW. Since it need supply only one-third of the overall capacity of the backup system, one would require – for this part of the backup plant – only one-third of $300 million, that is *100 million $/GW*.

We thus obtain for the *complete backup system* on the basis of CCGT power plants a total of $615 million plus $100 million, that is, *715 million $/GW* (2002-$).

---

1) E&M, 2007: "... start-up time from a hot or semi-hot condition following night or weekend standstill ..."
2) The considerably lower efficiencies of simple gas turbines (36%) are not significant for their use as an *emergency* backup component.
3) See also NTC (2008); there, for gas turbines, costs in the range of 200 $/kW are quoted.

10.1.2
Gas Costs

The contribution of gas costs to the final cost of power is influenced by the two factors *gas price* and *efficiency*.

The gas price was in the past subject to strong fluctuations and can hardly be predicted with any certainty. In particular, between 2004 and 2007, it increased sharply, and in the year 2007 (according to AEO, 2008) in the USA for gas delivered to power plants, it was 6.9 $/MMBTU (2006-$), higher heating value (HHV)), which corresponds to 2.62 ¢/kWh$_{gas}$ (lower heating value (LHV)); in 2002-$, this gives 6.16 $/MMBTU (HHV) = 2.34 ¢/kWh$_{gas}$ (LHV).[4] As a *base gas price*, we assume (for Germany also) 2.5 ¢/kWh$_{gas}$ (LHV) in 2002-$, both for base-load power plants and for backup plants. For simplicity, we have neglected the fact that owing to fluctuations in the amount of gas used by the backup plants, their price for gas tends to be higher than that for base-load plants. Since the gas price, considered in the long view, is in any case a very uncertain parameter, we dispense with distinguishing between the levels of gas prices in Europe and in the USA. In Germany, corresponding to the gas price in the year 2007 for power plants of 210 €/t of coal equivalent (tce) (LHV) (Kohlenstatistik, 2008) and recalculated to 2002 $, we find 2.48 US ¢/kWh$_{gas}$ (LHV)).[5] For comparison: The natural-gas import price in 2007 (Germany) was 5.5 €/GJ = 2.0 €-cent/kWh (HHV) (BMWi, 2008); converted to 2002-$ and expressed in terms of the real purchasing power of the dollar, this corresponds to 2.13 US ¢/kWh (LHV).

For the CCGT base-load power plants which will be constructed in future years, we assume a *gas-to-electric efficiency* of 60% (LHV). This corresponds to the best value for plants currently being built and would appear to be a conservative value in the long view. For the coming 10 years, in the literature, higher values are often quoted. The Chicago study gives 65% as "*R&D target.*" For *base-load power plants*, at the gas price given above, we would thus find a gas cost of *4.1 ¢/kWh$_{el}$* (2.5 ¢/kWh$_{gas}$/0.6).

For gas power plants operating as backup systems, we assume a somewhat lower efficiency of 58%.[6]

---

4) BTU: British thermal unit; 1 MMBTU = 10^6 BTU = 293 kWh = 1.055 GJ; for natural gas: LHV = 0.9 HHV. (The higher heating value includes the heat of vaporization of the water vapor which is formed on combustion, but cannot be utilized in power plants.) Inflation factor (US Consumer Price Index 2002 → 2006): 1.120; purchasing power parity 2002: 1$ = 0.96€.

5) 1 tce (ton of coal equivalent) = 8140 kWh; inflation factor (2002 → 2007 Consumer Price Index Germany): 1.084; purchasing power parity (PPP) 2002: 1$ = 0.96€ (since the imported gas in Europe is paid for in Euros, we use here the PPP, whereas for the oil imports (to Europe), which are paid for in dollars, we use the exchange rate.)

6) If *base-load* natural-gas power plants are to be replaced by solar power plants, we assume that the natural-gas CCGT plants that are in operation today will be used as backup plants for the solar-power system. If the total plant capacity is to be increased, that is, new backup plants are to be

We consider as examples the supply of solar electric power to Germany, on the one hand from Spain, and on the other from North Africa. The latter corresponds at the same time roughly to the situation in the USA. The expected average capacity factor for the Spanish solar plants of 70% means that the backup system operates 2630 h per year at full power, that is, per gigawatt of plant output power, 2630 GWh of electrical energy are generated. At an efficiency of 58%, this corresponds to a gas consumption of 4535 GWh and thus (at the gas price quoted above) to gas costs of 113 million $. Referred to the overall quantity of electrical energy generated, that is, for one gigawatt of nominal output power which corresponds annually to 8760 GWh, this gives a cost of 1.3 ¢/kWh$_{el}$. Expressed differently: Each kilowatt-hour generated by the backup system corresponds to a gas cost of 4.2 ¢ (see above). The fraction of 30% backup power in the overall power produced by the solar system then corresponds simply to a weighting factor of 0.3, so that we again find 4.2 ¢/kWh$_{el}$ × 0.3 = 1.3 ¢/kWh$_{el}$. In the case of Morocco (or the USA), with an expected capacity factor of 80% (fraction of backup power = 20%), the weighting factor is 0.2, so that we now find 4.3 ¢/kWh$_{el}$ × 0.2 = 0.8 ¢/kWh$_{el}$.

In summary:

Contribution of the gas cost for "Spain": 1.3 ¢/kWh
Contribution of the gas cost for "Morocco/USA": 0.8 ¢/kWh

10.1.3
Operating and Maintenance Costs

For operation and maintenance, 0.26 ¢/kWh, that is, ca. *0.3 ¢/kWh* are quoted in the Chicago study, which also corresponds to the value given by the AEO (2007, p. 79). We adopt this value for *base-load power plants*.

For the *backup power plants*, this value is adjusted by employing a breakdown into fixed and variable O&M costs, as found in AEO (2007). There, per gigawatt and year, $10 million is quoted as fixed cost and 0.17 ¢/kWh as variable cost. Referring the fixed costs for the backup plants to 8760 GWh and weighting (cf. the section on "gas costs" above) the variable costs for the case of Spain with a factor of 0.3, for Morocco/USA with a factor of 0.2, we obtain 0.17 or 0.15 ¢/kWh, respectively. In the tables, we use the rounded value (to compensate for rounding errors) of *0.1 ¢/kWh*.

constructed, these can be "exchanged" with the previous base-load plants as already described in Section 4.1. The plants currently in operation have on the average a lower efficiency than the newest plants; therefore, we use here an efficiency of 58%. Also, for the construction of new CCGT plants, owing to the relatively low annual operating time for backup power plants, the optimization of efficiency, and investment costs could be different from that for base-load plants. The assumed efficiency of 58% should be typical of this case, also.

10.2
Conventional Coal-Fired Plants

10.2.1
Investment Costs

The Chicago study lists typical costs for coal-fired plants of 1165 million \$/GW (2002-\$, "pulverized coal combustion"), along with a construction time of 4 years. Together with interest during the construction period, this yields overall investment costs of 1260 million \$/GW. (Compare AEO (2007): 1290 million \$/GW (2005-\$).) As the investment cost for *base-load power plants*, we adopt the plausible value of *1200 million \$/GW*.

When the backup system is based on coal-fired plants, it must be taken into account that to some extent, older plants can be employed, since solar power delivery will only seldom be completely disrupted. Often, there is only a more or less serious reduction of the power supply, so that the total backup system is in general seldom required. For this purpose, older coal-fired plants could be used, that is, plants that already have been in use for a good portion of their operating lives, and even those that would otherwise be shut down within the foreseeable future. Therefore, in the case that the overall power-generating capacity must be increased – this corresponds to the case of "annex construction" in Section 4.1 – we consider a mixed system consisting of one-half new and one-half older coal-fired plants.[7] For the older plants, the calculation presumes a cost of 50% of the cost of a new plant. In the mixed calculation, this leads to an average cost of 75% of the specific investment costs of new power plants. This gives – in the case of annex construction of plant capacity – investment costs for the *coal-fired plants* in the *backup system* of *900 million \$/GW*. In the case of *replacement* of operating coal-fired plants by the solar power system, the replaced coal plants are available so to speak "for free." This case was already discussed in Section 4.1.

With coal-fired plants, when there is a sudden interruption of power from the transmission line, several hours would pass before they could supply replacement power at full output. As with the steam-turbine part of CCGT plants, a fast-startup reserve is necessary, which in this case must supply the entire output power of the backup plant. For this purpose, either gas turbines or distributed diesel-powered generators could be used. The latter are to be sure more expensive than gas turbines, but they have the advantage that they can be started in ca. 1 min.[8] In

---

7) To be sure, new coal-fired plants with the same output capacity as the solar power plants would have to be constructed, but half of them could be exchanged for "used" coal plants.

8) According to the MAN Corporation (1986), large diesel engines can be started from a preheated condition, with oil and water temperatures of 60 °C, within a startup period of 80 s to full output power. Allowing a somewhat higher wear rate than would be acceptable for routine startup, they could be brought on line even faster. (The emergency power generators at nuclear power plants are kept in a prewarmed condition and can be put into operation almost "immediately.")

contrast to gas turbines, they can also be installed decentrally, and they then offer additional security against failures in the intermediate-voltage transmission system, which is currently not guaranteed. For *peak-load diesel generators* (in the power range > 3 MW), for example, Schwaegerl and Thieme (1987) cite investment costs of 330€/kW; starting from this number, in an earlier study (1998), we adopted a value corresponding to 375 $/kW (1995 $). In 2002-$, this would be *440 million $/GW*.

For the *overall backup system* based on coal-fired plants and fast-startup generators, we thus obtain investment costs of *1340 million $/GW* (in the case of "annex construction" of the entire power-plant capacity).

10.2.2
The Price of Coal

In the *USA*, according to AEO (2008), the average *price* of bituminous coal in the year 2007 was 1.73 $/MMBTU (2006-$, probably HHV); converted to the LHV (with the usual conversion factor for coal of HHV = 1.05 LHV), this gives 1.82 $/MMBTU (LHV) (2006-$) or 50.5 $/tce (LHV). In 2002-$, this would be 43.9 $/tce, that is, around *45 $/tce* or *0.55 ¢/kWh$_{coal}$ (LHV)*.[9]

For *Europe*—corresponding to the situation in the year 2007—one can assume a coal price of roughly twice that in the USA, that is, *90 $/tce (2002-$)* (= 1.10 ¢/kWh$_{coal}$ (LHV)). Justification: The price of imported coal increased in the period from January to December 2007 from 78 $/tce to 149 $/tce[10] (EURACOAL, 2008). Ninety dollars per tce (2002) at the dollar value of *2007* is 104 $/tce; rounded off, this gives *100 $/tce*.

Note concerning the assumption of ca. 100 $/tce (2007): The increase in the price of imported coal in the year 2007 was not due to a corresponding increase in the cost of coal production, but rather to a scarcity of coal on the world market. Once the production rate and especially the transport capacities have adjusted to the new demand, the price will again fall, so that in the long term, the current rather high price is not applicable. However, if the worldwide consumption of coal continues to increase in the future, it will be necessary to tap coal reserves that are more costly and have thus far not been developed; for this and other reasons, the coal price will probably never drop again to its previous low level. As is shown in the following section, the assumed price of 100 $/tce (2007-$) should reflect these developments fairly accurately.

Today's coal prices are *market prices*; they increased in recent years as a result of the increasing oil price, but they do not reflect the *production and transport costs* for coal. In contrast, the prices from the period before 2003, when a sufficient supply of coal was available in the world market (and competition was therefore strong), do reflect the actual production costs. Table 10.1 shows the

---

9) 1 tce (ton of coal equivalent) = 8140 kWh, 1 MMBTU = 293 kWh.
10) In 2002-$ (US inflation 2002–2007 = 1.152), this corresponds to an increase from 68 to 129 $/tce.

Table 10.1 Coal prices in recent years.

	1	2	3	4	5
Source	BP 2005, 2008a, 2008b	BP, 2005	BP, 2008a, 2008b	BP 2005, 2008a, 2008b	RWE, 2005 (p. 132)
	Northwest Europe Marker Price	US Coal Prices (Power Plants)	US Central Appalachian (CAPP)	Japan Steam Coal Import (cif)	EU Import (Average)
	Average prices	Average prices	Spot-price index	Average prices	Average prices
	US $/t[a]	US $/t[a]	US $/t[b]	US $/t[a]	US-$/t[a]
2000	36.0	27.1	29.9	34.5	37.9
2001	39.3	27.6	49.7	38.0	46.2
2002	31.6	27.9	32.9	36.9	41.4
2003	42.5	28.3	38.4	34.7	45.0
2004	71.9	29.9	64.3	51.3	69.4
2005	61.1		70.1	62.9	
2006	63.7		63.0	63.0	
2007	86.6		51.1	69.9	

a) Price per ton of coal (not tce!), monetary value of the corresponding year, heating value not given.
b) "Price is for CAPP 12 500 BTU."

cost trends in the years 2000–2007 in the monetary value of the corresponding year (however in tons of coal without taking its heating value into account). If we for simplicity take Illinois coal to be representative of American coal (heating value (LHV): ca. 6.7 kWh/kg), then the prices given above can be converted to tce (=8140 kWh (LHV)) by using the factor 1.22. The value for 2002 in column 2 would then be increased from 27.9 to 34.1 $/t. In Table 10.2, where power prices for various coal prices are given, a US price of – among other values – 35 $/tce (2002 $) is thus employed.

While for countries with their own coal production (USA, Canada, Australia, and so on), the costs of coal extraction are necessarily of significance (from the macroeconomic point of view), this aspect is irrelevant to the coal-importing countries. There, only the *price* that must be paid is decisive. Table 10.1 shows, however, that the costs of exporting coal lie far below the revenues that can be realized from it. It is therefore fundamentally conceivable that the price of imported coal will in the long run drop to below the value assumed here of 90 $/tce (2002). (As can be seen from the table – taking into account the factor of 1.22 for converting to tce – the costs of *mining* and *transport* of the imported coal are, however, *at* least 50 $/tce (2002 $).[11] This is thus the lower limit to which the price could – theoretically – again fall.) Whether such a major price decrease will in fact occur in the long term depends on the future market competitive situation, whose trends cannot be predicted with certainty by anyone.

In this connection, we must, however, consider the following:

1) In the long term, the oil price will be higher than in the past.

2) It can safely be assumed that coal consumption and production will increase over a period of some decades. This will be accompanied not only by the above-mentioned increases in production costs, but will also affect the situation of market competition; a tendency to an increasing influence on the part of multinational energy corporations is to be feared – basically similar to that on the oil price at present, although perhaps not so strongly pronounced.

Therefore, the price for imported coal assumed here of 90 $/tce (2002-$) or ca. 100 $/tce (2007-$) appears to be not implausible, especially since the prices were recently considerably higher, for example, in December 2007 at 149 $/tce (129 $/tce in 2002-$). In Table 10.2, as mentioned, we list various coal prices in order to make their influence on the cost of electric power clear to the reader.

10.2.3
Plant Efficiencies/Contribution of Coal Price to Power Costs

If we assume an *efficiency* for future coal-fired power plants of 45% (referred to the LHV),[12] we obtain for *base-load* power plants in the USA (coal price 45 $/tce) a contribution to the power cost due to the coal price of $1.3 ¢/kWh$ ($0.55 ¢/kWh_{coal}/0.45$); in Europe (90 $/tce = $1.10 ¢/kWh_{coal}$), this contribution is $2.5 ¢/kWh$.

For coal-fired plants in the *backup* system, we assume that 50% of the backup plants are new and 50% are used, with an average efficiency of only 42%. If only used plants are employed in the backup system (substitution of current

---

11) If we take as starting point for imported coal in Table 10.1 a price of 40 $/t (e.g., in the year 2001 in column 1, or the year 2002 in column 5), this corresponds to 48.8 $/tce (in 2002-$). This is the price at the border of the producing country, not the price at the power plant in the importing country. The coal price *including transport to the power plant* would then be at least 50 $/tce.

12) The German EWI study (EWI, 2007) presumes an efficiency of 45% for coal-fired plants presently under construction. The Chicago Study (2004, p. 6-3), in contrast, quotes a value of only 36% for currently operating American plants (probably based on the HHV, that is, 38% based on the LHV). In future, making use of supercritical steam, 45% is apparently likewise considered there to be feasible within a short time. "Newer units employing supercritical steam may reach efficiencies of 45 percent." However, this is associated with higher investment costs which are not precisely specified. Concerning future perspectives, the study states, "As materials advance, government R&D programs hope to reach efficiencies as high as 50 percent." In the EWI study, efficiencies of 51–52% by the year 2020 are held to be feasible.

Table 10.2 Coal-fired power plants.

Coal-fired power plants 2002-$		Base load	Base load			
		Modern conventional power plant	IGCC (from EIA AEO 2007)			
			Without CO_2 capture		With CO_2 capture	
Plant capacity	MW	600–700	550	550	380	380
Investment costs (including interest during construction 4%/a, 2a)	M$/GW	1200	1485	1485	2120	2120
Annual energy production (per GW)						
Capacity factor		91.3%	91.3%	91.3%	91.3%	91.3%
Hours of full-load operation	h/a	8000	8000	8000	8000	8000
	GWh_{el}	8000	8000	8000	8000	8000
Capital costs						
Real interest	%/a	4.0	4.0	4.0	4.0	4.0
Interest rate factor		1.040	1.040	1.040	1.040	1.040
Operating lifetime	a	45	45	45	45	45
→ real annuity	%/a	4.83	4.83	4.83	4.83	4.83
(Annuity at 2% real interest)	%/a	3.39	3.39	3.39	3.39	3.39
→ Capital costs per year	M$/a	58	72	72	102	102
Divided by electrical energy/year:						
→ Capital costs per kWh	¢/kWh	0.72	0.90	0.90	1.28	1.28
(Capital costs at 2% real interest)	¢/kWh	0.51	0.63	0.63 nth of a kind	0.90	0.90 nth of a kind
Efficiency		45.0%	41.1%	47.4%	35.2%	43.1%
Fuel costs (at the given efficiency)						
Bituminous coal (1 tce = 8140 kWh LHV) for a coal price of (at power plant):						
	$/tce	¢/$kWh_{coal\,LHV}$	¢/kWh_{el}		¢/kWh_{el}	
USA	35	0.43	0.96	1.05	0.91	1.22	1.00
USA (2007)	45	0.55	1.23	1.35	1.17	1.57	1.28
USA	60	0.74	1.64	1.79	1.56	2.09	1.71
Europe (import)	80	0.98	2.18	2.39	2.07	2.79	2.28
Europe (import 2007)	90	1.11	2.46	2.69	2.33	3.14	2.57
Europe (import)	100	1.23	2.73	2.99	2.59	3.49	2.85
Germany (bituminous coal)	150	1.84	4.10	4.48	3.89	5.24	4.28
Germany (lignite)	35	0.43	0.96	1.05	0.91	1.22	1.00

Table 10.2 Continued.

Coal-fired power plants 2002-$				Base load	Base load			
				Modern conventional power plant	IGCC (from EIA AEO 2007)			
					Without CO_2 capture		With CO_2 capture	
Operating and maintenance (O&M) costs								
Fixed costs per GW (base-load plant)			M$/a	24	33.5	33.5	39.4	39.4
Divided by electrical energy/year:			GWh/a	8000	8000	8000	8000	8000
→ Fixed O&M costs			¢/kWh	0.30	0.42	0.42	0.49	0.49
Variable O&M costs			¢/kWh	0.40	0.25	0.25	0.39	0.39
Total O&M costs			¢/kWh	0.70	0.67	0.67	0.88	0.88
				—	—	—	—	—
Energy cost (at 4% interest)								
	Coal price $/tce	¢/kWh$_{coal}$		¢/kWh$_{el}$	¢/kWh$_{el}$			
USA	35	0.43		2.4	2.6	2.5	3.4	3.2
USA (2007)	45	0.55		2.7	2.9	2.7	3.7	3.4
USA	60	0.74		3.1	3.4	3.1	4.3	3.9
Europe (import)	80	0.98		3.6	4.0	3.6	4.9	4.4
Europe (import 2007)	90	1.11		3.9	4.3	3.9	5.3	4.7
Europe (import)	100	1.23		4.2	4.6	4.2	5.6	5.0
Germany (bituminous coal)	150	1.84		5.5	6.1	5.5	7.4	6.4
Germany (lignite)	35	0.43		2.4	2.6	2.5	3.4	3.2
At 2% interest:								
USA (2007)	45	0.55		2.4	2.6	2.5	3.3	3.1
Europe (import 2007)	90	1.11		3.7	4.0	3.6	4.9	4.3

Note: The costs of CO_2 storage are not included (cf. Table 4.3)

relatively new coal-fired plants by solar plants), then we take the efficiency to be 40%.[13]

For the *USA*, we then find for the backup plants (with a weighting factor of 0.2 – cf. "gas power plants") that the contribution to the power price due to the

13) This efficiency should be typical of currently operating coal-fired power plants in Europe. In the USA, the value would be lower. Given the lower coal price, this difference has, however, only a minor effect on the results of our estimates, so that for simplicity, we use the value of 40% there, also.

cost of coal would be 0.25¢/kWh. For the power supply to central Europe, we again distinguish between solar power sites in Spain and Morocco. From the above coal price of 90 $/tce, for *Spain*, with a weighting factor of 0.3, we find a contribution of 0.8¢/kWh, and for *Morocco*, with a factor of 0.2, the contribution is 0.5¢/kWh.[14]

10.2.4
Operating and Maintenance Costs

The Chicago study gives a value for base-load plants of 0.75¢/kWh. In AEO (2007), fixed costs of $24 million (converted to 2002-$) are quoted, which for 8000 h of full-power operation correspond to 0.3¢/kWh; in addition, a contribution of 0.4¢/kWh due to variable costs is given. We thus employ a value of 0.7¢/kWh for *base-load* plants.

In the case of the backup system (as for gas-fired plants), the fixed costs of about 24 million $/GW (AEO, 2007) are referred to the entire solar power system per gigawatt with an annual output energy of 8760 GWh, giving 0.27¢/kWh. The variable costs of 0.4¢/kWh are, for the solar sites in *Spain* and *Morocco/USA*, weighted according to the fraction of backup power of 30% or 20%, respectively (giving 0.12 or 0.08¢/kWh). All together, we then find operating and maintenance costs for the *backup system* corresponding to a contribution to the power cost of 0.39 or 0.35¢/kWh, that is, in each case about 0.4¢/kWh.

10.3
Coal-Fired Plants with CO_2 Sequestration

The necessity of CO_2 sequestration, if coal is to be used in the future on a large scale for power generation or gas production, was already pointed out in Section 4.1 and elsewhere.

At present, the most promising design for a "CO_2-free" coal-fired power plant is the so-called integrated gasification combined cycle (IGCC) plant. It consists of a coal gasification facility (which will be described in more detail in Chapter 11) and a CCGT power plant, which converts the energy from the coal gas into electrical energy. The gasification plant and the power-plant process are closely coupled, so that heat from the gasification facility is fed into the power plant and the gas turbine provides pressurized air for the oxygen producing installation. One therefore refers to an "integrated" gasification. With the CO that – along with hydrogen – is the primary product of coal gasification, in a second step involving the reaction with water, H_2 and CO_2 are produced. The CO_2 is separated out and the hydrogen is fed into the gas turbine as fuel.

---

14) The calculation using efficiencies of 42% or of 40% leads to the same result, owing to rounding off of the values.

10.3.1
Cost Estimates According to EIA AEO 2007 (Without Storage Costs): The Cost of Power

As will be explained in Section 11.2, the gasification facilities for IGCC power plants – like coal gasification plants in general – are still under development. Such plants can, to be sure, already be constructed using present-day technology, but their efficiencies would be relatively low.

The US Energy Information Administration in its "Annual Energy Outlook" (EIA AEO, 2007) gave the expected costs for IGCC plants. (Similar investment costs, but notably higher O&M costs, were cited by STE (2006).)[15] In Table 10.2, the AEO values with and without "CO_2 capture" are listed. The next-to-last and the last columns give the values with CO_2 capture. The next-to-last column shows the expected costs for a power plant using today's technology, and the last column shows the predictions for plants with *future technology* (so-called *n*th of a kind). The efficiency is presumed to improve as a result of the new technologies, from 35.2% to 43.1%.

The resulting power cost is given as a function of the coal price. It contains the costs of CO_2 capture, but *not* the costs of its *storage* (which also includes the cost of transport of CO_2 to the storage depot). One can readily see that the power cost with today's technology is noticeably higher than that from conventional power plants, and even with the proposed future technology, it remains perceptibly higher. The cost of CO_2 storage must still be added in.

As was already shown in the cost overview in Section 4.1, the power cost from the IGCC plants with advanced technology and at the coal and CO_2 storage costs given there (10 $/t of CO_2) is even somewhat higher than the price from the solar power system (for the case of *replacement* of the conventional coal-fired power plants). If the currently estimated solar power cost is confirmed in the course of the required development program, then construction of such CO_2-free coal-fired plants would not make sense economically.

Furthermore, their development time would certainly be longer than that for solar power plants, so that this technology ("advanced" and with CO_2 sequestration) would probably not be available, for example, for the necessary worldwide construction of power plants in the coming decade (i.e., before 2020). This upgrading would thus have to be carried out with existing technology; even for the

---

15) STE (2006, p. 139 and pp. 50/51) quotes specific investment costs for IGCC power plants to be built in the year 2020 (in year-2000 €!) of 1900 €/kW (2050 $/kW), fixed O&M costs of 82 €/kW a (88 $/kW a), and variable O&M costs of 0.5 €-cent/kWh$_{el}$ (0.54 US ¢/kWh). The numbers in parentheses are in 2002-$. Conversion was performed with 1 € (2000) = 1.034 € (2002); 1 $ = 0.96 € (purchasing power parity in the year 2002); this gives all together a factor of 1.077. Thus, although the investment costs are nearly the same, the fixed O&M costs are more than twice as high (factor 2.2) and the variable costs are 40% higher than estimated in the AEO. Using these higher values, the overall O&M costs in Table 10.2 would be increased (from the value 0.88 ¢ given there) to 1.64 ¢/kWh.

replacement of currently operating plants, it would not be available within a comparably short period.

10.3.2
The Cost of Storing the Separated CO_2 (Including CO_2 Transport)

It is important to distinguish between storage on land and under the sea.

10.3.2.1 Storage on Land

The storage of CO_2 on land has been proposed using geological formations such as empty gas fields, but especially in water-bearing rock layers, so-called aquifers. Very diverse estimates of the cost of this type of storage are to be found in the literature. The lowest values apply to first projects in planning. For such initial trial applications, one can choose very favorable conditions with respect to distance and depot properties, resulting in lower costs. Whether these cost estimates are typical of the case of increased coal usage (strong increase in power production and beginning oil and gas substitution by coal), with the usual long distances from the plants to the storage depots and the need to make use of less favorable depots, is a question for whose answer no information is available. The literature accessible to the present authors contains cost estimates which (converted from $/t-$CO_2$ into $/tce)) range from 5 $/tce up to 70 $/tce.[16] The lowest value – for a power plant using future technology (*n*th of a kind) – would increase the price of energy from coal-fired plants by 0.14 ¢/kWh; the upper value by 2.0 ¢/kWh. As an *example*, we assume a value of *10 $/t of CO_2*[17]; this corresponds to 27 $/tce, and would increase the price of electrical energy by *0.8 ¢/kWh*. The question of whether there is all together a sufficient existing depot volume to accept the large amounts of CO_2 in the future for the case of the postulated increase in coal consumption will be treated in Section 11.3.

10.3.2.2 The Cost of CO_2 Storage at Sea

The potential of the oceans for CO_2 storage is practically unlimited. Apart from the still not clarified questions of what ecological effects would be associated with this type of storage and whether it could guarantee long-term retention of the CO_2 at all, we must consider here also the question of its cost. In contrast to the transport of CO_2 via pipelines on land, which represents the state of the art at present, there are apparently no reliable cost estimates for its sea transport.

---

16) The costs of transport and storage of the separated CO_2 must be added to these values: 2.7 $/t$CO_2$ = 7.5 $/tce (Stiegel and Ramezan, 2006), 1.9–6.2 €/t CO_2 = 5–17 €/tce (STE, 2006, p.57), 10 €/tCO_2 = 27.5 €/tce ("assumption" by Meyer, 2003, p. 19), 10–24 €/tCO_2 = 27.5–66 €/tce (COORETEC, 2003).

The spread in values given thus ranges from *2 $ to 25 $/t-$CO_2$*, that is, from *5 $ to 70 $/tce*. For conversion from $/t-$CO_2$ into tons of coal (tce), see Section 11.2.3. (Purchasing power parity 2002: 1$ = 0.96€.)

17) This corresponds to Meyer (2003), who also chose as an exemplary assumption the value 10€/t-CO_2 – cf. the previous footnote.

Table 10.3 Cost of CO_2 transport and injection into the deep ocean according to WBGU(2006)[a].

CO_2 transport and injection costs			WBGU, 2006[a]			Example[c] Coal-fired power plant – efficiency 45%
	Distance		$/t CO_2	= $/t C[b]	= $/t coal[b]	= ¢/kWh$_{el}$[b]
CO_2 pipeline on land	1000 km		4–30	15–110	11–82	0.3–2.2
CO_2 transport by ship	5000 km		15–25	55–91	42–70	1.1–1.9
Linearly extrapolated to the shorter distance[c]	per 1000 km		(3–5)	(11–18)	(8–14)	(0.2–0.4)
	per 2000 km		(6–10)	(22–36)	(18–28)	(0.4–0.8)
Injection into the deep ocean			0.5–8	1.8–30	1.4–22	0.04–0.6
			–	–	–	–
Totals, example: land 1000 km, sea 2000 km[c]			10.5–48	39–176	30–132	0.74–3.6

a) Quoted from IEA–International Energy Agency (2004): Prospects for CO_2 Capture and Storage. Paris. And from IPCC-Intergovernmental Panel on Climate Change (2005): Special Report on Carbon Dioxide Capture and Storage. Cambridge, New York: Cambridge University Press.
b) 1 t carbon = 3.66 t CO_2, assumption: 1 t coal (tce) = 0.75 t carbon.
1 tce (ton of coal equivalent) = 8140 kWh (LHV).
(To convert t-CO_2 into tce, see Section 11.2.3)
c) The columns and lines computed here for the "Example Power Plant," "extrapolated to shorter distances," and "Totals, example" are not contained in (WBGU, 2006).

In Table 10.3, one can see the wide range of the cost estimates. It is broad even for transport by pipeline over land (whose costs are in principle rather precisely known); probably the spread in values there is simply due to the varying transport capacities of different pipelines. In WBGU (2006), the amounts of CO_2 on which the cost estimates are based are not given. It can be assumed that for very large amounts, the lower cost limit is applicable; the cost might possibly then be even lower. The spread in cost estimates in the table is not so great for transport by ship. However, it shows that even these costs cannot be quoted with certainty, since suitable tanker ships have yet to be constructed. With regard to the costs of sea transport, the main open question is how great the relevant transport distances are. They will depend on the locations of the deep-ocean injection sites, where long-term storage is believed to be feasible.

The overall transport costs in each individual case will thus depend on the distance to the coast (pipeline) and on the oversea distance to the storage site. Although we are dealing here with cost estimates for technologies that should be familiar (e.g., CO_2 tanker ships), evidently up to now, no clear-cut statements have been made. The costs have probably simply not yet been estimated – insofar as they concern conventional pipeline transport, where they need only be recalculated for larger transported quantities than have been usual in the past.[18]

If we take the lower values within the estimated cost ranges and assume as an example a transport distance of 1000 km on land and 2000 km at sea, the transport costs would finally lie in the range of *10 $/t CO_2*. (This corresponds to the cost of storage on land assumed in the example above and leads with an IGCC power plant to an additional energy cost of 0.8 ¢/kWh.) In addition, we must consider the cost of injection into the deep ocean (Table 10.3: 0.5–8 $/t CO_2). If the lower limit is applicable here, also, its influence on the final energy cost would be negligible; with the upper limit, leading to a total of 18 $/t CO_2, the final storage cost would be nearly doubled.

---

[18] There are larger CO_2 pipelines in the USA, where CO_2 is transported to oil fields to be used in tertiary oil production.

11
Other Technologies for Backup Power Generation and Alternatives for Future Energy Supplies

The subjects that we discuss in this chapter go well beyond the topic of backup power generation for solar thermal power systems. At the same time, they contain, together with Chapter 12 on nuclear power, the most important facts regarding the costs of various other potential components of the future energy supply. The reader can find here in addition to an overview of the possible restructuring of the energy supply–also a compact treatment of the major options available (in particular with regard to their costs), so that he or she can better judge the significance of the solar-energy option. First, however, we shall take up the problem of backup power generation.

11.1
Generating Backup Power Without Natural Gas and Coal-Fired Power Plants

11.1.1
Overview

In our discussion of the overall power costs from a solar power system, we assumed as the baseline case that the generation of backup power would be accomplished using natural-gas combined cycle gas turbine (CCGT) power plants. Such plants can be regarded as an optimal solution as long as gas prices are not too high, due to their low investment and operating costs and their ability to start up rapidly. However, in case the oil price (and as a result also the gas price) rises drastically, the overall power costs from such a combined system would increase due to the contribution of the natural-gas costs. Where backup power plays a major role, as would be the case for example for plant locations in Spain, with up to 30% backup power, this cost increase would be all the more important. As an alternative, coal-fired backup plants can be envisaged, as already discussed, especially when obsolescent coal plants are available which would otherwise be shut down (cf. Section 4.1). There are, however, other alternatives:

Large-Scale Solar Thermal Power. Werner Vogel and Henry Kalb
© 2010 WILEY-VCH Verlag GmbH & Co. KGaA, Weinheim
ISBN: 978-3-527-40515-2

1) Gas from *coal gasification* as fuel for the CCGT plants:

    This is possible since in larger power plants, the necessary separate gas lines could be constructed at a relatively modest cost. As we shall show, gas as fuel for backup plants could probably be made available for ca. US 2.5 ¢/kWh; this holds for the coal prices both in the USA and also in Germany (lignite). If the natural-gas price should rise to above this value, one could convert to gas from coal gasification, provided that CCGT backup plants were used. For this reason, also, a reference price of 2.5 ¢/kWh was chosen in our cost estimates for the solar power system (Section 4.1).

    As we shall discuss in more detail below, coal gasification is in general a relatively low-cost alternative for producing a substitute for natural gas.

2) *Coal-fired ancillary boilers* in the solar power plants:

    At comparatively modest additional investment costs, *a portion* of the backup power could be generated directly at the solar plants by making use of coal.

    Solar power plants that produce their *entire* backup power with such an ancillary coal-fired system are usually termed *solar-coal hybrid power plants*.[1] They can deliver power without interruptions (except for technical downtimes). This concept is however relevant only when the solar power plants are in the immediate neighborhood of the consuming regions. When they are located some distance away, the possibility of failure of the transmission lines makes separate backup plants in the consuming region imperative, in spite of the coal-fired boilers at the solar-plant sites.

3) Solar power plants, at least in Europe, will probably be operated together with other renewable-energy plants, in particular with *offshore wind power installations* in the North Sea. Wind and solar energy are complementary with respect to their seasonal variations; wind plants generate more electrical energy in winter than in summer, while the reverse is true of solar plants. Combining the two allows the proportion of fossil-fuel backup power to be reduced.

    If the relatively low costs predicted for offshore wind energy are confirmed in practice, this renewable energy source will certainly provide an alternative or a complement to solar power. Owing to its relatively low capacity factor (offshore wind ca. 50% as compared to solar energy in Morocco at 80%, or in Spain at ca. 70%), wind energy is less attractive as a single renewable energy source. In a combination of the two systems, the duty-cycle problem is reduced (wind as a "conditional alternative"), and wind energy can nevertheless assume a considerable portion of the overall renewable energy supply. This portion will depend strongly on the actual cost ratio between wind and solar energy sources.

    Wind energy is also particularly important in Europe, since here there are no large favorable locations for solar plants (only limited regions in Spain).

---

1) With the ancillary fossil-fuel-fired boilers discussed here, such power plants could be termed "partial hybrid plants" instead of the concept "hybrid." In order to make a clear-cut differentiation we use the terms "ancillary firing equipment" or "coal-fired boilers."

In the case of a combination with wind energy, the potential area for plant locations in Spain would be utilized to a proportionally lesser extent. This would seem to be a fundamentally favorable solution. The main goal remains the exploitation of solar energy in North Africa, which is sunny and practically unlimited in terms of available area. For negotiations with the countries there, however, Europe needs alternatives. This aspect was discussed already in Section 4.3.4: "Political Costs – North-African solar energy as a 'relative' alternative for Europe."

11.1.2
Gas from Coal Gasification for Backup Power Plants

In the process of coal gasification, a mixture of gases containing H_2 and CO is produced from the coal. This gas mixture is generally referred to as "synthesis gas" or for short as "syngas." By means of an additional transformation step, it can be converted into pure hydrogen gas ($CO + H_2O \rightarrow H_2 + CO_2$; separation of CO_2). The terms syngas or hydrogen thus represent two possible end products of coal gasification. When in the following we refer in general to gas from coal gasification *in contrast to natural gas*, we will often simply use the general term *coal gas*.[2]

Supplying the backup power plants of the solar power system with coal gas is possible because power plants, in contrast to general gas consumers, can be fed by a separate large gas main connecting them directly and economically to the gas production or storage facilities. The coal gas cannot be transported in the same gas mains as natural gas, but instead, separate lines could be constructed for backup power plants and other large-scale consumers. Only if natural gas is to be substituted on a very large scale by coal gas in the future can the currently existing gas mains networks be used in their entirety. As we shall show, with production on a large, technical scale, coal gas could be made available at the backup plants for roughly 2.5 ¢/kWh$_{gas}$.

This is just the same gas that would be used in the planned coal-fired plants with integrated gasification (IGCC). In contrast to that concept, however, the gas production facilities and the CCGT power plants would be at separate locations. Thanks to the separation of the production facilities from the consumer (power plants), the gas could be produced at a continuous, constant rate in spite of the irregular consumption by the backup power plants, and stored in the interim in gas storage tanks which would deliver it on demand to the power plants. Power plants with *integrated* coal gasification would be uneconomical as backup plants due to their low capacity factors. The advantages of integrated gasification would be lost with separate gasification facilities, but this can be accepted for backup plants.

---

2) In the narrow sense, the term "coal gas" stands for the gas mixture which is obtained from bituminous coal by destructive distillation (i.e., heating under the exclusion of air, corresponding to coke production). Its principal components are hydrogen, methane, and carbon monoxide.

With separate gas production, the capacity of the gas production facilities could be adjusted to meet the annual *average* of the backup power requirements, or it would need to be only slightly higher. If the gas requirements in a particular year were to be greater than the capacity of the storage tanks due to a particularly low yield of solar energy in that year, the backup plants could switch their fuel source. Initially, one would switch to natural gas, which is readily stored. If natural gas should also become scarce in years with a poor insolation at the locations of the solar plants and simultaneously high gas consumption (e.g., due to a long, especially cold winter), the backup plants could be switched to light fuel oil.[3] The main portion of the fuel requirements would still be fulfilled by coal gas.

11.1.3
Smaller Coal-Fired Installations in the Solar Plants – Solar-Coal Hybrid Power Plants

Complementing a solar power plant with a coal-fired boiler requires considerably less in terms of investment funding than the construction of a separate coal-fired plant. In the following, this will be discussed using the example of a solar-tower plant with a molten-salt thermal circuit.

The steam circuit, the electrical components (among others the generators), and many of the switching and control mechanisms which would have to be provided for a separate coal-fired plant would already be at hand in the solar plant, so that the corresponding investment costs can be saved. The high-pressure steam generator, operated with molten salt, would also already be available. While for a separate coal-fired plant, a coal-fired steam generator and a *high-pressure* piping system would have to be constructed, for an installation within the solar plant, only a coal-fired *low-pressure* molten salt heater would be required. Although here, also, the coal firing system including flue gas purification facilities would be needed, the construction of a low-pressure system would be considerably less expensive than a high-pressure boiler as used in a conventional coal-fired plant (with its enormous size).[4]

The supply of solar energy drops completely to zero only on a few days each year. On most days with reduced insolation, it decreases only partially. For this reason, it would not be reasonable to design the coal-fired facility to supply the entire output capacity of the power plant. It would for example, be limited to one-third of the plant capacity. As soon as the daily insolation decreases, this relatively small coal-fired ancillary system would be activated. Only if the insolation were to

---

3) The possibility of operation of the plants using light fuel oil guarantees backup-power generation even in the case of a long-term interruption of the natural-gas supply, for example, if the supply of coal gas was insufficient and simultaneously the delivery of natural gas from foreign suppliers was to be interrupted. Power generation would still be secure even in such an extreme situation.

4) Thus, we read in (cav, 1989): "Heat transfer systems have been in use for over 50 years on a technical scale. In particular, the chemical industry recognized early on the specific advantages of indirect heat transfer and 'pressureless' operation at temperatures above 300 °C."

drop by more than one-third would the natural-gas CCGT plants, which would be located near the consumers, have to come on line.[5] Since the coal-fired boilers are switched on first, they cover more than one-third of the overall requirements for backup power on average. The precise value depends on the progression of the daily insolation at the location in question. Thus, Figure 5.10 shows the daily dependence of the insolation in Almeria (southern Spain) in the year 1990, a roughly average year. To the extent that the insolation is reduced, the backup plants would have to fill in. It can be readily seen that the backup plants would have to supply only a small portion of the overall power output on most days.[6] If the coal-fired boilers are switched on first, the fraction of the total backup power generation which they contribute would be relatively high. A numerical evaluation of the insolation data for Almeria shows that at this location, the fraction of power from the coal-fired system (designed to supply one-third of the overall capacity of the plant) would be 54% over the 12 years considered. It could thus be expected that this fraction would be *over 50%* for most locations in Spain.

Generating backup power in the solar power plants has four disadvantages:

- Backup power, like solar power, would have to be transported over the long-distance transmission lines, leading to losses.

- When dry cooling is used for the solar power plants, as in Morocco or the USA, the efficiency of coal-fired power generation would be lower than in conventional coal-fired steam power plants.

- The coal would have to be transported to the solar power plants, giving rise to additional costs.

- When backup power is generated at the solar power plants, it is still possible to experience a failure in the transmission lines, so that additional backup plants near the consuming regions would be required. For this reason alone, it makes sense to limit the ancillary coal-fired systems to a smaller fraction, for example, one-third of the output capacity of the solar plant. This condition can be relaxed only when the solar plants are located near the consuming regions and are connected to them – as with current power plants – by a redundant network of transmission lines. Failure of one transmission line then does not result in the loss of the entire output capacity of the solar power plants. When the consuming region is nearby, then, all of the backup power

---

5) When coal-fired ancillary heating systems are installed in solar power plants, operation of CCGT plants (near the consumers) with gas from coal gasification is less economical, since the gas consumption of the backup power plants would then be notably less uniform over time than in the case described above (without coal-fired ancillary boilers); the result would be a lower utilization of the gas transport pipelines and a still higher required storage capacity.

6) In interpreting Figure 5.10, it should be kept in mind that the mirror fields of the solar power plants in Spain would be designed to be so large (with a solar multiple of 4.4) that the plant would already attain full output capacity over 24 h with an insolation of 6.7 kWh/m^2 d. Only when the insolation drops below this value would there be any need for backup power.

can be generated at the solar plant itself (solar/coal or solar/natural gas hybrid power plants[7]).

Insofar as one limits the coal-fired system to a small fraction of the overall power generated, the disadvantages listed above are not too weighty for solar power plants in Spain. This is because there – provisionally, subject to further investigations – it can be assumed that wet cooling can be used. The distances for coal transport from the harbors to the plant sites near the coast are short, and the transmission lines to Central Europe are notably shorter, at 2000 km, than from locations in Morocco (or than in the USA for transmission of power from the Southwest to the East Coast). If the development of superconducting power transmission proves successful and the transmission losses can be further reduced (e.g., by half), coal-fired ancillary systems would also be feasible for locations in the US, at least for those locations where wet cooling is possible, as could be the case for some sites in the far Southwest (cf. Section 4.3.7).

Table 11.1 shows the energy costs which would result from an ancillary coal-fired system within the solar power plants. These energy costs are to be compared with those of backup power generation exclusively by natural-gas plants, as given in Table 4.2 for a gas price of 2.5 ¢/kWh (corresponding to an oil price of 40 $/barrel). (In Table 11.1 various values of the gas price are shown. The reader can also carry out such a variation in Table 4.2 by recalculating from the value of 2.5 ¢/kWh used there.)

The results shown in Table 11.1 are based on the following assumptions:

- Coal-fired ancillary boiler corresponding to one-third of the solar power-plant capacity;

- 50% of the backup power is generated by the coal-fired boiler. The CCGT power plants need supply only half of the backup power, leading to a reduction of their gas consumption by half. (To guarantee the supply of electrical energy, the full *generating capacity* of the CCGT plants is still required, e.g., in Germany.)

- Investment costs for the coal-fired components are two-thirds of the investment costs of a complete coal-fired power plant (corresponding to the assumptions

---

7) Solar-*natural-gas* hybrid power plants are of interest today at most only in those countries which have a low-cost supply of natural gas, such as those in North Africa or in the Near East. The currently operating parabolic-trough plants in California were designed as natural-gas hybrid power plants. At the time of their construction, the price of gas was considerably lower than today, and these plants also do not generate power for the base load, but rather for peak-load and medium-load demands. A simple ancillary gas heater, as in the case of these parabolic-trough plants, has a low efficiency for gas-fired power generation in comparison to modern natural-gas-fired CCGT plants. This is because the gas is utilized in the solar power plant only within a steam process and not with a combined-cycle process. In the past, there were to be sure numerous plans to combine the steam power cycles of solar-tower plants with gas turbines. However, evidently no convincing solution was found (at least not for tower plants with molten-salt receivers). The reason is that in that case, the temperatures and pressures in the solar steam circuit (of the tower plant) are different from those required in the waste-heat part of the CCGT process. For tower plants with air receivers, the conditions are possibly more favorable. This is still more applicable to parabolic-trough plants, with their lower steam temperatures.

Table 11.1 Energy costs from the solar power system with coal-fired ancillary boilers and CCGT backup power plants.

				Kalb/Vogel large-scale scenario in Spain SM = 4.4	Kalb/Vogel large-scale scenario in Morocco SM = 3.7, dry cooling	Kalb/Vogel large-scale scenario in the USA SM = 3.7, dry cooling
At a coal price of (US $/tce): →					90 $/t (1.11 ¢/kWh coal) Europe: imported coal	45 $/t (0.55 ¢/kWh coal) USA
At a gas price of: ↓						
Corresponding to an oil price of: ($/barrel)	$ per million BTU (I II IV)	US-¢/kWh (LI IV)	¢/cu.ft. 1030 BTU I II IV 0.273 kWh LHV	US-¢/kWh$_{el}$		
20	3.31	1.25	0.34	5.0	4.7	4.5
32	5.30	2	0.55	5.2	4.8	4.7
40	6.62	2.5	0.68	5.3	4.9	4.7
48	7.95	3	0.82	5.4	5.0	4.8
60	9.93	3.75	1.02	5.6	5.1	5.0
64	10.60	4	1.09	5.7	5.2	5.0
80	13.25	5	1.37	5.9	5.3	5.2
100	16.56	6.25	1.71	6.3	5.6	5.4

made in Chapter 10, this would be 1200 million $/GW with wet cooling). For the coal-fired portion of the plant, we thus find specific investment costs of 800 million $/GW; for one-third of the overall power output, this makes $266 million with wet cooling (referred to a solar plant with 1 GW output power), or $290 million with dry cooling.

- The assumed efficiency of the coal-fired portion of the plant is 36% with wet cooling and 33% with dry cooling. Compare: The efficiency for power generation in the solar plant is 39% (with wet cooling), and the combustion efficiency of the coal-fired portion is assumed to be 92% (39% × 92% = 36%). With dry cooling, we assume that 8% less power would be generated, that is, the electrical efficiency would then be 33%.[8]

8) Considering the reduction of efficiency due to dry cooling, it would in fact have to be taken into account that the generation of backup power by the coal-fired boiler would mainly occur on poor days for solar power (clouds, rain) and especially in the cooler winter half of the year, that is, at times when the air temperature would be lower than the annual average. The reduction of efficiency due to dry cooling would thus be smaller than expected on the basis of the annual average temperature. For simplicity, however, here we have used the average values.

- Price of coal:

  Europe (imported coal): 90 $/tce = 1.10 ¢/kWh coal (LHV)
  USA: 45 $/tce = 0.55 ¢/kWh coal (LHV)
  (1 tce = 8140 kWh)

- Operating and maintenance (O&M) costs (similar to those of coal-fired plants);

  Assumptions:

- Fixed O&M costs for coal-fired plants (base-load operation): 24 million $/a/GW.

- Of these, for the coal-fired boiler (base load): 60% = 14.4 million $/a/GW.

- No reduction of the fixed O&M costs of the coal-fired system with backup-power operation instead of base-load operation.

  This then yields for the coal-fired system, which delivers only one-third of the overall output power of the plant: $14.4 \times 0.33 = 4.8$ million $/a.

- Variable O&M costs for the coal-fired plant: 0.4 ¢/kWh$_{el}$.

- Assumption: The variable costs of the coal-fired plant are attributable to the extent of 100% to the coal-fired boiler system. Then, for the coal-fired portion of the solar power plant, 0.4 ¢/kWh is due to *coal power*. The power from the coal-fired ancillary boiler is, by assumption, equal to half of the overall backup power required; in Spain, it would thus be 15%, in Morocco 10% of the overall power output of the solar power system (that is of the solar power plant plus backup power plants). The variable costs for the coal-fired system, referred to the overall power output of the solar power system, would then be 0.06 ¢/kWh in Spain and 0.04 ¢/kWh in Morocco.

As is shown by a comparison of these values with those of Table 4.2, the energy cost would rise slightly when the gas price was very low, due to the additional investment costs for the coal-fired system. Starting with solar plants in Spain and a typical coal price for Europe of 90 $/tce, one however finds approximate price parity for a gas price of 2.5 ¢/kWh. The higher costs of the coal-fired system are thus already compensated by the gas costs saved. At still higher gas prices, the ancillary coal-fired power generation reduces the overall energy cost. It would thus drop from 7.2 ¢/kWh (gas only) to 6.3 ¢/kWh at a high gas price of 6.25 ¢/kWh (corresponding to an oil price of 100 $/barrel). Increasing gas prices thus would have less effect on the energy cost from the hybrid system.

In the USA, the cost situation is similar, given the low price of coal there (45 $/t) and in spite of the need for dry cooling and the greater distances for power transmission: at a gas price of 2.5 ¢/kWh, we find an identical energy cost; at a higher gas price of 6.25 ¢/kWh, the energy cost decreases from 6.0 to 5.4 ¢/kWh$_{el}$. In case the above assumptions prove to be applicable in the course of further developments, the construction of ancillary coal-fired boilers would thus be definitely worth considering.

We should point out two further aspects:

- If, in the power-importing region (e.g., Germany), due to the danger of power interruptions along the transmission lines, backup power plants are in any case necessary, then the backup power could be generated there initially, as long as the gas price remained low. With increasing gas prices, the solar power plants could be retrofitted with ancillary coal-fired boilers. For this purpose, however, a sufficient amount of space would have to be reserved for the combustion chambers, boilers, coal storage area, etc.

- It must be clarified whether the exhaust gases from the coal-fired system would give rise to corrosion of the heliostat reflecting surfaces. In a dry climate, only modest demands on corrosion resistance would normally be made. The coal-fired boilers would however be in operation mainly during the winter and during periods of bad weather, that is, when it would be cool and wetter. It must therefore be investigated whether the exhaust gases from coal combustion, with the usual degree of desulfurization and in combination with the wet-cooling towers (operating also during the cooler nighttime hours) could contribute to rapid corrosion of the mirror surfaces. Should there be indications that this would be the case, then the desulfurization system would have to be designed to ensure the necessary low emission values.

Ancillary coal-fired systems of this type were already investigated during the early development of solar power plants following the first oil crisis. (In the subsequent time period, characterized by decreasing energy costs, no need for a continued development was perceived.) At that time (1979), Rockwell designed a prototype coal-fired system with a power of 35 MW$_{therm}$. However, they did not publish cost estimates, so that here, we must resort to a comparison with the costs of a conventional coal-fired plant. An ancillary coal-fired system for a solar power plant however differs from a conventional power-plant boiler, so that this comparison can at best yield a rough reference value. The Rockwell system made use of a sodium thermal circuit. Instead of water (as in a coal-fired plant), sodium was to be heated and used to operate the steam generator of the planned solar power plant (thus using a molten sodium instead of a molten-salt thermal circuit). The advantage of a thermal circuit using liquid sodium or salts is its pressure-free operation, with the resulting lower investment costs for the complex piping system. As mentioned above, heating units of this type with pressure-free thermal transport media are in use in the chemical industry. However, the high input temperature of the liquid circuit must be kept in mind. In a solar power plant, the molten salt is heated from 290 °C to ca. 570 °C. The exhaust gases from the coal-firing system can therefore give up their heat only down to a temperature of 290 °C. (In a conventional steam power plant, the exhaust gases are cooled down to ca. 120 °C, since this heat at lower temperatures can be used for preheating the feed water.) In order to avoid heat losses from the molten-salt circuit, the exhaust gases are used for *preheating* the combustion air after they have been cooled to 290 °C. This however requires large-scale preheaters. In (cav, 1989), a coal-fired thermal oil circuit[9] is described: The input temperature of the liquid in the heater is given there as 365 °C, and the output temperature as 444 °C; the system is outfitted with a regenerative air preheater (to make use of the low-temperature heat from the exhaust gases). In the course of further research, such heating systems must also be designed for molten-salt circuits and for the temperature range typical of solar power plants. A precise thermotechnical design would shed light on the approximate construction costs to be expected. Regarding the adaptation to the higher temperatures of a solar molten-salt circuit (up to 570 °C), and with a view to a high degree of availability and a long operating lifetime, it can be expected that detailed research work will have to be carried out. A preliminary result could however be obtained quickly, although the final results would have to await conclusion of the development work.

11.1.4
The Combination of Solar Thermal and Offshore Wind Plants – Offshore Wind Power as a Conditional Alternative to Solar Energy for Europe

The combination of solar power with wind energy is particularly relevant for Europe. Here, there is a large and economically favorable potential for offshore wind energy in the North Sea. For offshore wind power plants, one can assume a comparable energy cost to that of solar plants, probably even a somewhat lower cost. In contrast, the capacity factor of the wind plants is poorer, and the supply of power thus less steady. If offshore wind plants are installed on a large scale, energy costs ranging from 3 to 5 €-cent/kWh[10] can be expected (BMU, 2004, p. 31).[11]

If this cost range proves correct, their potential could be utilized on a large scale, in spite of the lower capacity factor. (To this end, however, it would be required that the costs remain within this range even for greater water depths than have been considered thus far, so that sufficiently large areas of the North Sea can be made use of.) A combination with solar energy would be advantageous since the two systems are complementary with respect to their seasonal variations in power production: Solar power plants yield their maximum energy production in the summer, while wind plants deliver their maximum electrical energy in winter. Input from these two energy sources into the European power network would thus lead to a more uniform delivery of power, resulting in lower requirements for backup power from *fossil-fuel* sources.

In the North Sea, one can evidently assume a capacity factor for the offshore wind plants (depending on their design) of ca. 50%. On the basis of the wind

---

9) Thermal oil: Diphenyloxide/biphenyl. This synthetic heat-transport medium is also used in parabolic-trough power plants.
10) At a domestic purchasing power of the dollar (according to the OECD) of US $1 = €0.96, this corresponds approximately to 3–5 US-¢/kWh.
11) As is indicated by the large range of costs cited, this is only a rough estimate. It must therefore be verified by further development of wind-energy technology and marine power-transmission technology.

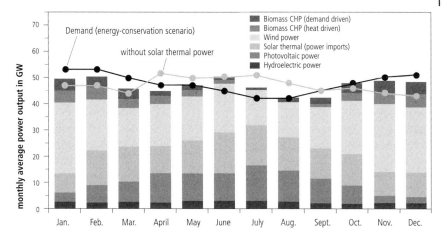

Figure 11.1 The combination of solar power imports and wind energy (here onshore!) in Germany – the monthly averaged power yield in GW (Quaschning and Geyer, 2000).

velocities determined for two years (2004/2005) at the offshore measurement platform Fino 1 (45 km north of the island of Borkum) at an altitude of 100 m, one finds *4500 full-load hours* annually (Neumann and Riedel, 2006)[12] (under the assumption that the wind plant reaches its full output power at a wind velocity of 12–13 m/s, as is typical of wind-energy plants, and with a cut-in wind speed of 4 m/s. This corresponds to a capacity factor of 51%. In contrast, solar-energy plants in Spain (SM 4.4) attain ca. 70%;[13] in Morocco and the USA (SM 3.7), they reach ca. 80%.

Fundamentally, wind energy could be utilized even without a *renewable* "partner," for example, with natural-gas backup power plants as for a solar power system. The low capacity factor however makes this appear less expedient. It would be relevant only if wind energy proves to have a *substantial* cost advantage (even at greater water depths) compared to the importation of solar energy. Offshore wind energy would thus seem to be a "conditional" alternative to solar energy, that is, in the sense of a "partner" instead of a "replacement."

In Figure 11.1, the monthly averaged power output from wind and solar power plants is shown. Along with imported power[14] the figure also assumes power generation in Germany using photovoltaics, and wind energy refers to plants on land, not offshore. In spite of these constraints, the figure demonstrates the *in*

12) The annual continuous power output is not given, only the wind velocities. At a cut-in wind speed of 4 m/s, the wind-energy plants would however be at a standstill during only about 10% of the year.

13) In the cost estimates for the solar power system, in estimating the gas costs a rounded value of 70% was employed.

14) Regarding solar thermal power generation, various locations in different countries are assumed here, that is, a location in Spain, one in Morocco, and three in Algeria. This permits a more uniform rate of power generation than would be possible with only one location (for example only in Spain).

principle opposite annual periodicity of solar imported power and wind power. The monthly outputs given here are meaningful in only a limited sense in terms of the actual overlap from various sources (day by day). On particular days, the solar-energy supply from Spain (or North Africa) and simultaneously the wind-energy supply in the North Sea could both be high, or both low. The daily overlap could indeed be readily evaluated from the daily data (wind velocity and insolation – in each case analogously to Figure 5.10); however, such a combination of data for solar plants in Spain and wind plants in the North Sea (or for Morocco–North Sea) is not yet available.[15] (This also indicates the level at which solar-power research has up to now been operating.) It is in any case clear that owing to the strong seasonal opposite periodicity of the two energy sources, even for the daily values a clear-cut compensation would be obtained. More precise figures must be forthcoming. Along with the wind-energy potential in the North Sea, further locations for wind-energy plants should be carefully investigated.[16]

11.2
Coal Gasification as a Gas Source for Backup Power Plants and as an Important Component of the Future Energy Supply

11.2.1
Gasification versus Direct Power Generation Using Coal – Solar Energy for Coal Replacement in Power Generation and for Hydrogen Production

As was explained at the beginning of this book, production of gas as fuel by means of coal gasification is interesting not only in view of its use in gas-fueled backup power plants. The following discussion of its costs makes it clear that it is also an

---

15) In (Nitsch and Trieb, 2000), the topic of overlap of wind and solar energy is discussed – in principle – but not applied in fact to the case of wind power from the North Sea and solar thermal power from Spain or Morocco. This is also true of Czisch (2005), although the topic is treated very thoroughly there.

16) Within reach of Europe, there is – among other places – in Morocco a considerable potential for wind energy, with likewise favorable economic prospects thanks to high and relatively constant wind speeds (which are roughly comparable to those in the North Sea; cf. for example, Czisch, 1999). However, these winds reach their maximum in the summer and are not suitable as a complement to solar energy, but instead to wind power from the North Sea. Their capacity factor is about the same as that of the latter, and is estimated to be 4200 hours annually (Czisch *et al.*, 1999). The trend however is that this complementary nature is not as complete as that of the combination of *solar power plants* with wind power from the North Sea. This potential for wind energy represents a certain alternative to *solar power generation* in Morocco, although the considerably less uniform supply of power is a disadvantage. The economic promise of this energy source should thus be appropriately investigated. In any case, some authors have presumed that it offers the potential of producing 35 GWa per year (300 TWh/a: Czisch *et al.*, 1999; this would however be only ca. 10% of the current total power consumption of Western Europe, which in 2004 was 364 GWa). At another point, though, an even considerably higher potential energy supply is held to be possible (Czisch, 1999).

important option for the general energy supply, so that this technology could play a definitive role in the long-term substitution of natural gas and petroleum. In the near term, it can in particular serve as a price hedge on imported natural gas; this is especially important for Europe. As likewise already emphasized, nonfossil power-generating technologies (solar energy, wind energy, or possibly nuclear energy) have, in terms of the substitution of oil and natural gas, the primary goal of freeing up large amounts of coal for gas production, which was previously used for electric power generation or would have to be employed in the future to supply the worldwide increasing demand for electric power.

The other great potential application of solar energy is for hydrogen production, especially through electrolysis. This will become economically interesting when the cost of solar hydrogen produced in this manner becomes acceptably close to the cost of coal gasification (with CO_2 sequestration). In the medium term, this appears relevant especially for a conceivable production of synthetic fuel (methanol) from solar hydrogen *and* gas from coal gasification. Through the production of solar fuel, in particular the USA with its enormous solar regions would be in a position especially to produce motor fuel if necessary for the entire world without an unduly intense exploitation of the reserves of low-cost coal and without the need to store CO_2–at a cost of ca 90 \$/barrel of oil equivalent. This could in turn represent an opportunity from the viewpoint of the USA to produce liquid energy carriers that would be independent of other countries and which–in the medium term–would define an effective barrier against price increases for petroleum for all of the oil-importing countries.

11.2.2
The Cost of Coal Gasification (for H_2 Production)

Figure 11.2 gives a schematic view of the process steps which take place within a gasification plant.

The costs discussed in the following are based on those given by Stiegel and Ramezan (2006). They, in turn, cite a fundamental study by Mitretek Systems (Gray and Tomlinson, 2002). On the origin of these data, Stiegel, Ramezan state: "Many have estimated the cost of producing hydrogen from coal, but the reported costs vary considerably. The variations in costs are due to different process configurations and process conditions as well as to different assumptions for economic and financial parameters. ... To obtain a consistent set of costs for the production of hydrogen from conventional as well as advanced technologies, both with and without the sequestration option, the U.S. Department of Energy's National Energy Technology Laboratory commissioned a study with Mitretek Systems to investigate the cost of producing hydrogen under various scenarios (Gray, Tomlinson 2002)." This study by the way forms the basis of the statements in the major American hydrogen report (NAE BEES, 2004), insofar as these relate to hydrogen from coal.

Table 11.2 shows the required capital expenditure and the cost of producing gas. One can distinguish among three different technologies (Cases 1 to 3). The cost of H_2 is shown as a function of the price of coal. For the USA, today's typical coal price is 45 \$/tce; in Europe, it is 90 \$/tce (cf. Chapter 10). The costs of gas

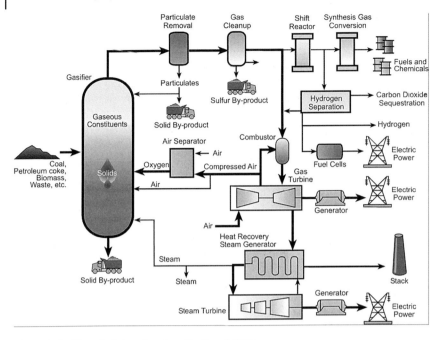

Figure 11.2 The principles of coal gasification (NETL).

Table 11.2 The cost of coal gasification (H_2 production); according to Stiegel and Ramezan (2006), but here as a real cost calculation and demonstration of the macroeconomic costs (without taxes, insurance. etc.).

		Case 1: Conventional coal-gasification Simple gas cleaning (option 50% of CO_2 is separated)	Case 2: Conventional coal-gasification Elaborate gas cleaning: 87% of CO_2 is separated	Case 3: Advanced gasification Advanced gas cleaning: >90% of CO_2 is separated	Comparison: As in Case 1, but with prod. of syngas (without $CO \rightarrow H_2$ conversion and without CO_2 separation)
Coal used (given) Illinois coal: moist ca. 7.1 kWh/kg	t/d	3000	3000	3000	
Coal used (rough computation)	MW	887	887	887	
H_2 output (given)	10^6 scf/d	131	119	158	
H_2 output HHV	$MW_{H_2\ HHV}$	520	472	627	

Table 11.2 Continued.

		Case 1: Conventional coal-gasification Simple gas cleaning (option 50% of CO_2 is separated)	Case 2: Conventional coal-gasification Elaborate gas cleaning: 87% of CO_2 is separated	Case 3: Advanced gasification Advanced gas cleaning: >90% of CO_2 is separated	Comparison: As in Case 1, but with prod. of syngas (without $CO \rightarrow H_2$ conversion and without CO_2 separation)
H_2 output LHV	$MW_{H_2\,LHV}$	440	399	531	
Additional power output	MW_{el}	20	27	25	
Investment costs – 1998 (given)	Million $	367	416	425	
Investment costs – 2002	Million $	405	459	469	
Investment per GW H_2 (LHV)	Million $/$GW_{H_2}$	920	1150	883	
Efficiency (HHV) (Stiegel/Ramezan)		63.8%	59.0%	75.5%	ca. 3–6% higher than Case 1
Efficiency (LHV)		54%	50%	64%	ca. 8–12% higher than Case 1
Capacity utilization (not given, assumed as in Staege, 1980)		8000 h/a	8000 h/a	8000 h/a	
H_2 costs					
Capital costs:					
Real interest rate		4%	4%	4%	
Operating lifetime		25 a	25 a	25 a	
Annuity at 4% int./25 a		6.4%	6.4%	6.4%	
(Annuity at 2% int./25 a)		5.12%	5.12%	5.12%	
Capital costs/a (4%)	Million $/a GW	58.9	73.6	56.5	
Capital cost/kWh_{H_2} at 4%	¢/$kWh_{H_2\,LHV}$	0.74	0.92	0.71	
(at 2% real interest)	¢/$kWh_{H_2\,LHV}$	0.59	0.74	0.57	
Cost of coal: (at above efficiency, LHV)	(1 tce = 8140 kWh)	¢/$kWh_{H_2\,LHV}$	¢/$kWh_{H_2\,LHV}$	¢/$kWh_{H_2\,LHV}$	
USA (as in Stiegel) 35 $/tce	0.43 ¢/kWh_{coal}	0.80	0.86	0.67	

Table 11.2 Continued.

		Case 1: Conventional coal-gasification Simple gas cleaning (option 50% of CO_2 is separated)	Case 2: Conventional coal-gasification Elaborate gas cleaning: 87% of CO_2 is separated	Case 3: Advanced gasification Advanced gas cleaning: >90% of CO_2 is separated	Comparison: As in Case 1, but with prod. of syngas (without $CO \rightarrow H_2$ conversion and without CO_2 separation)
USA (price 2007) 45 $/tce	0.55 ¢/kWh$_{coal}$	1.02	1.11	0.87	
USA 60 $/tce	0.74 ¢/kWh$_{coal}$	1.37	1.48	1.15	
Europe (import) 80 $/tce	0.98 ¢/kWh$_{coal}$	1.82	1.97	1.54	
Europe (import 2007) 90 $/tce	1.11 ¢/kWh$_{coal}$	2.05	2.21	1.73	
Europe (import) 100 $/tce	1.23 ¢/kWh$_{coal}$	2.28	2.46	1.92	
Europe (import) 150 $/tce	1.84 ¢/kWh$_{coal}$	3.41	3.69	2.88	
German lignite 35 $/tce	0.43 ¢/kWh$_{coal}$	0.80	0.86	0.67	
O&M costs:					
For above plant (Gray, p. B2)	Million $/a (2002)	21.5	22.6	19.3	
= per GW$_{H_2}$ LHV (2002)	Million $/a GW	49	57	36	
	¢/kWh$_{H_2\ LHV}$	0.61	0.71	0.45	
Overall H_2 cost (at 4% real interest)					
USA (as in Stiegel) 35 $/tce	¢/kWh$_{H_2\ LHV}$	2.1	2.5	1.8	Gas production costs are lower by ca. 15–25% than in Case 1
USA (price 2007) 45 $/tce	¢/kWh$_{H_2\ LHV}$	2.4	2.7	2.0	
USA 60 $/tce	¢/kWh$_{H_2\ LHV}$	2.7	3.1	2.3	
Europe (import) 80 $/tce	¢/kWh$_{H_2\ LHV}$	3.2	3.6	2.7	
Europe (import 2007) 90 $/tce	¢/kWh$_{H_2\ LHV}$	3.4	3.8	2.9	
Europe (import) 100 $/tce	¢/kWh$_{H_2\ LHV}$	3.6	4.1	3.1	
Europe (import) 150 $/tce	¢/kWh$_{H_2\ LHV}$	4.8	5.3	4.0	
German lignite 35 $/tce	¢/kWh$_{H_2\ LHV}$	2.1	2.5	1.8	
Comparison at 2% interest					
USA (45 $/t 2007)	¢/kWh$_{H_2\ LHV}$	2.2	2.6	1.9	

Table 11.2 Continued.

		Case 1: Conventional coal-gasification Simple gas cleaning (option 50% of CO_2 is separated)	Case 2: Conventional coal-gasification Elaborate gas cleaning: 87% of CO_2 is separated	Case 3: Advanced gasification Advanced gas cleaning: >90% of CO_2 is separated	Comparison: As in Case 1, but with prod. of syngas (without CO → H_2 conversion and without CO_2 separation)
Europe (import 90 \$/t)	¢/kWh$_{H_2\ LHV}$	3.2	3.7	2.8	
Storage of the separated CO_2 Example: 10 \$/t CO_2	¢/kWh$_{H_2\ LHV}$		+0.65	+0.5	

Notes:

- without CO_2 storage costs – cf. the last line; see also Table 4.3, there including storage costs with transport to the storage depot
- here for a small plant size
- The probable rise in coal prices in the case of a massive substitution of oil and natural gas by coal is not taken into account here.

Legend:

- Hydrogen: higher heating value (HHV[a]): 3.54 kWh/Nm³
  lower heating value (LHV[a]): 3.00 kWh/Nm³
- CO: heating value (HHV = LHV): 3.51 kWh/Nm³
- 1 scf at 60°F (= 15.5°C) => ideal gas: 0.02685 normal cubic meters (Nm³) at 0°C (32°F) (Note: 1 cf = 0.02832 m³) (Hydrogen: 1 million scf/24 h = 41666 scf/h = 1119 Nm³/h = 3.97 MW-HHV or 3.36 MW-LHV)
- US-inflation (consumer price index) 1998 to 2002: × 1.103
- Price of coal: from (Stiegel and Ramezan, 2006) "Illinois #6 coal, delivered cost (at plant) of \$1.26/10⁶ BTU." (\$1.26/10⁶ BTU = \$1.26/293 kWh = 0.43 ¢/kWh$_{coal}$ = 35 \$/tce). The heating value of Illinois #6 coal is not given by Stiegel and Ramezan. According to Maurstad et al. (2006), it can be taken to be ca. 7.1 kWh/kg (HHV).

Note: especially in Germany, the "coal equivalent" (ce) is used: 1 kg ce = 8.14 kWh.

[a] HHV is the energy content of the fuel including the heat of condensation of the water vapor contained in the combustion gases. LHV is the energy content without the heat of condensation. Water vapor is formed by oxidation of the hydrogen contained in the fuel. The difference between the HHV and the LHV is therefore greatest in the case of pure hydrogen (18%); for methane (CH_4), it is notably less (11%), and for CO (which contains no hydrogen), there is no difference. This difference thus has practical significance (condensing boiler!) in particular for H_2 and natural gas (CH_4) fuels (and less so for methanol (14%), liquid hydrocarbons (6–8%) and coal (ca. 5%)).

production were already shown in Table 4.3; there, also the cost of storing the separated CO_2 with exemplary costs of 10 \$/t CO_2 was given. The fact that the production of syngas is less expensive than that of hydrogen (Case 1) was also mentioned already. This is indicated in the table in the right-hand column (which is not from Stiegel and Ramezan), and will be discussed further below.

11.2.2.1 Conventional and Advanced Gasification

As Cases 1 and 2, Stiegel and Ramezan have taken the conventional Texaco process (entrained-flow gasification with pressurized oxygen). Both these cases refer in addition to conventional gas-cleaning methods: In Case 1, a simple gas treatment (conventional amine technology) is assumed; in Case 2, a more elaborate gas treatment with CO_2 separation (two-stage Selexol unit) is considered. In Case 3, "advanced technologies" are assumed. This refers in particular not only to the gas treatment, but also a further development of the gasification process is assumed.[17] This case illustrates the potential for further development as compared to Case 2. For the successful development of this technology, however, serious obstacles must be overcome; we discuss this point below.

The gasification process developed by Texaco and mentioned above (now General Electric) is economically attractive only when high-grade coal (bituminous coal) is used. (This is assumed from the outset by Stiegel, Ramezan: Illinois #6 coal.) In contrast, the quality of the coal is relatively unimportant for the Shell process discussed below. For the gasification of lignite (found among other places in Germany), along with the Shell process also the high-temperature winkler (HTW) (fluidized bed) process is suitable; cf. for example, Meyer and Lorenz (2004).[18] A number of other processes (many of them suitable for coal with a low heating value) are currently under development (Stiegel, 2008).

11.2.2.2 Operation and Maintenance (O&M) Costs

O&M costs are given by Gray and Tomlinson (2002) for the plants described above (in 1998-$, MW_{H_2}) (see Table 11.3).

11.2.3
The Assumed Cost of CO_2 Storage

In Cases 2 and 3 considered by Stiegel and Ramezan (2006), CO_2 is separated (Case 1 see footnote).[19] The effort required for this is included in the cost estimates, but not the cost of transporting CO_2 to the storage depot (only the cost of compressing is included) and, in particular, not the storage itself. As a rule, the injection of CO_2 into former gas fields and other underground gas depots (especially in water-bearing geological formations, so-called aquifers) is planned. To this

---

17) Case 3 (advanced technologies):

    a) improved gasification process: advanced entrained E-Gas gasifier;

    b) "high-temperature gas cleanup," consisting of hot gas sulfur removal and hot gas dust removal with candle filters.

    c) A porous ceramic *membrane* operating at 600 °C for the separation of H_2 from the H_2/CO gas mixture.

18) This fluidized-bed process presumes the use of reactive coal (lignite), and should be roughly equivalent in terms of efficiency and costs to the entrained-flow processes owing to its low gasification temperature, in spite of the high ash content of the coal used.

19) In Case 1, as a result of the gas cleaning, ca. 50% of CO_2 produced would also be separated out, and this portion could in fact be stored. However, for this partial yield of CO_2, Stiegel does not presume that storage would be carried out, so that in Case 1, no costs for compression of the gas to high pressure are included. The remaining 50% of CO_2 cannot be economically separated out of the gas stream, and is passed along with a small fraction of the hydrogen into the combustion chamber where energy for the process is produced.

Table 11.3 Operating and maintenance costs of coal gasification.

		Case 1	Case 2	Case 3
Capacity	MW_{H_2}	440	399	531
Consumables	Million $/a	1.7	1.6	6.8
Labor/Overhead	Million $/a	7.7	7.7	8.0
Admin. personnel	Million $/a	1.2	1.2	1.3
Other	Million $/a	8.9	10.0	1.5
		–	–	–
O&M 1998 $[a]	Million $/a	19.5	20.5	17.6
O&M 2002 $[a]	Million $/a	21.5	22.6	19.3

a) The local taxes and insurance (each amounting to ca. 8 million $/a) included by Gray and Tomlinson (2002) are not taken into account here, since we wish to give the macroeconomic costs. The bonus listed there for the sulfur obtained as a byproduct (2 million $/a) is also not included here; it can be assumed that with widespread use of coal gasification (among others IGCC), an oversupply of sulfur would result.

end, CO_2 as assumed by Stiegel/Ramezan – must be compressed to 200 bar. The investment and energy costs for this compression are included in the costs listed in Table 11.2, as mentioned (Cases 2 and 3). For transport and storage, Stiegel and Ramezan take a flat rate of 10 $/t of carbon (note: *not* per ton of CO_2).[20] These costs were however not included in Table 11.2. 10 $/t C corresponds to *2.7 $/t CO_2* or, recalculated to the carbon content of the coal, *7.5 $/tce*.[21] The cost assumptions

20) In the article of Stiegel, Ramezan, there is possibly a typographical error; in the hydrogen report (NAE BEES, 2004, p. 87), which no doubt referred to the same sources as Stiegel, an express reference to "$37/t C" is made. This is $10/t CO_2. Furthermore, NAE BEES (2004, p. 86) states: "The cost of storage is highly uncertain at this time and has not been a focus of this committee's analysis. The committee assumed $37/tC, which is consistent with the range of current estimates."

21) The carbon content of coals (referred to water- and ash-free hard coal!) is assumed here to be ca. 85%. Compare: The coal types listed in (Dubbel, 1987, p. L2) cover a range of 78–93% (the average value is thus 85%). The average water and ash content (both together) is ca. 10%, according to Dubbel (1987, p. L2) for the types of coal considered there. This yields a *carbon content* for the coal of roughly 75%.

The German coals used here as examples (Dubbel) have an average heating value of 8.56 kWh/kg and correspond rather precisely to the defined value of 8.14 kWh per kg coal equivalent (kg ce); the difference is only 5%. Thus, for our rough estimate, 1 t of coal can be considered as equivalent to 1 tce. With a carbon content of 75%, we then find approximately: 1 tce = 0.75 t *carbon* or 1 t *carbon* = 1.33 tce. Referred to the *energy content* of the coal, this ratio holds in general to a good approximation, so that for conversions, the actual carbon content of the particular type of coal (per kg of coal!) need not be considered. By way of explanation of the latter point: For coals with a higher water or ash content (relative to the German coals considered, which are practically identical per ton to one tce), this higher ballast content lowers both the heating value per kg of coal and the carbon content per kg of coal. The quotient "heating value per ton of C" is not dependent on the ballast content. (This ratio would be changed by other components of combustible substances such as carbon, methane, and hydrogen,

of Stiegel and Ramezan are, in comparison to other sources, extremely low.[22] In Chapter 10, the various values given in the literature were discussed. The range of values stretches roughly from 5 to 70 $/tce. As an *example*, we assumed in Chapter 10 a value of *10 $/t CO_2*. Referred to coal, this corresponds to *27 $/tce*. This would make hydrogen in Case 3 (future technologies) more expensive by *0.52 ¢/kWh*, and in Case 2 by 0.66 ¢/kWh. These values are listed in Table 11.2 on the last line and are included in Table 4.3.

11.2.4
Syngas as a Particularly Inexpensive Substitute

Stiegel and Ramezan (2006) investigated the costs of *hydrogen production*. In coal gasification, a gas mixture is obtained which consists of roughly 57% CO, 32% H_2, and 11% CO_2 (Texaco process). If this mixture is used directly as fuel, it is termed medium-BTU gas or synthesis gas (syngas).[23] Only in the next step, the so-called *CO conversion* (or CO shift conversion) is CO converted to CO_2, yielding additional hydrogen (CO + H_2O → CO_2 + H_2). After this step, CO_2 must be washed out of the gas stream. If, however, CO_2 is not to be sequestered – as in Case 1 – then one can dispense with the CO conversion and CO_2 separation. The raw gas (CO, H_2, plus a small concentration of CO_2) is then merely cleansed of sulfur and other contaminants, which can be accomplished by a relatively simple gas purification step, and then used directly as a fuel. This gas is considerably cheaper due to elimination of

e.g., in the case of lignite.) Then, even for the high-ballast hard coal types, with their lower heating values, the expression 1 t carbon = 1.33 tce holds to a good approximation. We thus use this conversion factor generally here. Furthermore, there is no information available on the precise carbon content of coals averaged over the world, and not even on those values in individual countries. The IEA Coal Information (2007) also gives no references to these values (average).

Since 1 t of carbon yields 3.66 t of CO_2 (molecular masses: $M(C)$ = 12, $M(CO_2)$ = 44), and 1 t C corresponds to 1.33 tce, we find: *1 tce = 2.75 t CO_2*. For the cost estimates, an analogous result is obtained: *1 $/t CO_2 = 3.66 $/t C = 2.75 $/tce*. – For comparison: In Germany, for computing CO_2 yields from hard coal, the Federal Environmental Agency uses the factor 92 t CO_2/TJ or *2.68 t CO_2/tce*; this is practically the same value as quoted above (2.75 t CO_2). For lignite the Environmental Agency uses 3.22 t CO_2/tce, and for subbituminous coal, the value 2.84 t CO_2/tce.

22) The low costs quoted by Stiegel/Ramezan probably referred to planned *initial* projects, for which one can presently choose favorable conditions regarding distance and storage-depot characteristics, resulting in lower costs. This however allows no conclusions to be drawn about the costs of large-scale CO_2 sequestration.

23) A note on the terminology: If the raw product of coal gasification is used as fuel gas, it is termed "medium-BTU gas" (owing to its relatively low heating value). If instead this gas is to be used as a raw material for synthesis purposes, it must be further purified of contaminants, and the ratio of CO/H_2 must be precisely optimized for the desired synthesis by reacting an appropriate amount of CO with water to give H_2 and CO_2. This final CO/H_2 mixture (after separation of CO_2) is then in fact what is termed "synthesis gas." These terms are however not used consistently in the literature; in particular, the short form "syngas" is often applied to the raw gas, that is, to medium-BTU gas. We also use it here in this sense.

the process steps mentioned above. Or, expressed differently: If the use of coal (gasification) is to be carried out in a "CO_2-free" manner (i.e., with CO_2 sequestration), the conversion of CO initially formed into CO_2 is indispensable. In this case, the product of coal gasification is pure hydrogen. If, on the other hand, one is willing to accept CO_2 emissions, one can dispense with the conversion, with the advantage that the product costs and the consumption of coal will be lower.

Staege[24], (1980) emphasized the advantages of dispensing with the conversion process: not only can the investment and operating costs of this process be saved, but also the associated reduction of efficiency[25] is avoided. If one dispenses with the conversion of the raw gas, a sulfur-free fuel gas with roughly the same heating value per unit volume as hydrogen is obtained. The investment costs for the plants are about one-third lower (Staege, 1980, p. 547). Although the consumption of coal does not decrease to the same extent,[26] the costs of producing this gas should be *perceptibly* lower than those for hydrogen as given above in Case 1; they could – depending on the price of coal – be around 15–25% lower.

The hydrogen costs quoted by Stiegel und Ramezan (2006) refer to relatively small gasification plants with a capacity of only 0.4 to 0.5 GW_{H_2}. The influence of the size of the plant was shown by Staege (1980). Thus, the investment costs increase only from €168 to €382 million (1980 monetary value), that is, by a factor of 2.3, on increasing the plant capacity from 0.6 GW to 2.6 GW_{gas} (by a factor of

---

24) After the first oil crisis, there was an intensive development of coal gasification processes, including investigations of the feasibility of large-scale plants. As energy prices again decreased, these developmental efforts were abandoned. There are evidently no newer estimates of the costs of large-scale installations, so that here, we must cite the earlier results. In the case of Staege (1980), of the Krupp Koppers firm in Germany, the process likewise involves an entrained-flow gasifier, namely the Koppers–Totzek process, which was at that time widespread. In contrast to the Texaco process, it operates at ambient pressure and with dry coal as input material.

25) This reduction of efficiency is especially noticeable in the decrease of the efficiency with respect to the lower heating value (LHV). The LHV depends on the H_2 content of the gas, the latter is only roughly 33% in the raw gas (relative to the combustible portion of the gas mixture); but in converted gas, it is 100%. (The difference between LHV and HHV for pure hydrogen is 18% (HHV = 3.55 kWh/Nm^3, LHV = 3.0 kWh/Nm^3); compare CO: LHV = HHV = 3.5 kWh/Nm^3 – see also the legend of Table 11.2.) For most applications, only the lower heating value can be utilized. An exception is space heating using condensing boilers; in that case, the advantages of syngas would not be so great.

26) In Staege (1980), the difference in efficiencies between syngas and hydrogen production is unfortunately not quoted (since Staege is concerned above all with the production of syngas *versus* SNG (synthetic natural gas), and only secondarily with pure hydrogen). From the rough estimate of the present authors, the efficiency of syngas production as compared to Case 1 should increase by ca. 3–6% (HHV) or by ca. 8–12% (LHV).

The amount of coal consumed per unit of gas produced could also be further reduced by using solar power in addition to supply the plant's own energy requirements (including those of the air separation system); this could be done at practically no additional cost. See the notes in Section 11.4.3.2 (Sun methanol: Coal consumption). The power consumed by the plant is normally generated by combustion of a portion of the gas produced in a small combined-cycle gas-steam turbine installation.

4.3). The specific investment cost thus decreases to 54%. The operating and maintenance costs should also decrease strongly for larger plants, especially due to the large proportion of personnel costs which they contain. The size of the plants also possibly plays a role in considerations of the advantages and disadvantages of integrated gasification (namely in power plants with integrated gasification) as opposed to pure gasification plants.[27]

Many participants in the current energy debate hold the opinion that the construction of new coal-fired power plants, even if they are designed for CO_2-emission-free operation, carries the risk of being seen as a poor decision in terms of energy policy. Such a construction program would bind up the use of coal in electric power generation for decades to come, although it can be expected that coal will be needed on a large scale for other purposes in the not-too-distant future. (If the oil price should increase in an uncontrolled manner, the industrialized nations could become entrapped in a situation in which a high proportion of petroleum and natural gas would have to be substituted very quickly by "coal gas"). At the same time, it can be seen that electric power cannot be generated with nuclear energy alone, but rather that in the near future, renewable energy sources will have to take over this function at similar costs to those of advanced – but expensive – coal-fired power plants with CO_2 sequestration. A long-term commitment of coal reserves for power plants would be less serious if the gasification systems of the integrated power plants could be used separately if needed, that is, without electric power generation, as gas production plants for the general gas supply. The question of how much additional effort would be required to construct power plants with integrated gasification in such a way that a subsequent retrofitting as gas-producing plants is feasible, should therefore be promptly investigated.

11.2.5
Backup Power Plants as Consumers of Gas – Gas Transport and Storage Costs

Hydrogen and syngas have the disadvantage that they cannot be transported in existing natural-gas mains together with the natural gas. For this reason, plans for

---

[27] Within the framework of high-priority research, the influence of the size of a plant on its efficiency needs to be investigated. Pure gasification plants could probably be made considerably larger (with capacities of up to e.g., $5\,GW_{gas}$) than installations which are integrated into electric power generating plants. Today, the development is restricted to the relatively small *integrated* gas generators for IGCC power plants (with capacities of roughly $1\,GW_{gas}$). If this topic is investigated, it would then become clear how the tradeoff between "integrated, but small" as opposed to "not integrated, but large" can be optimized. Even pure gasification plants include a small integrated electric power plant; it is, however, used almost exclusively to supply the plant's own requirements for electrical energy. In large gasification plants, this internal power plant can be of a similar capacity to current CCGT power plants (or at least "nearly so"), but with the advantage of a higher efficiency, particularly in the gas turbine. The gas turbine at the same time supplies pressurized air for the air separation system; it is thus an important plant component for determining the efficiency of the overall gasification process. (The air separation system is one of the most expensive components of a gasification plant.)

coal gasification in the past were directed toward the production of synthetic natural gas (SNG). This, however, requires not only the additional process steps already mentioned (CO conversion and CO_2 separation), following the production of synthesis gas in the first step, but also a methanization step, which is particularly complex.[28] The transport and distribution problem is thus an important obstacle to the goal of large-scale substitution of natural gas. Either entire regions must be switched over completely to the new gas (or to a gas mixture consisting of natural gas and syngas – see below); or else new gas mains (for the separate transport of different gases to consumers) would have to be constructed.[29]

In case only certain large consumers are to be supplied with gas, this disadvantage has only a limited effect. This is especially applicable to the backup power plants of the solar power system. They can either be constructed near to the gasification plants or gas storage installations, or – as would be feasible for the example of Germany – along a larger pipeline. (In Germany, this pipeline might run from north to south). Other consumers, also, in particular larger industrial customers, could be supplied from this pipeline. Nevertheless, the efforts required for the transport and storage of hydrogen or syngas are greater than for natural gas.[30] This must be taken into account in a complete cost comparison. In order to obtain more precise cost data, the future distribution network should be designed in detail: the locations of the gas production plants and of the large-volume storage facilities and power-input stations for the backup power plants. (For the gas storage facilities, salt domes could be considered. These large reservoirs could be combined with smaller above-ground storage tanks which would be near the power plants.) Regarding the possible locations for electric power input, a future dc power grid within the consuming region (e.g., Germany) should be planned. This would permit considerably greater distances between the power input points and the consumers, so that the locations of the backup power plants could be determined more by proximity to the gas supplies.

According to Staege (1980), the transport cost for syngas using a 2 GW gas line and over a distance of 300 km corresponds to 0.28 €-cent/kWh$_{gas}$ (1980). At the monetary value of the year 2002 (consumer price index: ×1.70; compare producer price index: ×1.36), this would be 0.5 €-cents. In US dollars, we find (converting

---

28) The methanization requires an especially effective gas purification process, due to the catalysts employed; this is similar to the methanol synthesis.

29) Because of the differences regarding their heating values and other properties between natural gas (CH_4) on the one hand, and H_2 or syngas on the other, a mixture of these gases would have different properties from either of them alone. Such a mixture has up to now not been permitted in terms of gas burners, gas meters, compressors, etc. Just how great the effort required to allow the use of mixed gas in these components would be, especially with variable mixing proportions, would also have to be determined in the course of the required research and development program.

30) The heating value per cubic meter of H_2 or syngas is only about one-third of that of natural gas (i.e., they would require a storage volume three times as great!). Furthermore, the noncontinuous nature of backup power-plant operation requires larger gas storage capacities than for base-load gas power plants.

by using the national purchasing power according to the OECD: ×1.05) a value of 0.53 US-¢/kWh (300 km) or 0.88¢/kWh per 500 km; this is for a gas line with a relatively small capacity. To supply a number of CCGT backup power plants (each for example, with an output power of 600 MW$_{el}$ and a gas consumption of ca. 1000 MW), a gas line with a capacity in the range of 10–20 GW$_{gas}$ (or even larger) would be necessary. The transport cost decreases correspondingly. Thus, the Enquete Commission (1990) quotes a cost of 1.02 €-cent/kWh for a 12.5 GW$_{H_2}$ pipeline (1.4 m in diameter, 80 bar, 6000 h of operation per year) and a distance of 2000 km. Converting to the monetary value of 2002 (×1.31) and linearly to the shorter distance (500 km), this yields 0.33 €-cent or *0.35 US-¢/kWh per 500 km*. The additional transport costs are thus small.

11.2.6
Backup Power Plants: Switching to Other Fuels When Gas is in Great Demand – Development of Combustion Chambers

As we have already mentioned, regarding the required storage capacity (and also the capacity usage of the transport pipelines), it must be remembered that the CCGT backup power plants can be operated with different fuel gases. Industrial consumers can also change fuels as required. The backup power plants would thus supply only a portion of their needs with gas from coal gasification; to be sure, this would be the major portion, for example, three-fourths. At times when the demand for gas is very high, when the supply capacity of the storage system is limited (due to the limited storage capacity), they could switch to natural gas, which is easier to store, or, if it were also in short supply, to light fuel oil.[31]

The precondition for such a fuel switchover is that the combustion chambers of the gas turbines be adaptable to operation with different fuels. Thus, current natural-gas CCGT power plants can be switched over during their operation to light fuel oil. Even in a gas turbine power plant fueled by syngas (medium BTU gas) at an Italian refinery, a switchover can be made to light fuel oil, which is available there at a favorable price.[32] The development of combustion chambers

---

31) The gas storage tanks thus need not be dimensioned for years with an especially high demand for gas, that is, for years in which the production of solar power is particularly low in the winter so that an unusually high demand for backup power occurs. The tanks can be designed to fill the long-term average demand *roughly* (only slightly larger). This also applies to the transport pipelines, which in the example of Germany would run from the gas storage regions in northern Germany down to the South. They would not have to be designed to supply the peak demand for gas. This peak demand would occur in the cold winter months, insofar as the pipelines would also supply gas for heating. On such peak-demand days, the backup power plants could interrupt their gas consumption. The gas which they usually require would then be at the disposal of other consumers, for whom a switchover to another fuel would not be possible (this applies in particular to the public gas supply).

32) In other power plants with integrated coal gasification (IGCC) which have already been built – for example, Buggenum (the Netherlands) – the fuel can be switched between gas from the gasification system and natural gas. These plants are also started up using natural gas.

for burning *pure* hydrogen is however still underway (STE, 2006). For these, also, a switchover to light fuel oil should be possible, but probably only at the price of higher NO_x emissions. Here, targeted R&D work must be carried out. In the course of the development program, such combustion chambers will have to be designed and brought to production readiness, and the possibilities for a similarly low-NO_x technology as is currently available for natural gas combustion must be explored.

Great progress has been made in the past two decades in the design of natural-gas combustion chambers, especially with respect to their NO_x emissions. Present-day combustion chambers can be regarded as nearly "ideal" in this respect. They give rise only to exceedingly low emissions. Such favorable NO_x-emission values may not be achievable with H_2 gas turbines, owing to different combustion characteristics of hydrogen. Thus for example, the premixed combustion technology cannot be used with pure hydrogen, at least not without considerable further development. The technology which was current ca. 20 years ago will have to be revived. The NO_x emissions were already at a fairly low level at that time (only the nearly "zero emission" level of today's natural gas burners had not yet been achieved). Development of H_2 combustion chambers is an ongoing project. Fundamentally, nothing would seem to prohibit that what is currently routine for syngas-fueled turbines, namely switching to light fuel oil as required, will also be possible with pure hydrogen fuel (possibly with the limitations mentioned for NO_x emissions; cf. Karg, 2008). In the longer term, there is hope that for hydrogen combustion, the premixed combustion technology, which is very favorable in terms of NO_x emissions, can be applied here as well. Furthermore, concerning the NO_x emissions, the short operating times of the backup power plants as compared to base-load plants must also be kept in mind.

11.2.7
Development of "Advanced Technology" with a View to a General Gas Supply and IGCC Power Plants – Barriers to Development

11.2.7.1 Gas Purification and Separation

As can be seen from Table 4.3, it is hoped with the aid of "advanced technology" to be able to develop the coal gasification process to the point that in spite of the need for CO_2 sequestration, the same cost for gas can be achieved as in the current technology without CO_2 sequestration. This further development is less concerned with the gasification reactors themselves as with the systems for *gas cleaning* and *gas separation*. Gas cleaning refers in particular to the removal of hydrogen sulfide (H_2S), which – depending on the sulfur content of the coal used – will be present at various concentrations in the raw gas. Gas separation refers to the removal of CO_2. Most cleaning agents absorb H_2S and CO_2, although with a strongly varying effectiveness. Therefore, gas cleaning and separation are carried out in many conventional processes using the same medium, but in different steps during the overall purification process.

In Case 1 above (Table 11.2), the relatively simple amine process is envisaged for the removal of H_2S. (Here, CO_2 is removed in an additional process step (pressure swing adsorption (PSA)) and only ca. 50% is separated out; it would have to be compressed to high pressure if it were to be stored in a depot). The content of H_2S and other sulfur compounds is reduced to the point that the hydrogen can be used as a fuel according to current environmental standards. This purification is however not sufficient for a subsequent methanol synthesis, since the catalysts used for that purpose are extremely sensitive to sulfur contamination.

In Case 2 as described in the table, the Selexol process is presumed for gas cleaning; it has been technically proven for some time, but is more complex. With this process, H_2S is almost completely removed and at the same time, 87% of CO_2 is separated out. This is a process in which the gases are dissolved in a liquid at near to room temperature. On heating and pressure reduction, they are expelled from the solution. This effective method is however technically complex, and it reduces the overall efficiency of the gasification to a considerable degree. For a similar but still more effective and more elaborate process (Rectisol), cold methanol is used as the solvent. The gases are dissolved at $-15\,°C$ (or sometimes down to $-40\,°C$). The high-purity gas which results is then suitable as a raw material for methanol synthesis.

The development efforts are now aimed at improving the efficiency of the Selexol process (or even the Rectisol process) while keeping costs low and especially while maintaining the overall efficiency of gas production at a high level. Such future technologies are presumed in Case 3 (Table 11.2). The central point is a novel gas separation process using a porous membrane, which is permeable to hydrogen but not to other gases (Stiegel and Ramezan, 2006, p. 188). In these ceramic membranes which were developed at the Oak Ridge National Laboratory (ORNL), by variation of the fabrication method, the pore diameter can be specified rather precisely and maintained within narrow tolerances. It is chosen so that hydrogen can pass through the pores, but not the other, larger gas molecules. Expressed simply, the membrane is a sieve which operates on the molecular level.[33] Relatively high H_2 flow rates with a low pressure difference can be achieved, but the operating temperature is quite high, at $600\,°C$.

The gas has to be purified of sulfur before reaching the membrane. This is carried out (at $600\,°C$) using the well-established so-called dry desulfurization process (adsorption onto solid particles). Then the gas has to be freed of dust (before it can be allowed to pass through the membrane). Since cooling off and reheating of the gas is to be avoided, the dedusting must also be carried out at ca. $600\,°C$. For this purpose, so-called hot gas dedusting using candle filters is applied; this is a process which is still not yet completely established. Here, likewise further development efforts must be made.

Whether or not this membrane technology will make the leap to a practical application is hard to judge at this point; no one can predict this with any certainty

---

33) The hydrogen does not in fact diffuse in the form of H_2 molecules through the membrane, but rather these dissociate (cf. Stiegel, 2008: Eltron Hydrogen Membrane). The dissociation and recombination occur in catalyst layers on either side of the membrane.

at present. The development is in any case still in its early stages. Thus, Stiegel and Ramezan write (2006, p.188): "The DOE and ORNL are currently initiating an effort to develop a large scale module for performance testing on coal-derived shifted synthesis gas." Quite possibly this is more of a goal than a prognosis. This development is of great importance. On p. 180, we read: "Because of the process's ability to operate at higher temperatures and the absence of the energy requirements associated with the amine absorption process, the efficiency of the process is about 11 efficiency points higher than Case 1 [as in Table 11.2] even with the capture and sequestration of carbon." (Compared to Case 2 (Selexol process), the efficiency (LHV) even increases by 14 points.) It must however be remembered that along with this process, a number of other development routes are being followed (Stiegel and Ramezan, 2006, p. 188). None of these, to be sure, can promise a comparable increase in efficiency.

11.2.7.2 Advanced Technology for IGCC Power Plants

IGCC power plants are also based on hydrogen production via coal gasification. They were already discussed in Chapter 10. In Table 10.2, two versions according to AEO (2007) are reproduced, both with CO_2 sequestration: the one represents the technical state-of-the-art in 2006, the other uses "advanced technology," referred to there as "nth of a kind." As shown in the table, an increase of the electrical efficiency from 35.2% to 43.1% is envisaged. If one for simplicity assumes an efficiency for power generation in the CC part of the power plant of 60%, this corresponds to an efficiency for the gasification process of 58.6% or of 71.8%, respectively; this is an increase by 13.2 efficiency points as compared to the situation in 2006. For the "nth of a kind" technology, comparable progress was thus predicted as in Stiegel and Ramezan, 2006 (Case 3 there). AEO (2007) unfortunately gives no information on the technology which they assumed for this case. Possibly, they presume an efficiency increase for the gas turbines, which however could make up only a part of the overall predicted increase. What was stated above thus applies on the whole to these power plants as well: the required development will be challenging, and it is by no means certain that the hoped-for results will in fact be achieved in the future.

A similar evaluation is to be found in the major American hydrogen study (NAE BEES, 2004, p. 208): "For new gasification technologies, the best opportunities for R&D appear to be for new reactor designs (entrained bed gasification), improved gas separation (hot gas separation) and purification technologies. These technologies, and the concept of integrating them with one another, are in very early development phases and will require longer-term development to verify their true potential and to reach commercial readiness."

11.2.7.3 Development of Gasification Facilities – The Higher Efficiency of the Shell Process

In the gasification process assumed by Stiegel and Ramezan (2006) (Table 11.2, Cases 1 and 2), originally developed by Texaco (now General Electric), a paste is prepared from pulverized coal and water, so that the coal can be pumped continuously into the gasification chamber, which is under pressure.[34] A disadvantage of this procedure is the large amount of water entering the chamber. In the Shell process—also an entrained-flow gasification under pressure—the coal is injected into the reactor chamber through pressure locks in the *dry* state.

The Texaco process thus suffers particularly when the heating value of the coal used for gasification is low and it already contains a quantity of hygroscopically bound water. The large amount of water entering the chamber must be heated as ballast to the elevated temperature required for gasification; this requires a large quantity of oxygen and lowers the overall efficiency of the process. According to Maurstad *et al.* (2006), the process is not at a disadvantage only if (i) high-grade coal is used and (ii) the resulting syngas is further reacted to give pure hydrogen. If—as in the case of the Shell process—insufficient water is contained in the syngas, the following CO conversion ($CO + H_2O \rightarrow H_2 + CO_2$) requires the addition of steam; this can be dispensed within the Texaco process. Only in this case is the Texaco process roughly equivalent.

This conversion is however only relevant when the gasification is to be combined with CO_2 sequestration, that is, for IGCC power plants or for a general supply of fuel gas in the form of hydrogen without CO_2 emissions. This step is dispensed with when the syngas is used directly as fuel gas; likewise when it is to be used as raw material for *sun methanol* (there, the syngas from coal gasification is mixed with solar hydrogen).

In syngas production using the Shell process, the disadvantage of the Texaco process, that is, the addition of water, is avoided. The Shell process is furthermore characterized by a more effective utilization of the waste heat. While in the Texaco process, the hot gas is cooled down by water quenching, with the result that most of its thermal energy is wasted, the hot gas in the Shell process is passed through heat-recovery boilers, where high-pressure steam is generated. According to Maurstad *et al.* (2006), for the production of syngas by the Shell process, considerably higher overall efficiencies can be obtained (they may possibly reach 70%). When low-grade, water-rich coal is to be used, a drying process which has already been tested in Germany with lignite must be applied.[35]

The Shell process would thus be usable for most types of coal with a high efficiency. According to Maurstad *et al.* (2006), roughly half of the worldwide "coal reserves" are unsuitable for the Texaco process—here, possibly only "reserves" in the narrow sense, that is, not including "resources," are meant. The Texaco process, however, according to Stiegel and Ramezan clearly has a certain edge in terms of development. Thus, up to 2004 roughly twice as many gasification projects using the Texaco process were under construction or planned as those using the Shell process (Meyer and Lorenz, 2004, p. 6). This however was no doubt due to

---

34) Texaco developed this process out of an earlier technique originally intended for the gasification of heavy oils.

35) Fluidized-bed coal drying with steam compression: The steam which is driven out of the coal is condensed after compression within a system of piping lying within the fluidized bed; it thus heats the drying chamber (heat-pump principle).

the fact that most of these newer projects were aimed at IGCC power plants. Both processes evidently do not guarantee a sufficiently high degree of reliability as yet, which, in particular for power plants, is an indispensable criterion. Both must still be perfected by detailed development (cf. Stiegel and Ramezan, 2008, p. 186).

Furthermore, alongside these two processes which have been in the foreground up to now, there are a number of other developments (Stiegel, 2008). One of these is the "advanced E-gasifier," which is considered as a reference process in Table 11.2, Case 3. Many of these processes are adapted to types of coal with low heating values, such as the German HTW fluidized-bed process mentioned above. A great step forward for all gasification plants would be a less complex process for oxygen separation. Evidently there is some hope of accomplishing this with the help of oxygen-permeable membranes (Stiegel and Ramezan, 2006, p. 188 and Stiegel, 2008, p. 46)

We can summarize the following conclusions:

- The gas-separation processes for efficient hydrogen production (advanced technology) are not required for syngas or for sun methanol production. For these two applications, the Shell process also yields a high efficiency.

- In the course of a large-scale energy research program, coal gasification technology must be developed as a matter of course more intensively and quickly than it has been thus far. A high-priority goal for energy policy is, however, to make available an alternative to importation of natural gas as soon as possible. Therefore, perfecting the most highly developed gasification techniques (Shell, Texaco, HTW fluidized-bed process for lignite) should be carried through initially with high priority. This is particularly important in Europe, where gas imports are predominant. In addition, certainly all the other types of gasification processes with high efficiencies should be further developed, even if they are most suitable for poorer grades of coal in some cases.

- In the medium term (and on a truly large scale), only "CO_2-free" hydrogen can be accepted as a substitute for natural gas and oil. With the "advanced technologies," it could potentially be produced for about the same cost as today's syngas. Therefore, the development of gas separation and gas cleaning processes is a central point in future process development.

- Syngas manufacture could also become very important for the USA, particularly with the background of "sun methanol" production. For this application, the syngas must be desulfurized in an especially efficient manner, for which the improvement of gas-cleaning technologies is likewise very important.

In view of the oil-price crisis of the year 2008 and probable similar crises in the years to come, the development of gasification technology should now be *intensified* in order to perfect it as quickly as possible and then, also with CO_2 sequestration, to attain an economically feasible position. The former nearly exclusive focus on IGCC power plants is thus obsolete. The goal must now be, in particular, to guarantee a gas supply and possibly also to substitute liquid fuels. This research program must be restructured in response to the new, much greater and especially urgent challenges – to enable it to react quickly and flexibly to all the recognized developmental approaches, in each one of the many branches of this field (at least in all of the more important ones).[36]

Fundamentally, the same conclusion holds as for solar energy: A continuation along the development routes followed up to now, which placed hope mainly on initiatives from industry, is totally insufficient. This method has yielded only very slow progress in the past 35 years, since the first oil crisis. Continuing with it could mean that we will need another 20 years, even with an intensified R&D program, to arrive at our goals. The approach used hitherto has failed to guarantee a systematic plan that includes all the options available. For this purpose, a central organization is indispensable. It must be in a position to carry out all those missions which are not subjects of initiatives by industrial partners. This of course presumes that changes in the system of support for research and development will be made. In many cases, governments will have to finance these missions 100%.

11.2.8
Preconditions for the Substitution of Natural Gas by H_2 or Syngas: Modification of the End-User Appliances and the Transport Networks

Both hydrogen and syngas (H_2/CO) differ from natural gas (methane) in terms of their heating values and their combustion and transport properties. Utilization of these gases therefore requires that components of consumer appliances be adapted, such as combustion chambers, burners (including the associated regulation and control devices), and gas meters. In the network of gas mains, modifications of the compressors are necessary.

If the substitution of natural gas on a large scale is planned, these modifications will have to be undertaken.[37] Thus, one is faced with the choice: either low-cost substitute gas *with* the modifications, or expensive natural gas *without* adaptations. In spite of the efforts required for making the modifications, the possible price decreases compared to natural gas will lead to overall lower costs in the general gas supply.

We have already discussed the fact that the substitute gas – in particular syngas – could take on the role of a "price brake" in the short term in negotiations of the price of natural gas with the exporting countries. This role is of course effective only when the new gas can indeed serve as a substitute, if need be for the entire natural-gas supply, in case the negotiations fail to produce an agreement. The new gas must thus represent a real (and not just a theoretical) competitor to natural gas. The intention would indeed not be to substitute more than a small portion of the natural gas initially. (A major substitution would not be

---

36) An essential part of this research and development program is the entire area related to the long-term storage of the separated CO_2 (see Section 11.3).

37) Without these modifications, mixing of hydrogen into natural gas would be possible only up to a H_2 concentration of ca. 15% (Ludwig-Bölkow-Systemtechnik, 1994). With syngas, its permissible concentration in the mixture would be in the same range.

reasonable at the outset, as long as large reserves of natural gas lie within reach of Europe. In the USA, the national reserves and those of neighboring countries are, relative to gas consumption rates there, considerably smaller.)

From the supply side, an ideal case would be a *variable* mixing ratio for the mixed gas, which can consist of the components methane and syngas or of methane and H_2. This would make it possible to operate the coal gasification plants, or later the installations for production of solar hydrogen or hydrogen from offshore wind power, continuously throughout the year, corresponding to the availability of the particular renewable energy source. Natural gas could then be used mainly for storage, given its high heating value per unit volume, and for meeting peak demands (or the medium consumption load).[38] The gas obtained from coal would then be produced continuously, and also consumed nearly continuously (to meet base-load demands), while natural gas reserves would be tapped only when the demand rose sharply. Furthermore, the somewhat reduced transport capacity of the gas pipelines for H_2 or syngas must be kept in mind. More precisely: transporting the same energy content as with natural gas requires a higher input of compressor power for these gases. With variable mixing ratios, one could switch mainly to natural gas during peak-load periods, so that the existing gas pipeline system would suffice to carry the required amounts of gas.[39]

In view of the required capability of gas consuming installations to operate with variable concentrations of the gas constituents, the following modifications or developments would be necessary:

- Suitable burners and regulators for all the relevant gas appliances in households, businesses, and industry will be needed.[40] For households, this means in particular the gas burners for space heating, water heaters, etc.[41]

- Gas meters would have to be able to register the amount of gas consumed as a function of time. The meters thus would have to be combined with clocks. With the data from the gasworks on the heating value of the gas at the time of its use (day and hour), the amount of energy delivered could then be computed.

- The compressor stations must be adapted to the varying properties of the gas mixture.

- Possibly, additional gas storage tanks will be required. As an alternative, one could implement the concept of storing mainly natural gas. Smaller and larger

---

38) Natural gas has approximately three times the heating value per Nm^3 of hydrogen or syngas. Seasonal energy-carrier storage is thus about three times less expensive using natural gas.

39) In the course of time, the gas-main network can be outfitted with pipes of larger diameter as new sections are constructed or older sections are renovated. The network must in the long term be modified so that when using pure hydrogen, all the limitations (such as higher required compressor power) would be eliminated.

40) Quite possibly, it will not prove possible to maintain the NO_x emissions at such a low level as is currently achieved with natural gas fuel.

41) Whether or not gas kitchen stoves can be operated with variable gas mixtures is still to be determined. If not, in the future, one would have to dispense with stoves designed for natural gas fuel.

gas storage tanks must then be integrated into the gas pipeline network. In Europe, locations for large underground storage reservoirs, which can be constructed economically by solution mining in salt domes, are mainly to be found near the German North Sea and Baltic coasts. Here, storage reservoirs of nearly unlimited volumes could be constructed at relatively low cost.

In the framework of a project supported by the EU and which is still underway, NaturalHy (2004), the question is being investigated as to whether the currently used plastic gas mains in urban areas are compatible with hydrogen as fuel or whether the use of hydrogen would shorten their operating lifetimes. The preliminary results indicate that there should be no major problems. This seems to apply also to the hydrogen compatibility of those types of steel used today for the construction of larger pipelines.[42]

All of these developments – in the case of the appliances, they include a multiplicity of different models from various manufacturers – should be carried out quickly with participation by the manufacturers involved, within a large-scale R&D program. The burners, regulators, gas meters, and compressors must be redesigned and a small number of prototypes manufactured and tested. The testing could be carried out within a limited region where the new gas mixture would be tried out. This includes of course an enormous number of tasks; in comparison to the whole breadth of the developments required, projects such as NaturalHy (2004) cannot be considered to represent a serious contribution toward solving the problems (although that is implied by the information available from NaturalHy).

11.2.9
The Possible Extent of Coal Gasification Using Substitutable Power-Plant Coal

In the "Preliminary Remarks and Summary" to this book, we pointed out that enormous quantities of gas could be obtained from the coal saved through the substitution of coal-fired power plants by solar plants. The numerical result was already quoted there. In the following section, we derive and justify these numbers. Along with the quantity of gas mentioned in the "Preliminary Remarks ...," which could be produced from the coal that would have to be used for power generation by the year *2030*, we also discuss in the following the quantity of gas which could be obtained from the coal used in *today's* power plants. Furthermore, we derive the amounts of coal which would be required to continue operation of coal-fired power plants (i.e., without their substitution by solar power plants), insofar as then, an additional demand for coal for the production of gas would have to be met. We deal in this section mainly with deriving the corresponding numerical data.

---

42) In any case, both these questions have thus far not been identified as major problems within the NaturalHy (2004) working group (Stolzenburg, 2008). Clearly, however, it will be necessary to monitor the gas lines more intensively at potential weak points such as valves and joints, making use of a pipeline management program (Müller-Syring, 2008).

Today's worldwide consumption of coal corresponds to 3700 GWa (2004), cf. Appendix C. Of this, ca. 2330 GWa of coal is used for electric-power generation. (The coal consumption by power plants is not specified in the statistics. It can be found from the power generated by coal-fired plants (770 GWa$_{el}$) and the plausible assumption of a worldwide average electrical efficiency of 33%.[43]) Then, 1370 GWa of coal are used for other purposes than power generation, i.e., for steel production, for space heating, etc.

In the coming years, we can expect a serious increase in electric-power consumption, in particular owing to the development of industrial nations in Asia. If we adopt the increases predicted by 2030 and suppose that all of the increased power consumption must be supplied by coal-fired plants, assuming modern plants with an efficiency of 45%, then the coal consumption for power generation would increase to 5050 GWa of coal; see Table 11.4. (Regarding the assumption that the increased power demands would be supplied from coal-fired plants alone: Considering the expected future high oil and gas prices, these other fossil energy carriers would not make a significant contribution in the medium to long term. In contrast, the present generating capacity supplied by these fuels would also have to be replaced.) The total consumption of coal, including its use for other purposes (1370 GWa), would then rise to 6420 GWa of coal. This is an increase compared to today's consumption (3700 GWa) by 75%. If, as is probable, that part of the current generating capacity which uses oil and gas would also have to be replaced by coal-fired plants, the consumption of coal for power generation would increase to 6100 GWa of coal (Table 11.4), and the total consumption of coal (including other applications) would be 7470 GWa of coal (8040 million tce). The quantity of coal mined worldwide (2004: 3700 GWa = ca. 4 billion tce) would thus have to be doubled.

Solar thermal power plants cannot substitute the whole amount of coal, owing to the need for backup power generation. The fossil-fuel base-load plants replaced by solar plants would then be available as backup power plants. Depending on the locations of the solar plants, the fraction of solar power would be 70% to 80% of the annual base-load power generated. The rest would thus have to come from coal-fired plants. On an average, roughly 75% of the coal which was thus far consumed by the coal-fired base-load plants could therefore be substituted. Relative to the annual coal consumption for future increased power generation (including substitution of plants hitherto fueled by oil and natural gas) of 6100 GWa of coal, only 4600 GWa of coal (5000 million tce) could be substituted.[44]

---

43) According to RAG, Steag (2001), the worldwide average efficiency of coal-fired power plants in the year 2001 was "of the order of 30%." Today (2008), owing to modernization of the power plants, it is no doubt somewhat higher, so that we have assumed a value of 33% here.

44) For this rough prediction of the *future* coal consumption by electric power generation, for simplicity we dispensed with making a

Table 11.4 Electric power generated (worldwide) in the year 2004, and predictions up to 2030 from (EIA–International Energy Outlook, 2007), along with our assumptions regarding possible future coal-fired power generation.

	Electrical energy generated (GWa$_{el}$)[a]	Electrical energy generated (TWh$_{el}$)	Coal required (GWa$_{coal}$)
Petroleum	110	966	
Natural gas	370	3 237	
Coal	770	6 746	2330 (efficiency: 33%[b])
Nuclear energy	310	2 678	
Renewable energies, especially hydroelectric	350	3 085	
Total (2004)	1900	16 712	
Estimated worldwide electrical energy consumption in the year 2030 (EIA)	3400	30 000	
Compare: Coal-fired power generation under the assumption that the increase in power demand by 2030 would be generated exclusively by coal-fired plants.	770 (2004) +1500 ——— 2270		5050 (efficiency: 45%[b])
Compare: Coal-fired power generation under the assumption that the increase in power demand by 2030 would be generated exclusively by coal-fired plants and the power currently generated with petroleum and natural gas would be replaced by power from coal-fired plants.	110 370 770 (2004) +1500 ——— 2750		6100 (efficiency: 45%[b])

a) 1 gigawatt year (GWa) = 8760 GWh = 1.076 million tce = 0.752 million t of crude oil.
b) Assumptions regarding coal requirements: The efficiency of the coal-fired power plants today is 33% (see the text). For 2030, worldwide modern coal-fired plants with 45% efficiency were assumed.

distinction between base-load power on the one hand and medium- or peak-load power on the other. For the estimate given below of *today's* substitutable quantities of coal, for clarity a blanket amount of 80% of the overall power generated was assumed as the base-load fraction. Furthermore, it can be assumed that a portion of the medium-load power could also be supplied from solar energy. This applies especially to "sunny" countries where power consumption increases during the summer (due to air conditioning).

11.2 Coal Gasification

The overall efficiency for coal gasification should (with future technology) be ca. 66% (LHV).[45] From the coal substituted (4600 GWa), gas with an energy content of 3000 GWa (LHV) could then be produced. This would be more than today's total consumption of natural gas (2900 GWa (LHV) = 3200 GWa (HHV)) and 60% of the current consumption of petroleum (5000 GWa). But remember: These numbers apply to a theoretical worldwide substitution of *all* future coal-fired power plants by solar power systems and are based upon the *future* coal-fired power generation which would be needed if oil- and gas-fired power plants were nearly all replaced by coal-fired plants, using neither solar nor nuclear power.

11.2.9.1 Gas Quantities Made Available by the Substitution of Current Coal-Fired and Gas Power Plants

If only currently operating coal-fired plants were to be replaced, less gas could be produced by the coal thus made available. Together with the natural gas saved by partial substitution of the current natural-gas-fired plants, however – again assuming a *theoretical*, that is, *total replacement* of all the base-load power plants – then still around half of the worldwide consumption of natural gas (outside of power plants) could be replaced by gas from coal gasification. Table 11.5 gives the numbers for the world, for the USA, and for Europe. We assume here generally that 80% of the power generated in coal-fired plants is base-load power, which could be substituted by power from the solar power systems.[46]

World:

As already explained, today's coal consumption by power plants is ca. 2330 GWa. According to the assumption above, about 80% of this power, i.e., 1865 GWa, is required to meet base-load power demands. Subtracting the consumption of the backup power plants (25%), then 1400 GWa of the coal remains for gasification. From this coal, 920 GWa of gas can be produced by gasification (at an efficiency of 66%). As can be seen from Table 11.2, under the assumptions made there, an additional 280 GWa of natural gas could be conserved (from electric-power generation) by using solar power. (The current consumption of natural gas for power plants was estimated roughly here by using their efficiency as well as the fraction of

---

45) From Table 11.2, the efficiency of hydrogen production using the future technology should be 64%. By using solar power for the energy supply of the gasification plants, the consumption of coal can probably be reduced still further (cost-neutrally), so that the yield of gas could be well over 66%. For simplicity, however, we have used the round value of two-thirds, that is, 66%.

46) Neither for the world as a whole, nor for the USA and Europe, are there data in the usual overview statistics on the division of electric-power demand into peak-load, medium-load and base-load power (especially for the individual types of power plants). Since coal-fired plants are frequently utilized for base-load power generation, and the peak- and medium-load demands are met for the most part by natural-gas plants, the assumed fraction of base-load power of 80% for the coal-fired plants should be approximately correct.

316 *11 Other Technologies for Backup Power Generation and Alternatives for Future Energy Supplies*

Table 11.5 Quantities of gas which could be produced from substitutable coal and substitutable gas from power plants (at today's power consumption).

	Fuel used in power plants in 2004	Assumption: Base-load fraction	Yields: Fuel used in base-load power plants	Fraction of backup power	Yields: Amount of substitutable fuel	Efficiency of coal gasification	Yields: Amount of gas	Compare: oil consumed in 2004	Compare: gas consumed in 2004	Compare: gas consumed in 2004 outside power plants
	GWa		GWa		GWa		GWa	GWa	GWa	GWa
World										
Coal	2330[a]	80%	1865	25%	1400	66%	920 LHV	Oil: 5000	3200 HHV	2170 LHV
Natural gas	740 LHV[b]	50%	370 LHV	25%	280 LHV		280 LHV		2910 LHV	
Total:							1200			
USA										
Coal	670[c]	80%	536	20%	430	66%	280	Oil: 1240	770 HHV	[d]530 LHV
Natural gas	170 LHV[c]	50%	95	20%	76		76		700 LHV	
Total:							360			
EU-25										
Coal	280[e]	80%	224	25%[f]	170	66%	112	Oil: 920	560 HHV	[g]370 LHV
Natural gas	140 LHV[e]	50%	70	25%	50		50		510 LHV	
Total:							160			

a) Coal requirements for power generation: cf. Table 11.4.
b) Power generated with natural gas (worldwide, 2004): 370 GW$_e$. Assumption: average efficiency of the gas-fired power plants (worldwide), referred to the LHV: 50%; this gives a gas consumption of 740 GWa (LHV).
c) USA: Gas used in power plants: 190 GWa (HHV) (cf. Appendix C) = 173 GWa (LHV). Natural gas: $HHV = 1.1 * LHV$.
d) 700 – 170 GWa (first column) = 530 GWa.
e) EU: Fuel consumption in power plants not given in the statistics; computed from the power generated (2004: coal 107 GWa, natural gas 69 GWa). Assumed average efficiency in Europe: for coal-fired power plants 38%; gas-fired power plants 50% (LHV).
f) Fraction of backup power for the solar plants in Spain 30%, for plants in Morocco 20%; Assumption: equal generating capacities in Spain and Morocco.
g) 510 – 140 GWa (first column) = 370 GWa.

base-load power, which was assumed to be 50% of the total power generated with natural gas fuel.) All together, according to this very rough estimate, ca. 1200 GWa$_{gas}$ (LHV) could be produced or conserved. This corresponds to 41% of the total worldwide gas consumption, or 55% of the gas consumed outside power plants, or 24% of the world's consumption of petroleum.

USA:
There, making similar assumptions, 51% of the total current natural gas consumed could be replaced, corresponding to 68% of the gas consumed outside power plants or 29% of the petroleum used.

Europe:
Here, we would have 31% of the overall gas consumed, or 43% of the gas used outside power plants, or 18% of the petroleum consumption.

If power generation using coal is to be maintained in the future *and*, for example, the quantity of gas mentioned above (3000 GWa)[47] is at the same time to be produced by coal gasification, then in addition to the 7470 GWa$_{coal}$ for power plants, a further 4600 GWa$_{coal}$ for gas production would be needed. Coal mine output would have to be increased in this example to 12070 GWa$_{coal}$ (13 billion tce); compared to today's output, this is a factor of 3.3 greater.

One must then ask the question as to whether such an increase by more than a factor of three in coal production would even be possible within the roughly 20 years remaining until the year 2030; or whether, in view of the difficulties of such a rapid increase, with the probable lag of production behind demand that must be feared, the price of coal would not rise sharply?

Furthermore, the associated CO_2 emissions must not be forgotten. An additional consumption of coal of this order of magnitude is unthinkable without sequestration of CO_2 emitted. Since the future demand for electric power and the substitution of the present gas-fired power plants alone would cause a doubling of coal consumption for power generation (and for the other applications), the necessity of CO_2 separation would present itself also for the power plants (at least for those *new plants* built), and not only for the gasification plants. For new construction, only CO_2-free power plants could be considered, that is the expensive IGCC plants with integrated coal gasification. At the same time, an extensive CO_2 disposal system (CO_2 sequestration) would have to be set up. Leaving off the question of *when* these power plants will be ready for construction (future technology!), we must still consider the fact that the construction of such expensive power-plant parks (IGCC) would lock in coal as the major energy source for a long time to come. The available supply of coal would then have to be *permanently* reserved for electric power generation and would not be available in the foreseeable future for

---

[47] That is the quantity of gas which would be producible in the year 2030 from the coal freed up by the substitution of coal-fired power plants.

other uses (substitution of oil and gas). This would thus represent a long-term, irreversible decision.

11.2.9.2 Limitations of the Natural-Gas Reserves in the USA

While Europe is within range of the large gas deposits in Russia and the Near East, and therefore gas production with coal, from the pure supply point of view, will not be necessary for some time to come, it could become essential in the USA on a large scale within the not-too-distant future. The proven natural-gas "reserves" (2004), according to the BP Statistical review of world energy (2005, p. 21), comprise $7.3 \times 10^{12} \, m^3$ (given in m^3, not in Nm^3) in North America (including Mexico); and in Central and South America, they are $7.1 \times 10^{12} \, m^3$. At a heating value of $9.5 \, kWh/m^3$ (LHV) (as for Nm^3), the North American "reserves" would correspond to $8000 \, GWa$ (LHV).[48] These reserves can be compared to the annual consumption in the USA of $700 \, GWa$ (LHV)[49] (2004). The BP Statistics therefore give the depletion time of the reserves for North America (not just the USA) as 10 years (cf. p. 24 there).

In BGR (2007, p. 18), for North America (including Mexico), the natural gas "reserves" are stated to be around $8 \times 10^{12} \, m^3$, and the "resources" (without the "unconventional resources") to be $27.3 \times 10^{12} \, m^3$.[50] The North American "reserves" and "resources" ($34.6 \times 10^{12} \, m^3 = 37\,000 \, GWa$ (LHV)) would then have a range of 53 years until their *complete* exhaustion – insofar as the "resources" can in fact be extracted to the full extent of the amount quoted, and insofar as they would offer any cost advantage at all over coal gasification with CO_2 sequestration, which is not the case for all of the natural-gas deposits (depending in particular on their geographical locations). Since a switchover can however not be initiated only when the reserves are already exhausted, it is quite likely that a massive buildup of alternative gas sources must begun within 15 to 20 years (in particular if the quoted "resources" cannot be completely exploited owing to economic factors). Furthermore, the accumulated consumption up to 2006 in North America was, according to BGR, $35 \times 10^{12} \, m^3$, just equal to today's presumably remaining "reserves" and "resources" all together.

11.3
Coal as the Only Major Alternative to Oil and Gas? – The Scope of the Coal Resources for Power Generation and Gasification on a Large Scale – the Potential for Sequestration of CO_2

Could the future energy supply, including the substitution of oil and gas, be based to a major extent on coal as energy carrier, that is, without the large-scale utilization

---

48) Compare: $10^{12} \, m^3 \times 9.5 \, kWh/m^3$
= 9500×10^3 GWh = 1084 GWa
($1 \, GWa = 8760 \, GWh$). $7.3 \times 10^{12} \, m^3$ then corresponds to 7920 GWa.
49) This corresponds to 770 GWa (HHV), cf. Table 11.5 and Appendix C.
50) For Central and South America, in BGR reserves of $7 \times 10^{12} \, m^3$ and resources of $9.9 \times 10^{12} \, m^3$ are quoted.

of other energy sources such as solar energy? With this question in mind, we take a closer look at the world's supply of coal and the potential for CO_2 sequestration.

11.3.1
Coal Reserves

The supplies of coal have thus far usually been claimed to be "nearly unlimited." Such statements are based on the so-called static lifetime, that is, the lifetime at today's consumption rate. However, if oil and natural gas were to be substituted in the future exclusively by coal, and the expected increase in power demand were also to be filled by coal-fired power plants, without utilization of solar energy,[51] then the supply situation would appear quite different. Then, the coal "reserves," that is, the known supplies which can be extracted economically under current conditions (the so-called proved recoverable coal reserves) would be exhausted already after 35 years (see below). Even the "resources," of which it is unclear what portion can be recovered at an economically feasible cost, could be exhausted within one to two centuries, depending on the consumption rates in the later decades; that portion of the resources which can be recovered at competitive prices would be used up even sooner (see below). Without an additional energy source such as solar energy, it is thus by no means the case, as is often claimed, that coal would suffice "for many generations."

The total available coal supply is however not even the most important aspect. Two others seem more significant: One of these is the geographical distribution of the reserves or the resources over various countries. The "resources" are located namely to the extent of 73% in Russia and China. For all the other countries in the world, coal will become rather expensive much sooner than one would expect on the basis of the overall known reserves. Thus, the perspective for the *economic* utilization of coal, in the case of a future energy supply based mainly on coal as fuel source, is *much* shorter for many countries than it would seem to be, given the global coal reserves. Considering the future price of imported coal, one must remember that this price was held down in the past by competition from low-cost oil. This will cease to be the case in the medium term. In addition, a rapid switchover from oil and natural gas to coal would mean that the amounts of coal mined could lag behind the demand, owing to the time required to open up new mines; this would likewise drive up the price of coal.

The other important aspect is the question as to whether sufficient locations for storage depots to store the corresponding enormous amounts of CO_2 exist, or whether such locations are available at acceptable costs in the regions where coal is consumed. There are several indications that this is not the case in many regions (see below). In that case, the available coal reserves are not the limiting factor for

---

51) Regarding a potential utilization of nuclear energy: Here, the fuel-supply situation is much more precarious than for coal, as long as uranium cannot be extracted from seawater (cf. Chapter 12). Thus, the only really serious alternative to coal can be considered to be solar energy.

History of Assessment of world coal resources

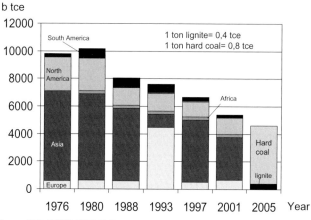

Figure 11.3 Differing estimates of the coal resources by the BGR over the years (Ludwig-Bölkow-Systemtechnik, 2007).

the particular region, but rather the availability of storage depots for CO_2 – quite apart from the question of whether CO_2 sequestration will even prove to be a realistic option.

Table 11.6 shows the "reserves" and "resources." (The amounts are given in tons of coal equivalent (tce).[52]) The current (2006) "reserves" amount to *730 billion tce*. Thus far, 240 billion tce have been consumed (BMWi, 2006). "Resources" are those quantities of coal which are either known but whose extraction costs would be too high (coal deposits at great depths, in very thin seams or in poorly accessible regions), or else speculative, that is, their existence is only presumed.[53] The worldwide resources were estimated up to 2006 by the BGR to comprise ca. 4000 billion tce (cf. BMWi, 2006). They were then suddenly increased to *8600 billion tce* due to an increased estimate of the resources in China and Russia. In past years, the resources have been repeatedly estimated with differing results, causing considerable fluctuations in the presumed total amount of coal; see Figure 11.3.

Before we consider in the following the length of time remaining until the whole reserves of coal are exhausted, we cast a glance at the situation of individual countries:

The coal supply situation is very different in different countries. Thus, *India* – at present the country with the second-largest population and in future the

---

52) 1 tce is defined as 8140 kWh.
53) BMWi (2006): "Resources are deposits which cannot yet be recovered economically or are not identified with certainty, but are expected on the basis of geological indicators. Price increases on the world raw-materials markets and the results of new exploration can convert resources into reserves."

Table 11.6 Coal reserves as of 2006 (BGR, 2007).

	"Reserves"			"Resources" (geological inventory[a])			Total
	in 10^9 t coal equivalent (tce)[b]						
	Hard coal[c]	Lignite[c]	Total	Hard coal[c]	Lignite[c]	Total	Reserves + Resources
PR China	142	9	151	3578	221	3799	3950
Russia	59	32	91	2268	450	2718	2809
USA[d]	181	12	193	612	138	750	943
Australia	34	13	47	130	62	192	239
India	81		81	133		133	214
Poland	10	1	11	142	14	156	167
Great Britain				161		161	161
Canada	3	1	4	119	18	137	141
Germany	1	14	15	72	12	84	99
Kazakhstan	7		7	80		80	87
Ukraine	27	1	28	42	2	44	72
Iran				38		38	38
Indonesia	2	1	3	20	12	32	35
Brazil	9		9	13		13	22
Czech Republic	3		3	18		18	21
Mozambique	2		2	13		13	15
Mongolia	3	0	3	8	3	11	14
Japan				10		10	10
North Korea				9		9	9
Venezuela				5		5	5
Vietnam					70	70	70
Pakistan		1	1		63	63	64
Republic of South Africa	41		41				41
	–	–	–	–	–	–	–
Above countries:	605	85	690	7471	1065	8536	9226
World total	627	100	727	7512	1082	8594	9321

Table 11.6 Continued.

	"Reserves"			"Resources" (geological inventory[a])			Total
	in 10^9 t coal equivalent (tce)[b]						
	Hard coal[c]	Lignite[c]	Total	Hard coal[c]	Lignite[c]	Total	Reserves + Resources
World without China and Russia	426	59	485	1666	411	2077	2562
EU-25	14	18	32	405	29	434	466
USA, Canada, Australia, EU-25	232	44	276	1266	247	1513	1789

a) Resources: In BGR (2007) (Federal Institute for Geosciences and Natural Resources, Germany), the geological inventory of the specific coal deposit ("in situ" amount) is listed under the coal resources (in contrast to oil and gas), that is, the overall amount of coal without taking the recoverable portion into account (p. 37). (In the case of oil and gas resources, on the other hand, only that portion which is considered to be recoverable in a presumed deposit is listed.)

b) In BGR (2007), the amounts of coal are specified in tons of coal. Conversion into t coal equivalent (tce) after BGR, p. 20/21: 1 t hard coal = 0.852 tce; 1 t lignite 0.352 tce (1 tce = 8140 kWh).

c) *Hard coals*: Anthracite, bituminous coal, and subbituminous coal, with an energy content (ash-free) >16.5 MJ/kg (>4.6 kWh/kg). *Lignite*: coal with an energy content of the raw coal (ash-free) <16.5 MJ/kg (<4.6 kWh/kg). From BGR, pp. 69 and 77: For Russia and the Ukraine, hard coals include only bituminous coal and anthracite, according to the national classification scheme; lignite is included there together with subbituminous coal. For definitions, see also IEA (note: not EIA) Coal Information (2007). (In some countries, lignite and subbituminous coal are referred to as *brown coal*.)

d) In Roan *et al.* (2004, p. 22), the US coal resources are listed according to data from the EIA (cf. also EIA (coal) 1999, 2006 and 2008). There a somewhat different distribution and overall quantity are given: The "reserves" (estimated "recoverable reserves") are quoted as 216 billion tce (6000 quadrillion BTU), which roughly corresponds to the value in Table 11.6 (193 billion tce). Of the all together 400 billion tce ("reserve base," in place!) (i.e., 11 100 quadrillion BTU), which have been "demonstrated" to be in the ground, the 216 billion tce mentioned are held to be recoverable under current conditions. Demonstrated means: "measured" (i.e., with a high degree of geologic assurance) or "indicated" (with a moderate degree of geologic assurance). In addition, there are the "inferred resources" (with a low degree of geologic assurance); they are given as 970 billion tce (in place), namely 1370 billion tce minus the 400 billion tce mentioned already. Furthermore, 1800 billion tce are thought to be conceivable but are "undiscovered" (in place) (49 000 quadrillion BTU). The potentially recoverable amounts of coal are thus evidently only partially known. However, what was already stated probably holds here also: The readily exploitable coal deposits have for the most part already been discovered, so that of the "undiscovered" and "inferred resources," no doubt only a smaller portion will prove to be recoverable under economically feasible conditions.

largest – has only relatively limited "resources." India is to be sure currently a coal-exporting country, but it will not be able to base its energy supply in the *long term* on coal, if it wishes to avoid becoming dependent on imports (which would mean surrendering to the international market prices).[54] With a view to a national energy policy, solar energy will therefore have a special significance in India, in particular because solar power plants can be constructed relatively quickly with the means available in developing countries. India should therefore be one of the countries in which solar thermal power plants are utilized on a large scale in the middle term; since it has – in contrast to China, for example – practically no alternatives for its national energy production.

A similar situation holds in the European Union (EU-25), as well as for the countries in South America. The EU has "reserves" of only 32 billion tce (of which nearly half is German lignite). For lignite, there are only small "resources." (These were, however, listed in earlier statistics at a higher value than is shown in Table 11.6. This indicates that the overall supplies – including deposits that would be more expensive to mine than those currently in use – are larger than shown in the table. Whether or not they would still have a substantial cost advantage over imported *hard coal* remains uncertain.)

As is already the case in Europe, many other regions and countries will in future be dependent on imported coal to a great extent if they are unwilling to develop their own relatively expensive "resources." Germany, like Great Britain and Poland, certainly has "resources" of hard coal (anthracite and bituminous coal).[55] The German example however makes it clear that – owing to their geological locations – these are *much* more expensive to exploit than the "reserves." Generating electric power from this coal (including CO_2 sequestration) would be uneconomical compared to solar power. Whether the costs of extracting German hard coal are typical for all coal "resources" is however not clear from BGR (2007). If one assumes extraction costs of around *150 $/tce* (2002) for German hard coal,[56] then the energy cost from an IGCC plant with future technology (including separation,

---

54) If India (1.05 billion inhabitants) were to be industrialized to a major extent and attained – with a frugal use of energy – half the per capita energy consumption of Europe today (EU-25: 450 million inhabitants, consumption of primary energy 2300 GW, without coal conversion for substitution of oil and gas), then according to this simple computation, its primary-energy consumption would be 2700 GW (2.9 billion tce); in fact, however, due to losses in the coal conversion processes (production of gas and liquid fuels), it would be higher, for example 4 billion tce/a. The "reserves" of 81 billion tce would then be used up at this consumption rate after only two decades. At that point, India would already have to fall back on the *more expensive* "resources," and even these would be exhausted after a further ca. three decades, if one assumes for simplicity that they could all be 100% exploited. Thus, a low-cost self supply with coal is not possible in the long term.

55) In some countries "hard coal" also includes "subbituminous coal."

56) We assume a typical price for German hard coal in the 1990s (thus before the recent increases in coal prices); this then reproduces to a good approximation the extraction costs: 138 €/tce 1995 (270 DM/tce) = 152 €/tce (2002) (consumer price index 1995–2002: 1.101) = *158 US $/tce (2002)* (purchasing power parity 2002: 1$ = 0.96€); thus, rounded off: *150 $/tce*.

but without storage of CO_2) would be 6.4 ¢/kWh[57] (compare Table 10.2). This is *considerably* higher than the energy cost from the solar power system, whereby the cost would increase still more if the coal price were higher.[58] The use of this costly coal for power generation would thus not be reasonable in view of its price disadvantage. A similar conclusion holds for coal gasification; it would then be practically cost equivalent to solar hydrogen production.[59]

11.3.2
The Future Consumption of Coal – Depletion Time of Resources

The depletion time given above as a result for the "reserves" and the "resources" will be derived and justified in this section. This time depends on the available supply and the future consumption of coal. Regarding the shorter time perspective, as is typical of the "reserves," one can make use of today's energy consumption or the predictions (substitution scenario). Regarding the very long-term consumption (this refers to the "resources"), one must resort to a great extent to speculation.

Concerning the time for exhausting the "reserves": The worldwide annual consumption of coal in the year 2004 was *4.0 billion tce*. What proportion of this was related to electric-power generation is not shown in the statistics; it is presumably ca. 2.5 billion tce. With this consumption of coal, the "reserves" would last for 180 years (static lifetime). The worldwide electric-power demand will however increase considerably in the future. If this enormous consumption is not covered by solar or nuclear power, the amount of coal used for power generation would increase roughly threefold by the year 2030; this to be sure does not include the higher efficiencies of modern power plants. The coal consumption for power plants would still more than double at this higher efficiency. Oil and natural gas will have to be substituted (at least partially) sooner or later owing to the increases in their prices. If this is accomplished by gasification of coal, at an efficiency of 66% (LHV, including separation of CO_2), we have $1.5\,\text{kWh}_{coal}$ per kWh_{gas}. From this we find the

---

57) For the USA, we have likewise assumed here a coal price of 150 $/tce. Quite possibly, those "resources," considering the very low-cost coal deposits found there up to now, were defined with a lower cost basis than in Europe. In BGR, this is not mentioned. The costs of CO_2 storage must be added, cf. Table 4.3. At 10 $/t CO_2, they would increase the electrical energy price by 0.8 ¢/kWh, to 7.2 ¢/kWh.

58) For the solar power system, this refers to the coal-fired backup power plants as well as to those CCGT-backup plants operated with gas from coal gasification. Due to the small proportion of backup power, an increase in the coal price however causes only a subproportional rise in the electrical energy cost. Utilizing coal-fired backup power plants (the case of "annex construction") and at a coal price of 150 $/tce, the electrical energy cost in Europe for solar-plant locations in Spain would increase to 5.8 ¢/kWh (compare 5.3 ¢/kWh at a coal price of 90 $/tce); in USA, the electrical energy price would increase to 5.3 ¢/kWh (compare 4.7 ¢/kWh at 45 $/tce; Table 4.2).

59) As seen in Table 11.2, in Case 3, the cost of H_2 would increase to 4.0 ¢/kWh at a coal price of 150 $/tce. Adding the cost of CO_2 storage (0.5 ¢/kWh assuming 10 $/t CO_2), this yields 4.5 ¢/kWh for H_2.

following amounts of coal consumed: If the coal is used to supply the increasing future demand for electric power and in addition only 20% of the present oil and gas consumption (coal gasification), the coal consumption would increase to ca. *10 billion tce/a*. In the case that 40% or even 100% of the oil and gas consumed would have to be substituted, then the consumption of coal would rise to 12 billion tce/a or 20 billion tce/a, respectively. The "reserves" of coal (730 billion tce) would then be already exhausted after 73 years (in the case of increased utilization, with 40% or 100% oil/gas substitution, within 60 or only 35 *years*).

If 40% or even 100% of today's oil and gas consumption were to be substituted, in addition to gas from coal gasification, also liquid energy carriers would be required, in particular methanol produced from coal gas. For the rough estimates given here, we have assumed the same conversion efficiency as for coal gasification, namely 66%, although it (with the same gasification technology) would in fact be nearer to 50%. Furthermore, a minor portion of the current liquid and gaseous energy carriers would be substituted by electrical energy.

Regarding the quantities listed (per year):

The total worldwide coal consumption in 2004 was 3690 GWa = 4.0 billion tce (compare: 1 GWa = 1.076 million tce).

Electric power generation from coal: The total electrical energy generated in 2004 worldwide was 1900 GWa$_{el}$. Of this, 770 GWa$_{el}$ was generated from coal. The coal consumed in power plants is not listed separately in the statistics. The worldwide average efficiency for electric power generation from coal should be roughly 33%. (Compare RAG, Steag (2001, p. 56): "The average efficiency of coal-fired power plants is presently (2001) of the order of 30%." In the meantime, the efficiency will have risen somewhat, so that the assumed value of 33% seems plausible today.) Assuming this efficiency, we find for the year 2004 a coal consumption by power plants of 2330 GWa$_{coal}$, corresponding to 2.5 billion tce. According to the predictions of the American Energy Information Administration (EIA–International Energy Outlook, 2007), by the year 2030, an increase of the worldwide electrical energy demand from 1900 GWa$_{el}$ to 3400 GWa$_{el}$ (30 000 TWh) is to be expected. The International Energy Agency (Paris) cites an increase to 3600 GWa (31 500 TWh) by the same year (IEA: World Energy Outlook – quoted in COORETEC, 2003, p. 7). If this increased demand (1500 GWa$_{el}$) were to be met exclusively by coal-fired plants – natural gas will not make a noticeable contribution due to its rising price; on the contrary, the present gas-fired power plants will have to be substituted to some extent – then power generation from coal would be tripled from 770 GWa$_{el}$ to 2270 GWa$_{el}$. With modern coal-fired power plants operating at an efficiency of 45%, this would correspond to a coal consumption of 5050 GWa$_{coal}$ = 5.4 billion tce.

Other coal demands: In addition, we must consider that quantity of coal which is consumed today in excess of what is used for power generation, i.e., for steel production, process heat, space heating, etc. This amounts to ca. 1.5 billion tce (overall coal consumption (4.0 billion tce) minus consumption in power plants (2.5 billion tce)). Including this amount, the overall coal consumption would then rise to around 7 billion tce.

Substitution of oil and gas: The worldwide demand for oil was 5000 GWa in 2004, and that for gas was 3200 GWa, together giving 8200 GWa. Assuming for the future an efficiency of 66% (LHV) for coal gasification, to produce 1 GWa of coal gas, 1.5 GWa $_{coal}$ would be required. For the substitution of, e.g., 1600 GWa of natural gas by coal gas – corresponding to one-half of today's consumption of natural gas, but only 20% of the present oil *and* gas consumption – we find an additional coal demand of 2400 GWa $_{coal}$ = 2.6 billion tce. Furthermore, replacement of 40% of today's oil and gas demand would give an additional requirement of 5.2 billion tce. For the complete substitution of oil and natural gas by coal gas, one would require 12.3 billion tce of coal.

Together with the future coal demand for power generation and other uses (a total of 7.0 billion tce, see above), and with substitution of 20% of the current oil and gas consumption, the coal demand would rise to 9.6 billion tce annually. With 40% substitution, it would increase to 12.2 billion tce per year, and with 100% substitution, to 19.7 billion tce per year.

Our considerations of the coal "resources" naturally extend far into the future. Concerning the worldwide energy demand at that point, we can only speculate.

We limit our further considerations of the depletion time of coal to that portion of the resources which are *economically recoverable*. We have already emphasized that "resources" are defined by two categories: (i) known deposits, which are however too expensive to recover (by today's standards); and (ii) still undiscovered deposits. This second group will include both low-cost and – presumably the major portion – expensive deposits. Only here can we expect any additional supplies which are "economically recoverable under current conditions." Thus only this less-expensive subgroup of the second category is of interest. The first subgroup (known but expensive) would without doubt not be competitive with solar energy, taking German hard coal as a yardstick, so that it does not represent an *economically* attractive resource. Unfortunately, this (known) subgroup is not listed separately in the statistics. Therefore, we are dependent on speculations concerning the overall quantity represented by the second subgroup (unknown), not to mention its less-expensive fraction. Regarding this less costly portion, one is forced to make educated guesses, so that only rough tendencies can be quoted, as given below. However, it is in any case clear that the time remaining until the exhaustion of this portion of the resources is *by far* not as long as has often been claimed.

11.3 Coal as the Only Major Alternative to Oil and Gas?

Only a small fraction of the thus-far undiscovered "resources" is likely to be economically recoverable. It is reasonable to assume that the coal deposits which are still undiscovered lie as a rule either at great depths or in remote and inhospitable areas (such as the polar or desert regions), that is, in locations where systematic exploration has not yet appeared to be profitable (especially since the exploration itself would be relatively expensive there).

Since however not even the known portion of the resources is listed in the statistics, the information needed for a rational estimation is completely lacking. (Even simply mentioning this known portion would allow a much better estimation of the overall situation.) Furthermore, there are no comments at all from the responsible authorities regarding the question of what fraction of the hitherto unknown resources might be economically recoverable.

The "resources" as a whole (known and unknown) are however listed as being roughly a factor of 10 greater than the "reserves"; thus the less-expensive fraction – even though it is probably only a small part of the total – could still be "worthy of mention." As an example, we assume here a tripling of the "reserves" through economically recoverable deposits which are still to be discovered (speculation); that is an increase of the economical "reserves" from 730 billion tce known today to 2100 billion tce.

Concerning the very long-term consumption of coal, we can also only speculate. In the long term, however, we must assume on the one hand that substitution of oil and natural gas to a large extent will be necessary, and on the other, that a considerably increased worldwide demand for energy will occur, owing to the increasing population and global industrialization. While the annual coal consumption, assuming complete substitution of the current oil and gas demands and coal-fired generation of electrical energy, would be 20 billion tce by the year 2030, it could in the longer term reach for example, 40 billion tce. At that consumption rate, the presumed economically recoverable total amount of coal reserves of 2100 billion tce would last only 55 years.[60] In fact, even the *total* "resources," that is, the theoretical quantity of 9300 billion tce (assuming it to be completely recoverable), would be consumed within 230 years. Even if the estimate given above as an example (2100 billion tce) should prove to be too pessimistic, one still has to presume in any case that the economically recoverable portion of the resources would be exhausted within *at most* 100 to 150 years.

As we have already discussed, given the worldwide distribution of coal supplies, it must be assumed that international coal prices would reach a high level even long before this global exhaustion of the low-cost supplies. Of the "reserves" (730 billion tce), only 277 billion tce are to be found in the USA, Canada, Australia, and Europe (EU-25), and thus in the presently industrialized nations which could theoretically be in a position to oppose a price diktat by the future major supplier

---

60) The high coal consumption has thus for simplicity been presumed to apply during the entire period of time. As stated, this is the depletion time of the coal reserves under the assumption that coal would have to supply the entire energy requirements *after* the oil and gas supplies are used up.

countries, Russia and China, by increasing their own coal production.[61] If in fact coal were the only alternative to oil in the future, and if all of the coal consumption rates estimated above were to be realized, these Western nations would certainly not be able to maintain their moderating influence on the pricing policies of the two large supplier countries for very long.

It has already become clear that the information available from the responsible agencies concerning the supply situation for coal is completely insufficient; this is certainly true of the German agency (BGR) (and is in all likelihood not very different in other countries). Not only is the known portion of the "resources" not given in the available statistics, and no professional commentary is provided on the probable undiscovered portion; but also there is no explanation of the numerous revaluations of the "resources" in the past, and thereby also of the possible uncertainties in the data.[62] The German Federal Institute for Geosciences and Natural Resources (BGR) has been criticized for some time because of these shortcomings, among others by Ludwig-Bölkow-Systemtechnik (2007).[63] The agency has thus far restricted itself to a simple reproduction of the statistics from individual countries. This however hardly requires a professional agency for raw materials; statistics of this kind could also be provided for example, by the (German) Federal Statistical Office. The technical expertise expected of the BGR is thus completely lacking. Today, when the pressing and necessary substitution of oil forces us to look to the future, this situation must be quickly corrected.

Even though many of the above considerations necessarily remained speculative, they still clearly permit the following conclusion: An additional major energy source besides coal is a compelling necessity. Whether it might be nuclear energy (considering the *much* smaller uranium reserves in comparison to coal) will be discussed in Chapter 12. Insofar as the costs of solar power estimated in this book

---

61) Of course, we must reckon with a certain fraction of economically recoverable, still-to-be-discovered "resources" in these Western countries also. The "resources" (known and not known) are however not very great here, all together 1500 billion tce. Furthermore, it must be assumed that particularly in these countries, coal deposits have been relatively well explored, so that the greater part of the "resources" is probably already known, but not economically recoverable.

62) How is it possible that the overall "resources" are suddenly – from one year to the next – listed as being twice as large as before, without any explanation of this great quantum leap?

63) In this connection, Ludwig-Bölkow-Systemtechnik (LBS) (2007) wrote: "A clear-cut example of the uncommented change in statistical assertions is the reduction of the proven German bituminous-coal reserves by 99 percent (!) from 23 billion tons to 0.183 billion tons in 2004. According to the World Energy Council, major reserves which previously were held to be proven were newly evaluated and are now classed as speculative. The responsible German agency has published no explanation at all of this action. Thus, this reduction – in spite of the active public debate at the time concerning the future of coal production in Germany – remained to a large extent unnoticed."

Regarding the supplies of *lignite coal* in Europe, one must keep in mind that in previous statistics, a larger amount of "resources" were listed than at present. Thus, in (BGR, 2002) for Europe without the Russian Federation countries, all together *60 billion tce* (1779 EJ) were listed. This is twice as much as in Table 11.6 (29 billion tce). "Resources" are however more costly, as mentioned, compared to the

are essentially confirmed in the course of further developments, there is no real alternative to the application of solar thermal (i.e., concentrating-solar) power-plant technology—worldwide and on a very large scale.

As part of a general program of research and development—as a reaction to the most recent oil crisis and for the avoidance of future crises—all of the major energy options, including a precise investigation of their potential, should be considered. In terms of coal usage (for electric power and gas production), this also includes the question of economically favorable reserves and—still more importantly—the topic of the potential for CO_2 sequestration as discussed below.

Concerning statistical evaluations of known and estimated supplies, in future the following basic principles should be observed:

- Known, but expensive "resources" should be listed separately.

- For each individual country, it should be noted what fraction of inexpensive coal can be expected within the still undiscovered "resources." When data of this type are missing, the professional agencies (e.g., in Germany the BGR) should use average values obtained from those countries for which data are available; they should thus give at least rough estimates.

- The goal should be (in the ideal case) to classify the coal "resources"—as in the case of uranium supplies—into cost categories (with estimates of the amounts still to be discovered).

- The uncertainties in the quantities listed should be discussed.

- All of the parameters which influence the results should be listed.

11.3.3
The Potentially Limited Capacity for Economical Storage of CO_2

On the one hand, we wish to investigate the potential for the storage of CO_2 *on land* and its costs in the case of large-scale utilization of coal. (We have already considered the question of costs in Section 10.3.2.) Within the extensive territory of the USA, there would seem in principle to be sufficient potential sites (CBO, 2007), even though it is unclear what the cost of transporting CO_2 to the regions where the storage depots are located would be. In other parts of the world, among others in Europe, the potential areas which can be developed economically are quite likely too small (cf. STE, 2006).[64] Then, not the coal supply, but rather the possibility of storing CO_2 produced by consuming coal would limit the timeline

---

(*very* inexpensive!) "reserves" utilized up to now. The later changes in the estimated "resources" are obviously connected with revaluations, in particular for Poland and other former Warsaw Pact countries. For the EU-15 (without Poland), in BGR (2007), only 33 billion tce (976 EJ) were given; thus 27 billion tce of the 60 billion tce total listed for Europe are attributed to Poland and the other Warsaw-Pact countries.

64) Additional literature on CO_2 sequestration: Meyer and Lorenz (2004), Ploetz (2003), COORETEC (2003).

for coal utilization. Thus for example, the deposition potential in the USA (mainly in "deep saline formations") was estimated in the CBO paper to be 1200–3600 billion t CO_2. In STE (2006, p. 28), for Europe, however, only 160 billion t CO_2 is quoted; however, this estimate appears to be incomplete.

On the other hand, the pressing research goals include the elucidation of the question of whether CO_2 deposition *at sea* is feasible in principle, or reasonable, in view of achievable storage times for CO_2. This involves firstly the diffuse injection of CO_2 into the ocean, and secondly, the concept of deposition of liquid CO_2 in sea-bottom sinks (at depths of 3000–4000 m); both must be intensively researched. Here, among other questions, it needs to be clarified whether the probably required cover layer (a so-called physical barrier) can in fact be constructed and whether it could be maintained over the long term. (Some notes on the subject of CO_2 deposition at sea are given below).

We cannot go into detail here concerning these aspects. Regarding deposition at sea, we however note the following: every method must guarantee not only the transitory retention of CO_2, but rather retention over a *very* long time: not just a century, but many centuries. (For this reason, for e.g., deposition of liquid CO_2 on the ocean floor, a covering layer seems indispensable.) Otherwise, a major part of the enormous amount of CO_2 stored could once again enter the atmosphere, for example, within a half-century or a century, and could no longer be recaptured, even though it previously (for climate protection) had to be sequestered "at all costs" and with great efforts (cf. e.g., WBGU, 2006). The climate would then be completely and irreversibly damaged.

Regarding the diffuse injection of CO_2 into the oceans:

In order to assess the question of whether the quantities of CO_2 dissolved in deep layers of water would remain there over long periods of time, one requires knowledge of the currents prevailing at those depths and locations. Thus far, only sparse information has been available about these currents; at least until a few years ago. This question could possibly be elucidated in the framework of the international ocean-current measurement program "Argo," which has been underway for several years, using so-called free-drifting profiling floats (cf. Argo, 2009; IFM, 2009; KDM, 2009). To this end, the program may need to be expanded. This should in any case be a part of the required large-scale energy research program. Around 3000 of these floats are in use today for investigations of the ocean currents; so far, however, only down to depths of ca. 2000 m. The floats sink to this depth and remain there with the aid of an automatic control system until they resurface after a certain time; they then report their new position by telemetry (including also data on measurements of the water temperature and salt content). After a series of such diving cycles, they are collected by a ship and taken to a new location. These systematic position determinations allow the path of each float at the preset depth to be reconstructed. In the case that the data obtained from this program for a water depth of 2000 m do not permit conclusions to be drawn about the currents near the ocean floor, the program

11.3 Coal as the Only Major Alternative to Oil and Gas?

would have to be extended to include floats for greater water depths. (Possibly, the proximity of the rough ocean floor would have to be taken into account, perhaps by including additional sensors in the floats which would maintain a constant distance from the float to the ocean bottom.)

Regarding the deposition of liquid CO_2 on the deep ocean floor:

Considering the required size of depots for liquid CO_2 on the ocean floor, we have to keep in mind the quantities of CO_2 which would need to be dealt with. The total worldwide CO_2 emissions from fossil-fuel power plants of all kinds in the year 2002 amounted to 11 billion tons of CO_2; in the USA, they were 2.4 (2005), in Europe 1.5, and in Germany 0.35 billion t CO_2 – compare Appendix C. The density of liquid CO_2 at a water depth of 4000 m corresponds nearly to that of water (it is only slightly heavier), so that 1 billion t of CO_2 just fills a volume of $1 km^3$. If, for example, the total CO_2 emissions of today's American power plants were to be deposited, this would require a depot volume of $2.4 km^3$ annually. The amount of CO_2 for the case of substitution by coal gas of one-third of the gas and oil consumed currently, and keeping the amount of electric-power generation in the USA constant, would increase to ca. 6 billion t annually. If all of it were to be stored on the ocean bottom, then $6 km^3$ of liquid CO_2 per year would have to be deposited. If furthermore the expected worldwide increase in electric-power demand were to be supplied by coal-fired plants, and additionally only 20% of today's oil and gas consumption were substituted by coal gas, then each year, 24 billion t of CO_2 would have to be deposited; in Europe (EU-25) – at the current rate of electric-power generation from coal and with 20% substitution of oil and gas – the amount would be 2.4 billion t, and thus only 10% of the world's total.

If we assume that CO_2 will be stored in depressions on the ocean floor with an *average* height of 0.2 km, this yields per km^3 a surface area of $5 km^2$ for the depots. (The "average" height depends of course on the shape of the depression; namely whether it is shaped like a crater (a flat bottom with steep walls) or like a cone. An average height of 0.2 km could thus require a maximum height of, e.g., 0.5 km, but possibly considerably more.) If, for example, the total CO_2 from currently operating coal-fired power plants in Europe and from future coal gasification (20% oil and gas substitution; $2.4 km^3$ per year) were to be stored in a single depot with this average height, this would require an area of $12 km^2$ on the ocean floor for each year. If this depot was required to hold the emissions from 30 years, the total area required would be 360, i.e., around $400 km^2$ (at 0.2 km height). The depot would thus have a volume corresponding to a square area of 20 by 20 km (0.2 km high). The amount of emissions worldwide would be ca. 10 times greater. These depots would have to lie at a depth of about 4000 m and to be covered with a protective layer, whose functionality would need to be guaranteed over many centuries. (Furthermore, the problem of possible expansion of the liquid CO_2 deposits due to the formation of CO_2 hydrates – and thus the potential instability of the depot—has apparently not yet been definitively clarified (Ploetz, 2003, p. 15).)

All of the necessary investigations should in any case be pursued emphatically. Thus far, the energy suppliers who support the construction of new coal-fired power plants have often touted future CO_2 sequestration as a real technical possibility on a large scale. This is correct, insofar as there are initially sufficient depot sites (on land), that is, for the first plants with sequestration. But if coal is to be utilized on a grand scale, this is no longer true. In that case—which we are considering here—fundamental questions still remain unanswered: The potential capacity of sites on land and their costs for storage of large amounts of CO_2 (e.g., when the distances from the depot sites to the power plants are great), and as an alternative to underground deposition, the basic questions regarding deposition in the oceans. This applies, as mentioned, not only to future coal-fired power plants, but also to coal gasification plants.

The necessary financial support for a more reliable determination of the supplies of coal available in the future, and for the much more involved studies of the potential feasibility and the costs of CO_2 sequestration, should be made available immediately. Only when the results are at hand will it be possible to make a conclusive comparison of various large-scale energy options. Such a comparison is however a necessary precondition for the decisions which must be taken immediately, or in any case very soon, concerning future energy supplies. Therefore, this research and development work should be carried out with a similar priority and on a similar time scale to that needed for solar energy. It is possible that we could thereby arrive at a reliable judgment of the large-scale coal option within ca. 4 years, and thus within the same time that will likely be required to clarify the fundamental questions regarding solar thermal power plants, if the work is begun soon and carried out rapidly; at least our knowledge of the probable *costs* of CO_2 sequestration would then be much more secure. It might become clear, as already suggested, that in certain regions of the world such as Europe, possibly also China, the country with the largest coal resources, the costs of CO_2 sequestration would be higher than assumed as an example in Table 4.3 (10 \$/t of CO_2); or also that the potential for economically feasible sequestration (and not the coal reserves) will prove to be the limiting factor for the future utilization of coal there.

11.4
Solar Hydrogen

11.4.1
Hydrogen Production from Electrolysis

In Section 4.1, we have already given a brief summary of hydrogen production by electrolytic decomposition of water. There, regarding a future electrolysis technology, which is still to be developed, that is, *high-temperature steam electrolysis*, we quoted an efficiency of 86% (LHV) and investment costs of *500 million* $/GW$_{H_2}$. This technology has also been described in several studies as a future possible variant for nuclear hydrogen production (NAE BEES, 2004, pp. 95 and 211; Verfondern, 2007, pp. 108–114), and it is under investigation in current projects (e.g., Hi2H2, 2005); in addition, the intensive development of similar designs for high-temperature fuel cells (HTFC) is also noteworthy. Present-day electrolysis installations have efficiencies of around 65%; it is possible that these low-temperature processes will achieve 75% (LHV) in the future (Nitsch, 2002, p. 8). A genuinely high efficiency, which could even be somewhat higher than assumed above, is however promised only by the newer technologies.

In recent years, there has been much talk of a hydrogen economy, and large sums have been expended on development projects. The most important aspect – in terms of renewable hydrogen – has however been practically neglected: the development of an efficient process for H_2 production. This is as if one would leave out the engine in a development project for a new type of automobile. One could even say: either it will be possible to develop an efficient electrolysis technology, or else hydrogen from renewable sources will have a hard time asserting itself against fossil fuels, even against hydrogen from relatively expensive imported coal.[65] (This holds however only for gas production; for the production of liquid fuels from solar hydrogen and coal, the situation is generally more favorable.)

Taking the cost of solar electrical energy at the plant to be 3.3 ¢/kWh$_{el}$ (Section 4.1), hydrogen – with the above values of efficiency and investment costs (including transport) – would cost 4.7 ¢/kWh$_{H_2}$ (LHV) (cf. Table 4.2). It would thus still be within a relevant cost range in comparison to hydrogen from coal gasification. (The latter, using imported coal, would cost 3.4 ¢/kWh$_{H_2}$ (LHV) (Section 4.1); of this, 0.5 ¢/kWh$_{H_2}$ is due to CO_2 storage.) At an efficiency of 75% for the electrolysis, the cost of solar hydrogen would rise to 5.3 ¢/kWh (LHV) (and with 70% efficiency, it would cost 5.7 ¢/kWh). With the assumed efficiency of 86%, for the production of 1 GW of hydrogen, a solar power-plant capacity of 1.16 GW$_{el}$ would be necessary; with 75% efficiency, 1.33 GW$_{el}$ would be required, and with 70%, 1.43 GW$_{el}$. The investment costs for these solar power plants ($3695 million per GW$_{el}$; see Table 2.1) would amount to $4520 million for 1.16 GW$_{el}$, but however $5180 million for 1.33 GW$_{el}$ (and $5550 million for 1.43 GW$_{el}$). Per GW$_{H_2}$, the increase in size of the power plants (for 75% instead of 86% efficiency) would thus require additional investments of $0.66 billion, which – referred to a fictitious overall capacity of, for example, 1000 GW for the hydrogen production installations[66] – would mean additional costs of $660 billion. This illustrates how important the efficiency of the electrolysis is, in particular, in connection with the question of which expenditures are justified for research and development.

---

65) The important option of the so-called thermochemical cyclic processes for hydrogen production cannot be treated within the framework of the present book; but see the note at the end of Section 4.1.

66) At an average capacity factor of 80%, this corresponds to a yearly H_2 production of 800 GW$_{H_2}$. This is not even 10% of present-day oil and gas consumption (2004: oil 5000 GW, gas 3200 GW).

Insofar as solar power can be generated at a cost within the range quoted above, the development of this electrolysis technology is thus of prime importance for a future economical H_2 production.[67]

We make the following assumptions concerning high-temperature electrolysis:

- It would be autothermal electrolysis, that is without supplying high-temperature heat, which would have to be at a temperature of 800–900 °C. This heat supply would be possible in principle and would further reduce the electric power requirements; theoretically, only 2.25 kWh of electrical energy per Nm^3 H_2 would then be needed. This however appears not to be relevant for practical reasons; instead, the power density of the installation will be increased to the point that resistive losses supply the heat required for the process (autothermal operation). Given the high current density, conditions are then favorable for lowering the investment costs of the installation. This operation mode corresponds to a theoretical electrical-energy consumption of 3.05 kWh/Nm^3.

Owing to heat losses which likewise must be replaced by electrical energy, the overall energy consumption would be somewhat higher than this theoretical value:

- Insulation losses, that is, radiation and other heat loss mechanisms; assumed additional electrical energy consumption: 2%.

- Losses during heat exchange, that is, incomplete heat exchange between the hot products (H_2 and leftover water vapor, O_2) and the media which have to be heated (steam plus a certain quantity of H_2): assumption of an additional energy consumption of 4%.[68]

For large-scale electrolysis processes, in terms of insulation and heat exchange we can safely assume a relatively high efficiency. With these assumptions, we find an electrical-energy consumption of 3.23 kWh/Nm^3 H_2 (3.05 × 1.06) and an electrical efficiency (without the energy required to vaporize the water) of 92% (LHV), which agrees in tendency with the values to be found in the literature.[69]

---

67) This situation would be different only if down-draft power plants prove to be feasible (cf. Section 3.5). In that case, there would seem to be some perspective in principle of achieving an electrical energy cost of only 2¢/kWh$_{el}$; their somewhat lower capacity factor is furthermore less of a disadvantage for hydrogen production (than for electric power generation).

68) The heat-exchange problem is discussed, for example, by Verfondern (2007, pp. 112 and 119) (but however no values for the resulting losses are given). According to this source, efficient utilization of the heat from the O_2 flow seems to be rather difficult (probably due to safety considerations concerning the nuclear power plants discussed there). In a solar power plant, for heating the electrolysis media (to temperatures of over 500 °C), solar heat at a molten-salt temperature of ca. 560 °C could however be utilized. (In *principle*, with solar power plants even higher temperatures could be reached. We shall however not discuss this point further here, since it would require high-temperature thermal circuits and high-temperature heat-storage systems, which have not reached a mature stage of development at present.)

69) 2.995 kWh (LHV): 2.995 kWh/3.23 kWh$_{el}$ = 0.927, that is, ca. 92% (LHV).

The energy losses mentioned (insulation and heat exchange), which yield an efficiency of 92%, have only been estimated here. Compare literature values: Nitsch,

11.4 Solar Hydrogen

As we shall show in the following, vaporization of water requires an equivalent electrical energy of ca. 6%. The overall efficiency including steam generation is then found to be 86%.

The electrical-energy equivalent of steam generation:

In the electrolysis of liquid water, the heat of vaporization of the water must be supplied as electrical energy. In the electrolysis of steam, the phase transition is effected using low-temperature heat. At ambient pressure (100 °C, vapor pressure 1 bar), vaporization of water requires a quantity of heat equal to 0.60 kWh/Nm^3 of H$_2$ (this includes the energy required to heat the water to its boiling point).[70] To limit the energy required for pressurizing the H$_2$, however, the electrolysis is carried out under pressure, at ca. 30 bar. This requires a vaporization temperature of 235 °C. The heat required is then 0.63 kWh/Nm^3 and corresponds to *21.0%* of the lower heating value (LHV) of hydrogen.

Inputting this energy in the form of heat (instead of electrical energy) allows

1) the avoidance of conversion losses in generating electrical energy from heat in the power plant, and;

2) permits this heat to be coupled out of the steam circuit at a relatively low temperature, after the steam has already performed a portion of its useful work.

The steam-circuit part of the solar power plant has an efficiency of around 40% (Kolb, 1996a): gross 41.9%, net ca. 39% – without the electrical energy consumed by the solar field. The energy required for the vaporization, corresponding to 21% (referred to the LHV of the H$_2$ produced), thus corresponds to an electrical-energy equivalent of 8.4% (21% × 0.4).

2002, p. 8: 3.2 kWh$_{el}$/Nm^3 = 93.6% (LHV). Enquete Commission, 1990, p. 709: 3.2 kWh$_{el}$/Nm^3 "... makes autonomous operation possible by heating the low—temperature steam from the electrical losses." (The energy consumption due to heating the media is thus contained in this value). Hi2H2, 2004: "The electrical efficiency demonstrated in the Hot Elly electrolyzer was close to 92%" (possibly including steam generation, but however at 1 bar). NAE BEES (2004, p. 222): "Efficiencies moving toward 95%" (probably without steam generation); but on p. 211: "85–90%" (probably with steam generation).

70) The energy required for vaporization and heating of the water up to its vaporization temperature: At 100 °C: 2676 kJ/kg H$_2$O, at 235 °C: 2802 kJ/kg H$_2$O.

1 kg water = 0.111 kg H$_2$. The density of H$_2$ is 0.090 kg/Nm^3.

Vaporization of water plus heating the steam to 1000 °C requires 0.99 kWh/Nm^3 of H$_2$; the heat content of the gases produced (H$_2$ + ½ O$_2$ at 1000 °C) is 0.59 kWh/Nm^3 of H$_2$ (from "Auf dem Wege zu neuen Energiesystemen," Part 3: Hydrogen, 1975, p. 19.)

> **Hydrogen**
>
> Higher heating value (HHV): $3.542 = 3.54\ kWh/Nm^3$* ($39.38\ kWh/kg$)
> Lower heating value (LHV): $2.995 = 3.00\ kWh/Nm^3$* ($33.32\ kWh/kg$)
>
> $$HHV = LHV + 18.3\%$$
>
> $$LHV = HHV - 15.4\%$$
>
> * HHV includes, and LHV does not include the condensation energy of the water vapor which is formed on combustion.
> Theoretical energy requirement for the electrolysis of water at 0 °C: $3.54\ kWh_{el}/Nm^3$ (corresponds to an efficiency for HHV of 100% or for *LHV* of 84.6%).
> Theoretical energy requirement for the electrolysis of water vapor at 900 °C, 1 bar (cf. Verfondern, 2007, p. 108) per Nm^3 of H_2:
>
> $2.25\ kWh_{el} + 0.8\ kWh$ of HT heat (at 900 °C) $+ 0.60\ kWh$ of LT heat (at 100 °C)
>
> Compare autothermal electrolysis: $3.05\ kWh_{el} + 0.60\ kWh$ of LT heat (at 100 °C)
>
> (Autothermal: without coupling in high-temperature (HT) heat; instead, this heat is produced electrically: $2.25 + 0.8 = 3.05\ kWh_{el}$.)

This heat is coupled out of the medium-pressure turbine at 235 °C. If we make the rough assumption that the steam in the power-plant circuit at this temperature has already generated one-third of the electrical energy which it will yield on being cooled to its final condensation temperature, then the electrical-energy equivalent is reduced by a further amount of one-third to 5.5%, i.e., to around 6%. The *thermal* power of the solar plant must therefore be increased by ca. 6%.[71]

In the electrolysis of *water* (at 0 °C, see info-box), the theoretical efficiency (LHV) is 84.6%. In practice, one will thus not be able to attain much more than the hoped-for 75%.

The high temperature of vapor-phase electrolysis (800–900 °C) not only allows an efficiency near to the optimum to be achieved (nearly 100% LHV), but also favors a high current density in the electrolysis cells (without using catalysts). This leads in tendency to lower investment costs.

The specific *investment costs* are assumed here to be 500 \$/kWh_{H_2} (LHV).[72] Note: These refer to an operating lifetime of the installation of 45 years, as generally

---

71) The electric power generation in the power plant remains unchanged, but tapping off the steam requires an increase in heat input by 6%, which would correspond to a 6% increase in electrical energy output (electrical-energy equivalent).

72) Czisch (2005) gives an overview of the specific investment costs cited in the literature for the various electrolysis processes considered to date (not high-temperature installations). They depend upon the size of the installation, the production rate assumed, the operating

assumed in this book. With shorter operating lifetimes, the investment costs would have to be correspondingly reduced.[73]

Given their high operating temperatures, frequent *switching on and off* of the installations, as would usually be expected for electric-power generation in a solar base-load power plant due to cloudy days, would have to be avoided. Therefore, either the electrolysis installations would be operated at minimum power even in the event of a complete stoppage of power generation in the local solar-power region (e.g., at 5% of their nominal power), or else the heat losses through the insulation would have to be compensated by an additional electrical heater. If no fossil-fuel backup power plants were located in the solar park, the required minimal amount of electric power could be transported back over the long-distance transmission line from the distant consumer region where it would be generated by the backup plants. (The power inverters for converting alternating current to direct current can always also be operated in reverse.)

The aforementioned advantages of HT-electrolysis are in opposition to the disadvantages of the high operating temperatures. A decisive precondition is the development of suitable materials. Just how great the probability is that this task can be successfully performed is difficult to estimate. To meliorate the materials and lifetime problems, a major goal of current development programs is the reduction of the operating temperature to ca. 600–800 °C. This should be possible by going to thin-film electrolytes, which allow high current densities even at lower temperatures. This development is in progress at least for the similarly constructed solid oxide fuel cells (SOFC) (cf. IEF-Jülich, 2007a, 2007b; see also Hauch *et al.*, 2007).

The development is on the whole complex and encompasses many paths (IEF-Jülich, 2007a): Depending on the temperature aimed at, different materials must

pressure, and – as estimates for the future – on the time of commissioning or the stage of development of the installation. Thus, on p. 142, he states: "For 30 bar high-pressure electrolysis apparatus, with production rates of ca. 3 MW$_{H_2}$, and for future mass production with over 1000 plants constructed for hydrogen service stations, (BWWZ01/Ludwig-Bölkow-Systemtechnik) we give a value of 400 €/kW$_{el}$" (Literature reference: Bussmann, Weindorf, Wurster, Zittel: *Geothermal Hydrogen – A Vision?*; Paper presented at the European Geothermal Energy Council's 2nd Business Seminar. L-B-Systemtechnik, Ottobrunn, 2001. cf. http://www.hyweb.de/Wissen/pdf/geoh22001.pdf). Regarding present-day installations, Czisch writes: "Norsk Hydro Electrolyzers quoted a price for their electrolysis apparatus with outputs of ca. 1.5 MW$_{H_2}$ and an efficiency of nearly 70% (Note: probably HHV) in 2004 of 1200 €/kW$_{H_2}$." While current installations (at low deployment rates) are thus still more expensive than assumed above, for future mass-produced series, lower costs are to be expected. These lower costs are also assumed here for the high-temperature installations, with very large-scale deployment (but referred to the relatively long operating lifetime assumed here; see footnote 73).

73) The shorter lifetimes which are to be expected apply most probably only to the electrolysis cells themselves, but not to the ancillary equipment such as power inverters, steam generators, or housings. At an operating lifetime of 20 years for the electrolysis cells, the equivalent investment costs could be, for example, 200 to 300 $/kW of H$_2$; that is, investments which correspond to 500 $/kW of H$_2$ referred to 45 years operating life for the overall installation.

be developed; there are several versions of the design of the cells (self-supported electrolyte, anode-supported); in particular, alongside the materials development, a high-quality fabrication is important, since defects in the electrolyte with the resulting gas penetration must be avoided. Different processes for fabrication and processing must be developed or perfected. It can be assumed that not only for materials development, but also in manufacturing processes different routes will be possible, each of which has to be investigated in terms of its feasibility. For the manufacturing procedures, this extends to the development of automatic fabrication techniques.

This spectrum of problems that ranges from basic research (materials development) to manufacturing technology naturally requires a particularly strongly interdisciplinary research organization (IEF-Jülich, 2007a). This in turn makes a particular development program with a suitable organization necessary. The situation is basically similar to that encountered in solar and coal research, as emphasized above. A tailor-made organization must guarantee that all of the new development tasks which arise, and which can be very diverse, are taken up quickly and effectively. In this field, one cannot count on rapid progress as in the case of solar power plants; not only because many development routes must be followed in parallel, but because even in the details, it will often be necessary to traverse new territory. If the work is begun soon, it is possible that within 10 or 15 years, tested and construction-ready plants of this type will be available (insofar as the overall development proves *in fact* successful). Within certain limits, the time required for such a program still depends on the intensity with which the research is carried out. If this research is classed as a key technology and in a correspondingly planned and broadly dimensioned special program *all* the possibilities for accelerating the development are utilized, it is possible that prototypes will be ready for construction even earlier.

In any case, the idea that one could essentially accelerate such a development with a 1.1 million-€ program (EU Project: Hi2H2, 2004) is at best naive. At the IEF in Jülich alone, 85 full-time staff members are involved in the development of the SOFC.[74] Whether for the development of electrolysis technology overall, for example, $100 million or possibly $500 million will be required (or – in connection with a real "crash program" – even more), is in the final analysis a secondary consideration; since here, on the one hand, we are dealing with the key technology for a long-term hydrogen economy; and on the other, the technology will possibly have an enormous significance in the short- to middle-term in connection with methanol production on a large scale from solar hydrogen and coal (sun methanol) as an alternative to increasingly expensive petroleum.

The cost of hydrogen could furthermore certainly be lower than estimated in Section 4.1 (Table 4.2; 4.7 ¢/kW$_{H_2}$) based on the solar power-plant costs assumed there. It is conceivable that the cost including transport could be as low as for example, 4 ¢/kWh$_{H_2}$, at least for the case of the use of hydrogen in methanol

---

74) The IEF has one of the largest research groups for solid oxide fuel cells worldwide. It is at the forefront of the development of fuel-cell materials and components as well as in the construction of installations. (IEF-Jülich, 2007a).

production. (If fundamentally different technologies should prove feasible, even lower costs could result.[75]) This value must be compared with the costs of coal gasification: 3.4 ¢/kWh$_{H_2}$ (Europe, imported coal) or 2.5 ¢/kWh$_{H_2}$ (USA).

Some reasons for possibly lower costs:

a) Taking full advantage of the *innovation potential* of solar power plants. This was discussed in Chapter 2. In the case that the cost framework cited above is basically confirmed in the course of further development – whereby making full use of the innovation potential was not, or only partially, taken into account – then the resulting power cost could in the end be lower than predicted above.

b) The *thermal losses* of high-temperature electrolysis cells will presumably be lower than assumed above, in particular for very large installations; owing to improved insulation and more efficient heat exchange (especially if the heat content of the oxygen could also be utilized in heat exchangers).

c) These *thermal losses* in the electrolysis cells (incomplete heat exchange) could be compensated to a great extent using solar heat instead of electrical energy (higher efficiency). The high-temperature thermal circuits required to achieve this have already been mentioned.Concerning the utilization of hydrogen for methanol production from coal, two additional facts should be noted:

d) The distances for transport to the coal fields in the USA, where the methanol production plants would be located, are considerably shorter than assumed in Section 4.1, there, e.g., for gas provision to countries in Europe or to the East Coast of the USA (3300 km including crossing the Mediterranean Sea near Sicily). Figure 11.4 shows the coal-producing regions in the USA. At an assumed average distance of only 1500 km, the cost of H$_2$ would decrease by 0.3 ¢/kWh. (The number of intermediate compressor stations would be reduced (power consumption and investment costs), while the primary compressor at the production plant would remain unchanged.)

e) With the application of solar power to hydrogen production, a portion of the unused solar heat from the power plants, assumed to be wasted

---

[75] If downdraft power plants can indeed be constructed at the cost given in Chapter 3 (resulting in electrical-energy costs of down to 2 ¢/kWh$_{el}$), then hydrogen could be supplied from, for example, Egypt, Mauritania, or Senegal, possibly for around 3 ¢/kWh$_{H_2}$. In the USA, the potential for down-draft plants is however more limited (cf. Figure 5.5). High-temperature electrolysis furthermore has some importance for the case that wind power in the North Sea will be able to supply electrical energy at a lower cost than solar-energy plants. From the European viewpoint, this would represent Europe's *own* energy source, while solar hydrogen in Europe raises the problem of dependence on the North African countries where the power plants have to be located.

11 Other Technologies for Backup Power Generation and Alternatives for Future Energy Supplies

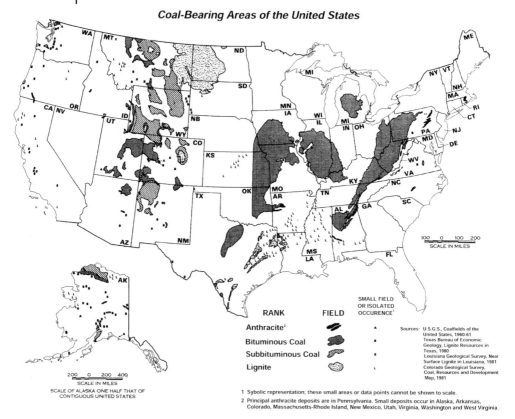

Figure 11.4 Coal bearing areas in the USA (Energy Information Administration).

due to the necessary overdimensioning when the plants are used for electric-power generation only, could be utilized. The resulting cost reduction can at present be only roughly estimated, since data for the annual trends in the USA are available for two years and for one site (Barstow) only: for the "poor year" 1984 and for the "good year" 1976; cf. Chapter 5. If we use these data as a guide (average values), the analysis shows that 9% of the solar heat is wasted. The solar power plants are assumed to have a capacity of a maximum of $8.0\,\text{kWh/m}^2\,\text{d}$ solar heat in their steam generators, while the insolation on peak days attains a value of more than $11\,\text{kWh/m}^2$. If instead, $9.6\,\text{kWh/m}^2\,\text{d}$ could be utilized, the power generated (in years corresponding to the average of these two reference years) would increase by 7.3%. If the heat is used only to generate *electrical energy*, the limitation to $8.0\,\text{kWh/m}^2\,\text{d}$ is reasonable, since otherwise not only the heat-storage reservoirs and the steam circuits of the power plants, but also the expensive power transmission lines and the backup power plants would have to be made larger. If, however, *hydrogen production* at a short distance from the power plants is envisaged, then only the heat reservoirs and the steam circuits would need to be larger; in addition, the electrolysis installations and the hydrogen pipeline would need to be correspondingly dimensioned. (The electrolysis installation could in principle even be operated in an overload mode, i.e., without being enlarged; this would be accompanied by a somewhat reduced overall efficiency). If the components mentioned were dimensioned for a heat input of 9.6 kWh/m² d (i.e., 20% larger than for electric power generation only; but for the electrolysis installation, only by an estimated 10%), then the total investment costs would be increased by 5.1%. The hydrogen production would however increase by 7.3%, according to the above simple calculation. A net advantage (cost reduction) of 2.2% would result.

Conclusion: If H_2 were to be used for methanol production, the cost would be reduced already by the shorter transport distances from 4.7 to 4.4 ¢/kWh of H_2. If in addition the power costs were to fall by only 10% due to the innovation potential of the solar plants and for the other reasons mentioned above, this would yield a final hydrogen cost of *4.0 ¢/kWh* of H_2 (LHV).

11.4.2
Transporting Hydrogen

The investment costs and losses due to long-distance transport of H_2, already mentioned in Section 4.1, will be discussed in more detail in this section. The data are based on the report of the Enquete Commission (1990, p. 727).[76] There, the following investment costs are cited for a gas pipeline of the assumed length of 3300 km (diameter[77]: 1.7 m, pressure: 100 bar) with a transport capacity of 25 GW_{H_2} (LHV)[78]: For the piping (3100 km on land, ca. 200 km undersea from Tunisia to Sicily), 16 000 million DM (1989) (i.e., 11 520 million in 2002-$);[79] for the compressor stations, 1950 million DM ($1405 million). The mechanical compressor power is given as 975 MW.[80] (Values for a pipeline of 2000 km length and

---

76) Since in the major American hydrogen study (NAE BEES, 2004), and also in the rest of the available literature, no numerical values are given for this topic, we make use of this older source.

77) In 1990, the diameter of the largest gas pipelines was 1.4 m (for a transported power of 12.5 GW of H_2). As needed, larger pipelines can however be constructed (of diameters 1.7 to 2.0 m). Thus, there on p. 727, the following diameters were assumed: for the "technology status" 1990 to 2005, 1.4 m; for 2005 to 2025, 1.7 m; and after 2025, 2.0 m.

78) This corresponds to 8.33 million Nm³/h or 29.6 GW of H_2 (HHV).

79) Conversion of 1989-DM into 2002-US $ using the factor 0.72 (inflation in Germany 1989–2002: 1.35; 1 € = 1.96 DM; 1 $ = 0.96 € (2002 purchasing-power parity)).

80) Entrance pressure at the first station: 30 bar (compression to 100 bar). The intermediate compressors raise the pressure from ca. 83 bar to 100 bar (pressure ratio 1.2, p. 725 of that reference).

conversion to other lengths are given in the footnote.[81]) The specific (with respect to the transport capacity) investment costs for this pipeline amount to *460 million $/GW* (for the undersea portion ca. 110 million $/GW), and for the compressor stations, *57 million $/GW*; all together, they are thus *517 million $/GW$_{H_2}$* (LHV).

The (mechanical) compressor power would be 40 MW/GW$_{H_2}$ (LHV), that is, *4%* of the overall H$_2$ power transported. With electrical operation of the compressors at a conservatively assumed efficiency for power transmission plus motor of 80%, the electric power (at the solar plant) required for the operation of the compressors would correspond to 50 MW$_{el}$/GW$_{H_2}$ or to 5.0% of the H$_2$ power transported. The transport efficiency would thus be *95%*.

Remarks on the electrical operation of the compressors: Here, we assume electrical operation of all the compressors, not only for the primary compression (at the electrolysis installation), but also at the intermediate compressor stations along the pipeline. This is possible if the consumer region is to be supplied with both base-load electric power and solar hydrogen. Then, it can be assumed that the hydrogen pipelines would be laid more or less parallel to the power transmission lines and would pass through countries or regions which were also supplied with base-load solar power. (An exception would probably be the sections of the pipeline passing through the Sahara.) The intermediate compressors could then be provided with solar electric power from the local public grid. Without a parallel solar-power supply in the regions through which the pipeline passes, providing the compressors with solar electric power would be difficult.[82]

---

81) For the 25-GW pipeline of 2000 km length from Spain to Germany also described in that study, the corresponding numbers per GW of H$_2$ (LHV) are: pipeline, $225 million (2002); compressor stations, $42 million (2002); mechanical compressor power: 29 MW.

With these data, only a rough extrapolation to other lengths is possible. To carry out a precise computation, it would be necessary to differentiate between primary and intermediate compressors and to consider separately the undersea portion of the pipeline, from Tunisia to Sicily. Comparing the data cited for the 2000-km pipeline (with only 2000 km for intermediate compression, but the same primary compression and without an undersea section), we can however arrive at the following rough conclusions:

a) The mechanical compressor power of the 3300-km pipeline (40 MW/GW) should be divided up approximately as follows: primary station ca. 13 MW, intermediate compressor stations ca. 27 MW.

b) The investment cost of the undersea pipeline is per km ca. *five times* higher than for the pipeline over land (cf. also p. 726); its losses are however roughly similar.

For H$_2$ production with *nuclear power plants*, a transport distance of 1000 km was assumed. For this case, the following values per GW of H$_2$ (LHV) (in 2002-$) are obtained: pipeline, $112 million; compressor stations, $30 million; all together *$142 million*. The mechanical compressor power would be 20.7 MW = 2.1%. With power-transmission and motor losses totaling 20%, we find a power requirement for compression corresponding to *2.6%*.

82) For natural-gas pipelines, the turbo compressors are usually driven by gas turbines (or by a combination of gas and steam turbines), which are supplied with natural gas fuel directly from the pipelines; in the past, the gas was very reasonably priced. (One could speak of "gas consumption for compression.") This concept was adopted in previous studies of

11.4 Solar Hydrogen

As already mentioned, the power-transmission losses over 3000 km with 800-kV DC transmission lines would amount to 11.5%, whereby we have assumed that all of the power would be transported to the end of the transmission lines. Some portion of the compressor stations would however be located near to the electrolysis installations, and the primary compression should consume a fraction of at least 30% of the total compressor power for the 3300 km long pipeline. The remaining compressor stations would be spaced at distances of several hundred kilometers along the pipeline, so that the power supplied to them would not have to be transmitted over the entire length, but on average over only half the distance (1650 instead of 3300 km). The losses due to power transmission would thus be reduced. The efficiency of electrical machines (generators and motors) in the power output class of a few 10 MW_{el}, typical of compressor stations (Enquete Commission, 1990, p. 726), is in the range of 95%, and the transformer losses are less than 2%. The overall losses (due to power transmission, motor, and transformer) would then be slightly more than 10%. For simplicity, we use the conservative value of 20%, that is, we assume that the electrical drive for the compressors has an overall efficiency of 80% (mechanical compressor power/electric power at the solar plant). In that study (Enquete Commission, 1990), operation of the compressors by gas turbines is assumed. As a back calculation shows, the efficiency of the gas turbines there was taken to be 35%, so that their "gas consumption for compression" (hydrogen) would correspond to 11.5% (p. 727 of that reference).

a future hydrogen supply. If solar power is generated by photovoltaic elements, as assumed by earlier solar-hydrogen studies, this construction is justified. Since photovoltaic plants generate power only during the daytime, they could not deliver base-load power for the uninterrupted operation of the compressors. A similar conclusion holds for concepts which in the past assumed isolated solar-hydrogen production, although utilizing solar thermal power plants. (Thus also in the report of the Enquete Commission (1990), operation of the compressors by gas turbines fueled from the H_2 pipeline was presumed; cf. p. 723. "The currently established method ..."). Without an accompanying provision of solar electric power to the northern countries, however, a small dedicated power transmission line would have to be constructed for operation of the compressors along the pipeline. If, in contrast (as in USA or in the case of a pipeline from North Africa to Europe), the hydrogen pipeline passes through countries or regions which are in any case provided with a solar electric-power supply, then the power could be taken from the local grid. Tapping off small amounts of power (e.g., for the compressors) from a high-voltage direct current transmission line, however, would require a considerable technical effort, since small inverter stations for converting the DC to AC have a much higher specific cost than large-scale inverters of the GW class; cf. Kanngießer (1998). In North Africa, at least the coastal region would have a supply of solar power (insofar as this energy concept is implemented); the same would hold for the European countries through which the H_2 pipeline would pass. Only in the section within the Sahara would the power for the first or second compressor station probably have to be supplied by its own power line. (There are however first indications of the development of methods for tapping off power from long-distance transmission lines which have some promise of achieving a more economical solution; cf. Kanngießer, 1998.)

Along the undersea portion of a proposed pipeline from Tunisia to Sicily, the *water depth* is less than 500 m. The question as to whether or not still greater ocean depths can be traversed with gas pipelines economically – they must withstand the outer hydrostatic pressure when their internal pressure is 1 bar – and if so, at which costs, still needs investigation. Thus for example, the 230 km long direct sea passage from the west border of Algeria to Spain is less than 1000 m deep. The 200 km long route from Tunisia to Sardinia lies at a depth range of 1000–1500 m.

Assuming that *superconducting power-transmission technology* progresses as hoped, it could furthermore be more favorable in future to transmit electrical energy, for example, from Africa to Europe and to generate the hydrogen there. The investment cost for the hydrogen pipeline (517 million $/GW$_{H_2}$ for 3300 km, as given above) corresponds rather closely to the current cost of an 800-kV overland transmission line (665 million $/GW$_{el}$ for 3000 km). The loss rate of the gas pipeline, at 5%, is however considerably lower than that of a conventional power line (11.5%). Using superconducting transmission, the investment costs and losses could possibly be still lower than those for a gas pipeline. Furthermore, with undersea cables, deep sections of ocean can be more readily crossed. Thus, possibly Egypt, which has favorable insolation (and is not an OPEC member) would come under consideration as a solar location (undersea transmission route to Greece). The decisive advantage of this concept would however be realized only if in addition to solar power also wind power (e.g., from the North Sea) were used to produce hydrogen. The electrolysis installation could then be operated with both primary energy sources, which have opposite seasonal tendencies. This would yield a higher capacity factor and more uniform H_2 production, and thus a reduced storage requirement. The electrolysis installations could also then be operated above their nominal capacities, which would not be possible with a hydrogen pipeline (fixed transport capacity). High-temperature vapor-phase electrolysis however presupposes that a certain portion of the input energy be in the form of heat (for steam generation). The operation of a large electrolysis complex in Europe would therefore be expedient only in combination with coal-fired power plants or coal gasification plants there; if the hydrogen were to be used for production of sun methanol, the methanol plant itself would however provide sufficient steam for the electrolysis. This concept should therefore be investigated more carefully within the framework of an energy research program.

11.4.3
Sun Methanol for Around 90 $/Barrel Oil Equivalent – An Effective Brake on the Oil Price. The USA as a Future Sun-Coal-Fuel World Power. "OPIC" as the Answer to OPEC

In the following section, the production of methanol from solar hydrogen and coal gas will be described. In particular, in the USA, with its large regions of high insolation and its considerable coal reserves, this synthetic fuel could be produced on a large scale. Before we consider the costs in detail, we begin with some basic preliminary remarks.

We have already explained in Section 11.2.7 that in the production of syngas (rather than hydrogen), the Shell gasification process offers clear advantages over the Texaco process, which we supposed above to be utilized for H_2 production (Table 11.2).[83] There are, however, no comparably detailed cost data available for the Shell process. Therefore, the following cost estimate is based for simplicity on the Texaco process (estimation of the upper cost limit). However, we adopt the data for the gas composition that have been reported for the Shell process. This allows among other things the requirements for solar hydrogen to be estimated. Regarding the costs, this provisional calculation is thus a mixed computation (Texaco: coal-gas costs; Shell: solar-hydrogen contingent). Regarding the determination of the solar hydrogen consumption and also the coal requirements, basing the calculation on the Texaco process would lead to an incorrect result, in view of the considerable differences between the two processes.

In terms of solar hydrogen, the cost considerations are based on two presuppositions which have already been discussed, and which we recall here: (1) That the cost data for solar power which we have used in this book will be verified in the course of further development; and (2) that an efficient process for producing hydrogen with the aid of high-temperature electrolysis can be developed. Then, a cost of ca. 4.0 ¢/kWh$_{H_2}$ (LHV) would appear to be achievable (where we have assumed among other things that the distance to the coal deposits is only 1500 km). Without the new electrolysis technology, but with advanced conventional technology, one would expect a hydrogen cost about 0.6 ¢/kWh higher.

Syngas[84] from coal gasification serves here as the carbon source for methanol (CH_3OH). In principle, all of the carbon contained in the coal could be transformed into methanol, so that in its production, no CO_2 would be released. In the currently used coal gasification and methanol synthesis, however, a small portion of the gas is diverted to generate energy for internal use in the process, so that correspondingly more coal is consumed and a certain amount of CO_2 must be separated. This additional coal consumption could be avoided for the most part by simultaneously utilizing solar electric power. By adding solar hydrogen as well as solar electric power, the coal requirements could be reduced to a minimum.

For the production of 1 kWh$_{methanol}$ (LHV), ca. 0.65 kWh$_{solar H_2}$ (LHV) plus 0.65 kWh$_{syngas}$ (LHV) from coal gasification are required. This product of solar energy and coal could thus be termed solar-coal methanol, or simply "sun methanol." We make use of this abbreviated notation here.

---

83) Most of the gasification projects being investigated today however utilize the Texaco process, because they are planned for application to CO_2-free IGCC power plants (utilizing H_2 as fuel).

84) Concerning nomenclature: "syngas" was used above as a synonym for medium-BTU gas (the raw product of coal gasification, merely purified of sulfur). We retain this usage in the following. Synthesis gas, which is required for methanol production, with the necessary stoichiometric composition (and also higher purity in comparison to medium-BTU gas), will be denoted by the complete term "synthesis gas."

11.4.3.1 Costs

For the methanol synthesis, an H_2/CO volume ratio of 2.03 is required (Hilsebein, 2009). In the Shell gasification process, the fractions of the combustible part of the gas derived from coal are 30% H_2 and 70% CO. Compare the raw gas composition according to Esquivel (2007) from Illinois #6 coal (the same coal as specified above for coal gasification, cf. Table 11.2): (Vol. %: H_2 27%, CO 63%, CO_2 1.5%, CH_4 0.03%, H_2S 1.3%, COS 0.1%, N_2+Ar (inert gases) 5.2%, HCl 0.03%, H_2O 2.0%). In contrast to hydrogen production from coal, for the manufacture of sun methanol, CO resulting from the gasification need *not* be "converted" to H_2 (using the shift reaction: $CO + H_2O \rightarrow H_2 + CO_2$). The ratio of H_2/CO required for the synthesis is instead adjusted by adding hydrogen externally. Furthermore, a fraction of ca. 3% of CO_2 is required in the synthesis gas (Hilsebein, 2009). It is also converted to methanol following the reaction $CO_2 + 3H_2 \rightarrow CH_3OH + H_2O$.

Methanol production in large-scale plants is characterized by relatively low investments as well as operating and maintenance costs. Its overall cost is therefore determined by the cost of the synthesis gas used.

As mentioned above, and as will be detailed in the following, for the production of $1.0\,kWh_{methanol}$ (LHV), $1.30\,kWh_{synthesis\ gas}$ (LHV) is required; it can be composed, based on the Shell gasification process and the type of coal specified, of $0.65\,kWh_{coal\ gas}$ (LHV) and $0.65\,kWh_{solar\ H_2}$ (LHV).

Materials Properties

H_2: $3.542\,kWh/Nm^3$ (HHV); $2.995\,kWh/Nm^3$ (LHV); density: $0.0899\,kg/Nm^3$
CO: $3.510\,kWh/Nm^3$; density: $1.250\,kg/Nm^3$
Methanol: $6.27\,kWh/kg$ (HHV); $5.42\,kWh/kg$ (LHV); density (at 25 °C) $0.79\,kg/l$.

According to data from the Lurgi corporation, for the production of 1 t of methanol, $2300\,Nm^3$ of synthesis gas (H_2/CO ratio = 2.03) are required (Hilsebein, 2009). (In addition, 1 t of steam is generated.) This gas has an energy content of 7875 kWh (HHV) or of 7060 kWh *(LHV)* – see below. Then the efficiency of the methanol synthesis referred to the lower heating value of methanol and synthesis gas is 5420/7060 = 76.8% (LHV) (and referred to the upper heating values, it is 79.6%). Per *kWh* of methanol (LHV), an amount of synthesis gas corresponding to 1.30 kWh (LHV) is thus required. Since the prices of gas and methanol in this book are given per kWh *(LHV)*, for the *cost estimation*, the efficiency referred to the LHV is relevant. (In contrast, the computations relating to the required amounts of gas are always referred to the higher heating value, i.e., to the total energy of the gas or of the methanol.)

Concerning the energy content of the synthesis gas:

Total volume 2300 Nm^3; of this, 3% is CO_2 (vol%); this yields for H_2 and CO *2231 Nm^3*.

At a H_2/CO ratio of 2.03 (vol), this can be divided up into 1495 Nm^3 of H_2 (with an energy content of 5292 kWh (HHV) or *4477 kWh (LHV)*), and 736 Nm^3 of CO (*2583 kWh*). The energy content of the total amount of gas is thus together 7875 kWh (HHV) or *7060 kWh* (LHV).

During gas cleaning, the sulfur components are washed out and water vapor is condensed from the raw gas product of coal gasification (with the composition given above). Apart from H_2, CO, and 3% of CO_2, only the inert gases remain. The above value of 2300 Nm^3/t of methanol according to Lurgi (Hilsebein, 2009) refers to a synthesis gas produced from *natural gas*, which contains practically no inert gases. Therefore, the gas ratios were computed here without the inert gases. (These cause no noticeable disturbance of the synthesis process (Hilsebein, 2009).) As shown by the computation, the gas from the Shell gasification process, in terms of its combustible fraction, is composed of 30% hydrogen and 70% CO.

This calculation presumes *Case 2* for the estimation of the gasification efficiency and the costs, as given in Table 11.2, that is, the Texaco gasification process (now General Electric). The Shell process is more promising for syngas production (and therefore also for methanol synthesis); but since no comparable cost data are available, we use Case 2 as the reference case. This suggests itself also because Case 2 includes a relatively intensive gas cleaning procedure (Selexol process). The methanol synthesis however requires a still more involved procedure, the so-called Rectisol gas cleaning. Other differences compared to the production of synthesis gas for methanol synthesis as in Case 2 are: the CO conversion used there, which – owing to the smaller LHV of hydrogen compared to CO – leads to a reduction of the energy yield (LHV), is not required. However, the more elaborate gas cleaning needed for methanol production (Rectisol) is associated with higher investment costs and a somewhat greater reduction in the overall efficiency than the Selexol process. In the following, for simplicity we assume that these two effects just compensate each other. Therefore, in the cost estimation, we assume as in Case 2 efficiencies of 59% (HHV) and 50% (LHV).

Regarding the gas composition (H_2/CO ratio), we take as stated the results for the Shell process, using the same type of coal as in Case 2 (Illinois #6 coal). The volume fractions of the raw gas have already been given. Referred to energy (instead of volume), we then find the following results: For a *coal input of 1 kWh*, the energy content of the gas mixture at a gasification efficiency of 59% (HHV) is 0.59 kWh (HHV) and is divided as indicated by the heating values into 0.412 kWh from CO[85] and 0.178 kWh (*HHV*) or 0.15 kWh (*LHV*) from H_2. Expressed in terms

---

[85] For CO, there is no difference between the higher and the lower heating values. This difference applies only to energy carriers which contain hydrogen.

of the lower heating value, the energy content of the coal gas is thus $0.562\,kWh$ *(LHV)* (0.412 + 0.150).

Since for the methanol synthesis, 1.256 kWh of synthesis gas (HHV) is required per kWh of methanol *(HHV)* (this corresponds to a synthesis efficiency of 79.6% *HHV*), while the gas mixture from coal gasification contains only 0.59 kWh (HHV), a quantity of solar hydrogen corresponding to 0.666 KWh (HHV) (1.256 minus 0.59) must be added, or $0.563\,kWh$ *(LHV)*. This can be compared with the required amount of coal gas, 0.562 kWh *(LHV)*. Thus, the amount of solar hydrogen added must have exactly the same energy content (LHV).

For the cost of solar hydrogen, we use the estimate of 4.0¢/kWh (LHV; in 2002-$) discussed in the final part of Section 11.4.1.[86] For the coal gas, we take the cost from Case 2 (Table 11.2), namely 2.7¢/kWh (LHV).[87] With the same relative amounts of the gases, we then find the cost of the synthesis gas to be 3.35¢/kWh (LHV). Since each $kWh_{methanol}$ (LHV) requires 1.30 $kWh_{synthesis\,gas}$ (LHV), the cost of the synthesis gas corresponds to $4.35¢/kWh_{methanol}$ (LHV) (i.e., 3.35¢/kWh × 1.30). (The additional amount of solar power required should not lead to a cost increase as compared to the energy supply assumed here to be supplied by combustion of a portion of the coal gas. This cost is already included in the coal-gas price taken here – see below.)

In addition, the relatively low capital costs as well as O&M costs of the synthesis plant are favorable. The specific investment costs should be in the range of 250 million $/$GW_{methanol}$ (LHV), which leads to capital costs of $0.2¢/kWh_{methanol}$ (LHV).[88] Here, an annual operational period of 8000 h/a is assumed for the plant (see

86) Compared to the cost for solar hydrogen of 4.7¢/kWh (LHV) quoted in Section 4.1, here we have assumed a shorter transport distance from the American solar sites to the coal fields where the gasification plants would presumably be located (only 1500 km instead of 3300 km as assumed for the general gas supply). This decreases the hydrogen cost to 4.4¢/kWh (LHV). In addition, a further cost reduction is presumed due to the higher capacity factor possible here and due to a certain exploitation of the innovation potential for the solar power plants (giving an overall reduction of the cost to 4.0¢/kWh LHV).
87) CO_2 storage costs do not accrue here.
88) Hilsebein (2009): "Investment costs are in general not quoted at present, because they depend very strongly on the size of the plants and furthermore are subject to frequent changes. For a rough estimate, however, one can assume that a large-scale plant constructed today (a mega-plant producing 5000 t of methanol per day = 1.12 GW of methanol (LHV)), in which the synthesis gas is obtained from natural gas, would cost all together 600 million Euros. Of this, roughly 35–40% would be due to the methanol synthesis and distillation." 40% of the overall investment (of €600 million) corresponds to €240 million per 1.12 GW (LHV) = 2.14 million €/GW (in 2008-€) = 197 million € (2002; inflation factor 1.084) = 206 million $/GW (in 2002-$). In the following, adding a security supplement, we take a value of 250 million $/GW of methanol (LHV, in 2002-$). These investment costs, taking 4% real interest and an operating lifetime (as for coal gasification plants) of 25 years (annuity 6.4%), lead to an annual capital cost of 16 million $/GW. With an annual operating period of 8000 h/a, corresponding

below). The O&M costs should be around $0.3\,¢/kWh_{methanol}$ (LHV).[89] Compared to the cost of the synthesis gas, both of these cost items are insignificant. Even if they should turn out to be somewhat higher due to the uncertainties in the data (even twice as high), the cost of the methanol produced would not change dramatically. The *methanol cost* is found to be all together $4.85\,¢/kWh$ (LHV) (in 2002-$), or, converted to 2008-$, $5.75\,¢/kWh$ (LHV). This corresponds to an oil price of 91 $/barrel (cf. Table 4.3, footnote i), or as a rounded-off value, *90 $/barrel oil equivalent* (in 2008-$).

If it should turn out that high-temperature electrolysis cannot be developed successfully (to achieve an efficiency of 86% LHV), but that further development of conventional electrolysis leads to an efficiency of 75% (LHV) (for the same plant cost), then the cost of solar hydrogen would increase by $0.6\,¢/kWh$ (LHV) (Section 11.4.1). The resulting methanol cost would correspond to nearly *100 $/barrel oil equivalent*.[90] If we also assume that the higher capacity factor of the solar plants, as given in Section 11.4.1 for hydrogen production, as well as the possible advantage due to exploitation of the innovation potential of the solar plants would not play a role, then the cost of the solar hydrogen would further increase by $0.4\,¢/kWh_{H_2}$ (LHV) (we have discussed this scenario already); then the methanol cost would correspond to *105 $/barrel oil equivalent*.

Since methanol production is a completely conventional technology, its costs, which we have provisionally estimated here (i.e., by making global assumptions for the gasification process, gas cleaning, investment costs of the synthesis plant and for its O&M costs), could be more precisely determined by plant engineering and construction firms within a few months; in particular for concrete projects, that is, for a specified type of coal and gasification process, and the associated conventional gas-cleaning technologies. This also holds for the coal-consumption balance, which we still have to discuss.

---

to a yearly methanol production of 8000 GWh (LHV) per GW, we then find a capital cost of exactly $0.2\,¢/kWh$ of methanol (LHV).

89) The O&M costs for methanol production, which are not quoted by Hilsebein (2009), were adopted for this rough estimate from Roan et al. (2004):

In *US-¢/gallon of methanol* (1 gallon of methanol (3.79 l) = 16.2 kWh (LHV)):

catalyst and chemicals	2.6 ¢/gallon
utilities	0.9 ¢/gallon
other fixed costs	4.0 ¢/gallon
intermediate sum	7.5 ¢/gallon
minus exported steam	−2.9 ¢/gallon
total	4.6 ¢/gallon = 0.28 ¢/kWh of methanol (LHV)

(The cost bonus for the steam generated (2.9 ¢/gallon) corresponds to $0.18\,¢/kWh$ of methanol (LHV).)

90) The cost of solar hydrogen would then increase from 4.0 to $4.6\,¢/kWh$ (LHV). The synthesis gas (50% solar hydrogen, 50% coal gas) would be $0.3\,¢/kWh$ more costly. Per kWh of methanol, 1.3 kWh of synthesis gas is required; per kWh of methanol, the cost of synthesis gas would thus increase by 0.39 = ca. $0.4\,¢/kWh$ (LHV), and therefore the methanol cost would increase from $4.85\,¢/kWh$ (as above) to $5.25\,¢/kWh$ (in 2002-$). In 2008-$, this corresponds to an oil equivalent price of 99 $/barrel.

First, we still make an explanatory remark on the assumed annual operation period of 8000 h/a, which can be skipped over in a first reading:

> Because of the storage of solar hydrogen in large *gas reservoirs*, methanol production plants could operate longer than the solar plants, which at a capacity factor of 80% (USA) would operate for 7010 full-output hours per year. (Since only a small portion of the gas would need to be stored, this should not increase the methanol price significantly.) This item in the cost calculation, which can be determined exactly only by exemplary planning of specific storage facilities, needs to be clarified, as thus far no data are available. To this end, determining potential sites for hydrogen storage facilities (in particular salt-dome caverns) and designing transport pipelines for the hydrogen (to the coal fields in the West and the East of the USA) would elucidate the question of how much additional infrastructure would be required to include the storage reservoirs, i.e., how much longer the pipelines would have to be. In addition, the conditions for preparation of salt-dome caverns need to be examined in typical cases, since their cost – which in general however should be rather low – depends on this factor. Figure 11.5 shows the locations of the salt-dome regions in the USA. Comparison with the locations of the coal fields in Figure 11.4 makes it clear that the salt-dome areas are nearly always in the neighborhood of coal fields, or – in the case of the Eastern coal regions – on the route from the solar sites to the coal fields. (The hydrogen could also be stored in caverns in rock layers, which would however be more expensive. Former gas fields are probably *not* suitable, since the hydrogen would be contaminated there by other gases.) These cost factors are still lacking in the overall cost balance for methanol production, so that its final cost will be somewhat higher, but only by a small amount.
>
> An alternative to H_2 storage (to cover interruptions in the solar hydrogen supply) would be a switch to operation with *coal alone*. Owing to the lack of solar hydrogen, the yield of methanol would be reduced by half. This emergency production would however be completely independent of the supply of solar power. A precondition would be the presence of those components of the plant which are necessary for converting a portion of CO from coal gasification into H_2 (CO conversion and separation of the resulting CO_2). It would have to be considered that these components, required for emergency operation with coal only, would not be used during normal operation, in order to maintain a high efficiency. The coal gasification plants would operate at full capacity; only the methanol synthesis plants (which in any case have much lower investment costs) would then operate with a reduced output, so that the utilization of the whole installation would not be too drastically impaired. Since the methanol production would decrease by half, in applying this strategy, methanol (instead of hydrogen) would have to be stored in some quantity, which could be carried out

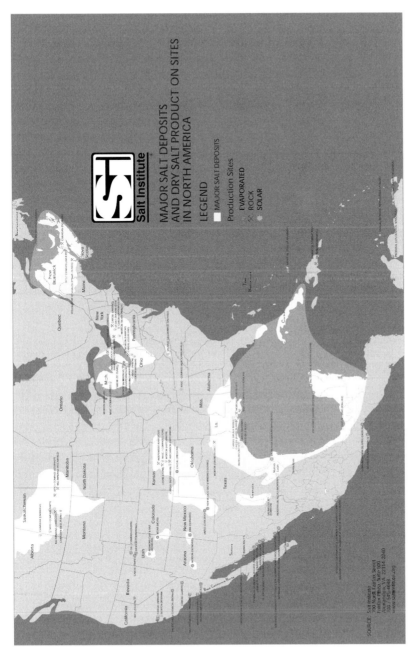

Figure 11.5 US salt production sites (potential locations for gas storage reservoirs in salt caverns) (Salt Institute, Alexandria, VA, USA).

without too great an effort. The investment costs for these additional plant components are in principle already included in the costs for Case 2, there however only for the less complex Selexol gas cleaning procedure, and not for Rectisol cleaning, so that with this strategy, certain additional costs would have to be borne. (The additional demand for electric power would be guaranteed by the backup power plants of the solar power system.)

11.4.3.2 Coal Consumption

What amount of coal would in fact be required for the production of 0.65 kWh of coal gas (corresponding to 1 kWh$_{\text{methanol}}$ (LHV))? With an efficiency for hydrogen production (Case 2) of 50% (LHV), for the manufacture of 0.65 kWh$_{\text{coal gas}}$, a coal requirement of 1.3 kWh is found. To estimate the *real* coal consumption, this comparison is not valid, as mentioned above.

In view of the lack of data on the coal consumption of operating or planned gasification plants using the Shell process (with the type of coal specified above), we estimate the coal demand on the basis of simple assumptions, starting with the known gas composition (from the type of coal specified). In terms of the estimates of the depletion time of the US coal reserves as given below (in Section 11.4.3.4), it is unimportant whether the true number is 10% higher or lower.

Regarding coal consumption by the Texaco gasification process for H_2 production, and by the Shell process for the production of syngas to manufacture sun methanol, the following differences can be noted:

a) For the Rectisol gas cleaning process (whose costs we estimate as stated by comparison with Case 2), the decrease in efficiency (increased electric power required) is only a minor portion of the additional expenditures.[91] The overall cost as in Case 2 thus reflects here in particular the increased investment cost. Without the decrease in overall efficiency due to the CO shift reaction, the coal consumption amounts to only 1.154 kWh per 0.65 kWh of coal gas (i.e., per 1 kWh of methanol produced (LHV)).[92]

---

91) Comparison of a methanol production plant using coal gasification (but a fluidized-bed process) according to Meyer and Lorenz (2004):

Electrical energy used in % of the methanol output:

coal drying* and processing	3.0%
gasification	0.3%
air separation facility	7.0%
Rectisol gas cleaning	1.8%
methanol synthesis	1.7%
CO_2 compression (not required here)	3.2%
miscellaneous	0.6%
Total:	17.6%

* Coal drying would be accomplished using the heat-pump process developed in Germany.

92) As already discussed above in the derivation of the quantity of coal gas resulting from coal gasification, the efficiency of gas production in Case 2, without the CO shift reaction, would be 56.3% (LHV). This leads to a consumption of 1.154 kWh of coal per 0.65 kWh of coal gas produced (LHV) (0.65/0.563).

b) The efficiency of the Shell gasification process is notably higher for the production of syngas (instead of H_2) than that of the Texaco process (Case 2). With the latter, due to the input of water (i.e., the increased energy requirement), a larger fraction of the carbon is converted to CO_2. (From this process, the raw gas composition according to Esquivel (2007) with Illinois #6 coal is (in vol.%): H_2 30.3%; CO 39.6%; CO_2 10.8%; H_2O 16.5%; N_2+Ar 1.6%, ...). Thus, a portion of the carbon input into the process is separated out as CO_2, resulting in an increased coal consumption. This loss is avoided to a great extent in the Shell process (with a CO_2 content of only 1.5% – see Section 11.4.3.1). In addition, the latter exhibits a superior utilization of heat energy, resulting in a lower gas requirement for energy production.

c) The large amount of steam resulting from the methanol synthesis can be used to generate electric power for the needs of the plant. This is not taken into account in Case 2 (where only gas production is considered).

d) Combustion of a portion of the coal gas for electric-power production to cover the energy requirements of the plant itself is not necessary. Instead, an additional quantity of solar power could be utilized. In Case 2, the electric power generation is accomplished by burning a special gas which is obtained from the gas separation; it would not be otherwise used at that point. This is a CO_2-rich partial gas flow from the PSA. This stage is not required for syngas production, since no gas separation is necessary in that case. Then there is also no need to burn this quantity of gas (insofar as the electric power requirements can be supplied from some other source). For the generation of electric power for the plant itself (foregoing the combustion of raw gas), on the one hand the additional steam (at medium pressure) from the methanol synthesis reactor can be used; on the other, the *purge gas* which accumulates from the synthesis,[93] which must likewise be utilized by burning it for

---

93) Regarding the so-called purge gas: in the methanol synthesis, the synthesis gas is injected into the gas circuit; in the process, the inert gases are concentrated. In order to limit their concentration, a small portion of the gas flow must be released (purge gas). If all the carbon were to be converted to methanol, only *0.75 kWh* of coal per kWh of methanol (LHV) would be required; see below. Including the purge gas released, this amount however increases to 0.86 kWh. This number can be justified as follows: For the production of 1 t of methanol, the required quantity of synthesis gas is 2300 Nm³ (stoichiometric mixture including 3% CO_2) and contains 430 kg C. 1 t of methanol (5420 kWh LHV) contains only 375 kg C. Thus, ca. 13% of the carbon contained in the gas is released as purge gas (375/430 = 87%). (The purge gas is usually burned in a CCGT plant.) The required energy is thus increased from 0.75 to *0.86 kWh* (0.75 × 430/375).

Derivation of the coal consumption for *complete* conversion of the carbon: Using the assumed amount of carbon from Section 11.2.3, that is, 0.75 kg of C contained in the coal per kg of *coal equivalent* (thus referring to the energy content of the coal; C: atomic mass 12),

power or heat production, can be exploited. No information is available regarding the question of whether these two sources together would suffice to cover the electric power requirements of the whole plant (gasification plus synthesis). Therefore, we assume here that an additional *0.1 kWh* of *solar energy* (per 0.65 kWh of coal gas produced) would be required.

In the following, we assume that the *coal consumption* will be reduced *at least* to $1\,kWh_{coal}$ (per $0.65\,kWh_{coal\,gas}$) for the reasons mentioned under items (b) through (d) from its original value of 1.154 kWh (item (a)). This upper limit can be expected to hold with certainty, at least when *0.1 kWh* of *solar electrical energy* is provided for the power supply of the plant.

In the Shell gasification process (and using the specified grade of coal), the real coal consumption could be expected to lie between ca. 0.85 und $1.0\,kWh_{coal}$ (per $0.65\,kWh_{coal\,gas}$). Then, given the nearly complete avoidance of CO_2 formation in the gasification, and omission of tapping off of raw gas to generate energy for the plants, carbon is removed from the process circuit only by the purge gas. With the assumptions made here, we then obtain a coal consumption of only 0.86 kWh.[94] Using other grades of coal, it could be higher, and would therefore lie on the average somewhere between 0.85 and 1.0 kWh. (The precise value could be readily determined in the course of detailed planning of the installations.) Here, for simplicity we take the conservative value of 1 kWh.

The demand for solar energy for hydrogen production and the additional electrical-energy requirements of the plants are found together to be $0.9\,kWh_{el}$ per $kWh_{methanol}$ (LHV) produced. (For the production of $0.65\,kWh_{H_2}$ (LHV), at the quoted efficiency for the electrolysis and H_2 transport (all together 81.7%; cf. Table 4.3, footnote *d*), $0.80\,kWh_{el}$ of electrical energy are required. We add to this the energy for the plant operation, equal to $0.1\,kWh_{el}$.)

The additional solar electrical energy supplied need not be taken into account separately, since it merely substitutes the energy that would otherwise be produced in a H_2-fueled CCGT power plant, which is already included in the gas costs (Case 2). The costs of this electrical-energy generation for use in the plant itself should be roughly the same as those of the solar electrical energy.[95] The assumed amount

then 1 kg *coal equivalent* would suffice to produce 2.0 kg of methanol (CH_3OH: molecular mass 32) (0.75 × 32/12). Recalculated in terms of the energy content (1 kg *coal equivalent* = 8.14 kWh (LHV), 2.0 kg of methanol = 10.84 kWh (LHV)), this then yields a coal consumption of 0.75 kWh (LHV) per kWh of methanol produced (LHV) (8.14/10.84 = 0.75).

94) See footnote 93.
95) Here, we are in fact dealing with an IGCC power plant (integrated coal gasification plus CCGT). As discussed in Section 4.1, electrical energy from this type of plant using H_2 fuel (including CO_2 separation, but without the cost of CO_2 deposition) and starting with American coal, would cost 3.4 ¢/kWh. This cost however holds only for very large IGCC plants with highly efficient, large-scale gas turbines, and for gasification using advanced technology (*n*th of a kind). In Case 2 (Table 11.2), we are considering small gas and steam turbines and not advanced technology for the H_2 production. The power costs would then be higher than the 3.4 ¢/kWh quoted above.

of solar electrical energy required, 0.1 kWh, would likewise seem to be conservatively estimated.[96]

Methanol production from coal alone (i.e., without the use of solar hydrogen) would require roughly twice as much coal. One can assume an efficiency of around 50% (LHV), so that $2\,kWh_{coal}$ per $kWh_{methanol}$ (LHV) would be required. An older study on methanol production from coal (AFAS, ISI-FhG, TÜV-Rheinland, 1984), which was likewise based on the Texaco process, however using high-grade German hard coal, quotes an efficiency of 48% (LHV). In the USA, there is only one plant which currently produces methanol from coal gasification (Eastman Corp., Kingsport, TN; cf. Olah *et al.*, 2006); however, no efficiency data are available for this plant.

11.4.3.3 A Price Brake on Petroleum – The Potential of Sun Methanol in the USA

Since for the production of sun methanol, only a relatively small amount of coal is required and since in the USA, enormous areas with a high insolation are available, sun methanol represents a great option there for the production of a liquid energy carrier. Probably the worldwide petroleum demand could in theory be substituted by methanol production in the USA.[97] The USA is the only industrial

---

These higher costs are to be compared with the cost of solar electric power at 4.7 ¢/kWh (Table 4.2), there for power transmission over 3000 km and including backup power plants. (The distance from the solar sites to the US coal fields is on the average less than half as great.) The cost difference between power generated in the gasification plant itself and solar electric power is thus very small, if there is in fact any difference at all. Switching from self-generated power to solar power thus has no noticeable effect on the gas production costs. (The cost of the solar power is compensated by the lower coal consumption and the avoidance of the small CCGT power plant within the facility, with its high specific cost, as well as its operation and maintenance costs.)

96) This number presumes, for simplicity, that a reduction of the coal consumption from the value listed in item (a) of 1.156 kWh to at least 0.9 kWh (i.e., by 0.256 kWh) would be due exclusively to the substitution by solar power, that is that power generation using the steam from the process and the purge gas makes no contribution. The amount of coal conserved, 0.256 kWh (the difference between 1.156 and 0.9 kWh) would, if consumed in an IGCC power plant at an efficiency of 43% (cf. Table 4.3), generate 0.11 kWh of electrical energy, which nearly corresponds to the additional quantity of solar electrical energy quoted, 0.1 kWh. Since the steam and purge gas in reality would also be utilized, we are again dealing with an upper limit here. This small fraction of electric power is furthermore not a sensitive quantity. Neither in the estimation of the required area for solar power plants for methanol production, nor in the estimation of the consumption of uranium in the nuclear variant, does this fraction of the overall power consumed play an essential role in the results, which are estimated only roughly here, in any case.

97) Regarding the required land area: As just mentioned, for the production of hydrogen (and for the direct electric power supply of the gasification unit), $0.90\,GW_{el}$ *of solar energy* would be required *per GW of methanol* (LHV) produced. At a typical capacity factor of 80% for sites in the USA, the annual generation of this amount of power implies a solar-plant output power of $1.11\,GW_{el}$. As described in Section 4.3.5, in the USA (at a solar multiple of 3.7), if the solar plants are designed for minimal area use, one can reckon with an area requirement of ca. $80\,km^2$ *per* GW_{el}. (Depending on the design, the land requirement lies between 75 and $90\,km^2$/GW. In Chapter 5, where we were dealing only with the *electric power supply* for the

country with truly favorable and large-area solar regions. The supply of solar energy is on the whole even better than in North Africa (at least in the western portion of North Africa) and in the Near East.[98]

This enormous US potential could – on the scale of the national energy economy – even be developed rapidly (insofar as the development of high-temperature electrolysis technology does not lead to major delays); for example, within 25 years. Therefore, this alternative production could lead "relatively" quickly to an effective price brake on the world petroleum market. Considering the time aspect: If we leave possible limitations due to electrolysis technology out of consideration, in any case the construction of the solar power-plant parks could begin for example, already in around 6 years (cf. Section 2.4), and the power-plant capacity could rapidly be increased on a very large scale – in the course of a broadly applied, high-priority construction program (crash program); for example, within 10 years up to $2000\,GWa_{methanol}/a$. From today, this would take 16 years.

The supply potential of the USA as provider of an oil substitute would then be immense. It would correspond to the fictitious case that an enormous oil well had been found there, which could supply the whole world with oil for a very long time (at an extraction cost of 90 \$/barrel).

The value of 90 \$/barrel oil equivalent quoted here refers by the way to the cost of fuel (as in the case of methanol), that is, to *processed* oil, not crude oil. The fuels obtained from crude oil are more expensive than the starting material, since it contains fractions which cannot be converted into fuels (or only with a great effort), such as liquid gases, heavy oils, and bitumen. The *crude-oil equivalent* of the metha-

---

USA, we used the higher value (90 km^2). In the case of a very large demand for solar energy for methanol production, the power plants could however be designed for a lower area usage.) Then for 1.11 GW$_{el}$, that is, for *1 GW of methanol*, 90 km^2 would be required. As we discussed in Section 5.3, according to the NREL map for the years 1998–2005 (NREL, 2008), the solar regions with a high insolation in the USA (most of them more than 7 kWh/m^2 d, some of them 6.5–7 kWh/m^2 d) comprise an overall terrain of roughly 500 000 km^2. Here, a tolerable slope of the terrain of 4% is assumed. (This is relevant only for solar-tower plants, but not for parabolic trough plants, which require virtually flat areas). Then in the USA, an annual production of sun methanol equal to (500 000/90 =) 5500 GW would be possible. (For comparison: The entire worldwide annual consumption of petroleum is currently 5000 GW.) The more extensive investigations regarding sites and solar insolation proposed in Chapter 5 would then yield the exact numbers relevant to this rather theoretical case (the complete substitution of the world's oil supply by methanol from the USA).

98) Apart from the climatic factors, the direct insolation which is relevant for solar thermal power plants depends on the altitude of the sites (lower absorption losses in the atmosphere). In the US Southwest, a large portion of the solar sites lie at a relatively high altitude. Similar insolation values are found in the Western Sahara only in a few mountainous regions, where land use for solar plants would be difficult or impossible due to the unfavorable topography. In the Central and Eastern Sahara, there are similarly favorable sites to those in the USA; the best of these however also lie in high-altitude regions or in mountainous areas, as one can readily confirm by comparing the solar map of Africa in Chapter 5 with a physical map.

nol price would thus be lower than 90 $/barrel. (For simplicity, we have however used this price for comparison here.)

The alternative supply (methanol) would define an absolute upper limit for the oil price. It would not only fix a price of 90 $/barrel (2008-$) as upper limit, but also the oil suppliers would have to offer oil at a price somewhat lower than the methanol; otherwise there would be no reason for the importing countries to continue using oil instead of the new energy carrier. Under these changed conditions, the power of the consumers to force the suppliers to offer lower prices would be still greater if they were to act as a united block; as *one* negotiating partner, who would face the oil cartel. Then, if the negotiations should fail, the supplier countries would have only a few possible customers; see below: "Organization of Petroleum Importing Countries" (OPIC). In combination with such an organization, the influence of the new energy supplier on the price policy of the OPEC would be enormous.

Considering its possible effect as a price brake, it is not important whether the alternative energy carrier (methanol) is in fact utilized; but rather, that it *could be* utilized. If this is possible (the USA as a potentially sufficiently large supplier for the world), then the actual production could be spread over other countries such as Australia which likewise possess large land areas with a very good insolation, as well as coal reserves; or also Mexico or the Republic of South Africa together with Namibia (there using imported coal). Since the USA could in principle supply the world demand (or at least a major part of it), these countries would have to offer their solar energy and – if present – their coal at attractive and stable prices over the long term. The solar potential of the USA would thus in reality need be exploited only to a certain (lesser) extent. (As one can see by considering the OPEC cartel, it is not necessary to be able to supply the entire demand. Thus, Saudi Arabia, with only 13% of the world's oil production and only 22% of the proven oil reserves, can dominate the cartel.)

11.4.3.4 Liquid-Fuel Production from Coal Alone? – Sun Methanol to Conserve US Coal Reserves

The consequences of methanol production on the depletion time of the American coal reserves will be demonstrated by the following hypothetical computation: Supposing the *total* current (2004) world petroleum consumption of exactly 5000 GWa (5.38 billion tce, i.e., around 5 billion tce) to be replaced by methanol produced from coal – at the efficiency already quoted of ca. 50% – and thus annually 10 billion tce of coal would be required. The proven and economically recoverable coal *reserves* of the USA amount to ca. 200 billion tce (Table 11.6), and would be exhausted within 20 years. In the case of sun methanol production, the coal demand would be roughly halved (cf. above: "at most" 1 kWh$_{coal}$ per kWh$_{methanol}$ (LHV) – possibly only 0.85 kWh$_{coal}$) and the depletion time would then be 40 years. If we furthermore assume that an annual methanol production of 2000 GWa (about 2 billion tce) would be sufficient to effectively influence the price policy of the OPEC, then for the sun methanol, only around *2 billion tce* of *coal* per *year* would be necessary. The reserves cited would then suffice for *100 years* (with methanol production exclusively from coal for 50 years).[99] A hundred years is a time horizon which reaches back to before the Age of Petroleum. US methanol as a price brake on oil would then be no longer relevant; by then, the global coal supplies could be utilized for methanol production.

The annual demand of 2 billion tce cited above should be compared with the present-day coal production in the USA, which is 0.8 billion tce (750 GWa). (This coal, which is mainly used as fuel in power plants, would, as suggested above, most reasonably be substituted by solar energy for power generation.)

We must furthermore consider the annual *US consumption* of *natural gas*, which amounts to 0.83 billion tce (770 GWa). If it – hypothetically – had to be completely substituted by coal gas (hydrogen) in the coming decades, at an efficiency (LHV) of 64% (Table 11.2, advanced technology), then annually *1.3 billion tce* of coal would be needed. In Central America and Venezuela, as well as in the rest of South America, there are also natural gas reserves of a considerable quantity, which could be used as a primary energy source (whereby the price would be limited by that of coal gas). Initially, natural gas would thus be only *partially* substituted by coal gasification. Depending on the actual quantity of natural gas "resources," possibly within 10–20 years, a massive substitution will however have to begin – compare the discussion of the lifetime of the US natural-gas supplies in Section 11.2.9. If natural gas was to be completely substituted, coal production would have to be increased to 3.3 billion tce/a (lifetime: 60 a), unless a part of the gas requirements were to be supplied as solar hydrogen.

11.4.3.5 Methanol Production using Nuclear Hydrogen

As we shall show in Chapter 12, at an assumed annual nuclear energy generating capacity of 3000 GWa (compare the expected worldwide electrical-energy consumption by the year 2030: 3400 GWa), the lifetime of proven and speculative uranium reserves under a maximum extraction cost of 130 \$/kg U as given in the statistics would be only 28 to 36 years. (Regarding the speculative reserves, two different values are given in the literature, so that here, we quote the corresponding two different times until the reserves are exhausted.) We have already explained in Section 4.1 that the available amount of uranium would possibly be doubled on increasing the price limit to *400 \$/kg U*. These reserves would then last for 56 or *72 years*. This is however not much more than one generation of nuclear power plants at their presumed operating lifetime of 45 years. For a uranium price of 800 \$/kgU, the reserves might be tripled. At that price, however, the cost of nuclear power[100] would already be just as high as that of solar power (at the power plant, for H_2 production), so that these quantities of uranium do not represent "economi-

---

99) If, after using up its low-cost coal reserves in around 100 years at the current rate of consumption, USA wanted to maintain its high rate of methanol production, then more expensive coal resources would have to be tapped. One has to keep in mind, however, that for sun methanol, the cost of coal gas affects only 50% of the final price. This product is thus relatively tolerant with respect to coal price increases.

100) This is only the microeconomical cost, and assuming the very low cost estimates given in Chapter 12.

cally recoverable" reserves for the support of an additional production of H_2 which might be implemented.

Since in the utilization of nuclear energy (instead of solar energy or coal), the figure quoted of 3000 GWa would already be required for the worldwide supply of electric power, an additional energy supply of more than the same amount again (3600 GWa) for hydrogen production would more than halve the lifetime of the reserves, so that the uranium at a price of up to 400 \$/kg U would last for only 25 or 33 years. Here, we have assumed that the same amount of methanol would have to be produced using nuclear power as would be available using the solar energy from the prime solar regions in the USA (4000 GWa$_{methanol}$, see above: "US production potential" and the footnote there; for this amount of methanol, 3600 GWa of nuclear energy would be needed yearly[101]). If only 2000 GWa$_{methanol}$ (LHV) were to be produced – this would require 1800 GWa/a of nuclear power – then the depletion time of uranium (price class ≤400 \$/kg U) would increase to 35 or 45 years. A depletion time much longer than this will not be feasible (insofar as the very rough estimates used here for this uranium price category prove to be approximately correct). The limit of what is possible would be reached,[102] so that a challenge to the oil-producing countries of a large-scale (or even complete) substitution of oil would not be credible. The increasing risks, including economic risks, accompanying a massive expansion of nuclear power capacity must also be considered. The estimated power cost involves numerous uncertainties, especially in terms of the investment costs and the evolution of the uranium price on the world market, which is not necessarily determined by the extraction cost of uranium, on which the statistics for the reserves are based (cf. Chapter 12).[103]

11.4.3.6 OPIC

As with every attempt to limit the prices demanded by a supply monopoly, with large-scale methanol production there would be a danger of a later price *decrease* by the monopoly. This would make the investments – which appeared economically rewarding at the higher price level – no longer profitable. Thus, if after a longer period of time, methanol could finally be produced as an alternative product to oil, then a sudden decrease in the oil price could push the methanol production below the level of profitability. In view of this fundamental risk in an attempt to

---

101) Compare: For solar power, the requirement is 0.9 kWh$_{el}$/kWh$_{methanol}$.
102) The power cost from nuclear energy would then be higher than that from solar energy.
103) Furthermore, this large number of nuclear power plants would have to be constructed in parks near large lakes or the seacoasts, owing to the need for cooling water. Otherwise, the additional cost of dry cooling would have to be taken into account. The power cost would then rise by ca. 8%, similarly to the case of solar power with dry cooling. To be sure, for nuclear power, the ambient air temperature would on the average be lower than at the desert sites of solar plants. However, nuclear power plants have a lower efficiency than solar plants (in their conventional steam circuits), so that an increase in output temperature would have a stronger effect on the overall efficiency. This cooling problem would also be present if the nuclear plants were built (for hydrogen production), for example, nearby to the coal fields in order to minimize the transport cost for the H_2.

force a price limit, investors frequently are reluctant to engage themselves in such a project, so that the necessary investment funds would be available only if the cost difference between the monopoly product (petroleum) and the alternative product (methanol) were very great. Thus, if the USA were to establish a production capacity of 1000 to 2000 $GW_{methanol}$ as a brake on the oil price and were successful in this attempt, so that the oil price in fact were to drop below the cost level of methanol, then they would have to bear the cost difference alone. The gain from lower oil prices would however be shared by all the oil-importing countries. The USA would thus be saddled with the costs of this strategy while the result (a limit on the oil price) lies in the interest of all the larger industrial countries. The USA will therefore not willingly take up such a burden. The solution to this dilemma would be a common engagement of all the oil-importing or industrialized countries for the construction of a solar alternative production in the Southwest of the USA. Each country would be assigned a contingent corresponding to its share of the oil consumption.

Such a coalition of nations, or the associated organization, might be termed the "Organization of Petroleum-Importing Countries" (OPIC). The contractually regulated cooperation would guarantee that all the countries would contribute their portion of the cost of the construction measures. The countries organized in the OPIC would then be able to decide on a common basis what fraction of the world's oil production should be replaced by sun methanol (for example, 1000 GWa/a). Even with decreasing oil prices, each country would have to maintain its share.

In the long term, within such an association of nations, a shift in the constellation of interests might occur. If namely one day the oil reserves were practically exhausted, the power of the supplier nations would gradually shift toward those countries with large reserves of coal, not least the USA itself. Price increases for coal would be the result. The member states of the OPIC would then be subject to a price diktat, for example, for American coal. In view of their large investments there (solar plants), they could hardly avoid such a diktat. The association of oil-importing countries would thus have to agree in advance on long-term price controls for coal. Without such guarantees, the non-American countries would hardly be willing to make a large-scale commitment in the USA.

11.4.3.7 The CO_2 Balance

Without gas consumption for power production for operation of the gasification plant (by making use of solar power), only the purge gas would have to be burned (cf. Section 11.4.3.2), and CO_2 emissions would be reduced to a minimum. Without these emissions from combustion of the purge gas, CO_2 would be formed only on later combustion of the methanol product, to the extent of 0.25 kg CO_2/$kWh_{methanol}$ (LHV) (compare the emissions from other energy carriers[104]). Including combus-

---

104) CO_2 emission factors (from the Federal Office of the Environment, Germany) in kg CO_2/kWh (LHV): Bituminous coal, 0.33; lignite coal, 0.40; subbituminous coal, 0.35; crude oil, 0.28; *light fuel oil, 0.27*; bunker oil, 0.30; diesel fuel, 0.27; *gasoline, 0.26*; methanol, 0.25; ethanol, 0.26; *natural gas, 0.20*.

tion of purge gas, this value increases – at the coal consumption given above – to 0.29 kg CO_2/kWh$_{methanol}$ (LHV) (i.e., 0.25 × 0.86/0.75).

Thus, for sun methanol, the lowest conceivable specific CO_2 emissions of any synthetic liquid fuel achievable in practice would be attained. They could be further reduced only by manufacturing methanol from CO_2 or by gasification of biomass (see below); and of course by dispensing with liquid carbon-containing energy carriers, in particular by an increased direct utilization of electrical energy or hydrogen in the transportation sector.

In this connection, one must keep in mind that in the course of a solar strategy, the previous high CO_2 emissions from coal-fired power plants would be eliminated or strongly reduced, so that the overall CO_2 emissions (relative to the current energy consumption) would be lower than they are today. This is particularly the case if natural gas were also to be substituted either by hydrogen from coal (with CO_2 sequestration) or by solar hydrogen. In the view of many scientists, a major decrease in the emissions of greenhouse gases will be necessary by the middle of this century in order to prevent climate change (the "climate catastrophe"). The goal of a reduction of CO_2 output by 80% or at least by 50% can be attained only through massive limitations on the use of (carbon-containing) liquid energy carriers. In the mid- and long-term, methanol thus does not represent a suitable replacement for gasoline and diesel fuels, at least not at today's consumption rates. Instead, it can be only a part of the final, that is, sustainable solution to the energy problem. The goal thus remains the construction of CO_2-free systems, especially for the transportation sector, which in future will probably continue to grow rapidly.

> *Biomass:* It is particularly well-suited for gasification, both under the technical as well as the economic aspect. Since the biomass removes CO_2 from the atmosphere during growth, it represents a closed CO_2 circuit (no net emissions). Its potential for the large-scale energy supply is however too limited. According to Angerer (2007), within the EU-25 by the year 2010 (with biofuel production from the biomass alone, without the addition of solar hydrogen) 105 GWa/a of biofuels could be produced, which corresponds to 21% of the motor fuel demand. In the middle term, with a corresponding preference for the utilization of biomass, replacement of up to 30% seems possible. (By the additional use of solar hydrogen for methanol synthesis – instead of internal hydrogen produced from CO conversion – the fraction of renewable motor fuels would be increased still further, but it would not double.) The motor fuel consumption in the EU-25 in the year 2010 is estimated to be 510 GWa (16 000 PJ) (Angerer, 2007, p. 25). Motor fuels account for more than half of the overall oil consumption (920 GWa in the year 2004), however only 35% of the oil and gas consumption (1480 GWa in 2004; cf. Appendix C). In this case, biofuels would be able to replace only around 7% of the current oil and gas consumption (with a share of 21% of the motor-fuel consumption). With mixing in of solar hydrogen before the methanol synthesis, this share could be increased, e.g., to ca. 10% (at most to 14%). In other regions, the numbers are somewhat

different. For the supply of the total energy requirements, which are expected to grow strongly worldwide, this energy source will thus be able to make only a relatively small contribution. (In view of the possibility of a supply of the worldwide demand for motor fuels (methanol) by the USA alone, the probable fraction of biomass there would in any case be *very* small.) Concerning the supply of individual countries with liquid energy carriers, the local share of biofuels could however be quite noteworthy in some cases.

In particular, within a CO_2-minimized renewable energy system, in which carbon is required only for the production of liquid energy carriers and in which at the same time its use is suppressed as far as possible, biomass could thus permit an additional, significant reduction in CO_2 emissions. The potential of biomass should naturally be utilized as much as possible before coal gasification is envisaged on a large scale, but only insofar as this utilization does not enter into competition with food production and does not lead to additional environmental burdens resulting from "industrial" agriculture (monoculture of energy plants, large-area use of pesticides, etc.).

11.4.4
Hydrogen and Coal for Liquid Energy Carriers in a Future Solar-Hydrogen Energy System

In order to reduce the consumption of coal as far as possible, not only replacement of coal-fired power plants by solar plants, as explained above, but also the use of solar hydrogen instead of natural gas or coal gas will be necessary. If all gas requirements were to be filled by solar hydrogen, coal would need to be used only as a carbon source for liquid energy carriers.

In this connection, we cast a glance at an energy flow diagram from an earlier publication of the present authors. It is taken from a study which had model character and which applied especially to Germany (Kalb and Vogel, 1980), where a solar hydrogen energy system with a minimal use of coal was described in detail (Figure 11.6) (here in english). In contrast to the production of sun methanol described above, here however, coal hydrogenation–likewise making use of solar hydrogen–is employed, that is, the direct "liquefaction" of coal by hydrogenation for the manufacture of gasoline, fuel oils and other liquid hydrocarbons.[105] For comparison, the energy-flow diagram for Germany in the year 1976 is shown (Figure 11.7). As one can see, using a hydrogen system and with the actual coal consumption for that year, the complete supply of the energy requirements of West Germany would have been possible.

---

105) The direct liquefaction of coal by hydrogenation is technologically more complex than the methanol synthesis (high-pressure processes, complicated product treatment), but it requires less hydrogen; cf. for example, Verfondern (2007, p. 34). On the whole, methanol production should be noticeably more economical.

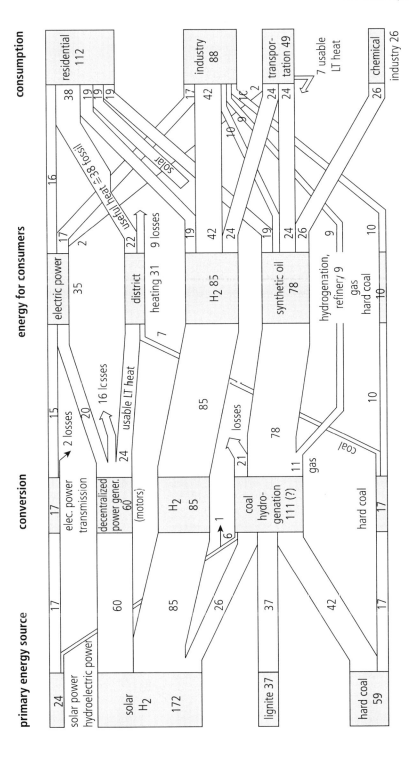

Figure 11.6 A solar-hydrogen-coal energy system for the Federal Republic of (Western) Germany (Kalb and Vogel, 1980) (assumption: overall energy consumption as in 1976).

364 | *11 Other Technologies for Backup Power Generation and Alternatives for Future Energy Supplies*

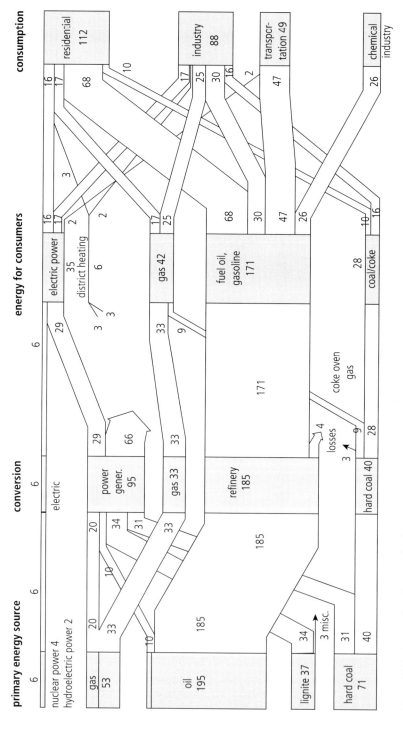

Figure 11.7 Energy flow diagram for the Federal Republic of Germany 1976 (Kalb and Vogel, 1980).

Such a reservation of coal as a carbon source for liquid energy carriers for transportation, for the chemical industry and for peak-load power generation for industry and households is not only desirable in view of the conservation of coal reserves. This would already become relevant early on, when the cost difference between hydrogen from coal (including CO_2 sequestration) – in particular with expensive imported coal – and solar hydrogen was not very great, that is, economically acceptable; then, solar hydrogen could in principle also become interesting for substitution of the general gas requirements. Compare Section 4.1 (Tables 4.2 and 4.3): H_2 from coal gasification in Europe (advanced technology, including 10 $/t CO_2 for sequestration) would cost 3.4 ¢/kWh; solar hydrogen (from North Africa, but without political costs) would cost 4.7 ¢/kWh. As described in Section 11.4.1, with a very large-scale utilization of solar energy (exploitation of its innovation potential), costs of down to nearly 4.0 ¢/kWh could be possible. (The solar-hydrogen option is also interesting for the case that the difficulties in sequestration of CO_2 should prove greater than expected, or if they should lead to higher costs than previously estimated.) For Europe, the solar-hydrogen cost quoted presumes, to be sure, a constructive offer on the part of the North African countries involved (with security guarantees and a rather low political premium) to allow the construction of solar plants by the European countries there. Without such an offer, one would certainly prefer the coal-based variant for the future gas supply (and the liquid energy carriers would in any case be produced in the USA).[106] Regarding the gas supply, one must in particular keep the large natural gas deposits in the "neighborhood" in mind. As long as these are present, the solar-hydrogen option will certainly be implemented in Europe only on a smaller scale.

As can also be seen from the flow diagram (Figure 11.6), in connection with the supply of electric power, already in this study, and for the first time in it (at least in the German-language literature) the import of solar thermal electric power, making use of high-voltage direct-current transmission (HVDC) was suggested.[107] The power requirements were to be fulfilled in roughly equal parts by

106) In the face of these alternatives, one can assume that corresponding offers from the North African countries would be forthcoming.

107) Transmitting power from the southerly solar regions to the distant consumer regions in Northern Europe was a very novel idea within the discussion of alternative energy sources at that time. Even after the presentation of the Kalb and Vogel study (1986a), the concept of a solar base-load electric power supply for Central Europe was still unheard of in professional circles, also, as the reactions of the time illustrate (see Appendix D). This extensive study (1986) was the first (and for a long time the only) publication in which the importation of solar thermal power was suggested as a real, economically feasible major option, capable of implementation within a manageable time for the energy supply, and in which its consequent development was called for. In addition to the publications in the press, at the time 80 copies of the detailed summary (which currently can be found on the Internet, see the reference for Kalb and Vogel, 1986b) were sent to all of the important institutions in the country: to the relevant research establishments, the appropriate industrial organizations, to institutes, but also to the media and the

imported (solar) power and decentralized generation using hydrogen–likewise produced on a solar thermal basis. For decentralized power generation, engine-powered block heating and power plants were planned. At that time, fuel cells were already under development, but the future economical operation of fuel-cell power generators was in general still regarded with great skepticism–especially in Germany, less so in the USA (cf. Part 2 of the double study initiated at that time by the present authors), and the efficient CCGT power plants in use today were not yet on the market. These distributed H_2-fueled power plants would have had at the same time the function of providing backup power to cover reductions in solar power production.

responsible political agencies. Up to that time, for unfathomable reasons, only hydrogen was considered to be an energy carrier which could be economically transported over long distances (at least within the European debate on renewable energy sources). An appendix to this circular gave special attention to the differences from the hydrogen concept. This part appeared later in Scheer 1987 (Kalb and Vogel, 1987: Solar electricity and Solar Hydrogen, pp. 213–229). The fact that the concept propagated up to that time (H_2 transport) was maintained is all the more surprising when one considers that already seven years earlier (1979) in Africa, the first HVDC transmission line (1400 km long) had been commissioned. The background of this behavior was described in the remarks on the organization of German solar research in Kalb and Vogel (1993).

12
The Large-Scale Use of Nuclear Energy

12.1
The Costs of Nuclear Power – Results

Table 12.1 gives the costs of new nuclear power plants of American design under mass production. The assumptions regarding investment costs, fuel-element processing, uranium costs, and operating and maintenance costs will be discussed individually below. The table lists the resulting energy price for three different assumptions concerning the specific investment costs. For an investment of 1100 $/kW ($1100 million per GW) and at a uranium price of 130 $/kg, as well as a real interest rate of 4% as generally assumed in this book, and an operating lifetime of 45 years, the resulting energy cost is 2.4 ¢/kWh. In the case of nuclear power-plant pools, which are located far from the regions where the power is consumed, the cost would be 2.9 ¢/kWh. In the lower part of Table 12.1, the accompanying investment costs including the transmission lines and backup power plants are also shown.

12.2
Investment Costs under Mass Production

12.2.1
Estimates According to the "Chicago Study"

At the beginning of Table 12.1, specific investment costs within the range of 900 to 1100 $/kW to 1400 $/kW are listed. In the following, we explain how these values were derived from the publication "Uni-Chicago (2004)." This study ("The Economic Future of Nuclear Power"), commissioned by the US DOE, will be referred to for brevity in the following as the "Chicago Study." It gives estimates for the cost of future nuclear power plants in the case of a certain degree of mass production, but at a deployment rate which is still small compared to that required for the conceivable complete substitution of oil and gas by nuclear energy.

The Chicago Study begins by estimating the investment costs without mass production. It gives a range with the three cost categories of 1200, 1500, and

Large-Scale Solar Thermal Power. Werner Vogel and Henry Kalb
© 2010 WILEY-VCH Verlag GmbH & Co. KGaA, Weinheim
ISBN: 978-3-527-40515-2

Table 12.1 Investment and energy costs of future nuclear power plants for large deployment series and for power plant pools. In the latter case, power transmission and backup plants are included.

Nuclear power plants (Large-scale scenario – capacity worldwide 1000–3000 GW$_{el}$)

2002-$						Nuclear pools, e.g., Scotland/Sweden (distance 1000 km)		
			900 $/kW	1100 $/kW	1400 $/kW	900 $/kW	1100 $/kW	1400 $/kW
Investment cost (1 GW plant)		M$	900	1100	1400	900	1100	1400
Investment with transmission + backup plant (see below)		M$				1684	1894	2208
Annual energy production								
Capacity factor		%	91	91	91	91	91	91
Operation at full output power		h/a	8000	8000	8000	8000	8000	8000
Annual energy yield (1 GW plant)		GWh	8000	8000	8000	8000	8000	8000
Capital costs								
	Real interest rate	%/a	4.0	4.0	4.0	4.0	4.0	4.0
	Interest factor		1.040	1.040	1.040	1.040	1.040	1.040
	Operating lifetime	a	45	45	45	45	45	45
	Real annuity	%/a	4.83	4.83	4.83	4.83	4.83	4.83

12.2 Investment Costs under Mass Production

(annuity at 2% real interest)	(%/a)	3.39	3.39	3.39	3.39	3.39	3.39	3.39
Capital cost per year Divided by annual energy output, gives:	M$/a	43.4	53.1	67.6	81.3	91.4	106.6
Capital cost	¢/kWh	0.54	0.66	0.84	1.02	1.14	1.33
(Capital cost at 2% real interest)	(¢/kWh)	0.38	0.47	0.59	0.71	0.80	0.94
Fuel cycle							
Fuel cycle cost without uranium	¢/kWh	0.50	0.50	0.50	0.50	0.50	0.50
including transmission losses	¢/kWh				0.52	0.52	0.52
Uranium price	in $/kg U:						
as in the middle category of U-reserves –>		80	80	80	80	80	80
as in the upper category of U-reserves –>		130	130	130	130	130	130
		200	200	200	200	200	200
(Cf. Spot price in mid-2007: 350 $/kg !)		300	300	300	300	300	300
Corresponding prices in $/lb U_3O_8:		31	31	31	31	31	31

Table 12.1 Continued.

Nuclear power plants (Large-scale scenario – capacity worldwide 1000–3000 GW$_{el}$)

2002-$		900 $/kW	1100 $/kW	1400 $/kW	Nuclear pools, e.g., Scotland/Sweden (distance 1000 km)		
					900 $/kW	1100 $/kW	1400 $/kW
(Cf. Spot price in mid-2007: 135 $/lb!)		50	50	50	50	50	50
		77	77	77	77	77	77
Future natural-U consumption (currently 25.5)	kg·U/GWh$_{el}$	115	115	115	115	115	115
		14.5	14.5	14.5	14.5	14.5	14.5
Energy cost due to Natural-U at 100 $/lb (future consumption)	¢/kWh	0.38	0.38	0.38	0.38	0.38	0.38
Including transmission losses	¢/kWh				0.40	0.40	0.40
	$/kg	80					
Energy cost due to natural-U at a price of ($/kg U) (future cons.)	¢/kWh	0.12	0.12	0.12	0.12	0.12	0.12
	¢/kWh	130					
		0.19	0.19	0.19	0.20	0.20	0.20

12.2 Investment Costs under Mass Production

			200	300		200	300	
Operation and maintenance (O&M) costs		¢/kWh	0.29	0.29	0.29	0.31	0.31	0.31
		¢/kWh	0.44	0.44	0.44	0.46	0.46	0.46
(In literature directly given in ¢/kWh)		¢/kWh	1.0	1.0	1.0	1.0	1.0	1.0
O&M costs including transmission losses		¢/kWh				1.05	1.05	1.05
Energy cost			–	–	–	–	–	–
4% interest	U-Price: 80 $/kg	¢/kWh	2.2	2.3	2.5	2.7	2.8	3.0
	130 $/kg	¢/kWh	2.2	2.4	2.5	2.8	2.9	3.1
	200 $/kg	¢/kWh	2.3	2.5	2.6	2.9	3.0	3.2
	300 $/kg	¢/kWh	2.5	2.6	2.8	3.0	3.2	3.4
2% interest	U-Price: 80 $/kg	¢/kWh	2.0	2.1	2.2	2.4	2.5	2.6
	130 $/kg	¢/kWh	2.1	2.2	2.3	2.5	2.6	2.7
	200 $/kg	¢/kWh	2.2	2.3	2.4	2.6	2.7	2.8
	300 $/kg	¢/kWh	2.3	2.4	2.5	2.7	2.8	3.0

Table 12.1 Continued.

Nuclear power plants (Large-scale scenario – capacity worldwide 1000–3000 GW$_{el}$)

2002-$		900 $/kW	1100 $/kW	1400 $/kW	Nuclear pools, e.g., Scotland/Sweden (distance 1000 km)		
					900 $/kW	1100 $/kW	1400 $/kW
Investment costs including power transmission and backup plants							
Investment costs, power plant (as above)	$/kW				900	1100	1400
Power transmission:							
Voltage					800 kV	800 kV	800 kV
Distance	km				1000	1000	1000
Transmission losses:							
Losses overhead line per 1000 km					3.3%	3.3%	3.3%
Losses overhead line of above distance					3.3%	3.3%	3.3%
Losses converter stations (two)					1.5%	1.5%	1.5%
Overall losses					4.8%	4.8%	4.8%

12.2 Investment Costs under Mass Production

Total investment costs:				
Required plant output power	MW	1048	1048	1048
Nuclear power plant including losses	M$	943	1153	1467
Overhead line				
Overhead line per 1000 km (without interest)	M$	100	100	100
Overhead line of above distance	M$	100	100	100
Interest rate	%/a	4	4	4
Interest during construction (4a of full interest)	M$	16	16	16
Overhead line (including interest)	M$	116	116	116
Undersea transmission line	M$			
Converter stations (two stations)	M$	225	225	225
Backup power plant (gas turbine)	M$	400	400	400
		–	–	–
Overall investment costs (nuclear power plant pools)	M$	1684	1894	2208

Table 12.2 Nuclear power plants: specific investment costs for a certain number of reactors constructed vby a particular manufacturer (according to the Chicago Study, 2004).

Number of reactors per manufacturer	Doublings	GW per manufacturer (at 1.35 GW per reactor)	Construction rate per manufacturer (total/30 a)	GW worldwide (with e.g., eight manufacturers)	Construction rate worldwide (total/30 a)	Cost reduction per doubling (%) (Chicago study)	Cost reduction factor per doubling
		GW	GW/a	GW	GW/a		
64	6	86	2.9	688	23	3%	0.97
64	6	86	2.9	688	23	5%	0.95
64	6	86	2.9	688	23	10%	0.90
256	8	345	11.5	2760	92	3%	0.97
256	8	345	11.5	2760	92	5%	0.95
256	8	345	11.5	2760	92	10%	0.90
1024	10	1380	46	11040	368	3%	0.97
1024	10	1380	46	11040	368	5%	0.95
1024	10	1380	46	11040	368	10%	0.90
2048	11	2765	92	22120	737	3%	0.97
2048	11	2765	92	22120	737	5%	0.95
2048	11	2765	92	22120	737	10%	0.90

a) Assumptions concerning interest payments: construction time 7 years. Full-interest period corresponds to one-half the construction time, that is, 3.5 years. At an interest rate of 4%, this gives for the interest during construction 14.7% ($1.04^{3.5} = 1.147$); at an interest rate of 2%, the value would be 7.2% ($1.02^{3.5} = 1.072$).
b) US inflation 2002–2003 = 2.3%.

1800 \$/kW (in each case without interest payments during the construction period, the so-called "overnight cost," in 2003-\$). The price under mass production is then estimated by making the assumption that the investment costs would be reduced by a certain percentage for each doubling of the *total* number of reactors of a particular type constructed (or of reactors constructed by a particular manufacturer), as a result of the "learning effect."

Table 12.2 gives an overview of this method of estimation: for the cost reduction per doubling of the number of reactors constructed (of a particular type or manufacturer), three different values were given as parameters: 3%, 5%, and 10%. On page 10-4 of the Chicago study, then, the probably relevant number of reactors

Cost reduction factor overall	Unit	Specific investment costs *without* interest during construction (2003-$)			Interest during construction at interest rate of (%)[a]		Specific investment costs *with* interest (1%) during construction (2002-$)[b]			Specific investment costs with interest (2%) during construction (2002-$)[b]		
		For cost category (Chicago study)					For cost category (Chicago study)			For cost category (Chicago study)		
	$/kW	1200	1500	1800	4%	2%	1200	1500	1800	1200	1500	1800
							With interest, category equates to:					
	$/kW				14.7%	7.2%	1345	1682	2018	1257	1572	1886
0.833	$/kW	1000	1249	1499	14.7%	7.2%	1121	1401	1681	1047	1309	1571
0.735	$/kW	882	1103	1323	14.7%	7.2%	989	1236	1484	924	1155	1387
0.531	$/kW	638	797	957	14.7%	7.2%	715	894	1073	668	835	1002
0.784	$/kW	940	1176	1411	14.7%	7.2%	1054	1318	1582	986	1232	1478
0.663	$/kW	796	995	1194	14.7%	7.2%	893	1116	1339	834	1043	1251
0.430	$/kW	517	646	775	14.7%	7.2%	579	724	869	541	677	812
0.737	$/kW	885	1106	1327	14.7%	7.2%	992	1240	1488	927	1159	1391
0.599	$/kW	718	898	1078	14.7%	7.2%	806	1007	1208	753	941	1129
0.349	$/kW	418	523	628	14.7%	7.2%	469	586	704	438	548	658
0.715	$/kW	858	1073	1288	14.7%	7.2%	962	1203	1444	899	1124	1349
0.569	$/kW	683	853	1024	14.7%	7.2%	765	957	1148	715	894	1073
0.314	$/kW	377	471	565	14.7%	7.2%	422	528	633	395	493	592

constructed in the USA is taken to be *64 reactors*. Calculated from the first reactor constructed, this corresponds to six doublings. Using the middle value for the learning effect, that is, a cost reduction of 5% per doubling, the cost decreases computationally to 73.5% of the cost of the first reactor; referred to the lower cost category (1200 $/kW), this gives 882 $/kW (without interest payments, in 2003-$).

Taking the interest into account and converting to 2002-$ yields 989 $/kW.

Using the lowest value for the learning effect (3%), the costs decrease only to 1121 $/kW. (The average of these two numbers amounts to 1055 $/kW.) In Table 12.1, the value of 1100 $/kW quoted, including interest (in 2002-$), is thus more or less typical of a production volume of this order of magnitude (64 reactors).

If 256 reactors were built by one manufacturer (eight doublings), the costs according to Table 12.2 at a 5% reduction rate per doubling would drop to 66.3%, that is, from 1200 to 893 $/kW (including interest, in 2002-$), thus to around 900 $/kW. (At a learning factor of 3% per doubling, they would drop only to 1054 $/kW.) In Table 12.1, therefore, only the lowest cost range (900 $/kW) should roughly represent this level of mass production. Below, we will discuss the topic of predictions of cost decreases with increasing construction volume in more detail.

In Table 12.1, the complete investment costs *including interest* (in 2002-$) are listed. Regarding the interest payments, we deviate from the Chicago study in order to maintain a unified description, and – as in all the other cost comparisons in this book – we use a real interest rate of 4% (baseline rate).[1] With a construction period of 7 years (Chicago Study, 2004, p. 5–17), and the assumption that the investment costs for the plant, which accrue successively during the construction period, are subject to interest on the average over one-half the construction period (i.e., 3.5 years), the investment costs are then increased due to interest payments by 14.7% (Table 12.2). (At an interest rate of 2%, they would increase by 7.2%.)

12.2.1.1 Conclusions from Table 12.1

The *middle cost range* (1100 $/kW) in the table represents *roughly* the rate of mass production (more precisely the production volume) which we are considering here, namely a total of 64 power plants of a particular type. It was, therefore, taken as the *baseline case* for comparison with other types of power plants. The lower cost range of 900 $/kW represents a larger number of plants constructed as a series. Comparing to the Chicago study, this corresponds roughly to the construction of 256 power plants of the same type. This cost range thus represents the deployment of *very large* numbers of plants. The construction of 256 power plants – at 1.35 GW per reactor, giving a total of 350 GW – within the assumed time of 30 years is equivalent to an *annual* construction of 8.5 reactors of a particular reactor type or manufacturer. All together, this would correspond worldwide to a *very large-scale scenario*: if one assumes, for example, that there would be eight manufacturers worldwide, then a total of 2760 GW (1.35 GW × 256 × 8) or roughly 3000 GW output capacity would be constructed in 30 years (with 64 instead of 256 reactors, the total would be 700 GW). However, 3000 GW would mean nearly a factor of 10 more than today's capacity. In the year 2004, the worldwide nuclear-plant capacity was 350 GW, with an annual energy output of 306 GWa. If 3000 GW would have to be installed within 30 years (and if we for simplicity assume a constant construction rate during this time), this would correspond to 100 GW per year of additional generating capacity.

For comparison, the estimates of the mass-production costs for *solar power plants* assumed considerably lower deployment rates. Thus, SunLab (quoted in S&L, 2003) assumed an annually increasing construction rate up to the year 2020 (beginning in 2006); in the last years, one would then have a deployment rate of *1 GW*

---

[1] This applies to the interest payments during construction, but even more to the computation of the capital cost share of the electricity costs – cf. Table 12.1.

per year (at a solar multiple of 3.7). This would require the production of nearly 100 000 heliostats per year, each with an area of 150 m², as assumed in this book, and as already presumed by Kalb and Vogel (1998) (100 000 heliostats of 150 m² each per year correspond to 0.9 GW/a at an SM of 3.7). In contrast to SunLab, Sargent & Lundy assumed for the last years of the same time period (i.e., up to 2020) only *0.2 GW per year* (at SM 3.7; cf. (S&L, 2003), pp. E-22, E-34, 3–7, B-3, 5–13). In comparison, the above assumption of 64 nuclear reactors constructed per year (86 GW) over 30 years yields a rate of 3 GW per year. This construction rate (which results from the 64 reactors and the time of 30 years that we have arbitrarily assumed) would, however, apply only if the deployment rate were constant. In reality, the rate – similarly to the case of solar power plants – would begin at a low level and would increase over the years so that in the last year of the period considered, the construction rate would be much higher. This must also be taken into account in making comparisons with the construction rates of solar power plants.

The Chicago study was based (like EIA AEO, 2007) in the main on American power plant types.[2] In the case of the European pressurized water reactor (EPR), which is currently under development for the European market and which is technically more complex (it has among other features a core-catcher and a double-walled containment vessel), we could expect higher investment costs; according to Uni-Chicago (p. 3-2), they could be 20–40% higher. For a construction series of all together 64 plants, they would – according to that study – cost, for example, 1400 $/kW instead of 1100 $/kW.

The recently completed single nuclear power plants in Japan and South Korea, in contrast, cost 2400 and 2300 $/kW, respectively (Uni-Chicago, p. 2–14). It is noticeable that also the US Energy Information Administration (EIA) in its "Assumptions to the Annual Energy Outlook 2007" (EIA AEO, 2007) lists a cost of 1982 $/kW (valid from the 5th power plant on, that is, after overcoming the initial problems in the first plants built, without interest payments, in 2005-$). Converted to 2002-$, this would be 1880 $/kW.

In contrast to the Chicago study, the EIA does not quote a cost *range*. The cost value given by the EIA lies at the upper limit of the range quoted in the Uni-Chicago report (1200–1800 $/kW for individual power plants without mass production and interest payments). Nevertheless, we follow here the extensively documented Chicago study, with its lower value. This lower value (1200 $/kW) represents – according to the Chicago study – on the one hand the newest American reactor types; on the other hand, as is emphasized there, the range of 1200–1800 $/kW also expresses a range of uncertainty in the cost estimates (p. 9-5: "the cost

---

2) According to the Chicago study, in particular the following types are to be considered: among the American power plant types are the ABWR (advanced boiling water reactor, 1350 MW) from General Electric (here, there is a cooperation with Japanese firms), the AP 1000 (pressurized water reactor, 1150 MW) from Westinghouse, and the French SWR 1000 from the former Framatome (now Areva) in cooperation with the German company Siemens, which is also aimed at the US market (boiling water reactor, 1010 MW) (pp. 9-5, A4-10, A4-21).

range also allows for uncertainty in cost estimates for reasons other than reactor type"). The investment costs, which we have derived here from this value, thus represent a *lower limit* to the costs to be expected in mass production.

The deployment numbers assumed in Table 12.2 (e.g., the 64 plants per manufacturer or per type) should be quite relevant. Thus, the president of Toshiba was quoted in the year 2008 in the German business newspaper Handelsblatt (2008a) as stating that his company would construct 31 nuclear reactors by 2015 (i.e., within only 7 years). By the year 2020, the construction rate would already be 65 (probably referring in each case to the beginning of construction). Toshiba took over the nuclear plant manufacturer Westinghouse in 2006 and is now the market leader, ahead of the French firm Areva.

That these plants will be sold for *considerably* less than the above-mentioned first power plants in Japan und Korea (2300 $/kW) is hardly to be expected. Therefore, in Section 4.1, we presumed a price of 1800 $/kW without interest payments for nuclear plant types that are being constructed for the first time, whereby, in addition, the more complex European reactor types such as the EPR were assumed; with interest, this would correspond to 2020-$, that is, around 2000 $/kW. The value 1800 $/kW corresponds to the upper cost category in the Chicago study and represents there also the more expensive, safer reactor types (EPR: with among other things a core-catcher and stronger armoring to protect against airplane crashes). The operating and maintenance costs and the price of uranium consumed were taken to correspond to current values.

The energy costs listed in Table 12.1 do not contain the *external costs* associated with nuclear power plants, that is, the risk of a *major accident* and the related damage in a larger region around the power plant. The renowned Swiss PROGNOS Institute in 1992 took it upon itself to quantify the external costs related to today's reactors under European conditions, arriving at a value of 2.3 €-cent/kWh (Enquete Commission, 1994), corresponding to 3.0 US-¢/kWh in 2002 currency.[3] Even though such risks can hardly be numerically quantified (in the sense of a reliable prediction), the cost range in this estimate is worthy of note. The nuclear power-plant operators, to be sure, predict that future reactors will be more secure than those currently in operation (on which the PROGNOS study was based). Newer evaluations of the external costs, therefore, obtain lower numbers. But here, we must consider that these numbers are derived from the accident risk without consideration of the *danger of terrorist activities*. Given this fundamental change in the risk potential as compared to earlier times, one must conclude that the potential for future damage is *in total* not less than previously estimated, but–possibly *quite seriously*–even greater.

This potential danger by itself makes it seem out of the question to deploy the very large number of nuclear power plants mentioned above, which would be required for the replacement of fossil energy carriers to a large extent, in densely populated regions of the relevant countries. In Table 12.1, we therefore also give the energy costs for the case that the nuclear plants would be built far from popu-

---

3) Inflation (Germany): × 1.20; 1 = 1.96 DM; purchasing power parity (OECD) 2002: 1 = 1.043.

lation centers and – like the solar plants – would have to be connected to them via long-distance transmission lines. (Such suggestions have already been made in the past, beginning in the 1960s.) They would then also require a backup system near the power-consuming region for the case of an interruption in power transmission, for example, gas-turbine plants.[4] The nuclear plants would then reasonably be combined into large pools at their distant sites; for example, 5–10 plants could be built at one location. This concept of distant plants with a long transmission line is referred to here under the generic term "nuclear power-plant pools." (In the general energy discussion, however, the term "pool" as a rule means only a group of power plants in the neighborhood of a consuming region, that is, without a long transmission line; this terminology is used also in the Chicago study – see below.)

In the next section, we shall take up the important question of whether or not the power plant manufacturers, if they can indeed produce the plants at the costs estimated above, will also *sell* them at the corresponding price, or at a close approximation to it; or whether the buyers would not have to pay considerably higher prices.

First, however, for the interested reader we want to take a closer look at the Chicago study, regarding the question as to whether the cost-reduction factor of "3–5%" per doubling of the number of plants constructed, as used above, is in fact a reasonable value for the case we are considering here:

> The topic of cost reduction with increasing deployment volume is complex; of course there can be no guarantee of the correctness of the prognoses. The estimates are oriented on the experience gained in building nuclear power plants from the 1960s to the 1980s. The Chicago study, therefore, does not quote a definite value, but rather it specifies a range from 3% to 5% up to 10% cost reduction per doubling of the number of reactors constructed – whereby these cases are at the same time denoted as typical in their trends for certain boundary conditions. In the Chicago study, these questions were unfortunately not dealt with in a consistent manner and are not clearly described. On the one hand, the percentage cost reductions are, as mentioned, attributed to certain boundary conditions. As stated on pages 4–24, briefly summarized, the 3% rate applies to the construction of individual reactors, irregular orders and an uncertain legal framework (in relation to the requirements to be fulfilled by the reactors, with resulting delays in the construction times). The value 5% applies to practically regular orders, whereby the power plants are, however, only occasionally combined

---

4) For a complete gas-turbine power plant, we assume a price of 400 $/kW (400 million $/GW); see Chapter 10. Reminder: for the CCGT backup plants in the solar power system, gas turbines were presumed to be the fast start-up facility for the slower-starting steam-turbines. Since the gas turbines would be integrated into the CCGT plants, and therefore many components of a stand-alone gas-turbine power plant would be unnecessary, we assumed there a cost of only 300 $/kW.

into "pools,"[5] and with a stable legal situation. The value 10% applies to the construction of pools, at a fixed, planned construction rate and a stable legal situation. On the other hand, the most important question for the case of a large-scale deployment scenario as considered in this book, namely up to what number of reactors constructed these values continue to hold, is not clearly discussed. This question is only hinted at; the conclusions of the Chicago study thus remain unclear on this point. As extensively discussed there on p. 4–9, the basic idea that for each doubling of the number of plants constructed, a *constant* cost-reduction factor can be expected, is controversial. For example, Zimmerman (1982), who is cited there, expects on the basis of his evaluation of the cost trends in the past that for the first doublings of the number of reactors constructed, the cost reductions will be considerably greater than for later doublings. According to his results, the cost-reduction effect after the first doubling is 7.4%, after the second 5.9%, after the third 4.0%, and after the fourth only 2.7%. As concluded in the Chicago study, he arrives, however, "on the average" likewise at a value of ca. 5% per doubling – but only for this limited number of reactors (16)! Thus, on p. 9-8, the high cost-reduction factors are discussed only in connection with a number of reactors constructed of up to 8. And in their summary (p. 10-4), where the cost reduction is referred to a total of 64 reactors constructed, only the factors 3% and 5% are mentioned. (The upper limit of the range of cost-reduction factors quoted in the Chicago study, 10% per doubling, will probably retain a limited validity only for the first few doublings.) In general, the cost-reduction factors can clearly not be employed continuously up to arbitrarily high construction volumes. For 64 reactors, the correct value should lie somewhere between 3% and 5%. In this case, the investment cost of 1100 $/kW assumed above should be roughly typical of this scenario.

The fact that the relative cost reductions cannot be extrapolated to arbitrarily high construction volumes is incidentally already clear from the cost examples given in Table 12.2. Thus, the factor 10% would lead with six doublings (64 reactors) to investment costs (without interest payments) of only 638 $/kW (in 2003-$) or, in 2002-$, of 622 $/kW; this is practically the same as the current cost of a natural-gas-fueled CCGT power plant (580 $/kW without interest in 2002-$)! After eight doublings, a nuclear power plant would, according to this calculation, be even noticeably cheaper than this technically simplest of all power plant types. This alone demonstrates the purely theoretical character of such considerations, when one tries to apply the same factor to arbitrarily large numbers of reactors constructed.

---

5) In contrast to the use of the term "nuclear power-plant pool" in this book (namely: large power plant parks located at some distance from the consumer regions, with correspondingly long transmission lines), in the Chicago study, "pool" is understood only as a large group of nuclear power plants somewhere near the consumer region.

12.2.2
A Problem: The Lack of Competition among System Manufacturers – The Contrast to Solar Energy

The numbers given above refer to the *production costs* of reactors. The *prices* demanded by the reactor manufacturers under unfavorable circumstances could – apart from a profit margin within the "usual" range – be quite different (i.e., considerably higher). This is true especially in the case of a high rate of power plant deployment.

As we have already mentioned, a situation could occur in which, due to continued major increases in oil prices, a *very* rapid switchover to other energy sources would be desirable or necessary. The resulting massive construction of new power plant parks would then have to be carried out within a very short time. It is not out of the question that an extreme situation would then arise, with massive price increases and as a result with the need to escape from this vicious circle as quickly as possible. Then the construction of new power plant capacity ought to be completed in the ideal case within 15 to 20 years. This is, however, by no means feasible in the case of nuclear energy. With respect to maintaining the energy supply and reducing the economic pressure from expensive imported oil, longer construction periods would not be justifiable; in the extreme case perhaps 20–25 years at most. Given the necessary volume of power-generating capacity which would have to be constructed, nuclear energy would however – even with a maximal effort – not be able to meet this goal. In order to achieve an appreciable substitution effect for fossil energy carriers, the worldwide nuclear-energy generating capacity would have to be increased roughly *tenfold* relative to the current level. In Section 12.6.1, on the lifetime of uranium resources, we discuss this in more detail. There, we presume that nuclear energy would be used only for electric power generation (substitution of coal in fossil-fuel power plants), but not for hydrogen production. Regarding the dimensions of the necessary construction rate, we must keep in mind that at present – in contrast to the 1980s – only a small reserve of industrial and personnel resources is available for the construction of nuclear plants. Increasing the construction capability within the manufacturing companies (personnel and material), and the actual construction of the plants, would have to take place nearly simultaneously. We note that here, we are speaking of setting up a construction capability roughly 10 times greater than what was available in the 1970s and 80s. In the past, 10 years were allowed just for the planning and construction of a single nuclear power plant, and this with tried and tested reactor types (but for the construction of single reactors, and not of reactor pools).

We must thus ask on the one hand the question as to whether such short times for replacing the power plant park using nuclear plants are at all feasible. On the other hand, we must keep in mind that with such a construction boom, the prices of nuclear plants might not remain nearly stable. (Here, it is unimportant whether the large number of plants were to be constructed within 20 or 30 years.) Since the task at hand (high construction rates and simultaneous buildup of their construction capabilities) would already be very difficult for the manufacturers to accomplish in a short time – under the important assumption that no shortcuts were taken in the planning and fabrication of all the components relevant to the security of the plants – the companies would have only a limited interest in accepting orders for still more plants. If these were nevertheless required, this would be reflected in corresponding price increases.

An additional fact to be considered here is the reduction in the number of manufacturers of nuclear power plants in comparison to the 1980s, which can be observed in at least some countries. In some cases, the manufacturers have gone out of business (in Germany, e.g., the AEG company), but more importantly, there were a number of mergers and takeovers. For example, the German company Siemens and the former Framatome (France) are currently working together (developing the EPR) so that in Germany and France, only this one system manufacturer still exists – insofar as one leaves out the Swedish firm ABB, which has some production facilities in Germany. ABB incidentally also cooperates in building nuclear plants with Westinghouse, while General Electric works with Japanese manufacturers,[6] so that worldwide, a reduction in the overall number of manufacturers can be ascertained; in any case, there are not more manufacturers active now than at earlier times, as would be required to satisfy a large increase in demand for nuclear plant construction.

In the face of these two aspects (large volume of demand and limited number of manufacturers), it can hardly be assumed that the suppliers would sell the power plants at prices which reflect only their production costs. As a result of limited construction resources and the enormous demand, the chance to "make a killing" (with very high profits) would open up. The conditions for a price diktat would thus be fundamentally different from those in the past. From the macroeconomical viewpoint, to be sure only the production *costs* are relevant; the profits remain within the national economy (insofar as the manufacturers are not globalized). But from the viewpoint of power consumers in households and industry, this interpretation is only partially relevant, since such profits can be socialized only to a limited extent. A certain portion of the increased expenses for the consumers (in the case of very high profits) represents a real loss for a major part of society.

The problem "boom in demand and limited time frame" of course would apply in a comparable situation also to solar energy. Here, however, there is an important difference: nuclear power plants represent a *complex* und *security-relevant* technology, which can be delivered ready-to-operate only by large *system manufacturers*. In the case of solar power plants, one is dealing with a technology with a modular structure. The resulting important consequences in terms of construction times were already discussed at the beginning of this book. Regarding the competition situation, it is interesting to note that parts of a solar plant (such as the mirror fields, or their individual components including tracking drives and mirror frames) could be produced by many industrial firms. Even the orders for individual com-

---

6) From the Chicago study, it is however not clear just how close these co-operations are (cf. pp. A4-10 and A4-16).

ponents could, if necessary, be spread over a number of suppliers. Here, all the resources available (such as those of the automobile industry or steel construction industry) could without major changes be mobilized for the production of the required components. A similar situation holds for solar tower plants regarding the tower thermal circuits (they could be fabricated by many firms engaged in plant engineering and construction) or for the thermal-storage systems (plant engineering, chemical industry). In contrast to nuclear plants, there are also no security-relevant construction steps which would make it essential to employ a competent system manufacturer. In awarding production orders for the components, the competition among the suppliers would thus be completely different from the situation for nuclear plants. Here, from the viewpoint of the buyers, the production costs would be decisive and would represent the principal factor in price negotiations so that very large profit margins or inflated oligopoly prices would not be a problem. In contrast, for nuclear plants, during a construction boom the prices could rise to levels which lie *considerably* above the production costs. Such price inflation is not taken into account in the tables. A comparable scenario could, by the way, also occur for the price of natural uranium (and already did so in the summer of 2007); see the following section.

12.3
Operation and Maintenance Costs; Fuel Costs

12.3.1
Operation and Maintenance (O&M) Costs

In the case of a major buildup of nuclear-plant capacity, it can be assumed that at each location, several plants would be constructed. Then a certain decrease in operating costs (in particular the personnel costs) can be assumed. The O&M costs will be estimated in this chapter for the case of a future massive deployment of nuclear plants – in contrast to the WNA data given in Table 12.3, which reflect *today's* costs (with a relatively small number of power plants); here, we take a value of 1 ¢/kWh (2002-$).

Comparison: in Uni-Chicago (2004, p. 5–17), annual fixed O&M costs of 60 $/kW (with 8000 h of full-output operation per year, this corresponds to 0.75 ¢/kWh) and variable costs of 0.21 ¢/kWh are given so that all together, 0.96 ¢/kWh

Table 12.3 Operating and maintenance costs for nuclear power plants according to the World Nuclear Association (WNA Report, 2005).

	1981	1985	1990	1995	2000	2003
	¢/kWh (2003-$)					
O&M costs	1.41	1.93	2.07	1.73	1.37	1.28

results. (The question of a possible dependence of the O&M costs on the future extent of deployment of nuclear plants is not discussed there.)

12.3.2
Enrichment and Other Fuel Costs, Not Including the Cost of Natural Uranium

Table 12.4 shows the nuclear fuel costs without the raw uranium cost. In the year 2003, they amounted to 0.36 ¢/kWh. This includes the cost of conversion into UF_6, enrichment, and the fabrication of the fuel elements (as well as the cost of spent-fuel disposal). According to UIC (2007), in the year 2007 per kg of uranium as reactor fuel, roughly $90 of the cost was due to conversion, $985 to enrichment, and $240 to the fuel-element fabrication. Enrichment thus makes up the greatest portion of the cost. As shown in Table 12.4, the "fuel cost without uranium" has

Table 12.4 Nuclear fuel costs without the cost of natural uranium (WNA Report, 2005, p. 11).

Year	1981	1985	1990	1995	2000	2003	2007
	¢/kWh (2003-$)						
Fuel costs–total[a] (¢/kWh)	1.06	1.28	1.01	0.69	0.52	0.44	[d]
Uranium price[b] $/lb U_3O_8	(44 1980)	(15)	(9)	(10)	(9)	(13)	135 Spot price mid-2007
Portion of energy costs due to uranium[c] (¢/kWh)	0.29	0.10	0.06	0.07	0.06	0.08	0.9 ¢/kWh
Fuel costs–without uranium[a] (¢/kWh)	0.77	1.18	0.95	0.62	0.46	0.36	

a) Total fuel costs according to WNA (2005). From these, the cost of natural uranium – from other sources – was subtracted (it depends on the uranium price in the particular year) so that in the last line, the fuel costs *without* uranium costs are obtained. (The line "fuel costs–total" includes spent fuel-element recycling for plants in the USA; in Germany, these costs are higher than in the USA – WNA Report, 2005, p. 11.)
b) Uranium price of the relevant year according to UXC and Uranium.info (quoted in Wikipedia, 2007).
c) Regarding the consumption of raw uranium per kWh_{el}, here we have used the worldwide average for currently operating power plants (2004). An increase of 10 $/lb in the uranium price yields an increase in the energy cost of 0.066 US-¢/kWh_{el} (see Section 12.4).
d) The right-hand column shows as an additional piece of information just how the extremely high uranium price, which in June 2007 reached 135 $/lb of U_3O_8 (= 350 $/kg U), would affect the price of electrical energy. (In mid-2008, the price was at a minimum (55 $/lb), and then rose again by August 2008 to 65 $/lb U_3O_8.) At the worldwide average uranium consumption rate per kWh_{el} of present reactors, the maximum price of 135 $/lb, which was reached for a brief period, would have caused an increase in energy price by nearly 1 ¢/kWh, while the price in August 2008 of 65 $/lb corresponds to an energy-price increase of 0.43 ¢/kWh (cf. also Section 12.6.2).

fallen sharply since 1990; this applies in particular to the *enrichment cost*. This reduction was due for one thing to the overcapacity for enrichment available at the end of the Cold War, and for another to the introduction of an improved centrifugal enrichment process. Thus, the enrichment firm URENCO, with its own development of centrifuge technology, was able to increase its production capacity in spite of the lower prices. (The immense and possibly devastating consequences of this technical development, which are already becoming clear, will be discussed below)

For comparison with the value quoted above (2003: 0.36 ¢/kWh), we refer to Uni-Chicago (2004, p. A5-4). This study cites fuel costs of 1420–2209 $/kg U (5% enrichment), corresponding to a contribution to the energy price between 0.36 and 0.55 ¢/kWh. Here, however, the raw uranium cost (222–353 $/kg of *enriched* uranium) and the interest payments are included. (The costs of fuel processing in each process step during the time up to the loading of the fuel elements into the reactor are subject to interest.) Without raw uranium and interest, the price would be 840–1295 $/kg U; this then yields for the contribution to the energy price 0.21–0.32 ¢/kWh (in 2003-$). In Uni-Chicago (2004), the fuel-cycle costs are not discussed in any detail, even though the study deals with a possible future use of nuclear power on a very large scale; only the numbers quoted by the Nuclear Energy Agency (NEA) from the year 1994 and those of the Nuclear Assurance Corporation (NAC) from the year 2000 are quoted (converted to 2003-$). In addition to the fuel costs, the cost of *spent-fuel recycling and disposal* of *0.11 ¢/kWh* must be taken into account, whereby long-term disposal at the Yucca Mountain site following short-term storage is presumed. The overall costs including waste disposal (but *without the raw uranium cost*) are thus found in the Uni-Chicago study to be 0.32–0.43 ¢/kWh.

Comparison: EWI (2007) gives the total fuel costs *including* raw uranium as 1 €-¢/kWh.[7] Converted using purchasing power parity according to the OECD (2006: $1 = 0.88 Euro), this is equivalent to 1.1 US-¢/kWh.

Corresponding to the results of the WNA report and the Chicago study, we thus find for "fuel costs – without uranium" for future large-scale deployment of new reactors, and *at the current level of depletion* of the natural uranium, a value of ca. *0.4 ¢/kWh*.

Through a future *increased depletion level* of the natural uranium – currently, it is depleted from 0.71% ^{235}U to 0.3%; in the future, with high uranium-ore prices, it will probably be depleted down to ca. 0.1% – the effort required for isotope separation will be increased by ca. 70% (Greenpeace, 2006, p. 25). Since it is not quite clear just how this will affect the price, in this book we simply assume an increase in the cost of enrichment by one-third. The enrichment costs, however, make up ca. 75% of the total "fuel costs – without uranium," so that the latter would increase *by 25%*. The "fuel costs – without uranium" are then found to be 0.5 ¢/kWh. We shall use this value in this book.

---

7) Note: In (EWI, 2007, table 4), these costs are not listed, due to a typographical error. Personal communication EWI: "The `variable costs' (0.2 €-cent/kWh) listed are only the *remaining* variable costs. To these must be added the fuel costs (including waste-disposal costs) of around 1 €-cent/kWh, which were not given there."

12.4
Consumption and Cost of Natural Uranium per kWh$_{el}$

For the computation of the contribution of the uranium price to the overall energy cost given above, one must assume a certain consumption of natural uranium per kWh$_{el}$. The starting point for estimating this consumption is the worldwide average from currently operating nuclear power plants. Schindler and Zittel (2007) give the ratio of current natural-uranium consumption by operating nuclear plants (2004: 67 320 t) to the electrical energy which they generate (2004: 2638 TWh). According to their numbers, the average specific consumption is *0.0255 g natural uranium/kWh$_{el}$*. Therefore, a price of 100 $/lb U$_3O_8$[8] for natural uranium corresponds to a contribution to the overall energy cost of 0.66 ¢/kWh$_{el}$.

When the uranium price is high and with newer reactors, the specific consumption of natural uranium will decrease. The reasons for this are as follows:

1) The more intensive depletion of the uranium ore by the enrichment plants: thus far, uranium from the ore is depleted from 0.71% ^{235}U, corresponding to its natural abundance, down to ca. 0.3%. In future (with higher uranium prices and in view of today's already relatively low enrichment costs), it will probably be depleted down to ca. 0.1% – cf. Greenpeace (2006), p. 25. The yield up to now was 0.71% − 0.3% = 0.41% ^{235}U; in the future, however, it would be 0.71% − 0.10% = 0.61%. For this reason, the consumption of natural uranium per kWh$_{el}$ would decrease relative to the present value to (0.41/0.61 =) 67%.

2) A lower percentage of remaining ^{235}U content[9] in spent fuel elements due to the higher degree of enrichment: at a higher degree of enrichment, the ratio of the ^{235}U remaining in the spent fuel elements after their removal from the reactor to their initial ^{235}U content is lower, that is, more favorable. (A higher enrichment is also advantageous due to the resulting lower waste-disposal costs.) At present, new fuel elements contain ca. 4%, and spent fuel elements ca. 1% of ^{235}U. This corresponds to a degree of utilization of 75%. In the future, new fuel elements will contain ca. 6% (cf. also Hospe, 1996, p. 40: for SWR 1000: 5.3% instead of 3.2%), and spent fuel elements will contain ca. 1% of ^{235}U. This yields a ^{235}U utilization of 83%. The consumption of natural uranium per kWh$_{el}$ would then sink relative to the current level to (75%/83% =) 90%.

3) A higher electrical efficiency of the power plants: the current efficiency is 34%; in the future, it could be 36% (Peter and Ständer 1995, p. 214); cf. also Hospe

---

[8] 1 lb = 0.4536 kg; 1 kg U$_3$O$_8$ = 0.848 kg U ⇒ 1 $/lb U$_3O_8$ = 2.60 $/kg U.

[9] This is not identical with the so-called burn-up. The latter specifies how much thermal energy per kg of uranium – or more precisely: per kg of the mixture of uranium isotopes employed – is produced in the reactor. Highly enriched uranium, with a higher fraction of fissionable ^{235}U, can deliver correspondingly more energy per kg of uranium; but more natural uranium is required to produce it, so that there is no net savings on natural uranium.

(1996, p. 40): 35.3% instead of 33.4%. The consumption of natural uranium per kWh$_{el}$ compared to the present value is thereby lowered to (34/36 =) 95%.

Summary: Owing to these three effects, *in future* the specific consumption of natural uranium as compared to the present worldwide average will decrease to (0.67 × 0.9 × 0.95 =) 57%; thus decreasing from today's value of 0.0255 g/kWh$_{el}$ down to 0.0145 g/kWh$_{el}$. Referred to a uranium price of 100 $/lb U$_3O_8$ (260 $/kg U), this then yields a contribution from the cost of natural uranium to the overall energy price of 0.38 ¢/kWh$_{el}$.

12.5
The Problems Associated with Nuclear Energy

12.5.1
Consequences of the Development of Centrifuge Technology

The URENCO concern mentioned above, a German/Dutch/British consortium (Urenco Ltd., Marlow/GB), the licensee of the low-cost centrifuge process, has a good chance of going down in history as the initiator of one the worst catastrophes for humanity: thus if terrorists or "terrorist regimes," making use of illegally copied centrifuge technology from URENCO–which would not be particularly difficult (cf. the example of Pakistan's nuclear weapons capability)–are able to acquire and use nuclear weapons against large numbers of people or at least threaten to do so, future political developments and the world order would become completely unpredictable. The cost advantage of the URENCO process for producing fuel elements as mentioned–a cost reduction which is completely insignificant on a worldwide scale–is in total disproportion to these potential repercussions. The USA alone spends annually $590 billion in order to maintain a certain "world order" and in particular to avoid itself being subject to extortion (defense budget in 2006: $410 billion plus associated items). These expenditures would be rendered worthless by the uncontrollable proliferation of nuclear weapons.

The fact that such developments, which are driven by the potent individual economic interests of minor players, cannot be prevented in practice makes it clear just how much danger is associated with the peaceful use of nuclear energy. Furthermore, every country that has a nuclear power plant could construct nuclear weapons with a moderate effort, even without the centrifuge technology. The future global use of nuclear energy on a large scale would thus be accompanied by a final proliferation of nuclear weapons throughout the whole world.

12.5.2
General Problems of Nuclear Power Generation

One could quite correctly take the standpoint that the further proliferation of nuclear weapons is already unavoidable today (in view of the centrifuge technology), and would therefore not be prevented by a worldwide ban on the construction of nuclear power plants, but at most would be delayed. However, it becomes clear in any case just what havoc could be brought by the use of nuclear power, or rather what results have been brought about by those who have played down its risks. The overall byproducts of the use of nuclear power, in view of these consequences which are already becoming all too clear, are devastating, as already pointed out. One should thus learn a lesson from the mistakes of the past (even if the problem that has resulted from them can no longer be completely avoided), and consider the risks in a realistic manner before taking the decisions, which are now imminent. For example,

- Regarding the problems of *technical* security associated with the construction of hundreds of new nuclear plants in densely populated regions of Europe, America, and Asia; even considering only those security failures which have already occurred, such a scale of construction would be hard to justify.

- Regarding the completely new security situation, that is, the current real danger of targeted *terrorist attacks* on nuclear facilities. In the past, in discussing nuclear security, the question was only how probable it is that the plants would fail *by accident*; the probability of purposeful damage or even attacks was never considered. Today, this danger has not only become real, but also the manner of possible attacks has fundamentally changed compared to what was previously conceivable. Not only is there a possibility of considerable destruction of the facilities outside the reactor containment vessel caused by massive explosive charges transported in automobiles or trucks, but also the *new* reactors will have to be protected against intentional airplane crashes. Some reactors (e.g., EPR) should be protected by appropriately thick concrete walls not only against civil aircraft, but also against the much faster military aircraft. But even this does not offer complete protection against possible terrorist tactics, for example, not against air attacks with bombs on board, and also not against attacks with concrete-breaking weapons. For the evaluation of protective measures against air attacks, up to now the rigid, concrete-penetrating turbine shafts of civil and military aircraft were taken to constitute the decisive threat. However, special penetrating warheads could be purposely carried on an aircraft and optimized for the purpose of causing maximum damage to the reactor. Modern concrete-breaking weapons as a rule contain depleted uranium; even for this, worldwide organized terrorists could readily find a number of supply sources.

- Regarding the massive changes in *uranium mining and extraction* brought about by the necessary use of low-grade ores. These mean not only that per kilogram of uranium extracted, at least 10 times (possibly up to 100 times) more ore must be processed than was required in the past, but also that in connection with the uranium extraction (dissolution of uranium out of the rock), correspondingly larger amounts of tailings will accumulate, whose radioactivity is only ca. 15% lower than that of the original ore. For the mining of low-grade ores, presumably to a large extent stockpile-leaching will be used, or for extraction, the *in situ* leaching process; with these mining and extraction processes, serious environmental problems can be expected. (In Greenpeace (2006), mining and extraction and their environmental aspects are described in detail; see also (WISE, 2006).)

- Regarding the problem of *long-term waste deposition*. We will not treat this important and much-discussed question in any detail here. It is clear that this already serious problem would be correspondingly enormously exacerbated if the current nuclear power plant capacity were to be increased by tenfold.

- Regarding the question of when older nuclear plants should be shut down; that is, the problem which will increasingly be on the agenda in the coming years, that with *increasing age*, the probability of plant failures increases, but at the same time, potent economic interests oppose the shutdown of the older plants.

- Last but not least, regarding the newer, more economical reactors such as those which are currently being developed in the USA.

For all these questions, in practice there is not only the matter of a factual assessment (which, given the potential dangers, must be carried out with great care), but also one must keep in mind that there are powerful forces which oppose such an assessment. Whether political controls can function in the face of such forces is more than questionable. It may, for example, be true that the new reactors which are currently under consideration for a large-scale build-up of nuclear energy, and which were taken as examples for the cost estimates given above (ABWR, AP 1000), will in spite of their simpler construction principles be just as secure or even more secure than the previous generation, as a result of their novel security concepts (more passive instead of active safety). (Such questions can, however, be answered only after a very thorough-going investigation of the subject.) But it must be kept in mind that this new generation of power plants was not developed primarily with the goal of improved security, but rather mainly for cost reduction – in the face of the overpowering price competition of natural-gas power plants at the time of low-cost gas in the 1990s. This applies even more to even cheaper reactors such as the simplified boiling water reactor (Uni-Chicago, 2004, p. S-20) or the economic simplified boiling water reactor (ESBWR). Who can judge whether the simultaneously improved security claimed is only wishful thinking? Certainly, nuclear power plants will become less expensive under mass production, and fundamentally, it is also conceivable that improvements in plant design could reduce costs *and* improve security. But for the objective observer, it appears just as clear that such (technical) improvements in security are more than compensated by the new danger of targeted attacks, and that this new danger is particularly acute for the often poorly protected older plants – still more than for the planned new ones – while the older plants are now expected to be kept on line for decades to come. Here too, there was a complete misjudgment in the estimation of the "probability" of a major nuclear accident in the past, as is clearly visible in hindsight.

Just how strongly certain interests enter into the "objective" evaluation of the situation can be seen especially clearly in the comparing the past estimates of uranium resources. On the part of the proponents of nuclear energy, there was a *complete* reversal of opinion. In the 1980s, when there was a discussion in Germany of the acceptance of nuclear-fuel reprocessing and of continuing the Fast Breeder project, the compelling need for both was justified by the supposedly limited stocks of uranium, which would otherwise not guarantee the long-term operation of the nuclear power plants. Today, these technologies are no longer an option. In the current discussion of a major expansion of the use of nuclear energy *without* fuel reprocessing, the argumentation has changed radically; according to current claims, the uranium stocks are so immense that there are no problems whatsoever to be expected on this count. It may be that the earlier estimation was simply false. It is, however, equally possible that the future problems of uranium mining and extraction are now being intentionally minimized. (See, e.g., Greenpeace (2006) for a discussion of these questions.)

For citizens and responsible politicians, the problem faced in most decision making takes the form of "it might be, or it might not be." The large-scale application of nuclear energy thus becomes a kind of all-or-nothing gamble. In the case of a *massive* deployment of nuclear plants, the risk would be correspondingly great.

12.6
Uranium Reserves

12.6.1
Lifetime of the Reserves in the Case of a Massive Increase in Nuclear Power Production

Classification of the reserves according to the degree of their exploration (Lübbert and Lange 2006):

RAR: "reasonably *assured* resources."

EAR I: "*estimated* additional resources," category I.
EAR I refers to resources in *known* deposits, whose extent has however not yet been adequately investigated by drilling boreholes so that the amount of uranium available can only be estimated.

EAR II: "*estimated* additional resources," category II.
EAR II refers to ore deposits whose existence is *presumed* in *the neighborhood* of known deposits, but has not yet been directly demonstrated.

SR: "*speculative* resources."
Sources whose existence in certain countries has been assumed on the basis of the general geological characteristics of the territory, but which have not been localized in any precise way.

12.6.1.1 Lifetime
Table 12.5 shows the reserves of uranium. The present worldwide consumption of uranium is 68 000 t per year. If, as a result of the future increase in consumption

Table 12.5 Uranium reserves, ordered by production cost (source: Framatome, 2005).

	Framatome, 2005			Comparison: BGR, 2007 (there, from NEA, 2006)
Extractable at mining costs of up to:	40 $/kg U	80 $/kg U[a]	130 $/kg U[a],[c]	[d]
This corresponds to a price of up to:	15 $/lb U_3O_8	30 $/lb U_3O_8	50 $/lb U_3O_8	
	Million tons of uranium			Million tons of uranium
RAR (demonstrated by boreholes)[b]	1.7	2.4	3.2	3.3
EAR I (estimated without boreholes within the ore-deposit region)[b]	0.8	1.1	1.4	1.4
	–	–	–	–
Intermediate sum	2.5	3.5	4.6	4.7
EAR II (estimated to be in the neighborhood of known deposits)[b]	–	1.5	2.3	2.5
SR (speculative – presumed only on the basis of geology)	No data	No data	4.4	7.5
	No data	No data	–	–
Overall sum			11.3	14.7
			Unconventional reserves and thorium – see below	

a) The overall amount in each case, including resources which can be extracted for a lower price. The third column thus contains the total amounts producible up to a price of 130 $/kg U.

b) Lübbert and Lange (2006): "The statements about uranium reserves in the various literature sources are practically the same concerning the categories RAR to EAR-II, since they are all based on the so-called "Red Book" from the Nuclear Energy Agency (NEA) of the OECD. There are, however, differing opinions concerning the speculative and especially the unconventional reserves." Unconventional reserves are discussed below. An overview of uranium reserves is also given in WEC (2007), pp. 200–205.

c) The uranium price in the year 2007: in June, 2007, uranium sold at a top price of 135 $/lb U_3O_8 = 350 $/kg U; by November 2007, it had again fallen to 62 $/lb U_3O_8; see also the section 12.6.2.

d) Up to 130 $/kg U.

of electric power, we assume as an *example* a *tenfold increase* in the nuclear plant capacity (see below) as well as a constant specific consumption of natural uranium (per kWh), we estimate a future consumption of 0.7 million tons per year. Through higher levels of enrichment, lower percentage contents of remaining uranium in the spent fuel elements, and a somewhat higher electrical efficiency, the specific consumption of natural uranium in the new reactors will, however, be notably lowered (see above), presumably to ca. 60% of its present value. We would thus find for a tenfold increase in nuclear power generating capacity a consumption of *0.4 million t/a*. The lifetime of the reserves at extraction costs of up to 130 $/kg, including the relatively secure resources (intermediate sum: 4.6 million t) would then be *12 years* and, including the speculative reserves (11.3 million t), it would be *28 years*; or with reserves of 14.7 million t as given in (BGR, 2007), it would be *36 years*. After this time, these price categories would be exhausted and one would have to access more expensive sources. What amounts of uranium would be available there is not clear. (In Section 4.1, we gave our own provisional estimate; it predicts that the reserves could possibly be doubled by going up to a price of 400 $/kg, and tripled up to a price of 800 $/kg.[10]) In contrast, it is clear that the environmental damage caused by the necessary mining and extraction of uranium from these low-grade ores would be much greater than it has been thus far. Furthermore, long before the exhaustion of the favorable deposits in the classical supplier countries (USA, Canada, Australia), a large margin for price increases would open up for new suppliers so that in a large-scale nuclear power scenario, price increases would be possible relatively early (due to potential supplier cartels) and could be *far* higher than the extraction costs (see also Section 12.6.2).

> A *tenfold increase* in the nuclear power generating capacity could indeed cover the expected world electric power consumption to a large extent, but it would make no contribution to the substitution of petroleum and natural gas. Compare:
>
> *Electric power.* The electrical energy generated annually by nuclear plants is currently (2004) 300 GWa$_{el}$ (2680 TWh), while the overall electrical energy generated is *1900 GWa$_{el}$*, with an expected increase by the year 2030 to *3400 GWa$_{el}$*. Subtracting energy from renewable sources (350 GWa$_{el}$) gives ca. 3000 GWa$_{el}$.

---

10) This attempt to make a prediction of the uranium reserves available at costs of more than 130 $/kg was based on the arbitrary and speculative assumption that the amounts extractable at higher cost would increase in the same ratios as seen in Table 12.5 for the reserves available at extraction costs between 40 and 130 $/kg (there, in the categories RAR and EAR I, for which amounts are available for all three price classes): a rough tripling of the cost corresponds to a doubling of the total amount of uranium available. Such a provisional estimate is naturally not at all sufficient for a reasonably secure appraisal of the situation regarding uranium reserves. It could be supplanted within a short time by evaluations from a number of geological institutes (possibly of a preliminary nature) of the probable reserves within the higher cost classes (e.g., 400 and 800 $/kg).

12.6 Uranium Reserves

Primary energy sources. If in the medium term oil and gas would need to be substituted by nuclear or solar energy, the required capacities would be of a much greater magnitude. The annual world consumption of primary energy is currently 13 600 GWa. Of this, 3700 GWa are due to coal, which is to a large extent used for electric power generation. *Oil* and *gas*, which potentially must be substituted, make up ca. *8200 GWa* of today's annual energy consumption. Since the requirements for utility energy (energy supply services) is increasing worldwide, in spite of intensive energy savings programs (in particular in the currently industrialized countries), it will not be possible to reduce the world's energy consumption *fundamentally*. On the contrary, with growing economic development of the emerging nations, it will increase. Thus, very large amounts of energy must be substituted. In the case of a tenfold increase in the nuclear energy production capacity, the annual amount of nuclear electrical energy generated would thus be 3000 GWa$_{el}$.

1 GWa (gigawatt-year) = 8760 GWh = 8.76 TWh.

A power plant with a nominal output power of 1 GW and a (theoretical) annual capacity factor of 100% would generate 1 GWa of electrical energy per year. The American power plants (1050-MW class) have an annual capacity factor of 91% (2004), according to Uni-Chicago (2004, p. 5–26); correspondingly, they generate 0.96 GWa of electrical energy per year. Note: 1 GWa corresponds roughly to the annual electrical energy output of a present-day nuclear power plant. For an annual production of 3000 GWa of nuclear electrical energy, one would therefore require ca. 3000 nuclear power plants of the current size.

In the case of a planned massive upgrading of the nuclear energy generating capacity – analogous to the expansion of solar energy proposed in this book – the question of uranium resources would thus have to be considered very carefully. With an operating lifetime of 45 years for the nuclear plants, the reserves given above (at a price of up to 130 $/kg), including the speculative reserves (11.3 million t), would, with a tenfold increase in capacity, be exhausted already after ca. two-thirds of the operating lifetime of the plants (cf. above: 28 years); even with the largest estimated amount of resources quoted in the table of 14.7 million t (timescale: 36 years), the reserves would not be sufficient to fuel even one whole power plant generation, so that toward the end of their operating life, uranium sources with notably higher extraction costs would already have to be tapped.

The amounts of reserves and their cost categories as given above hold for *current extraction technologies*. Some authors – in particular "Bunn *et al.*, 2003," as cited in (Uni-Chicago, 2004, p. 7–13) – hope that improved technologies will make it possible to extract uranium over the long term which previously could not be produced at the prices given above. They are certain that technical advances (in mining and extraction) will be made, and consider it possible that the reserves in the price category up to 130 $/kg could then be a factor of two to six times greater than the earlier estimates. The example of such a technical advance which they give, namely the greatly improved extraction of uranium as a by-product of copper production, is however not very convincing, since the annual amounts produced would be linked to copper production and are thus only marginal. Bunn *et al.* have thus not given an adequate justification of their hoped-for technical advances, and their statements are not quoted in any other studies that we evaluated in connection with this book, aside from (Uni-Chicago, 2004).

The above timescale for the exhaustion of uranium reserves does not take the use of *fast breeder reactors* or fuel-element reprocessing into account. In the case of the fast breeder, we are dealing with a completely different reactor technology, which has not only a different, and probably much greater risk potential, but also—as is generally expected—would have considerably higher costs (cf. the discussion in Section 4.1). This technology will, therefore, not be considered further here.

The *fuel-element reprocessing technology* would also save on uranium; it is however likewise accompanied by additional risks (danger of accidents and nuclear proliferation) and increased costs (UCS, 2008). In Section 4.1, the accompanying expected additional cost of electrical energy of ca. 0.6 ¢/kWh was already mentioned. Nuclear power would then increase in price from 2.4 ¢/kWh (Table 4.3) to 3.0 ¢/kWh, or, in the case of nuclear plant pools, to 3.5 ¢/kWh. With pools, the cost difference from solar energy (e.g., the USA in the case of "replacement" of coal plants (Table 4.2): 4.1 ¢/kWh; Spain: 4.8 ¢/kWh) would then not be very significant. According to Framatome (2005, p. 4), the yield of fissionable material can be increased by fuel-element recycling and the reuse of plutonium and unconsumed uranium by *up to 30%*, which is equivalent to an increase in the uranium reserves by 30%. The timescale of *36 years* estimated for the above example (tenfold increase in the nuclear capacity to an annual energy production of 3000 GWa) would then be lengthened to *47 years*, about the same as the assumed operating lifetimes of the power plants (45 a).

In the longer term—to the year 2030 and beyond—the world's annual energy consumption may in fact become notably higher than estimated above, namely if the most populous countries, in particular those in Asia, continue to develop economically as desired. Thus, if a *serious* substitution becomes necessary (e.g., half of today's oil and natural-gas consumption), one would not be able to go very far with nuclear energy, at least not by employing the favorable uranium resources from the classical supply countries. In terms of affordable uranium reserves, the situation would be different only if the less safe and more expensive fast breeder reactors were used on a large scale, or if uranium can in fact be extracted at a reasonable cost from seawater (see below).

12.6.1.2 Classification of Ores According to Their Uranium Content

Based on the amount of uranium contained, one can make the following rough classification (Greenpeace, 2006, p. 6):

Very high	>10% uranium	(>100 kg U/t ore)	<10 kg ore/kg U
High	1–10% uranium	(10–100 kg U/t ore)	=10–100 kg ore/kg U
Good	0.2–1% uranium	(2–10 kg U/t ore)	=100–500 kg ore/kg U
Moderate	0.1–0.2% uranium	(1–2 kg U/t ore)	=500–1000 kg ore/kg U
Low*	<0.1% uranium*	(<1 kg U/t ore)	>1000 kg ore/kg U

Assumption. In future, uranium will be extracted from ores with still lower concentrations: 0.01–0.1% uranium (0.1–1 kg U/t ore) = 1000–10 000 kg ore/kg U**

* In the past (at a low uranium price), uranium was considered to be worth extracting from ores with less than 0.1% only under special circumstances.

** At a concentration of 0.01%, that is, 10 t ore/kg, the amount of ore processed is a factor of 1000 greater, compared to the best deposits with "very high" U content, which were exploited in the past; and compared to the current "moderate" to "good" ores, it is greater by a *factor of 10–100*. At a U content of *0.02%*, the amounts of ore which must be processed are greater by a *factor of 5–10* compared to "moderate" ores, and by a factor of *10–50* compared to a "good" ore. The size of the mining and extraction facilities must be correspondingly larger, as also the volume of the tailings and waste leavings from processing.

12.6.1.3 Unconventional Uranium Reserves

Phosphates Along with the reserves mentioned above, it is sometimes pointed out that uranium is found in phosphate minerals at a total quantity of ca. 20 million tons (Framatome, 2005). Other sources quote 5–15 million tons (Greenpeace, 2006, p. 12) or 9 million tons (IAEA, 2001, p. 65); of this, 7 million tons is supposed to be in the large phosphate deposits in Morocco. The uranium from these deposits is, however, only partially extractable (if at all). Since its concentration is very low, the uranium can probably be extracted only as a by-product of phosphate production. Thus, the IAEA (2001) estimated an annual production of only 3700 t (compared to the worldwide uranium consumption of 68 000 t per year).

Differing statements are made about the uranium content of these ores. Greenpeace (2006, p. 12) gives an average uranium content of 0.005–0.02% (50–200 g/t), while the IAEA (2001, p. 65) cites a "typical average from 0.0006% to 0.012%" (6–120 g/t). The Greenpeace numbers yield 5–20 t of ore per kg uranium (from the IAEA numbers, the value would be 8–170 t). For a subsequent mining exclusively for the purpose of uranium production, according to the data of Greenpeace, to achieve a price of, for example, 200 $/kg U, a ton of ore would have to be able to be mined *and processed* at a cost of $10 to $40 (assuming that all the uranium could be extracted from the ore). Compare: American coal can be mined for less than 30 $/t. This is, however, only the cost of mining itself; for uranium, the processing, that is, extracting the uranium from the rock, must be added. Nevertheless, this comparison shows that depending on the type of deposit, production of uranium from phosphate deposits is not completely out of the question. It is, however, notable that the phosphates are not included among the known reserves – for example, in the category up to 130 $/kg. This indicates higher production costs (possibly considerably higher – including extraction of the uranium from the ore!). Thus, the IAEA (2001, p. 65) notes: "However, there are no rigorous estimates of phosphorite deposit resources, so this total (estimate of 9 million t) should be considered a mineral inventory rather than conforming to standard resource categories."

Regarding the amount of rock which must be handled, the IAEA (2001) published an exemplary treatment: for the present worldwide production of phosphates, annually 142 million tons of rock are mined, containing 66 million tons of phosphate concentrates. For the extraction of uranium, however, only the so-called marine deposits (ocean sediments from *earlier eras*) are usable; they constitute 80% of the phosphate deposits. Of these, currently 70% are processed with the phosphoric-acid technique; only in this case can uranium be extracted (currently as a by-product). We can thus conclude that 80 million t/a of rock and 37 million t/a of concentrate are processed. Assuming an extractable content of 0.01% uranium, we obtain the above-mentioned annual production of 3700 t/a. According to these numbers, in order to obtain 1 kg uranium, 22 t of rock with a content of 10 t of phosphate concentrate would have to be mined and processed. Whether or not the assumed concentration of 0.01% is typical of the overall phosphate reserves appears questionable, as indicated by the data above.

The question of a possible economically feasible exploitation of these phosphate resources due to increasing uranium prices should be answered definitively in the near future, in order to judge what potential this route in fact holds.

Seawater The uranium content of seawater has been estimated to be ca. 4000 million tons (Framatome, 2005, p. 3) and is thus practically unlimited (compare conventional reserves including the speculative resources: 11.3 million tons). To be sure, the uranium is present at an extremely low concentration in seawater. It is nevertheless hoped – by using adsorption techniques – that its extraction will be possible. Regarding the costs of such an extraction, "estimates" up to now range from 300 to 500 $/kg U. Considering that the spot price of uranium already went as high as 350 $/kg, extraction costs of that order would not represent a fundamental obstacle to the use of this resource. At the future expected specific consumption of natural uranium of 0.0145 g/kWh$_{el}$ (see above), a price of 500 $/kg U would contribute 0.75 ¢/kWh$_{el}$ to the cost of nuclear electric energy.

Regarding the estimates of the extraction costs for uranium from seawater, cf.:

- IAEA (2001, p. 65). "Research in Japan indicates that uranium could potentially be extracted from seawater at a cost of approximately US$ 300/kg U."

- Lübbert and Lange (2006). "Extraction of uranium from seawater would thus be technically and financially extremely laborious and is at present not being seriously considered."

- NRAW (2006, p. 11). "This source (seawater) could be tapped using a new absorption technique developed by Japanese scientists at an estimated price of 400 to 500 $/kg."[11]

11) Literature: N. Seko and M. Tamada: "R&D towards recovery of Uranium from Seawater," IUPAP Working group on Energy, Report on R&D of Energy Technologies, 2004.

These extraction costs should indeed not just be set out as an "expectation," but rather need to be practically demonstrated. To this end, the technology should be developed as quickly as possible. Thus, a doubling of the expected price (from 500 to 1000 \$/kg), which would result in a contribution to the energy cost of 1.5 ¢/kWh, would cast serious doubts on the economic feasibility of the large-scale application of nuclear power.[12] Since we are faced today with the need to decide on the future course of energy supplies over a long time to come, a precise knowledge of the long-term perspectives is indispensable. The research funding needed to clarify this question should, therefore, be made available without delay.

It appears dubious whether the statements made thus far are more than simply educated guesses. How could one expect to predict the costs more or less precisely without constructing a large-scale facility or at least a complete semitechnical installation? It therefore appears unlikely that these estimates can have an uncertainty of, for example, 10%, or even of only 50%. How easily can simple estimates – of the kind we are apparently dealing with here – be wrong by a factor of 2 or 3! In any case, it is not permissible to base decisions concerning the world's future energy supply on estimates made by individuals (probably only on the basis of laboratory experiments), which have hardly been assessed by the responsible institutions in the major industrial countries nor subjected to any other sort of critical evaluation. Here, a concrete verification must be presented soon; otherwise, statements about the uranium reserves in the oceans as an energy resource are pointless, since these "reserves" can quite possibly not be exploited at an economically feasible cost.[13]

12.6.1.4 Thorium Reserves

Thorium was not yet included in the above considerations of nuclear-fuel reserves. This fuel can however be used only in new reactors (e.g., high-temperature reactors). These are indeed under development, and prototypes exist; but they have not yet been operated on a commercial scale in practice, and have thus not been tested in a comparable way to light-water reactors. (The prototypes have thus far chalked up only an insignificant number of hours of operation.) Since this is a completely different type of reactor, uncertainties regarding safety remain. Whether they will turn out to be in fact comparable to light-water reactors in this respect is hard for nonexperts and for the political decision-makers to judge. We can, however, safely assume that if nuclear power is to be used on a large scale,

---

12) This refers to a purely *microeconomical* cost-effectiveness, that is, cost-effectiveness from the viewpoint of the operator, who in the main need not assume the cost of the (*external*) consequences of his actions. The *external* costs – in particular the possibly enormous damage in the case of a major nuclear accident – were not considered in the estimates given here.

13) Thus, the WEC (2007, p. 202) states: "The technology to extract uranium from seawater has only been demonstrated at the laboratory scale, and extraction costs were estimated in the mid-1990s at US\$ 260/kg U (Nobukawa *et al.*, 1994) [Authors' note: this corresponds to 315 \$/kg U in 2002-\$], but scaling up laboratory-level production to thousands of tons is unproven and may encounter unforeseen difficulties."

with high uranium prices and scarce resources, these reactors will find their way to the practical application and thereby somewhat relieve the demand for uranium fuel. The learning process, which has already been traversed for the currently used reactor types, still lies ahead for thorium reactors. Regarding the large-scale deployment of these reactor types, there are thus more open questions than in the case of the currently used types.

The known reserves of thorium are considerably smaller than the uranium reserves. In BGR (2007), "reserves"[14] of 2.2 million tons of thorium are quoted. In addition, "resources" of only 2.4 million tons of thorium are listed; together, this makes 4.6 million tons. (Compare uranium: according to BGR (2007), there are 14.7 million tons – see Table 12.5.) As a result, one cannot hope that thorium will offer a major improvement in the situation regarding reserves of nuclear fuels. However, exploration for thorium appears to have been rather incomplete up to now so that according to WEC (2007, p. 202), the reserves are probably larger than previously assumed.

12.6.2
The Present and Future Price of Uranium – Geographical Distribution of the Uranium Reserves

The uranium price increased dramatically in recent years due to competitive speculation, and in June 2007, it reached its maximum thus far, at 135 $/lb U_3O_8, corresponding to *350 $/kg U*.[15] By mid-2008, the price had fallen to 55 $/lb U_3O_8 (143 $/kg U), and it then rose again slightly. The level of June 2007 (135 $/lb) amounted to a factor of 2.7 more than the highest price category listed in Table 12.5 (50 $/lb). (The price trends of the past years in $/lb were as follows: 2003, 10; 2004, 15; 2005, 21; 2006, 35; early 2007, 75; June 2007, 135; September 2007, 85; November 2007, 62 (UIC – Newsletters Nos. 4 to 7, 2007); June 2008, 55; August 2008, 65 (Handelsblatt, 2008b); December 2008, 52 (Uranium Info 2008).) Speculation buying is evidently fed by the expectation of a temporary scarcity of uranium in the coming years due to a worldwide "nuclear-energy renaissance." Although the price dropped again after reaching a maximum in 2007, this speculation will, in the opinion of many observers, continue in principle as long as the production capacities, which were reduced in the past, have not yet been built up again. That may take several years, possibly up to 10 years. The reason for the high prices at present thus has little to do with the reserves or the extraction costs, but rather with the momentary production capacity and evidently also with intentional speculation, in which hedge funds also play a role (fondsexperte 24 2007). At 350

---

14) Thorium *reserves*: According to the definition in BGR (2007), these however include only the known resources corresponding to the category "RAR" for uranium, and only those available at extraction costs of up to 40 $/kg. Thorium is treated in this report as having the same energy content per kg as uranium. Thorium *resources*: These evidently correspond in BGR (2007) to the remaining resource categories of uranium, including the "speculative resources."

15) 1 $/ U_3O_8 = 2.60 $/kg U.

$/kg, the uranium price already reached two-thirds of the presumed cost of extracting uranium from seawater.

Regarding a possible future price diktat, we note that 60% of the reserves (at a cost of up to 130 $/kg) are to be found outside the classical supply countries (the USA, Canada, and Australia: Table 12.6). The danger of price fixing by a supplier cartel will arise not just at the moment when the low-cost reserves (≤130 $/kg) in the industrial countries are exhausted, but rather sooner. With a massive use of nuclear energy, this problem could already appear in the coming decades. As long

Table 12.6 Uranium resources by country.

Country	Verified	Not yet verified			
	RAR + EAR I[a] up to 130 $/kg U	EAR II[a] Estimated in the neighborhood of known deposits	Speculative[b]	Totals (absolute)	Totals (relative)
	1000 t of uranium				
USA	342	1273	1340	2955	20.0%
Canada	444	150	700	1294	8.8%
Australia	1143	0	0	1143	7.7%
EU-25	106	27	328	461	3.1%
				–	–
				5900	40%
Kazakhstan	815	310	500	1625	11.0%
Republic of South Africa	340	110	1113	1563	10.6%
Mongolia	54	0	1390	1444	9.8%
Brazil	279	300	500	1079	7.3%
Russia	172	105	545	821	5.5%
Uzbekistan	115	85	135	334	2.2%
Ukraine	62	15	255	332	2.2%
Namibia	282	0	0	282	1.9%
Nigeria	225	25	0	250	1.7%
India	65	12	17	94	0.6%
P.R. of China	34	4	4	42	0.3%
				–	–
				7900	53%
Other countries				1000	7%
World	4740	2520	7540	14800	100%

Source: Federal Institute for Geosciences and Natural Resources (BGR, 2007, p. 83/84), cited there from NEA (Nuclear Energy Agency) 2006: "URANIUM 2005: Resources, Production and Demand," 391 pp. (2006), OECD, Paris.
a) EAR I is now called "Inferred Resources" (IR)
   EAR II is now called "Prognosticated Resources."
b) Speculative Resources: assumed amounts, based on the geological structure of the territory of the country, without precise localization of the deposits. (EAR II and speculative resources should likewise refer to extraction costs of up to 130 $/kg U)

as a cartel must fear a corresponding increase in production by the noncartel countries when it precipitates a scarcity on the supply side, its freedom to increase prices is limited. This situation would change drastically at the moment when those noncartel countries were no longer able to increase their production in a sustainable manner without resorting to more expensive reserves. Likewise, in the case of oil, the price will increase not just when the reserves are finally exhausted, but rather as soon as alternative supplies for the non-OPEC countries, which could compensate for reduced deliveries over a period of time, are no longer accessible or are available only at higher cost. Here, an additional aspect must be considered: for the more expensive uranium resources, the time required to develop new deposits is considerably longer than for the favorable ore deposits in the past. A rapid reaction to a supply bottleneck initiated by a cartel is thus not possible. This also increases the possible influence of a cartel on the price level. It should furthermore be taken into account that precisely the more expensive extraction projects using low-grade ore are dependent on planning reliability, due to the large amount of capital required. Such reliability is however not present when there are strong price fluctuations (as is typical of markets in which cartels are at work). This means that investments will be forthcoming only when the price rises to *well above* the actual production costs. It then still requires several years before the increased production can in fact come on line. For this reason, also, it must be feared that price rises for uranium could occur rather early in a large-scale nuclear scenario.

In the case of such a large-scale nuclear power scenario (e.g., a tenfold increase in plant capacity), we would thus move relatively rapidly toward the point at which the classical supply countries could no longer react quickly, emphatically, and in a sustainable manner to compensate for supply scarcities. If in fact a large-scale increase in nuclear power generation is implemented, the future – and in fact possibly even the near future – would no longer experience a consistent price development, but rather, as with oil, price uncertainties, strong fluctuations, and – most probably – a high price level would predominate. The events of the year 2007 gave a foretaste of what could be expected.

Up to now it has been assumed that the main supplier countries *USA, Canada,* and *Australia* would offer their worldwide customers uranium at a reasonable price with a long-term price guarantee, close to the production cost. From the point of view of the USA, supplying the western industrial nations was a matter in the interest of national security. Military aspects also played an important role in the planning of uranium mining and extraction in the USA. In the past, the western European countries could be included in this provisioning with little additional effort. In contrast, the future supply of uranium – always assuming a major increase in the use of nuclear energy – will for most countries (in Europe, China, India, ...) be mainly dependent on supplier nations, which have no political or strategic interest in maintaining a reasonably priced supply of uranium to other countries.

The numbers given above should be regarded critically once more: if the future world consumption of uranium, in the case of an assumed tenfold increase in nuclear power plant capacity (0.4 million tons of uranium per year), has to be supplied exclusively from the resources of the above three industrial countries plus EU-25, their economically favorable resources including "speculative resources" costing up to 130 $/kg[16] (Table 12.6: 5.9 million tons of U) would be exhausted within 15 years so that in these countries, one would have to fall back on more expensive reserves.

For the remaining producing countries to join into a *de facto* cartel, it would by no means be necessary that their governments supply the driving force; this could also come from the mining companies. If it turns out that the reserves costing *more than* the category of 130 $/kg U are not distributed geographically in a different way from those listed in Table 12.6, such a cartel would soon be able to dictate the price of uranium. Even countries like Australia and Canada, which themselves have limited uranium requirements, would have a greater interest in increasing uranium prices than in keeping them low in favor of their national electrical-generating economies. This would in particular apply if uranium mining and extraction in those countries were in the hands of private financial investors and if mining in addition was limited to only a few companies which also might have worldwide interests. The consumers then might well find themselves at the mercy of producer cartels or multinational firms.

---

16) In BGR (2007) (which quotes a total amount of 14.7 million tons U), it is not stated whether the estimated amounts in the categories EAR II and "speculative resources" likewise refer to the extraction costs of up to 130 $/kg (this is mentioned specifically for the categories RAR and EAR I); in Framatome (2005), this cost category is expressly stated (cf. Table 12.5). However, it is noticeable that the "speculative resources" given by BGR are considerably greater than those given by Framatome. (In both reports, the "unconventional" resources and the thorium reserves are not discussed.)

Appendix A
Solar Tower Power Plants: Comparison of Kolb (1996), Kalb/Vogel[1], SunLab[2], S&L[2]

Solar power plant:

Basic data
Efficiencies
Investment costs

Solar power system:

Investment costs
Capital costs
Gas costs
O&M costs
Electric power costs

1) Kalb and Vogel, 1998
2) SunLab and S&L from (S&L, 2003)

Large-Scale Solar Thermal Power. Werner Vogel and Henry Kalb
© 2010 WILEY-VCH Verlag GmbH & Co. KGaA, Weinheim
ISBN: 978-3-527-40515-2

Appendix A Solar Tower Power Plants: Comparison of Kolb (1996), Kalb/Vogel, SunLab, S&L

Table A.1 Basic data.

Basic data		Kolb, 1996a 200 MW "Advanced receiver 200" SM 2.7 11.0 kWh/m² d	Kolb, 1996a 200 MW "Advanced Receiver 200" SM 4.4 6.7 kWh/m² d Spain	Kalb/V. (as Kolb96) 200 MW "Advanced Receiver" SM 2.7 11.0 kWh/m² d Mass production[a]	Kalb/V. (as Kolb96) 200 MW "Advanced Receiver" SM 4.4 6.7 kWh/m² d Mass production[a] Spain	SunLab 200 MW SM 2.9 10.3 kWh/m² d	S&L 200 MW SM 2.9 10.2 kWh/m² d	SunLab 220 MW SM 2.9 10.3 kWh/m² d (supercritical steam, advanced heliostats)
						pp. 5–37, E10	pp. 5–37, E10	pp. 5–38, E10
Electric power output	MW$_{el}$	200	200	200	200	200	200	200
Receiver output (maximum)	MW$_{th}$	1400		1400		1400	1400	1400
Heliostat field (S&L, p. E4)	10⁶ m²	2.477	4.070	2.477	4.070	2.600	2.667	2.642
Heliostat type		Glass/metal (1996-b)	=	=	=	Glass/metal	Glass/metal	Advanced heliostat
Heliostat size	m²	150	=	=	=	148	148	148
Heliostats (number)		16 500	27 130	16 500	27 130	17 608	18 021	17 851
Land area (S&L, p. E-3)	km²	13.8 (18%)	22.6 (18%)	13.8 (18%)	22.6 (18%)	(13.7)	13.7	(13.9)

Appendix A Solar Tower Power Plants: Comparison of Kolb (1996), Kalb/Vogel, SunLab, S&L

Area utilization (heliostat area/land area)		as Utility st. (22.6%) 18%[b]	=	=	=	(19.0%)	(19.5%)	(19.0%)
Receiver area (SL, p. 5–37)	m²					1870	1930	1650
Heat transfer medium		Solar salt	Solar salt	Solar salt	Solar salt	Solar salt	Solar salt	Solar salt
Maximum temperature (salt)	°C	565/288	=	=	=	574	574	650
Horizontal piping		–	–	Yes	Yes	–	–	–
Heat storage (hours)	h	13	15	13	15	13	13	16
Heat storage capacity (Kolb, p. 22)	MWh$_{th}$	6800						
Heat storage medium		Solar salt	Solar salt	Solar salt	Solar salt	Solar salt	Solar salt	Solar salt + O$_2$-blanket
Steam conditions (S&L, p. E-46)		538°C	=	=	=	180 bar 540°C 2 reheat?[c]	180 bar 540°C 2 reheat?[c]	300 bar 640°C 2 reheat

Table A.1 Continued.

Basic data	Kolb, 1996a 200 MW "Advanced receiver 200" SM 2.7 11.0 kWh/m² d	Kolb, 1996a 200 MW "Advanced Receiver 200" SM 4.4 6.7 kWh/m² d Spain	Kalb/V. (as Kolb96) 200 MW "Advanced Receiver" SM 2.7 11.0 kWh/m² d Mass production[a]	Kalb/V. (as Kolb96) 200 MW "Advanced Receiver" SM 4.4 6.7 kWh/m² d Mass production[a] Spain	SunLab 200 MW SM 2.9 10.3 kWh/m² d	S&L 200 MW SM 2.9 10.2 kWh/m² d	SunLab 220 MW SM 2.9 10.3 kWh/m² d (supercritical steam, advanced heliostats)
Power block efficiency (gross average)					42.8%	42.8%	46.1%
Wet or dry cooling	Wet	Wet	Wet	Wet	Wet[d]	Wet[d]	Wet[d]
Efficiency total (see Table A.2)	17.6%	=	=	=	17.93%	17.56%	19.27%
Electric power production per day (full load) GWh/d	4.8	=	=	=	4.8	4.8	5.28 220 MW
Solar energy required per day (at above efficiency) GWh/d	27.27	=	=	=	26.77	27.33	27.40 220 MW

Appendix A Solar Tower Power Plants: Comparison of Kolb (1996), Kalb/Vogel, SunLab, S&L

Design insolation[e] (Insolation/m² for 24-h operation)	kWh/m²d	11.01	6.7 Spain (Morocco[e])	11.01	6.7 Spain (Morocco[e])	10.30	10.25	10.37
Solar multiple		2.7 (given)	4.4 (Morocco 3.7)	2.7	4.4 (Morocco 3.7)	$2.89^{f)} = 2.9$	$2.90^{f)}$	$2.87^{f)} = 2.9$
Annual capacity factor (given) SunLab, S&L probably including plant outages		63%[g] (including plant outages)				(74%)[g] p. 5-1, 5-37	(74%)[g] p. 5-1, 5-37	(73%)[g] p. 5-1, 5-38

a) "Kalb/Vogel" – Mass production. Different assumptions concerning heliostat costs and "indirect costs"; horizontal piping is included. Heliostat and indirect costs do not affect the efficiency; heat losses, and parasitic power unknown for horizontal piping and thus neglected for calculating the efficiency.
b) Land utilization: Not given in (Kolb, 1996a,b); here, according to Utility Studies (1988): 22.6% (surround field). In the case of combining several surround fields and including interstitial areas: 18% (cf. Section 4.3.5).
c) Steam conditions: SunLab 200 and S&L 200: In S&L (2003) probably a typing error, most likely in fact only one reheat stage.
d) Wet cooling (S&L, p. G-3) Water cost as in SEGS parabolic-trough power plants (cf. Section 4.2.1.2).
e) Design insolation: From this and from the "Required amount of solar energy per day" (depending on efficiency), the total mirror area is obtained; see Section 4.3.2. Cf. Morocco: Design insolation: 8.0 kWh/m² d; resulting mirror area: 3.394 million m².
f) Solar multiple (SM). For SunLab and S&L, the SM was computed starting from the design insolation as compared to Kolb (1996a) (there, SM = 2.7 is cited). The SM and the design insolation are indirectly proportional. (On doubling the mirror area, i.e., doubling the SM, the daily insolation per m² required for 24-h operation (DNI) is halved.) In this conversion, the same ratio of daily maximum efficiency to daily average efficiency is assumed as by Kolb.
g) Annual capacity factor:
- Kolb (1996a): *without* plant outages 68.5%, referring to the insolation at Barstow (probably several-year average, that is ca. 2500 kWh/m² a, cf. Section 5.3).
- SunLab and S&L: referring to the insolation at Kramer Junction in the year 1999, that is a value of 8.0 kWh/m² d = 2920 kWh/m² a (S&L, 2003, p.5-2).

Table A.2 Efficiencies.

Efficiencies		Kolb, 1996a 200 MW "Advanced receiver 200" SM 2.7 11.0 kWh/m² d	Kolb, 1996a 200 MW "Advanced Receiver 200" SM 4.4 6.7 kWh/m² d Spain	Kalb/V. (as Kolb96) 200 MW "Advanced Receiver" SM 2.7 11.0 kWh/m² d Mass production[a]	Kalb/V. (as Kolb96) 200 MW "Advanced Receiver" SM 4.4 6.7 kWh/m² d Mass production[a] Spain	SunLab 200 MW SM 2.9 10.3 kWh/m² d	S&L 200 MW SM 2.9 10.2 kWh/m² d	SunLab 220 MW SM 2.9 10.3 kWh/m² d (supercritical-steam, advanced heliostats)
Heliostat field (K/V 98, p. 81; S&L, p. 5–37)	%	55.3	=	=	=	p. E-9 56.1	p. E-10 55.2	p. E-9 57.0
Reflectivity (clean mirror)	%					94.5	94.0	95.0
Field efficiency (geometric losses)	%					62.8	62.8	62.8
Field availability	%	99 (K/Vp81)	=	=		99.5	99.5	99.5
Mirror corrosion	%					100	100	100
Mirror cleanliness (water cf. p. G-4)	%					96	95	97
High wind outage (p. E-10)	%					99	99	99
Receiver (S&L, p. 5–37)	%	87.0[b] Advanced receiver	=	=	=	83.5	83.5	82.0 high temperature
Defocus, Dump, Startup, Clouds	%	(nonadvanced 94.2%[c])				93.4	93.4	93.4
Absorptance	%	(nonadvanced 93.2%[c])				94.5	94.5	94.5

Appendix A Solar Tower Power Plants: Comparison of Kolb (1996), Kalb/Vogel, SunLab, S&L

		Kolb	Kalb/Vogel	Kalb/Vogel Mass production	SunLab	S&L
Thermal losses (radiation, convection)	%	(nonadvanced. 92.9%[c])		94.7	94.7	92.9 high temperature
Vertical piping heat losses	%	99.96[c]	=	99.9	99.9	99.9
Horizontal piping[a),d)]			Heat losses unknown (neglected)			
Heat storage	%	99.5[c]	=	99.5	99.5	99.5
Power block (EPGS) (gross, average)	%	41.9[e] Wet cooling	=	42.8[f] Wet cooling	42.8[f] Wet cooling	46.1[f] Wet cooling
Parasitic power (efficiency)	%	88.9[g]	88.9[h]	(Roughly the same)[j]	90.0[j]	90.0[j]
(Power block net (including parasitic power for solar field)	%	(37.2)	(37.2)	(Roughly the same)[j]	(38.3)	(41.5)
Efficiency total[k] Daily average without plant outages	%	17.6[l] (1996-b) Wet cooling	=	17.93 S&L, p. E-9/10 Wet cooling	17.56 S&L, p. E-9/10 Wet cooling	19.27 S&L, p. E-9/10 Wet cooling
Plant outages[k] (plant wide availability)	%	0.92 (1996-b)	=	94.0	94.0	94.0
Efficiency total including plant outages	%	16.2 (1996-b)	=	16.9 (p. 5-27: 16.8%)	16.5	18.1 (p. 5-27: 17.8%)

a) "Kalb/Vogel" – Mass production. Horizontal piping is included. Heat losses, and parasitic power unknown for horizontal piping and thus neglected for calculating the efficiency.
b) Receiver: cf. *nonadvanced* receiver – expected 1993: 80.3% (Becker and Klimas, 1993; cf. Kalb and Vogel 1998, pp. 81, 67), "current" (1996): 78% (Kolb, 1996b).
c) SANDIA (Kolb): cf. *nonadvanced* receiver – expected 1993 (Becker and Klimas. 1993; cf. Kalb and Vogel 1998, p. 81):

Table A.2 Continued.

Startup	96.4%	
Min. flow	98.0%	
Defocus heliostats	99.8%	
	(Total: 94.2%)	
Absorptance	93.2%	
Thermal efficiency (radiation, convection)	92.9%	
	Total: 80.3%	

c) Thermal efficiency of vertical piping (tower) and pipes to storage system: 99.96%
   Heat storage efficiency (including steam generator startup): 99.5%
d) Horizontal piping: Molten-salt pipes for connecting six towers (each equivalent to 200 MW$_{el}$ at SM = 2.7) into a power plant with a 700 MW$_{el}$ power block (at SM = 4.4) and for thermal interconnection of several such 700-MW power plants (see Section 4.3.6).
e) Power block: 41.9% (Becker and Klimas, 1993; cf. Kalb and Vogel (1998, p. 81); cf. Kolb (1996b) (rounded off): 42%.
f) Power block: SunLab and S&L: annual average, that is including startup and partial load (S&L, 2003, p. 5–8).
g) Parasitic power for 200 MW: nonadvanced – expected 1993 (Becker and Klimas, 1993; cf. Kalb and Vogel, 1998, p. 81):

Balance of plant	4.3%	
Turbine plant	3.5%	
Heliostat array	3.8%	
Overnight	0.8%	
	Total 12.4%	

Note: As given in (Becker, Klimas): "total" 11.1%
(cf. total parasitic power in Kolb (1996b): ca. 9%).

h) "Kalb/Vogel". Additional power consumption of the *horizontal salt circuit* pumps and for electrical heating of the horizontal pipes is unknown and therefore neglected.
i) Parasitic power of the solar field at a higher Solar Multiple: The absolute electric power consumption of the solar field (heliostat drives and salt pumps for the tower circuit) increases proportionally with increasing SM. Since the overall power output does not in fact change. (If there is an oversupply of solar energy on especially sunny days, a portion of the solar field will be switched off.) At sites with less favorable insolation (e.g., Spain instead of the USA or Morocco), the *relative* parasitic power consumption for the tower and salt pumps for the tower is however somewhat higher (more frequent cloudiness over the mirror field). (The relative power consumption of the salt pumps for the towers and for horizontal piping does not increase; with cloudiness, the salt flow rate is lower). This effect was neglected in this book for sites in Spain. For a more precise calculation, it would have to be quantified in detail.
j) S&L (2003, p. 5–8) – Parasitic power:
   Power block: feedwater and cooling water pumps, cooling tower fans, miscellaneous.
   Solar field: Salt pumps for the tower circuits (without pumps for horizontal piping and power consumption for electric heating of pipes for starting), heliostat drives, instrumentation etc.
k) Efficiency total: daily average (i.e., average with respect to the mean daily insolation (DNI)) = annual solar-to-electric efficiency for yearly insolation (DNI).
   Without plant outages: Outage hours of the power block reduce yearly energy production, but not energy production on normal-operation days. For the electric power production on those days, the daily average efficiency "without plant outages" is relevant.
l) Kolb (1996b): 17.6%. Expected 1993 (Becker and Klimas, 1993; cf. Kalb and Vogel, 1998, p. 81): 17.8%.

Table A.3 Investment costs: 200 MW, 220 MW.

Solar power plant (200 MW or 220 MW) – investment costs

2002 $		Kolb (SANDIA) advanced receiver SM = 2.7	Kolb (SANDIA) advanced receiver Spain SM = 4.4	Kolb (SANDIA) advanced receiver Morocco SM = 3.7	Kalb/Vogel Mass production advanced receiver SM = 2.7	Kalb/Vogel Mass production advanced receiver Spain SM = 4.4 Land:Spain	Kalb/Vogel Mass production advanced receiver Morocco SM = 3.7
Output power	MW	200	200	200	200	200	200
Solar multiple (SM)		2.7	4.4	3.7	2.7	4.4	3.7
Design insolation (24 h full output)	kWh/m² d	11.0	6.7	8.0	11.0	6.7	8.0
Mirror area	km²	2.477	4.037	3.394	2.477	4.037	3.394
Land area	km²	13.8	22.5	18.9	13.8	22.5	18.9
Mirror area coverage	%	17.9	17.9	17.9	17.9	17.9	17.9
US inflation 1995–2002 (factor)		1.180	1.180	1.180	1.180	1.180	1.180
Investment costs (200 or 220 MW)	Million $						
Heliostats (per m²)	$/m²	138	138	133	83	83	83
Heliostat field	M$	342	557	463	205	333	280
Tower and receiver (total)	M$	59	96	81	59	96	81
Receiver	M$						
Tower + vertical piping	M$						
Heat storage capacity	h	13	16	16	13	16	16
Heat storage	M$	71	87	87	71	87	87

Table A.3 Continued.

Solar power plant (200 MW or 220 MW) – investment costs

2002 $		Kolb (SANDIA) advanced receiver SM = 2.7	Kolb (SANDIA) advanced receiver Spain SM = 4.4	Kolb (SANDIA) advanced receiver Morocco SM = 3.7	Kalb/Vogel Mass production advanced receiver SM = 2.7	Kalb/Vogel Mass production advanced receiver Spain SM = 4.4 Land:Spain	Kalb/Vogel Mass production advanced receiver Morocco SM = 3.7
Power block	M$	118	118	118 Wet cooling	118	118	118 Wet cooling
Land preparation	M$	12	19	16	12	19	16
		–	–	–	–	–	–
Subtotal	M$	602	878	771	464	654	583
plus (Kalb/Vogel):							
Horizontal piping	M$				21.8	35.6	29.9
Land price (Spain = Kalb/Vogel 98)	$/m²					1.25	
Land costs (Spain = Kalb/Vogel 98)	M$		0			28.1	
		–	–	–	–	–	–
Total (direct costs)	M$	602	878	771	486	718	612
Indirect costs:							
Construction time (as small volume production)	a	4	4	4	2	2	2
Full interest period (50% of construction time)	a	2	2	2	1	1	1
Interest rate	%/a	4.0	4.0	4.0	4.0	4.0	4.0

Appendix A Solar Tower Power Plants: Comparison of Kolb (1996), Kalb/Vogel, SunLab, SaL

Interest during construction	M$	48.1	70.2	61.7	19.4	28.7	24.5
Owner's costs (% of investment)	%	6	6	6	3	3	3
Owner's costs	M$	36.1	52.7	46.2	14.6	21.5	18.4
Planning and contracting (% of investment)	%	9	9	9	4	4	4
Planning and contracting	M$	54.1	79.0	69.4	19.4	28.7	24.5
Engineering, management, and development (% of investment)	%						
Engineering, management, and development	M$	7.0	7.0	7.0	0	0	0
Contingency cost margin (% of investment)	%	42.1	61.4	54.3			
Contingency cost margin	M$	–	–	–	–	–	–
Total (indirect costs)	M$	180	263	231	53	79	67
		–	–	–	–	–	–
Total (overall) (Wet cooling)	M$	782	1141	1002	539	797	680
% less power (dry/wet cooling)	%			8			8
Investment costs with dry cooling (includes higher investment costs for cooling towers)	M$			1083			739

Table A.3 Continued.

Solar power plant (200 MW or 220 MW)

2002 $		SunLab 200 MW (Spain) SM = 2.9	SunLab 200 MW Spain SM = 4.4 Land: Sp.	SunLab 200 MW Morocco SM = 3.7	S&L 200 MW (Spain) SM = 2.9	S&L 200 MW Spain SM = 4.4 Land: Sp.	S$L 200 MW Morocco SM = 3.7	SunLab 220 MW (Spain) SM = 2.9	SunLab 220 MW Spain SM = 4.4 Land: Sp.	SunLab 220 MW Morocco SM = 3.7
Output power	MW	200	200	200	200	200	200	220	220	220
Solar multiple (SM)		2.9	4.4	3.7	2.9	4.4	3.7	2.9	4.4	3.7
Design insolation	kWh/m² d	10.3	6.7	8.0	10.2	6.7	8.0	10.3	6.7	8.0
Mirror area	km²	2.600	3.945	3.317	2.667	4.046	3.403	2.642	4.009	3.371
Land area	km²	13.7	20.8	17.5	13.7	20.8	17.5	13.9	21.1	17.7
Mirror area coverage	%	19.0	19.0	19.0	19.5	19.5	19.5	19.0	19.0	19.0
Investment costs	Million $									
Heliostats (per m²)	$/m²	96	96	96	117	117	117	76	76	76
Heliostat field	M$	250	379	318	312	473	398	201	305	256
Tower and receiver	M$	61	93	78	70	106	89	59	90	75
Receiver alone	M$	37	56	47	46	70	59	34	52	43
Tower + vertical piping	M$	24	36	31	24	36	31	25	38	32
Heat storage capacity	h	13	16	16	13	16	16	??? 16	16	16
Heat storage	M$	56	69	69	56	69	69	57	57	57
Power block	M$	85	85 Wet cooling	85	91	91 Wet cooling	91	104	104 Wet cooling	104

Appendix A Solar Tower Power Plants: Comparison of Kolb (1996), Kalb/Vogel, SunLab, S&L

Land preparation	M$	7	11	9	7	11	9	7	11	9
Subtotal plus (Kalb/Vogel):	M$	459	636	559	536	750	657	428	566	502
Horizontal piping	M$	21.8	33.1	27.8	21.8	33.1	27.8	21.8	33.1	27.8
Land price (Spain=Kalb/V98)	$/m²	0.5	1.25	0.5	0.5	1.25	0.5	0.5	1.25	0.5
Land costs	M$	6.9	26.0	8.7	6.9	26.0	8.7	7.0	26.4	8.9
Total (direct costs)	M$	487	695	596	565	810	693	457	626	538
Indirect costs:										
Construction time	a				1	1	1			
Full interest period	a				(0.5)	(0.5)	(0.5)			
Interest rate	%/a				(4%/a)	(4%/a)	(4%/a)			
Interest during construction	M$				(4%/a = 11)	(4%/a = 16)	(4%/a = 14)			
Owner's costs	%									
Owner's costs	M$									
Planning and contracting	%									
Planning and contracting	M$									
Engineering, management, and development (% of interest)	%	7.8	7.8	7.8	15.0	15.0	15.0	7.8	7.8	7.8

Table A.3 Continued.

Solar power plant (200 MW or 220 MW)										
2002 $		SunLab 200 MW (Spain) SM = 2.9	SunLab 200 MW Spain SM = 4.4 Land: Sp.	SunLab 200 MW Morocco SM = 3.7	S&L 200 MW (Spain) SM = 2.9	S&L 200 MW Spain SM = 4.4 Land: Sp.	S$L 200 MW Morocco SM = 3.7	SunLab 220 MW (Spain) SM = 2.9	SunLab 220 MW Spain SM = 4.4 Land: Sp.	SunLab 220 MW Morocco SM = 3.7
Engineering, management, and development	M$	38.0	54.2	46.5	84.7	121.4	104.0	35.6	48.8	42.0
Contingency cost margin	%	7.4	7.4	7.4	14.3	14.3	14.3	8.1	8.1	8.1
Contingency cost margin	M$	36.1	51.4	44.1	80.8	115.8	99.1	37.0	50.7	43.6
Total (indirect costs)	M$	74 –	106 –	91 –	166 –	237 –	203 –	73 –	99 –	86 –
Total (overall) (Wet cooling)	M$	562	801	687	730	1047	896	529 220 MW	725 220 MW	624 220 MW$_{wet}$ 220 MW$_{dry}$
% less power (dry/ wet cooling)	%			8	Recalculated to 200 MW →		8			8
Investment costs with dry cooling	M$			746			974	481	659	678 617

Table A.4 Investment costs: 1000 MW.

Solar power plant – investment costs per 1000 MW

2002 $		Kolb (SANDIA) advanced receiver SM = 2.7	Kolb (SANDIA) advanced receiver Spain SM = 4.4	Kolb (SANDIA) advanced receiver Morocco SM = 3.7	Kalb/Vogel Mass production advanced receiver SM = 2.7	Kalb/Vogel Mass production advanced receiver Spain SM = 4.4 Land:Spain	Kalb/Vogel Mass production advanced receiver Morocco SM = 3.7
Output power	MW	1000	1000	1000	1000	1000	1000
Solar multiple (SM)		2.7	4.4	3.7	2.7	4.4	3.7
Design insolation (24 h ful output)	kWh/m² d	11.0	6.7	8.0	11.0	6.7	8.0
Mirror area	km²	12.39	20.18	16.97	12.39	20.18	16.97
Land area	km²	69.0	112.4	94.6	69.0	112.4	94.6
Mirror area coverage	%	17.9	17.9	17.9	17.9	17.9	17.9
Investment costs	Million $						
Heliostats (per m²)	$/m²	138	138	138	83	83	83
Heliostat field	M$	1710	2786	2343	1023	1667	1402
Tower and receiver (total)	M$	295	481	404	295	481	404
Receiver alone	M$						
Tower + vertical piping	M$						
Heat storage capacity	h	13	16	16	13	16	16
Heat storage	M$	354	436	436	354	436	436

Table A.4 Continued.

Solar power plant–investment costs per 1000 MW

2002 $		Kolb (SANDIA) advanced receiver SM = 2.7	Kolb (SANDIA) advanced receiver Spain SM = 4.4	Kolb (SANDIA) advanced receiver Morocco SM = 3.7	Kalb/Vogel Mass production advanced receiver SM = 2.7	Kalb/Vogel Mass production advanced receiver Spain SM = 4.4 Land:Spain	Kalb/Vogel Mass production advanced receiver Morocco SM = 3.7
Power block	M$	590	590	590 Wet cooling	590	590	590 Wet cooling
Land preparation	M$	59	96	81	59	96	81
		–	–	–	–	–	–
Subtotal	M$	3008	4389	3854	2321	3270	2913
plus (Kalb/Vogel):							
Horizontal piping	M$				109	178	150
Land price (Spain = Kalb/Vogel 98)	$/m²		0			1.25	
Land costs (Spain = Kalb/Vogel 98)	M$	–	–	–	–	140.6	–
Total (direct costs)	M$	3008	4389	3854	2430	3588	3062
Indirect costs:							
Construction time (as small volume production)	a	4	4	4	2	2	2
Full interest period (50% of construction time)	a	2	2	2	1	1	1
Interest rate	%/a	4.0	4.0	4.0	4.0	4.0	4.0

Appendix A Solar Tower Power Plants: Comparison of Kolb (1996), Kalb/Vogel, SunLab, S&L

Interest during construction	M$	240.6	351.1	308.3	97.2	143.5	122.5
Owner's costs (% of investment)	%	6	6	6	3	3	3
Owner's costs	M$	180.5	263.3	231.2	72.9	107.6	91.9
Planning and contracting (% of investment)	%	9	9	9	4	4	4
Planning and contracting	M$	270.7	395.0	346.9	97.2	143.5	122.5
Engineering, management, and development (% of investment)	M$						
Engineering, management, and development							
Contingency cost margin (% of investment)	%	7	7	7	0	0	0
Contingency cost margin	M$	210.6	307.2	269.8			
	M$	—	—	—	—	—	—
Total (indirect costs)	M$	902	1317	1156	267	395	337
	M$	—	—	—	—	—	—
Total (overall)	M$	3910	5706	5010	2697	3983	3399
(Wet cooling)							
% less power (dry/wet cooling)	%			8			8
Investment costs with dry cooling (includes higher investment costs for cooling towers)	M$			5446			3695

Table A.4 Continued.

Solar power plant (1000 MW)

2002 $		SunLab 200 MW (Spain) SM = 2.9	SunLab 200 MW Spain SM = 4.4 Land: Sp.	SunLab 200 MW Morocco SM = 3.7	S&L 200 MW (Spain) SM = 2.9	S&L 200 MW Spain SM = 4.4 Land: Sp.	S$L 200 MW Morocco SM = 3.7	SunLab 220 MW (Spain) SM = 2.9	SunLab 220 MW Spain SM = 4.4 Land: Sp.	SunLab 220 MW Morocco SM = 3.7
Output power	MW	1000	1000	1000	1000	1000	1000	1000	1000	1000
Solar multiple (SM)		2.9	4.4	3.7	2.9	4.4	3.7	2.9	4.4	3.7
Design insolation	kWh/m² d	10.3	6.7	8.0	10.2	6.7	8.0	10.3	6.7	8.0
Mirror area	km²	13.00	19.72	16.59	13.34	20.23	17.01	12.01	18.22	15.32
Land area	km²	68.5	103.9	87.4	68.5	103.9	87.4	63.2	95.9	80.6
Mirror area coverage	%	19.0	19.0	19.0	19.5	19.5	19.5	19.0	19.0	19.0
Investment costs	Million $									
Heliostats (per m²)	$/m²	96	96	96	117	117	117	76	76	76
Heliostat field	M$	1248	1894	1592	1560	2367	1991	913	1385	1164
Tower and receiver	M$	305	463	389	350	531	447	268	407	342
Receiver alone	M$	185	281	236	230	349	293	155	234	197
Tower + vertical piping	M$	120	182	153	120	182	153	114	172	145
Heat storage capacity	h	13	16	16	13	16	16	? 16	16	16
Heat storage	M$	280	345	345	280	345	345	259	259	259

Appendix A Solar Tower Power Plants: Comparison of Kolb (1996), Kalb/Vogel, SunLab, S&L

Power block	M$	425	425	425	455	455	455	473	473	473
			Wet cooling			Wet cooling			Wet cooling	
Land preparation	M$	36	55	46	35	55	46	33	50	42
		–	–	–	–	–	–	–	–	–
Subtotal	M$	2294	3181	2797	2381	3752	3283	1945	2573	2280
plus (Kalb/Vogel):										
Horizontal piping	M$	109	165	139	139	165	139	99	150	126
Land price (Spain = Kalb/V)	$/m²	0.5	1.25	0.5	0.5	1.25	0.5	0.5	1.25	0.5
Land costs	M$	34.3	129.9	43.7	34.3	129.9	43.7	31.6	119.8	40.3
		–	–	–	–	–	–	–	–	–
Total (direct costs)	M$	2437	3476	2980	2324	4048	3465	2076	2843	2447
Indirect costs:										
Construction time	a				1	1	1			
Full interest period	a				(0.5)	(0.5)	(0.5)			
Interest rate	%/a				(4%/a)	(4%/a)	(4%/a)			
Interest during construction	M$				(4%/a=56)	(4%/a=81)	(4%/a=69)			
Owner's costs	%									
Owner's costs	M$									
Planning and contracting	%									
Planning and contracting	M$									
Engineering, management, and development (% of investment)	%	7.8	7.8	7.8	15.0	15.0	15.0	7.8	7.8	7.8

Table A.4 Continued.

Solar power plant (1000 MW)

2002 $		SunLab 200 MW (Spain) SM = 2.9	SunLab 200 MW Spain SM = 4.4 Land: Sp.	SunLab 200 MW Morocco SM = 3.7	S&L 200 MW (Spain) SM = 2.9	S&L 200 MW Spain SM = 4.4 Land: Sp.	S&L 200 MW Morocco SM = 3.7	SunLab 220 MW (Spain) SM = 2.9	SunLab 220 MW Spain SM = 4.4 Land: Sp.	SunLab 220 MW Morocco SM = 3.7
Engineering, management, and development	M$	190.1	271.1	232.4	423.7	607.2	519.8	161.9	221.8	190.9
Contingency cost margin	%	7.4	7.4	7.4	14.3	14.3	14.3	8.1	8.1	8.1
Contingency cost margin	M$	180.4	257.2	220.5	403.9	578.8	495.6	168.2	230.3	198.2
Total (indirect costs)	M$	370	528	453	828	1186	1015	330	452	389
Total (overall) (Wet cooling)	M$	2808	4004	3433	3652	5234	4481	2406	3295	2836
% less power (dry/wet cooling)	%			8			8			8
Investment costs with dry cooling (includes costs for dry cooling towers)	M$			3731			4870			3083

Appendix A Solar Tower Power Plants: Comparison of Kolb (1996), Kalb/Vogel, SunLab, S&L

Table A.5 Investment costs: solar power system (1000 MW).

Solar power system (including transmission and backup plants) – investment costs per 1000 MW

2002 $		Kolb (SANDIA) advanced receiver SM = 2.7	Kolb (SANDIA) advanced receiver Spain SM = 4.4	Kolb (SANDIA) advanced receiver Morocco SM = 3.7	Kalb/Vogel Mass production advanced receiver SM = 2.7	Kalb/Vogel Mass production advanced receiver Spain SM = 4.4 Land: Spain	Kalb/Vogel Mass production advanced receiver Morocco SM = 3.7
Solar plant (1000 MW at the plant)	M$	3910	5706	5446 Dry cooling	2697	3983	3695 Dry cooling
Solar power for 1000 MW output of transmission line (transmission losses)	MW	1088	1088	1130	1088	1088	1130
Solar power plant (of above power) (for 1000 MW output of transmission line)	M$	4255	6209	6153 Dry cooling	2935	4334	4175 Dry cooling
Electric power transmission							
Distance	km	2000	2000	3000	2000	2000	3000
Type: Overhead line 800 kV DC							
Losses:							
Losses overhead line per 1000 km	%	3.3	3.3	3.3	3.3	3.3	3.3
Losses at above distance	%	6.6	6.6	9.9	6.6	6.6	9.9
Undersea cable:							
Cable (20 km: ca. 0.1%)	%			0.1			0.1
Special converters for cable (two)	%			0			0
Converter stations (two)	%	1.5	1.5	1.5	1.5	1.5	1.5
Losses (total)	%	8.1	8.1	11.5	8.1	8.1	11.5
Investment power transmission (1 GW):							
Overhead line:							
Investment per 1000 km (1 GW)	M$	100	100	100	100	100	100

Table A.5 Continued.

Solar power system (including transmission and backup plants) – investment costs per 1000 MW

2002 $		Kolb (SANDIA) advanced receiver SM = 2.7	Kolb (SANDIA) advanced receiver Spain SM = 4.4	Kolb (SANDIA) advanved receiver Morocco SM = 3.7	Kalb/Vogel Mass production advanced receiver SM = 2.7	Kalb/Vogel Mass production advanced receiver Spain SM = 4.4 Land: Spain	Kalb/Vogel Mass production advanced receiver Morocco SM = 3.7
Full interest period (electrical system growth)	a	4	4	4	4	4	4
Real interest	%/a	4.0	4.0	4.0	4.0	4.0	4.0
Interest total during system growth	%	17.0	17.0	17.0	17.0	17.0	17.0
Overhead line (above distance) without interest	M$	200	200	300	200	200	300
Interest	M$	34	34	51	34	34	51
Overhead line including interest	M$	234	234	351	234	234	351
Undersea cable:							
Cable (20 km, Gibraltar)	M$			12			12
Special converters for cable (two)	M$			0			0
Converters (two)	M$	225	225	225	225	225	225
Investment total (1 GW input)	M$	459	459	588	459	459	588
Power input for 1 GW output (as above)	MW	1088	1088	1130	1088	1088	1130
Investment total (1 GW output)	M$	499	499	664	499	499	664
CCGT power plant (Gas-fired, 1 GW) (CCGT 2002: 580 M$/GW + Interest 4%/a*1.5a)	M$	615	615	615	615	615	615
Fast start-up gas turbine (1/3 of total power)	M$	100	100	100	100	100	100
CCGT plant total	M$	715	715	715	715	715	715
		–	–	–	–	–	–
Investment costs total	M$	5469	7423	7533	4150	5548	5554
				Dry cooling			Dry cooling

Appendix A Solar Tower Power Plants: Comparison of Kolb (1996), Kalb/Vogel, SunLab, S&L

Solar power system (1000 MW)

2002 $		SunLab 200 MW (Spain) SM = 2.9	SunLab 200 MW Spain SM = 4.4 Land: Sp.	SunLab 200 MW Morocco SM = 3.7	S&L 200 MW (Spain) SM = 2.9	S&L 200 MW Spain SM = 4.4 Land: Sp.	S&L 200 MW Morocco SM = 3.7	SunLab 220 MW (Spain) SM = 2.9	SunLab 220 MW Spain SM = 4.4 Land: Sp.	SunLab 220 MW Morocco SM = 3.7
Solar plant (1000 MW at the plant)	M$	2808	4004 / Dry cooling	3731	3652	5234 / Dry cooling	4870	2406	3295 / Dry cooling	3083
Solar power for 1000 MW output of transmission line (transmission losses)	MW	1088	1088	1130	1088	1088	1130	1088	1088	1130
Solar plant (of above power)	M$	3055	4357 / Dry cooling	4216	3974	5695 / Dry cooling	5503	2618	3586 / Dry cooling	3483
Electric power transmission										
Distance	km	2000	2000	3000	2000	2000	3000	2000	2000	3000
Type: Overhead line 800 kV DC										
Losses:										
Overhead line per 1000 km	%	3.3	3.3	3.3	3.3	3.3	3.3	3.3	3.3	3.3
Losses at above distance	%	6.6	6.6	9.9	6.6	6.6	9.9	6.6	6.6	9.9
Undersea cable:										
Cable (20 km: ca. 0.1%)	%			0.1			0.1			0.1
Special converters for cable (two)	%			0			0			0

Table A.5 Continued.

Solar power system (1000 MW)

2002 $		SunLab 200 MW (Spain) SM = 2.9	SunLab 200 MW Spain SM = 4.4 Land: Sp.	SunLab 200 MW Morocco SM = 3.7	S&L 200 MW (Spain) SM = 2.9	S&L 200 MW Spain SM = 4.4 Land: Sp.	S&L 200 MW Morocco SM = 3.7	SunLab 220 MW (Spain) SM = 2.9	SunLab 220 MW Spain SM = 4.4 Land: Sp.	SunLab 220 MW Morocco SM = 3.7
Converter stations (two)	%	1.5	1.5	1.5	1.5	1.5	1.5	1.5	1.5	1.5
Losses (total)	%	8.1	8.1	11.5	8.1	8.1	11.5	8.1	8.1	11.5
Investment power transmission per 1 GW:										
Overhead line:										
Investment per 1000 km (1 GW)	M$	100	100	100	100	100	100	100	100	100
Full interest period	a	4	4	4	4	4	4	4	4	4
Real interest	%/a	4.0	4.0	4.0	4.0	4.0	4.0	4.0	4.0	4.0
Interest total during growth	%	17.0	17.0	17.0	17.0	17.0	17.0	17.0	17.0	17.0
Overhead line without interest	M$	200	200	300	200	200	300	200	200	300
Interest	M$	34	34	51	34	34	51	34	34	51

Appendix A Solar Tower Power Plants: Comparison of Kolb (1996), Kalb/Vogel, SunLab, S&L

		1	2	3	4	5	6	7	8	9
Overhead line including interest	M$	234	234	351	234	234	351	234	234	351
Undersea cable:										
Cable (20km, Gibraltar)	M$			12			12			12
Special converters for cable (two)	M$			0			0			0
Converters (two)	M$	225	225	225	225	225	225	225	225	225
Power transmission total (1 GW)	M$	459	459	588	459	459	588	459	459	588
Power input for 1 GW line output	MW	1088	1088	1130	1088	1088	1130	1088	1088	1130
Transmission total (1 GW output)	M$	499	499	664	499	499	664	499	499	664
CCGT power plant (gas-fired, 1 GW)										
(580M$/GW + interest 4%/a×1.5a)	M$	615	615	615	615	615	615	615	615	615
Fast start-up gas turbine	M$	100	100	100	100	100	100	100	100	100
CCGT plant total	M$	715	715	715	715	715	715	715	715	715
		–	–	–	–	–	–	–	–	–
Investment costs total	M$	4270	5571 Dry cooling	5595	5188	6909 Dry cooling	6882	3833	4800 Dry cooling	4862

Table A.6 Capital costs, gas costs.

Capital costs per kWh, gas costs (solar power system)

2002 $		Kolb (SANDIA) advanced receiver SM = 2.7	Kolb (SANDIA) advanced receiver Spain SM = 4.4	Kolb (SANDIA) advanced receiver Morocco SM = 3.7	Kalb/Vogel Mass production advanced receiver SM = 2.7	Kalb/Vogel Mass production advanced receiver Spain SM = 4.4	Kalb/Vogel Mass production advanced receiver Morocco SM = 3.7
Investment	M$	5469	7423	7533	4150	5548	5554
Electric power (transmission line output)	MW	1000	1000	1000	1000	1000	1000
				Dry cooling			Dry cooling
Capacity factor (solar power system)	%	100	100	100	100	100	100
Hours of full-load operation	h/a	8760	8760	8760	8760	8760	8760
Power production (solar power system)	GWh/a	8760	8760	8760	8760	8760	8760
Capacity factor (solar power plant)	%	55	70	80	55	70	80
Hours of full-load operation	h/a	4818	6132	7008	4818	6132	7008
Power production (solar power plant)	GWh/a	4818	6132	7008	4818	6132	7008
Capacity factor backup power plant	%	45	30	20	45	30	20
Hours of full-load operation	h/a	3942	2628	1752	3942	2628	1752
Power production (backup power plant)	GWh/a	3942	2628	1752	3942	2628	1752
Capital costs per kWh							
Interest rate (real)	%/a	4.0	4.0	4.0	4.0	4.0	4.0
Interest factor		1.040	1.040	1.040	1.040	1.040	1.040
Lifetime	a	45	45	45	45	45	45
→ Annuity factor	%/a	4.83	4.83	4.83	4.83	4.83	4.83
(Annuity factor at 2% real interest)	%/a	3.39	3.39	3.39	3.39	3.39	3.39

Annuity factor = $q^n \times (q-1)/(q^n-1)$, where q = Interest factor, n = years

Annuity			M$/a	264	358	364	200	268	268			
Power output			GWh/a	8760	8760	8760	8760	8760	8760			
Capital costs			¢/kWh	3.01	4.09	4.15	2.29	3.06	3.06			
(Capital costs at 2% real interest)			¢/kWh	2.12	2.87	2.92	1.61	2.15	2.15			
						Dry cooling			Dry cooling			
Gas costs (CCGT backup power plants)												
Efficiency (LHV)			%	58	58	58	58	58	58			
Share of backup power			%	45	30	20	45	30	20			
Gas costs/kWh$_{el}$ (solar power system):												
Gas price:												
Equivalent oil price $/barrel	$/MM BTU (HHV)	¢/kWh (LHV)	¢/cu ft 1030 BTU HHV 0.273 kWh LHV	¢/kWh$_{el}$	¢/kWh$_{el}$	¢/kWh$_{el}$	¢/kWh$_{el}$	¢/kWh$_{el}$	¢/kWh$_{el}$			
20	3.31	1.25	0.34	0.97	0.65	0.43	0.97	0.65	0.43			
32	5.30	2	0.55	1.55	1.03	0.69	1.55	1.03	0.69			
40	6.62	2.5	0.68	1.94	1.29	0.86	1.94	1.29	0.86			
48	7.95	3	0.82	2.33	1.55	1.03	2.33	1.55	1.03			
60	9.93	3.75	1.02	2.91	1.94	1.29	2.91	1.94	1.29			
64	10.60	4	1.09	3.10	2.07	1.38	3.10	2.07	1.38			
80	13.25	5	1.37	3.88	2.59	1.72	3.88	2.59	1.72			

1 barrel = 159 l; 1 liter oil = 10.0 kWh LHV; Intern.:1 cf gas = 1030 Btu = 0.302 kWh HHV = 0.273 LHV. 1 million Btu= 293 kWh (Germany 1 Nm^3 = 8.8 kWh LHV = 9.8 HHV)

Table A.6 Continued.

Capital costs per kWh, gas costs (solar power system)

2002 $		SunLab 200 MW (Spain) SM = 2.9	SunLab 200 MW Spain SM = 4.4	SunLab 200 MW Morocco SM = 3.7	S&L 200 MW (Spain) SM = 2.9	S&L 200 MW Spain SM = 4.4	S$L 200 MW Morocco SM = 3.7	SunLab 220 MW (Spain) SM = 2.9	SunLab 220 MW Spain SM = 4.4	SunLab 220 MW Morocco SM = 3.7
Investment	M$	4270	5571	5595	5188	6909	6882	3833	4800	4862
Electric power (transmission output)	MW	1000	1000	1000	1000	1000	1000	1000	1000	1000
				Dry cooling			Dry cooling			Dry cooling
Capacity factor (solar power system)	%	100	100	100	100	100	100	100	100	100
Hours full-load operation	h/a	8760	8760	8760	8760	8760	8760	8760	8760	8760
Power production (solar power system)	GWh/a	8760	8760	8760	8760	8760	8760	8760	8760	8760
Capacity factor (solar power plant)	%	55	70	80	55	70	80	55	70	80
Hours full-load operation	h/a	4818	6132	7008	4818	6132	7008	4818	6132	7008
Power production (solar power plant)	GWh/a	4818	6132	7008	4818	6132	7008	4818	6132	7008
Capacity factor (backup power plant)	%	45	30	20	45	30	20	45	30	20
Hours full-load operation	h/a	3942	2628	1752	3942	2628	1752	3942	2628	1752
Power production (backup power plant)	GWh/a	3942	2628	1752	3942	2628	1752	3942	2628	1752

Appendix A Solar Tower Power Plants: Comparison of Kolb (1996), Kalb/Vogel, SunLab, S&L

Capital costs per kWh										
Interest rate (real)	%/a	4.0	4.0	4.0	4.0	4.0	4.0	4.0	4.0	4.0
Interest factor		1.040	1.040	1.040	1.040	1.040	1.040	1.040	1.040	1.040
Lifetime	a	45	45	45	45	45	45	45	45	45
–> Annuity factor	%/a	4.83	4.83	4.83	4.83	4.83	4.83	4.83	4.83	4.83
(Annuity factor at 2% real interest)	%/a	3.39	3.39	3.39	3.39	3.39	3.39	3.39	3.39	3.39
Annuity	M$/a	206	269	270	250	333	332	185	232	235
Power output	GWh$_{th}$/a	8760	8760	8760	8760	8760	8760	8760	8760	8760
Capital costs	¢/kWh	2.35	3.07	3.08	2.86	3.81	3.79	2.11	2.64	2.68
(Capital cost at 2% real interest)	¢/kWh	1.65	2.16	2.17	2.01	2.67	2.66	1.48	1.86	1.88
				Dry cooling			Dry cooling			Dry cooling
Gas costs (backup power plants)										
Efficiency (LHV)	%	58	58	58	58	58	58	58	58	58
Share of backup power	%	45	30	20	45	30	20	45	30	20

Gas costs/kWh$_{el}$ (solar power system):

Gas price: Equivalent oil price $/barrel	¢/kWh$_{el}$	¢/kWh$_{el}$	¢/kWh$_{el}$	¢/kWh$_{el}$	¢/kWh$_{el}$	¢/kWh$_{el}$	¢/kWh$_{el}$	¢/kWh$_{el}$	¢/kWh$_{el}$
20	0.97	0.65	0.43	0.97	0.65	0.43	0.97	0.65	0.43
32	1.55	1.03	0.69	1.55	1.03	0.69	1.55	1.03	0.69
40	1.94	1.29	0.86	1.94	1.29	0.86	1.94	1.29	0.86
48	2.33	1.55	1.03	2.33	1.55	1.03	2.33	1.55	1.03
60	2.91	1.94	1.29	2.91	1.94	1.29	2.91	1.94	1.29
64	3.10	2.07	1.38	3.10	2.07	1.38	3.10	2.07	1.38
80	3.88	2.59	1.72	3.88	2.59	1.72	3.88	2.59	1.72

Table A.7 O&M costs, electric power costs.

O&M costs, electric power costs

2002 $		Kolb (SANDIA) advanced receiver SM = 2.7	Kolb (SANDIA) advanced receiver Spain SM = 4.4	Kolb (SANDIA) advanced receiver Morocco SM = 3.7	Kalb/Vogel Mass production advanced receiver SM = 2.7	Kalb/Vogel Mass production advanced receiver Spain SM = 4.4	Kalb/Vogel Mass production advanced receiver Morocco SM = 3.7
O&M costs							
Solar power plant							
Solar power plant output	MW	200	200	200	200	200	200
Solar multiple (SM)		2.7	4.4	3.7	2.7	4.4	3.7
Design insolation	kWh/m² d	11.00	6.70	8.00	11.00	6.70	8.00
O&M costs per year: (in 2002-$) per:	MW	200	200	200	200	200	200
For SM = 2.7 (for computation)	M$/a	7.1	7.1	7.1	7.1	7.1	7.1
for above SM:	M$/a	7.1	11.5	9.7	7.1	11.5	9.7
per:	MW	1000	1000	1000	1000	1000	1000
	M$/a	35.4	57.7	48.5	35.4	57.7	48.5

Appendix A Solar Tower Power Plants: Comparison of Kolb (1996), Kalb/Vogel, SunLab, SQL

per:						
Power production/a (solar power system 1 GW)	MW	1088	1088	1130	1088	1130
O&M costs (solar power plant)	M$/a	38.5	62.8	54.8	38.5	54.8
	GWh/a	8760	8760	8760	8760	8760
	¢/kWh	0.44	0.72	0.63	0.44	0.63
				Wet cooling		Wet cooling
Dry cooling: x% less power	%			8		8
O&M costs (dry cooling)	¢/kWh			0.68		0.68
				Dry cooling		Dry cooling
CCGT backup power plant						
Fixed O&M/1000 MW CCGT baseload	M$/a	10.0	10.0	10.0	10.0	10.0
Factor backup-/baseload mode (assumption)		1	1	1	1	1
→ Fixed cost for backup plant (1000 MW)	M$/a	10.0	10.0	10.0	10.0	10.0
Produced power/a (solar power system)	GWh/a	8760	8760	8760	8760	8760
→ Fixed cost per kWh (solar power system)	¢/kWh	0.11	0.11	0.11	0.11	0.11
Variable O&M cost per kWh backup power	¢/kWh	0.17	0.17	0.17	0.17	0.17
Variable cost per kWh (so or power system)	¢/kWh	0.08	0.05	0.08	0.05	0.03
O&M cost backup plant total	¢/kWh	0.19	0.17	0.19	0.17	0.15

Table A.7 Continued.

O&M costs, electric power costs

2002 $

| | Kolb (SANDIA) advanced receiver SM = 2.7 | Kolb (SANDIA) advanced receiver Spain SM = 4.4 | Kolb (SANDIA) advanced receiver Morocco SM = 3.7 | Kalb/Vogel Mass production advanced receiver SM = 2.7 | Kalb/Vogel Mass production advanced receiver Spain SM = 4.4 | Kalb/Vogel Mass production advanced receiver Morocco SM = 3.7 |

Electric power costs at gas price:

Equivalent oil price $/barrel	$/MM BTU (HHV)	¢/kWh (LHV)	¢/cu ft 1030 BTU HHV	¢/kWh$_{el}$	¢/kWh$_{el}$	¢/kWh$_{el}$	¢/kWh$_{el}$	¢/kWh$_{el}$	¢/kWh$_{el}$
20	3.31	1.25	0.34	4.6	5.6	5.4	3.9	4.6	4.3
32	5.30	2	0.55	5.2	6.0	5.7	4.5	5.0	4.6
40	6.62	2.5	0.68	5.6	6.3	5.8	4.9	5.2	4.7
48	7.95	3	0.82	6.0	6.5	6.0	5.2	5.5	4.9
60	9.93	3.75	1.02	6.6	6.9	6.3	5.8	5.9	5.2
64	10.60	4	1.09	6.7	7.0	6.4	6.0	6.0	5.3
80	13.25	5	1.37	7.5	7.6	6.7	6.8	6.5	5.6
						Dry cooling			Dry cooling
(Cf.: at 2% real interest:)		2.5	0.68	4.7	5.0	4.6	4.2	4.3	3.8

Appendix A Solar Tower Power Plants: Comparison of Kolb (1996), Kalb/Vogel, SunLab, S&L

O&M costs, electric power costs

2002 $		SunLab 200 MW (Spain) SM = 2.9	SunLab 200 MW Spain SM = 4.4	SunLab 200 MW Morocco SM = 3.7	S&L 200 MW (Spain) SM = 2.9	S&L 200 MW Spain SM = 4.4	S&L 200 MW Morocco SM = 3.7	SunLab 220 MW (Spain) SM = 2.9	SunLab 220 MW Spain SM = 4.4	SunLab 220 MW Morocco SM = 3.7
O&M costs										
Solar power plant										
Solar power plant output	MW	200	200	200	200	200	200	220	220	220
Solar multiple (SM)		2.9	4.4	3.7	2.9	4.4	3.7	2.9	4.4	3.7
Design insolation	kWh$_t$/m^2 d	10.30	6.70	8.00	10.20	6.70	8.00	10.30	6.70	8.00
O&M costs per year:										
per:	MW	200	200	200	200	200	200	220	220	220
	M$/a	4.7	4.7	4.7	9.1	9.1	9.1	4.7	4.7	4.7
For SM = 2.9 (for computation)										
for above SM:	M$/a	4.7	7.1	6.0	9.1	13.8	11.6	4.7	7.1	6.0
per:	MW	1000	1000	1000	1000	1000	1000	1000	1000	1000
	M$/a	23.5	35.7	30.0	45.5	69.0	58.1	21.4	32.4	27.3
per:	MW	1088	1088	1130	1088	1088	1130	1088	1088	1130
	M$/a	25.6	38.8	33.9	49.5	75.1	65.6	23.2	35.3	30.8

Table A.7 Continued.

O&M costs, electric power costs

2002 $		SunLab 200 MW (Spain) SM = 2.9	SunLab 200 MW Spain SM = 4.4	SunLab 200 MW Morocco SM = 3.7	S&L 200 MW (Spain) SM = 2.9	S&L 200 MW Spain SM = 4.4	S$L 200 MW Morocco SM = 3.7	SunLab 220 MW (Spain) SM = 2.9	SunLab 220 MW Spain SM = 4.4	SunLab 220 MW Morocco SM = 3.7
Power production/a (1 GW output)	GWh/a	8760	8760	8760	8760	8760	8760	8760	8760	8760
O&M solar power plant	¢/kWh	0.29	0.44	0.39	0.57	0.86	0.75	0.27	0.40	0.35
				Wet cooling			Wet cooling			Wet cooling
Dry cooling: x% less power	%			8			8			8
O&M costs (dry cooling)	¢/kWh			0.42			0.81			0.38
				Dry cooling			Dry cooling			Dry cooling
CCGT backup power plant										
Fixed O&M/1000 MW baseload	M$/a	10.0	10.0	10.0	10.0	10.0	10.0	10.0	10.0	10.0
Factor backup-/baseload mode		1	1	1	1	1	1	1	1	1
→ Fixed cost for backup plant	M$/a	10.0	10.0	10.0	10.0	10.0	10.0	10.0	10.0	10.0

Appendix A Solar Tower Power Plants: Comparison of Kolb (1996), Kalb/Vogel, SunLab, SQL

Power/a (solar power system)	GWh/a	8760	8760	8760	8760	8760	8760	8760	8760
→ Fixed cost/kWh (solar power system)	¢/kWh	0.11	0.11	0.11	0.11	0.11	0.11	0.11	0.11
Variable O&M/kWh backup power	¢/kWh	0.17	0.17	0.17	0.17	0.17	0.17	0.17	0.17
Variable cost/kWh (solar power system)	¢/kWh	0.08	0.05	0.03	0.08	0.05	0.03	0.08	0.05
O&M cost backup plant total	¢/kWh	0.19	0.17	0.15	0.19	0.17	0.15	0.19	0.17
Electric power costs at gas price:									
Equivalent oil price $/barrel		¢/kWh$_{el}$	¢/kWh$_{el}$	¢/kWh$_{el}$	¢/kWh$_{el}$	¢/kWh$_{el}$	¢/kWh$_{el}$	¢/kWh$_{el}$	¢/kWh$_{el}$
20		3.8	4.3	4.1	4.6	5.5	5.2	3.5	3.9
32		4.4	4.7	4.3	5.2	5.9	5.4	4.1	4.2
40		4.8	5.0	4.5	5.6	6.1	5.6	4.5	4.5
48		5.2	5.2	4.7	5.9	6.4	5.8	4.9	4.8
60		5.7	5.6	4.9	6.5	6.8	6.0	5.5	5.2
64		5.9	5.7	5.0	6.7	6.9	6.1	5.7	5.3
80		6.7	6.3	5.4	7.5	7.4	6.5	6.4	5.8
				Dry cooling			Dry cooling		
(Cf.: at 2% real interest)	40	4.1	4.1	3.6	4.7	5.0	4.5	3.9	3.7

Last column (rightmost): ¢/kWh$_{el}$ values: 3.6, 3.9, 4.1, 4.2, 4.5, 4.6, 4.9, Dry cooling, 3.3

Appendix B
Inflation, Purchasing Power Parities

Table B.1 Inflation in the USA and Germany (OECD, 2008).

Price indices

	Consumer price index				Producer price index			
	All items		All items – nonfood, nonenergy		Total (industrial products)		Manufacturing products	
	Germany	USA	Germany	USA	Germany	USA	Germany	USA
2007	112.1	120.4	109.2	116.2	119.1	130.1	112.6	1??
2006	110.6	117.1	107.2	113.6	116.8	124.1	109.7	117.5
2005	107.9	113.4	106.4	110.8	110.6	118.6	106.8	113.0
2004	106.2	109.7	105.5	108.5	105.8	110.5	103.9	107.1
2003	104.5	106.8	103.8	106.6	104.1	104.1	102.1	102.7
2002	103.4	104.5	102.9	105.1	102.4	98.8	101.5	100.1
2001	102.0	102.8	101.3	102.7	103.0	101.1	101.3	100.8
2000	100	100	100	100	100	100	100	100
1999	98.6	96.7	99.4	97.6	97.0	94.5	97.0	96.1
1998	98.0	94.7	98.9	95.6	98.0	93.7	97.2	94.5
1997	97.1	93.2	97.6	93.5	98.4	96.1	97.4	95.5
1996	95.3	91.1	95.8	91.3	97.3	96.2	96.8	95.2
1995	93.9	88.5	94.3	88.9	98.5	94.0	96.7	93.1
1994	92.3	86.1	92.4	86.3	96.8	90.7	94.7	90.4
1993	89.9	83.9	89.9	83.9	96.3	89.6	94.0	89.2
1992	86.1	81.5	85.3	81.3	96.3	88.3	94.0	87.9
1991	81.9	79.1	80.8	78.4	95.0	87.8	92.5	86.8
1990	78.7	75.9	77.9	74.7	92.7	87.6	90.5	85.7
1989	76.6	72.0	76.0	71.1	91.1	84.6	89.2	82.1
1988	74.5	68.7	74.0	68.1	88.4	80.6	86.3	78.2
1987	73.6	66.0	72.7	65.2	87.3	77.4	84.9	75.5
1986	73.4	63.7	71.7	62.6	89.5	75.5	85.3	73.7
1985	73.5	62.5	71.0	60.2	91.8	77.7	87.3	75.3
1984	72.0	60.3	69.5	57.7	89.6	78.1	85.5	74.6
1983	70.3	57.8	67.9	54.9	87.1	76.3	83.1	73.1
1982	68.1	56.0	65.3	52.9	85.9	75.3	81.9	71.9

Large-Scale Solar Thermal Power. Werner Vogel and Henry Kalb
© 2010 WILEY-VCH Verlag GmbH & Co. KGaA, Weinheim
ISBN: 978-3-527-40515-2

Appendix B Inflation, Purchasing Power Parities

Table B.1 Continued.

Price indices

| | Consumer price index | | | | Producer price index | | | |
| | All items | | All items – nonfood, nonenergy | | Total (industrial products) | | Manufacturing products | |
	Germany	USA	Germany	USA	Germany	USA	Germany	USA
1981	64.7	52.8	62.4	49.2	81.1	73.9	78.2	69.2
1980	60.8	47.9	59.1	44.6	75.2	67.7	73.7	63.3
1979	57.7	42.1	56.2	39.9	69.9	59.3	69.0	55.8
1978	55.5	37.9	54.5	36.1	66.8	52.7	65.5	50.2
1977	54.0	35.2	53.0	33.7	65.9	48.9	65.1	46.6
1976	52.0	33.0	51.0	31.7	64.2	46.1	63.3	43.7
1975	49.9	31.3	49.1	29.7	62.2	44.0	61.4	41.9
1974	47.1	28.6	46.5	27.2	60.2	40.3	59.4	37.8
1973	44.1	25.8	43.6	25.1	53.1	33.9	52.4	32.8
1972	41.2	24.3	41.0	24.3	49.7	30.0	49.1	30.0
1971	39.0	23.5	38.9	23.6	48.6	28.7	48.0	29.1
1970	37.1	22.5	36.8	22.5	46.6	27.8	46.0	28.3

Table B.2 Inflation (USA, Germany): Price increases 1995–2002.

Results for 1995–2002

| Consumer price index | | | | Producer price index | | | |
| All items | | Nonfood, nonenergy | | Total | | Manufacturing products | |
Germany	USA	Germany	USA	Germany	USA	Germany	USA
10.1%	18.0%	9.1%	18.2%	4.0%	5.1%	5.1%	7.5%

Note:
The *US Producer Price Index* exhibits an *anomaly* for the important year 2002, insofar as prices actually fell between 2001 and 2002. Therefore, in this book, for the USA the *Consumer* Price Index (all items) is used. The values for "all items" and for "nonfood, nonenergy" differ up to the year 2002 only slightly; we thus generally employ "all items" for comparisons. The conversion factor is particularly important for the transfer of the estimated costs from the studies of Kolb (1996a) and of Kalb and Vogel (1998), which were quoted at the monetary value of 1995, to the year 2002. This latter year is used as a reference in the present book, since the costs from the SunLab and S&L reports (both cited in S&L, 2003) are given in terms of 2002-$.

In Germany, the Producer Price Index and the Consumer Price Index likewise differ considerably. In order to avoid quoting costs that are too low due to the use of an overly small inflation factor, for Germany also the more rapidly increasing Consumer Price Index (all items) is used for calculations. The conversion of German cost data to Dollar (incl. inflation in Euro), however, applies to only a few and as a whole insignificant items in the cost calculation.

Table B.3 Purchasing power parities and exchange rates.

	Purchasing power parities (PPP) (OECD) Euro (Germany)/US $				Exchange rates Euro (Germany)/US $	
	Series 1980–2007[a] (revised with respect to series 1980–2006)	Series 1980–2006[a] (revised with respect to series 1980–2004)	Series 1980–2004[a]			
	€/$				€/$[d]	$/€[d]
2007	0.88		Note the significant differences in the years 2003 and 2004		0.73	1.370
2006	0.87	0.88			0.80	1.255
2005	0.89	0.88			0.80	1.242
2004	0.90	0.89	0.97	0.85[b]	0.80	1.243
2003	0.92	0.91	0.98		0.88	1.131
2002	0.94	0.96	0.98	Result 2002:	1.06	0.946
2001	0.96	0.97	0.99	1 $ = 0.96 €	1.12	0.896
2000	0.97	0.98	0.99	1 € = 1.043 $[c]	1.08	0.924
1999	0.97	1.00	1.00		0.94	1.066
1998	0.99	1.00	1.01		0.90	1.112
1997	0.99	1.01	1.01		0.89	1.127
1996	0.99	1.01	1.01		0.77	1.301
1995	1.00	1.03	1.02		0.73	1.364
1994	1.01	1.03	1.03		0.83	1.206
1993	1.00	1.03	1.02		0.85	1.182
1992	0.99	1.01	1.01		0.80	1.254
1991	0.96	0.99	0.98		0.85	1.177
1990	0.97	0.99	0.98		0.83	1.210
1989	0.97	0.99	0.99		0.96	1.040
1988	0.98	1.00	1.00		0.90	1.112
1987	1.00	1.02	1.02		0.92	1.088
1986	1.01	1.04	1.03		1.11	0.901
1985	1.00	1.03	1.02			
1984	1.01	1.04	1.03			
1983	1.03	1.05	1.05			
1982	1.04	1.07	1.06			
1981	1.06	1.08	1.07			
1980	1.11	1.14	1.12			

Before 2002:
1 Euro = 1.95583 Deutsche Mark (DM)
1 Deutsche Mark = 0.51129 Euro

a) OECD, 2008 (Statistics: Price Indices und PPPs).
b) Compare: purchasing power parity according to the Deka-Bank (2004) for the year 2004: 0.85. (The DekaBank (Germany) carries out its own computations of purchasing power parity.)
c) The revised value for 2002 from the series of years 1980–2007 was available only at the end of 2008. All conversions in this book were carried out up to that point using the value from the

Table B.3 *Continued.*

series 1980–2006 ($1 = €0.96). This value was, therefore, retained. The difference with respect to the series 1980–2007 is small. (A good description of the problems associated with purchasing power parities (their computation and predictive power) can be found in Terres (2000) (in German).)

d) Exchange rates 2005–2007: OECD StatExtracts (2009) 1986–2004: RWE Weltenergiereport (2005, p. 132).

Appendix C
Energy Statistics

Large-Scale Solar Thermal Power. Werner Vogel and Henry Kalb
© 2010 WILEY-VCH Verlag GmbH & Co. KGaA, Weinheim
ISBN: 978-3-527-40515-2

Units
1 J (− 1 Nm) − 1 Ws (1 W = 1 J/s)
==> 1 kWh = 1000 W × 3600 s = 3.6 MJ
   1 TWh = 10^{12} Wh = 3.6×10^{15} J = 3.6 PJ
   1 GJ = 277.7 kWh (1/3.6 = 0.27777) // 1 TJ = 0.277 GWh
1 cal = 4.1868 J
1 Btu (British thermal unit) = 0.252 kcal = 1055.056 J (10^6 Btu = 1.055 GJ)
(1000 Btu = 1055 kJ = (0.29307) ca. 0.3 kWh) (10^6 Btu = 293.0 kWh)
   1 barrel = 159.106 l, 1 US gallon = 3.79 l, 1 pound (lb) = 0.45359 kg,
   1 sh t (2000 lb) = 0.90718 t

1 GWa = 8760 GWh = 8.76 TWh = 31.536 PJ = (29.89) 30 trillion (10^{12}) Btu
   (1 EJ = 31.71 GWa)
   (10^{15} Btu = 33.45 GWa; i.e. 3 quads (10^{15} Btu) = 100.35 GWa)
Conversion:
1 million t ce (coal equivalent) = 0.93 GWa (1 GWa = 1.076 Mtce)
1 million t oe (oil equivalent) = 1.33 GWa (1 GWa = 0.752 Mtoe)
1 million barrels crude oil = 0.180 GWa
   (i.e., 5.5 million barrels crude oil = (0.99) 1.0 GWa)
   (i.e., 15,000 barrels/d = 0.99 GWa/a)
10^9 Nm^3 natural gas (Germany 8.81 kWh/Nm^3) = 1.01 GWa LHV (=1.12 GWa HHV)
10^9 cubic feet natural gas (international: BP 10.78 kWh/Nm^3 HHV) = 0.0344 GWa HHV
(= 0.0310 GWa LHV)
(i.e., 30 billion (10^9) cubic feet natural gas = 1.03 GWa HHV (= 0.93 GWa LHV))

For the coming era of electrical energy supply from power plants, the appropriate unit is the gigawatt-year: 1 GWa = 1 GW × 8760 h = 8760 GWh.
A power plant with a nominal output power of 1 GW and 100% capacity utilization produces 1 GWa of energy per year.
   Compare: a 1 GW solar power plant with 80% capacity utilization (Morocco, USA: SM = 3.7) produces 0.8 GWa of solar energy, plus 0.2 GWa from backup plants; this gives 1 GWa per year (100% overall utilization). A 1 GW offshore wind power park in the North Sea with 50% capacity utilization would yield 0.5 GWa. American nuclear power plants of the current 1050 MW class operating at 91% capacity utilization (2004) produce 0.96 GWa per year, while German plants of the 1250 MW class at 87.3% utilization (2004) yield 1.09 GWa per year.
   Note: 1 GWa is the annual yield of a 1 GW solar power system or roughly 1 nuclear plant.

Additional costs
Note: a difference in electrical energy cost of 1 US-¢/kWh, referred to an annual energy production of 10 GWa, corresponds to additional costs of 0.876×10^9 US $ = ca. 1 billion $ per year.

Heating values
Coal (ce) = 7000 kcal/kg LHV = 29308 kJ/kg = (8.141) ca. 8.1 kWh/kg LHV
   (ce = coal equivalent) 1 t ce = 8141 kWh LHV
   Bituminous coal (depending on provenance): 7 to 9.9 kWh/kg LHV. (USA kWh/kg HHV:
      Wyoming: 5.5; Illinois #6: 7.1; Upper Freeport PA: 8.6; Pocahontas-#3 VA: 9.6)

Crude oil (oe): 10000 kcal/kg LHV = 41868 kJ/kg = 11.630 kWh/kg = (9.944) ca. 10.0 kWh/l
   LHV (oe = oil equivalent). Usual American conversion factor: 1 barrel (159.1 l) crude
   oil = 136 kg crude oil (cf. BP 136.4) => density 0.855 kg/l
Domestic fuel oil: 41868 kJ/kg LHV D:0.86 kg/l =>36006 kJ/lit =10.0 kWh/l LHV
Gasoline: 12.90 kWh/kg HHV (46683 kJ/kg), 11.80 kWh/kg LHV (42496 kJ/kg)
   Density: 0.72 - 0.80 = 0.76 kg/l => 9.80 kWh/l HHV; 9.0 kWh/l LHV
Methanol: 6.27 kWh/kg HHV; 5.42 kWh/kg LHV (Density 4°C: 0.81, 25°C: 0.79 kg/l)

Natural gas: Heating values depending on provenance:
   Natural gas L: 9.77 kWh/Nm^3 HHV; 8.81 kWh/Nm^3 LHV <– German convention
   Natural gas H: 11.48 kWh/Nm^3 HHV; 10.37 kWh/Nm^3 LHV
   Germany, the Netherlands (ca.): 9.8 kWh/Nm^3 HHV; 8.8 kWh/Nm^3 LHV
   USA Example Panhandle: 12.2 kWh/Nm^3 HHV; 10.8 kWh/Nm^3 LHV
International convention: If heating values for statistical conversion not given, then 38 MJ/
Nm^3 (10.55 kWh/Nm^3) HHV
RWE (2005): 10.55 kWh/Nm^3 HHV (ca. 9.53 LHV) = 0.0295 kWh/scf HHV = (1008)
ca.1000 Btu/scf HHV (scf = standard cubic foot)
cf. BP2005: 1030 Btu/cf HHV =0.302 kWh/cf HHV (ca 0.273 LHV) = 10.78 kWh/Nm^3
HHV (ca. 9.7 LHV)
1 scf at 60°F (=15.5°C) = ideal gas: 0.02685 norm cubic meter (Nm^3) at 0°C (32°F)
(cf. 1 cu ft =0.02832 m^3). Natural gas is not an ideal gas; conversion depends on gas
composition. BP (2005) uses for conversion: 1 cubic foot of natural gas = 0.028 Nm^3
CH_4:   11.1 kWh/Nm^3 HHV; 10.0 kWh/Nm^3 LHV
H_2:   3.54 kWh/Nm^3 HHV; 3.00 kWh/Nm^3 LHV
CO: 3.51 kWh/Nm^3
   HHV: Higher heating value or "gross calorific value" or "gross energy"
   LHV: Lower heating value or "net calorific value" or "net energy"
      HHV: Energy content of the fuel including the heat of condensation of the water
      vapor contained in the combustion gases, which normally cannot be used.
      LHV: Without heat of condensation. American and British statistics use the HHV.

			Germany	USA
kJ = 10^3 J	TJ = 10^{12} J	10^6	Million (Mill.)	million
MJ = 10^6 J	PJ = 10^{15} J	10^9	Milliarde (Mrd.)	billion
GJ = 10^9 J	EJ = 10^{19} J	10^{12}	Billion	trillion
		10^{15}	Billiarde	quadrillion

Table C.1 Energy consumption per year.

Primary energy consumption 2004[a]	Germany (82 M inhabitants)			EU-25 (457 M inhabitants)				USA (290 M inhabitants)				World (6500 M inhabitants)			
	M tce	M toe	EJ	GWa	M tce	M toe	EJ	GWa	M tce	M toe	EJ	GWa	M tce	M toe	EJ	GWa
Crude oil	176	123	5.2	163	992	694	29.1	922	1339	937	39.2	1244	5383	3767	157.8	5003
Natural gas (HHV)	110	77	3.2	102	600	420	17.6	558	832	582	24.4	773	3458	2420	101.3	3214
Coal	121	85	3.6	113	439	307	12.9	408	806	564	23.6	749	3970	2778	116.3	3689
Hard coal[b]	66	43		57		239		317								
Lignite[b]	56	36		48		74		98								
Nuclear energy[c]	53	37	1.5	49	319	223	9.3	296	267	187	7.8	248	892	624	26.1	829
Hydropower[c]	9	6	0.3	8	104	73	3.1	97	84	59	2.5	78	906	634	26.6	842
Total	472	330	13.8	438	2455	1718	71.9	2282	3331	2331	97.6	3096	14609	10224	428.2	13577

a) Source: BP Weltenergiestatistik 2005, p. 38. Details for Germany see Energy flow chart (http://www.ag-energiebilanzen.de).
b) Hard coal: Anthracite, bituminous, and subbituminous coal. Germany: Arbeitsgemeinschaft Energiebilanzen 2005; EU: Eurostat 2006.
c) To obtain the quoted quantities of primary energy, BP uses a conversion efficiency for nuclear power and hydroelectric plants similar to that of fossil-fuel power plants (coal-fired), i.e., 38%. (That is, a calculational consumption of primary energy from nuclear and hydroelectric plants can be considered to be equivalent to the substitution of coal from coal-fired plants.).

Breakdown for Germany: Primary energy for electric power generation 2004 (GWa): Hard coal 39, lignite 48; gas 13, oil 2 (Arbeitsgemeinschaft Energiebilanzen 2007).
Breakdown for the USA: Primary energy for electric power generation 2004 (GWa): Coal 670; gas 190, oil 35. For details see US energy flow chart (2004).

From crude oil 2004:[a]	Germany		EU-25				USA				World			
	M tce	GWa	1000 barrel/d	kWh/l	GWh/d	GWa	1000 barrel/d	kWh/l	GWh/d	GWa	1000 barrel/d	kWh/l	GWh/d	GWa
Gasoline			3514	9.0	5029	210	9436	9.0	13503	563	25587	9.0	36615	1526
Domestic fuel oil, diesel, kerosene			6607	10.0	10505	438	6087	10.0	9678	403	28979	10.0	46077	1920
Heavy fuel oil			1652	13.1	3441	143	795	13.1	1656	69	9924	13.1	20671	861
Others?[b]			2810	7.0	3123	130	4199	7.0	4673	195	17409	7.0	19376	807
Total			14583		22102	921	20517		29511	1230	81898		122739	5114
Motor fuels	90	84												
Domestic fuel oil	35	33												

a) Source: BP Weltenergiestatstik 2005, p. 12; motor fuels and domestic fuel oil in Germany: Arbeitsgemeinschaft Energiebilanzen 2005.
b) Heating value for "Others" was not given and is estimated (7.0 kWh/l); resulting errors probably small because of the low quantity of "Others."

Table C.2 Electricity generation per year.

Electricity generation	Germany[a] (82 M inhabitants)			EU-25[b] (457 M inhabitants)			USA[c] (290 M inhabitants)			World[d] (6500 M inhabitants)		
2004	TWh$_{el}$	GWa$_{el}$		TWh$_{el}$	GWa$_{el}$		TWh$_{el}$	GWa$_{el}$		TWh$_{el}$	GWa$_{el}$	
Coal							1978	226	50%	6746	770	40%
Hard coal	141	16	23%	652	74	20%				Prognosis coal 2030: plus 800[d]		
Lignite	158	18	26%	289	33	9%						
Natural gas	61	7	10%	604	69	19%	709	81	18%	3237	370	19%
Other gases				31	4	1%	17	2	0.4%	Prognosis gas 2030: plus 500[d]		
Oil	10	1	2%	143	16	4%	120	14	3%	966	110	6%
Nuclear energy	167	19	27%	986	113	31%	788	90	20%	2678	306	16%
Renewable total										3085	352	18%
Hydropower	28	3.2	5%	337	38.5	11%	268	31	7%			
Other renewables:							90	10	2%			
Wind	26	2.9	4.1%	58	6.6	1.8%						
Biomass, waste	13	1.4	2.0%	68	7.8	2.1%						
Photovoltaics	0.6	0.07	0.1%	0.0	0.0	0.0%						
Others	12	1.4	2.0%	5	0.6	0.2%						
Total	616	70	100%	3190	364	100%	3970	453	100%	16712	1908	100%
										Prognosis for 2030[d]		
										30000	3425	

a) BMWi 2007b Energiedaten Tab. 22 (Primary energy consumption for electricity generation; see Table C.1 footnotes).
b) EU (25 countries): Eurostat (2006).
c) EIA 2006 (Primary energy for electricity generation; see Table C.1 footnotes).
d) Prognosis (coal and gas) in GWa$_{el}$. Source: EIA International Energy Outlook 2007 (Chap. 6 Electr. p. 61, 62) – Prognosis for 2030: Electricity ca. 30000 TWh (total) (p. 61) – Without nuclear and solar energy in the case of rising gas prices, the major portion of the increase of ca. 1500 GWa would have to be supplied by coal.

Electricity generation

2004 (RWE 2005)	France	Great Britain	Italy	Spain
Hard coal	5%	25%	43%	18%
Lignite	0%	0%	0%	5%
Gas	4%	49%	17%	23%
Oil	1%	1%	0%	8%
Nuclear	76%	20%	9%	23%
Hydro, others	14%	4%	31%	24%
GWa$_{el}$	62	43	32	29

Table C.3 Power plant capacity.

Power plant capacity 2005	Germany[a] GW$_{el}$		Western Europe GW$_{el}$	USA[b]			World	
				Plants	GW$_{el}$			GW$_{el}$
Coal			(Power plant capacity broken down according to fuel used is not given by Eurostat)	1522	335	31%		
Hard coal	29	22%						
Lignite	22	17%						
Natural gas	20	15%		5467	436	41%		
Other gases				102	2	0%		
Oil	6	5%		3753	65	6%		
Nuclear	21	16%	121	104	105	10%		367
Hydropower	10	8%		3993	77	7%		
Other renewable[c]	18	14%		1671	23	2%		
Total	132	100%		16807	1067	100%		

a) Source: BMWi 2007b Energiedaten.
b) Source: EIA 2006.
c) Other renewables: in Germany: wind.

Appendix C Energy Statistics

Nuclear power plants Capacity 2005[b]

		Plants	GW$_{el}$	Under construction[a]			Plants	GW$_{el}$	Under construction[a]
Germany[c]	France	59	63	–	North America	USA	104	99	–
	Germany	17	20	–		Canada	14	12	–
	Great Britain	23	12	–		Total	118	111	–
	Sweden	10	9	–	East Asia:	Japan	54	45	3
	Spain	9	8	–		South Korea	20	17	–
	Belgium	7	6	–		India	14	2.5	9
	Switzerland	5	3	–		PR China	9	7	2
	Total	130	121	–		Taiwan	6	5	2
Former Eastern Block countries:						Total	103	77	16
	Russia	31	22	2	World total				
	Ukraine	15	13	–	Countries here listed		409	350	18
	Czech Republic	6	3.5	–	Other countries		32	17	9
	Slovakia	6	2.5	–	World total (given)		441	367	2005[a] 27
	Total	58	41	2					
	Europe total	188	162	2					

a) 2005 plants under construction (cf. 2009 under construction world: 52 plants (IAEA)).
b) Source: IAEA (cited in Weltalmanach 2006, p. 658).
c) German nuclear power plants: Hours of full-load operation/a 2004: 7670 h = 87.3 % (VGB PowerTech 2005).

Table C.4 CO_2 emissions per year.

CO_2 emissions (anthropogenic, from energy consumption)

	Germany 2002[a]		EU-25 2002[a]		USA 2005 (see below)		World 2002[a]	
	Mt CO_2		Mt CO_2		Mt CO_2		Mt CO_2	
Total		841		3700		5933		22900
Power plants	42%	353	41%	1517	40%	2373	47%	10763
Transportation	21%	177	21%	777	33%	1958	21%	4809
Industrial	16%	135	16%	592	17%	1009	18%	4122
Residential, commercial	21%	177	19%	703	10%	593	14%	3206

In the year 2000 only ca. 60% of all greenhouse gases are caused by energy consumption – see Table C.6

a) Data for Germany, EU and World according to IEA and BMU (cited in STE 2006, p. 4).

Table C.5 US CO_2 emissions per year.

USA CO_2 emissions 2005 (anthropogenic, from energy consumption) million t CO_2				Total (direct CO_2 emissions of sector)		Indirect CO_2 emission of sector by electricity consumption	Total CO_2 emissions of sector (including CO_2 emission by electricity consumption)	
	Coal	Oil	Gas					
Electric power generation	1944	100	318	2362	40%			
Transportation	5	1921	31	1957	33%	16	1973	33%
Industrial	188	431	399	1018	17%	663	1681	28%
Commercial	8	55	166	229		821	1050	
Residential	1	105	261	367		885	1252	
Residential + commercial (total)	9	160	427	596	10%	1706	2302	39%
Total	2146 36%	2612 44%	1175 20%	5933		2385	5956	

Source: EIA 2007.

Table C.6 Emissions of greenhouse gases per year.

Greenhouse gas (GHG) emissions (anthropogenic) World 2000 (Total 42,000 million t CO_2 equivalent)

Sources		Gases		Gases (total)				From energy consumption		% of total anthropogenic GHG
Energy consumption	61.5%	CO_2	59.0%	CO_2	Energy consumption	59%		Transportation:		
		CH_4	1.5%		Deforestation	18%		Road		9.9%
		HFCs	1.0%		C in chemical process	3%		(Road USA)		(4.8%)
					Subtotal	80%		Air		1.6%
			61.5%	CH_4		11%		Rail, ship		1.6%
C in chemical processes	3.4%	CO_2	3.4%	N_2O		8%		Subtotal		13.5%
				HFCs, PFCs		1%		Electricity production		24.6%
Deforestation	18.2%	CO_2	18.2%		Total	100%		Other fuel combustion[a]		9.0%
								Industry		10.4%
Agriculture	13.5%	CH_4	5.1%	Livestock and manure						
		CH_4	1.5%	Rice cultivation				Fugitive emissions		3.9%
		N_2O	6.0%	Agriculture soils						
Waste	3.6%	CH_4	3.4%					Total		61.5%
Total	100%		100%							

Source: World GHG Emissions Flow Chart (World Resources Institute – WRI).
See this informative flow chart under: http://cait.wri.org/figures.php?page=/World-FlowChart (Access 07).
(WRI: "All calulations are based on CO_2 equivalents, using 100-year global warming potentials from the IPCC [1996]".)

a) E.g. for heating.

Appendix D
Comments on the Earlier Study (Kalb and Vogel 1986a)

Siemens–Interatom 1986

Following the publication of a summary version of the Kalb/Vogel study in the journal *Sonnenenergie* (DGS, February and April 1986 – see the DGS homepage: Kalb and Vogel, 1986b), the authors sent this summary in May 1986 as a circular to all the relevant institutions in Germany (Kalb and Vogel, 1986c). These included Prof. Beckurts, the chairman of the Board of Directors of Interatom, the leading German power plant manufacturer. The Interatom Division of Solar Energy then prepared a comment, which was also sent to the present authors. This 23-page document contained a four-page summary, which we quote as excerpts in the following. These comments make clear the general attitude to this topic that prevailed at the time – at least in Europe. Their conclusion was that the expected costs would be *three to four* times higher, possibly even *six to eight* times higher, than estimated in our study. (This, however, was revised in a second letter in April 1988, after in the meantime the Utility Studies (1988) had been published. Using the cost data of this extensive American report, upon which the follow-up study of Kalb and Vogel (1993) was based (although referring to mass-produced plants), the inflation-corrected costs proved to be even somewhat lower than estimated in the original study of Kalb and Vogel (1986a).)

"INTERATOM Note Id.-No. 60.14730.8; TVS-No: 324731; Date: *31.07.1986*

Summary

We make the attempt here, without knowledge of the overall study "The Spain Solar Program" by the above-named authors, to consider in detail and to comment on some essential points which are unclear, to some extent incorrect or which are of fundamental importance.

Summing up, we can make the following statements:

The suggested base-load electric power supply making use of solar tower power plants installed in Spain and overland transmission of electrical energy to Germany requires a critical examination in order to gain more definite knowledge about its feasibility and the costs to be expected.

At present, we know of no other studies which have investigated a solar base-load power supply. At Interatom, up to now no tower plants for 24-h operation have been considered in detail.

A higher proportion of solar power for supplying the annual base-load requirements would necessitate enormous thermal energy storage systems with corresponding high-efficiency solar-specific components, which would give rise to considerable cost increases. Such systems would operate far from the cost optimum. Systems optimized in terms of cost would, from the present-day viewpoint, involve at most 4 to 6 additional hours of full-power operation at the Barstow insolation and would thus operate in the peak- or middle-load range.

A considerable, but currently unknown, proportion of the annual base-load power requirements would have to be supplied by nuclear or fossil-fuel generating plants within Germany, since a thermal storage capacity limited by technical/economic considerations and climatic conditions would not permit the overall power needs to be supplied from solar energy. A more certain knowledge of the proportion of solar energy within the supply would necessitate a more precise investigation of the annual yields based on simulations using meteorological data for particular sites as well as models for energy storage.

The assumptions in this study regarding the use of solar energy, the availability of low-cost technologies, and the resulting cost of electric power appear to us to be extremely optimistic. ..."

A series of critical remarks[1] follows, which culminates in the following conclusion:

"A precise comparison of cost data is not possible for us at present due to a lack of knowledge of the overall study, and it would require involved

---

1) Some part of the criticism of the technical design in our study was related to the heat-storage system suggested by the authors, based on molten chloride salts. In the early period of solar power plant development in the second half of the 1970s, when the authors began their work on solar energy and hydrogen technology, the currently favored nitrate salts for heat storage were not yet under discussion. Great hopes were instead placed on latent-heat storage systems, which however proved to have major disadvantages. The authors, therefore, suggested developing a molten-salt storage system using low-cost chloride salts, combined with a high-temperature design for the power plant. Later, the feasibility of heat-storage systems using molten nitrate salts became evident. In the 1986 study (Kalb 1986), three storage concepts utilizing liquid heat-storage media were presented: with chloride salts, nitrate salts, or with sodium,

recalculations. Starting from available cost data / 5/[2], we would expect, roughly estimated, an increased cost factor of ca. 6 to 8 as compared to the investment costs given [by Kalb and Vogel], if a design for winter operation (solar multiple = 6–7) is presumed. Even for summer operation (solar multiple = 3), an increased cost factor of ca. 3 to 4 is to be expected. All together, we are of the opinion that the statement that the base-load power supply could be provided using solar power from Spain at acceptable costs, utilizing available technology, and on a manageable time scale is not sustainable, and is possibly even dubious."

Finally, the following assessment is given:

"The investigations suggested [by Kalb and Vogel, to obtain a] 'more certain knowledge of the costs which are to be expected' (p. 15) and 'making use of several mutually independent studies' (p. 16) appear reasonable. The 'specifying study' (p. 24), termed a 'first stage,' to be carried out by a qualified institution, e.g. by a university department, seems to us to be quite desirable.

Furthermore, Spain has been engaged in solar tower power plant development since the beginning (1977), and is operating its own 1 MW_{el} prototype installation. In effect, the Spanish themselves should be the first to be convinced of the possibility of a solar power supply for their own country – if the necessary conditions are in fact as favorable as suggested in the Kalb/Vogel report."

BMU

The German Ministry for the Environment, Nature Conservation and Nuclear Safety (BMU) wrote to the authors in January 1987: "… A fundamental problem of your suggestion involves the question of the transmission of power over long distances, as well as energy storage and backup power generation. One possibility for solving these problems is currently seen in the concept of a solar-hydrogen technology. Your suggested solutions (molten-salt heat storage system, high-voltage direct-current transmission) would have to be compared with this and with other concepts, in particular under the viewpoints of environmental impact, practicability, development potential and economic feasibility."

---

which is relatively expensive. As reference variant, however, the chloride-salt system was retained, as was likewise stated in the summary to which the Interatom comments referred. After the concept of nitrate-salt storage had been worked out in more detail in the USA (and later formed the basis for the Utility Studies (1988) concept), chloride-salt storage systems were no longer relevant so that this storage system was in fact obsolete by the time of the final publication of the study (only in early 1986).

2) Reference: *Technologieprogramm GAST, Analysis of the Potential*, June 1985.

BMFT

The department head of Solar Energy Research at the German Federal Ministry for Research and Technology (BMFT) made in June 1986 statements such as: "... of all your cost assumptions, I found most unusual your heliostat cost of 160 DM or 190 DM/m², respectively. [Authors' note: these values correspond to 80 or 95 €; corrected for inflation, that is $120–140 (2002), that is they lie well above the costs which are estimated today (less than $100 in 2002-$).] Whether or not such costs will be achievable would certainly be an important point for discussion – the USA after all uses exclusively the more costly (glass) technologies. ... It would however be especially interesting to me to know just why you insist on base-load operation; would not the whole system be less costly without heat storage – i.e. power from the sun, as long as it is shining, and power from local plants when solar power is not sufficient."

In August (1986), another staff member of the same ministry wrote: "... I am of the opinion that every new idea should first be introduced to the scientific community ... I thus request – regarding also your suggestion of carrying out discussions with me – that you instead seek a scientific interaction with researchers in industry and basic research institutions. ..."

Prof. Buckel

Prof. W. Buckel (University of Karlsruhe), at that time President of the European Physical Society, wrote in 1988[3]: "... the scientific and technical conclusions appear compelling to me. The support which I am giving to Mr. Kalb and Mr. Vogel is due to my annoyance at the lack of interest on the part of the responsible agencies. They were not even willing to have the study checked. ..."

Even before the actual publication of the study, there were several responses:

L. Bölkow / MBB

In response to a request by the authors regarding possible support for the publication of their study from Ludwig Bölkow[4] (as a well-known and respected public figure), he answered in a letter on October 26th, 1983: "... After reading through your letter with the outline of a proposed article, I want to assure you that I am in principle willing to support this publication with an accompanying comment. ..." He kept this promise, even though the department of MBB which dealt with solar power plants recommended otherwise (28th of November, 1983):

---

3) In a letter to H.-D. Harig (at that time Deputy Chairman of the Board of Directors of VEBA Power Plants Ruhr).

4) Aircraft designer and co-owner of the aircraft company Messerschmitt-Bölkow-Blohm (MBB, today a part of the EADS aerospace concern); initiator of the Ludwig Bölkow Foundation, which is one of the sponsors of the present book (see Acknowledgments).

"The importance of solar power plants in Spain to electric power generation for Germany, as suggested in the title, is to some extent not consistently reasoned out and presents no new theoretical concepts: The EURELIOS power plant in Sicily also feeds power into the European grid. It is however completely unsatisfactory to speak of the direct interaction of a solar power plant located in southern Europe with a coal-fired plant in Germany. ... The difficult position of solar power would only be further exacerbated by such a publication. I would suggest, dear Dr. Bölkow, that you not provide the author with any encouragement." Ludwig Bölkow nevertheless advocated publication in the journal *Sonnenenergie* (DGS; Nos. 1 and 2, 1986) in the face of a likewise reluctant stance on the part of the journal's editors so that the summary of the study finally appeared shortly before the reactor catastrophe in Chernobyl.

DLR

In a four-page response by the German Aerospace Center (DLR) on 18th April, 1985, the introductory remarks state: "the overall idea of integrating solar-tower power plants into the power grid is quite attractive. In detail, there are however numerous weak points in the suggested concept. On the other hand, we support the conclusion that increased research in this area will be rewarding. ..." Continuing, among other things they write: "Whether or not 'new' coal-fired power plants or 'out of service' plants are utilized (for backup power) is not a function of whether solar plants are deployed or not. Cheap power from 'out of service' plants would still be cheap without solar plants."

BMWi

Finally, we quote a statement that is typical of the situation prevalent even around the year 2000, and in particular of several responses to our study (Kalb and Vogel, 1998). In October 1999, the Parliamentary State Secretary in the German Ministry for Economy and Technology (BMWi), Siegmar Mosdorf, wrote to the Member of Parliament Ms. Marga Elser: "... The technology of conventional solar thermal power plants can today be considered to be ripe for use. For the implementation of this technology, however, we need a corresponding interest on the part of industry, power-plant manufacturers and the electric utility concerns. By renouncing all their patents, for example Siemens let us know at the beginning of this year that the industry does not expect any improvement in the economic chances of solar thermal power plants even over the longer term. No matter how much one may regret this situation, it will not be cleared up by a continuation or even an expansion of governmental subsidies. I would be genuinely happy if the industry would nevertheless commit its resources to this field. But I see no possibility of implementing the scenario suggested by Mr. Kalb and Mr. Vogel through the use of public funds. ..."

References

ABB (2008) Xiangjiaba–Shanghai ±800 kV UHVDC transmission project, www.abb.com (accessed January 2008).

AEO (2007) See EIA AEO (2007).

AEO (2008) Annual energy outlook (Projection Tables, Reference Case, Table 3 Energy Prices). US Energy Information Agency Report DOE/EIA-0383(2008). http://www.eia.doe.gov/oiaf/archive/aeo08/index.html (accessed 22 September 2009).

AFAS, ISI-FhG, TÜV-Rheinland (1984) Methanol für den Strassenverkehr (Methanol for Road Transportation, in German). Research Center Karlsruhe, Department of Applied Systems Analysis (AFAS), Fraunhofer Institute for Systems and Innovation Research (ISI), TÜV-Rheinland.

Altmann, T., Carmel, Y., Guetta, R., Zaslavsky, D., and Doytsher, Y. (2005) Assessment of an "energy tower" potential using a mathematical model and GIS. *Solar Energy*, **78** (6), 799–808.

Altmann, T., Zaslavsky, D., Guetta, R., and Czisch, G. (2006) Evaluation of the potential of electricity and desalinated water supply by using technology of "energy towers" for Australia, America and Africa. ECMWF Special Project "Evaluation of the global potential of energy towers" (Principal Investigator: G. Czisch, ISET, University of Kassel, Germany). www.ecmwf.int/about/special_projects (accessed 25 February 2008).

Altmann, T., Guetta, R., Zaslavsky, D., and Czisch, G. (2008) Evaluation of the potential of electricity production by using technology of "energy towers". ECMWF Special Project "Evaluation of the Global Potential of Energy Towers", Interim Report. http://www.ecmwf.int/about/special_projects (accessed 3 October 2009).

Amadeo, K. (2007) U.S. military budget– Understanding the US military budget and its impact on the US economy. http://www.useconomy.about.com/od/fiscalpolicy/p/2008_defense.htm (accessed December 2007).

Angerer, G. (2007) Zukunftsmarkt Synthetische Biokraftstoffe. Study of the Fraunhofer Institute ISI, Karlsruhe, commissioned bei the German Federal Environment Agency (UBA), Germany, FKZ 206 14 132/05.

Arbeitsgemeinschaft Energiebilanzen (AGEB) (2007) www.ag-energiebilanzen.de (accessed 13 September 2007).

Argo (2009) www.argo.ucsd.edu (accessed February 2009).

Bammert, K. and Deuster, G. (1974) Das Heliumturbinen-Kraftwerk Oberhausen–Auslegung und Aufbau. *Energie und Technik*, **26** (1), 1–6.

Bammert, K. and Seifert, P. (1981) Zur Auslegung berohrter Receiver von Solarkraftwerken. *Atomkernenergie Kerntechnik*, **39** (3), 163–173.

Baum, V., Aparici, R., and Garf, B. (1957) High power solar installations. *Journal of Solar Energy, Science, and Engineering*, **1** (1), 6–12.

Becker, M. and Klimas, P. (eds.) (1993) *Second Generation Central Receiver Technologies–A Status Report*, C. F. Müller, Karlsruhe, Germany.

BGR (2002) Reserven, Ressourcen und Verfügbarkeit von Energierohstoffen 2002 (Nr. 519). Federal Institute for Geosciences and Natural Resources (BGR), Germany. www.bmwa.bund.de (accessed 26 March 2008).

BGR (2007) Reserven, Ressourcen und Verfügbarkeit von Rohstoffen 2006. Brief study (Authors H. Rempel, S. Schmidt, U. Schwarz-Schampera), Federal Institute for Geosciences and Natural Resources (BGR), Germany. www.bgr.bund.de (acccessed 26 March 2008).

Blake, D., Moens, L., Rudnicki, D., and Pilath, H. (2006) Lifetime of imidazolium salts at elevated temperatures. *Journal of Solar Energy Engineering*, **128** (1), 54–57.

BMBF (1996) See Pilkington (1996).

BMU (2004) Ökologisch optimierter Ausbau der Nutzung erneuerbarer Energien in Deutschland (Authors: Nitsch, J., Krewitt, W., Nast, M., Viebahn, E., Gärtner, S., Pehnt, M., Reinhardt, G., Schmidt, R., Uihlein, A., Scheurlen, K., Barthel, C., Fischedick, M., Merten, M.). DLR, IFEU and Wuppertal Institute for Climate, Environment and Energy, study commissioned by the German Federal Ministry for the Environment, Nature Conservation and Nuclear Safety (BMU), FKZ 901 41 803. www.dlr.de/tt (accessed 1 October 2009).

BMWi (2006) Verfügbarkeit und Versorgung mit Energierohstoffen. Abridged Report, Arbeitsgruppe Energierohstoffe, BMWi Abt. III, 29 March 2006. www.bmwi.de (accessed 7 August 2009).

BMWi (2007a) Entwicklungsstand und Perspektiven von CCS-Technologien in Deutschland. Joint Report of the ministries BMWi, BMU, and BMBF for the German Federal Government. http://www.bmwi.de (accessed 7 August 2009).

BMWi (2007b) Energiedaten–Energieträger–Tab. 22: Stromerzeugungskapazitäten und Bruttostromerzeugung nach Energieträgern. www.bmwi.de (accessed 11 November 2009).

BMWi (2008) Entwicklung von Energiepreisen und Preisindizes, Energiedaten. Tabelle 26, November 2008. http://www.bmwi.de/BMWi/Navigation/Energie/energiestatistiken (accessed 2 February 2009).

Boeing Engineering and Construction Co. (1978) Solar central receiver prototype heliostat, Vol. 1: final technical report, June 1978, (SAN-1604-1, UNCLAS).

BP (2005) BP statistical review of world energy (weltenergiestatistik), June 2005. http://www.deutschebp.de/liveassets (accessed 15 October 2009, in German).

BP (2008a) BP statistical review of world energy, June 2008. www.deutschebp.de (15 October 2009, English or German).

BP (2008b) BP statistical review of world energy (single statistics, yearly actualized). http://www.bp.com/statisticalreview (September 2007 and November 2008).

Bradshaw, R. (2008) SNL project overview. Solar Energy Technologies Program (SETP) Annual Review Meeting, 22–24 April 2008, Austin, TX, Session: Thermal Storage. http://www1.eere.energy.gov/solar/review_meeting/index.html (accessed 3 October 2009).

Brösamle, H., Mannstein, H., Schillings, C., and Trieb, F. (2001) Assessment of solar electricity potentials in North Africa based on satellite data and a geographic information system. *Solar Energy*, **70** (1), 1–12.

Brosseau, D., Hlava, P., and Kelly, M. (2004) Testing thermocline filler materials and molten salt heat transfer fluids for thermal energy storage systems used in parabolic trough solar power plants. Sandia National Laboratories SAND2004-3207. http://www.nrel.gov/csp/troughnet (accessed 23 July 2009).

Brosseau, D. and Kolb, G. (2007) Salt freeze protection & recovery. NREL Trough Technology Workshop, 8 March 2007, Golden, Colorado. http://www.nrel.gov/csp/troughnet (accessed 4 October 2009).

Buckel, W. (1994) Die Verantwortung der Wissenschaftler gegenüber der Öffentlichkeit (The responsibility of scientists towards the public). Meeting of the German Physical Society (DPG), 14–18 March 1994, Hamburg, in Vorträge des Arbeitskreises Energie (ed. K. Schultze), DPG, Bad Honnef.

Buckel, W. (1996) Personal communication.

Bunn, M., Fetter, S., Holdren, J., and van der Zwaan, B. (2003) The economics of reprocessing vs. direct disposal of spent nuclear fuel. Report for Project on Managing the Atom, Belfer Center for Science and International Affairs, Harvard Kennedy School, Cambridge, Report DE-FG26-99FT4028 (quoted in Chicago Study 2004, p. 7–13). (See also: Bunn, M., Fetter, S., Holdren, J., van der Zwaan, B. (2005) The Economics of Reprocessing vs. Direct Disposal of Spent Nuclear Fuel. Nuclear Technology, 150 (3), 209–230.).

Burbridge, D., Mills, D., and Morrison, G. (2000) Stanwell solar thermal power project. Proceedings of the 10th Solar-PACES International Symposium, 8–10 March 2000, Sydney, Australia.

Butti, K. and Perlin, J. (1980) *A Golden Thread – 2500 Years of Solar Architecture and Technology*, Van Nostrand, New York.

Cabanyes, I. (1903) Projecto de motor solar. *La Energia Electrica*, (4), 61–65; (5), 81–84.

California Energy Commission (2002) Comparison of alternate cooling technologies for California power plants – economic, environmental and other tradeoffs, Final Report (Author J. Maulbetsch). EPRI, Palo Alto, CA, and California Energy Commission, Sacramento, CA, publication No. P500-02-079F. www.energy.ca.gov (accessed 27 September 2009).

Carlson, P. (1975) Power Generation through Controlled Convection (Aeroelectric Power Generation). Lockheed Aircraft Corp., US Patent No. 3,894,393.

cav (1989) Wärmeträgeranlagen im Großformat. *cav chemie-anlagen und verfahren*, 22 (6), 18–22.

Cayetano Hernandez (2009) Iberdrola Renovables y el Sector Solar Termolectrico, GENERA Madrid, 13 de Mayo 2009. http://www.ifema.es/ferias/genera/ponencias09/06/Cayetano_Hernandez_Iberdrola.pdf (accessed 20 July 2009).

CBO (2007) The potential for carbon sequestration in the United States. The Congress of the United States, Congressional Budget Office (CBO), Publication No. 2931. http://www.cbo.gov/publications (accessed 7 August 2009).

Chicago Study (2004) The economic future of nuclear power (directed by G. Tolley and D. Jones). Study conducted at the University of Chicago for the US Department of Energy/Argonne National Laboratory. www.anl.gov/Special_Reports/NuclEcon-Aug04.pdf (accessed August 2007).

China Physical Atlas (1999) *The National Physical Atlas of China*, China Cartographic Publishing House, Beijing.

Cohen, G., Kearney, D., and Kolb, G. (1999) Final report on the operation and maintenance improvement program for CSP plants. Sandia Report SAND99-1290, Sandia, Albuquerque, NM.

Committee on Science and Technology (2007) Solar energy research and advancement act of 2007. House Report 110-303, US House of Representatives, Committee on Science and Technology. science.house.gov (accessed 4 October 2009).

COORETEC (2003) Research and development concept for zero-emission fossil-fuelled power plants – summary of COORETEC. German Federal Ministry of Economics and Labour (BMWA), Documentation No. 527 (COORETEC: CO_2 Reduction Technologies). www.bmwi.de (accessed 20 September 2009).

CSP Industry (2000) A rebuttal of the national research council's review of the U.S. DOE concentrating solar power program, executive summary. U.S. Concentrating Solar Power Industry Review Panel, Appendix B in Morse, F. (2000) The Commercial Path Forward for Concentrating Solar Power Technologies, document prepared for Sandia National Laboratories, SAND2001-2520P. http://www.solarpaces.org/Library/docs/FMorse.pdf.

Czisch, G. (1999) Potentiale der regenerativen Stromerzeugung in Nordafrika. Meeting of the German Physical Society (DPG), 15–19 March 1999, Heidelberg, Germany, in *Energie, Plutonium, Strom und die Umwelt* (ed. W. Blum), DPG, Bad Honnef. http://www.iset.uni-kassel.de/abt/w3-w/projekte (accessed 3 October 2009).

Czisch, G. (2005) Szenarien zur zukünftigen Stromversorgung – Kostenoptimierte Variationen zur Versorgung Europas und seiner Nachbarn mit Strom aus erneuerbaren Energien. Dissertation, Department of Electrical Engineering /Informatics,

University of Kassel, Germany. http://kobra.bibliothek.uni-kassel.de (accessed 4 October 2009).

Czisch, G., Durstewitz, M., Hoppe-Kilpper, M., and Kleinkauf, W. (1999) Windenergie gestern, heute und morgen. Presented on the Conference "Windwirtschaft 2000 plus", September 22–24, 1999, Husum, Germany. http://www.iset.uni-kassel.de/abt/w3-w/projekte.

Deka-Bank (2004) Kaufkraftparitäten: Ein Blick auf gleichgewichtige Wechselkurse in den Industrieländern (Author J.-U. Wächter). Volkswirtschaft Spezial (Deka-Bank), 10 March 2004. http://www.dekabank.de/db/de/economics/publikationen/index.jsp (accessed 2 October 2009).

Dersch, J. and Richter, C. (2007) Water saving heat rejection for solar thermal power plants, NREL Parabolic Trough Technology Workshop, March 8–9, 2007, Golden, Colorado. http://www.nrel.gov/csp/troughnet (accessed 4 November 2007).

Desertec (2009) Red paper – an overview of the desertec concept. Desertec Foundation. http://www.desertec.org/en/concept/redpaper (accessed 4 October 2009).

Dinter, F., Geyer, M., and Tamme, R. (eds.) (1991) *Thermal Energy Storage for Commercial Applications*, Springer, New York.

DLR (2008) Deutsches Zentrum für Luft- und Raumfahrt (German Aerospace Center) – TerraSAR-X Satellit. Wikipedia. http://de.wikipedia.org/wiki/DLR (accessed 30 August 2008).

DLR IMF (2008) Das Institut für Methodik der Fernerkundung (IMF). Brochure V01, German Aerospace Center (DLR). www.dlr.de (accessed 20 August 2009).

DLR-MED (2005) Concentrating solar power for the mediterranean region, final report (MED-CSP). DLR Institute of Technical Thermodynamics, Division of Systems Analysis and Technology Assessment, Stuttgart, Germany. http://www.dlr.de/tt/med-csp (accessed 30 September 2009).

DLR-TRANS (2006) Trans-mediterranean interconnection for concentrating solar power (Trans-CSP). DLR Institute of Technical Thermodynamics, Division of Systems Analysis and Technology Assessment. http://www.dlr.de/tt/trans-csp (accessed 30 September 2009).

DOE (1997). Renewable energy technology characterization. Topical Report TR-109496, prepared by Department of Energy/OUT and EPRI. http://www1.eere.energy.gov/library (accessed 15 July 2007).

DOE (2007) Report to congress on assessment of potential impact of concentrating solar power for electricity generation. (EPACT 2005 – Section 934(c)). Report No. DOE/GO-102007-2400, US Department of Energy. http://www.nrel.gov/csp (accessed 7 August 2009).

Dubbel (1987) *Dubbel – Taschenbuch für den Maschinenbau*, 16th edn (eds. W. Beitz and K.-H. Küttner) (chapter L) Springer, Berlin.

E&M (2007) The great challenge – power generation technologies for the future (Author L. Balling). Energie & Management (Special: Power-Gen Europe), 15 June 2007.

Eck, M. and Zarza, E. (2006) Saturated steam process with direct steam generating parabolic troughs. *Solar Energy*, **80** (11), 1424–1433.

ECOSTAR (2004) European concentrated solar thermal road-mapping (ECOSTAR) (eds. R. Pitz-Paal, J. Dersch, and B. Milow). European Commission Roadmap Document SES6-CT-2003-502578.

Edison (2005) Major new solar energy project announced by Southern California Edison and Stirling Energy Systems, Inc. Press release, 9 August 2005. http://www.edison.com/pressroom (accessed 6 October 2009).

Ehrenberg, C. (1997) *Solarthermische Kraftwerke, Vol. VI of the book series "Regenerative Energien"*, VDI Society for Energy Technology VDI-GET, Düsseldorf, Germany.

EIA (2006) Electricity (data for 2006). http://www.eia.doe.gov/fuelelectric.html (accessed 5 November 2007).

EIA (2007) Environment – state carbon dioxide emissions by energy sectors. http://www.eia.doe.gov/environment.html (accessed 8 September 2007).

EIA – International Energy Outlook (2007) International energy outlook 2007

(Chapter 6 – Electricity). EIA Report DOE/EIA-0484(2007). http://www.eia.doe.gov/oiaf/ieo/ieoarchive.html (accessed 11 October 2009).

EIA (coal) (1999) U.S. coal reserves: 1997 update (Chapter 1: EIA Coal Reserves Data, there: Important Terminology). EIA Report DOE/EIA-0529(97) (published 1999). http://www.eia.doe.gov/fuelcoal.html (accessed 15 October 2009) or http://www.eia.doe.gov/cneaf/coal/reserves/front-1.html (accessed 15 October 2009).

EIA (coal) (2006) Annual coal report 2006. EIA Report DOE/EIA-0584 (2006).

EIA (coal) (2008) Coal reserves current and back issues. Report released October 29. 2008. http://www.eia.doe.gov/cneaf/coal/reserves/reserves.html (December, 2008).

EIA AEO (2007) Assumptions to the annual energy outlook 2007. US Energy Information Administration, Report DOE/EIA-0554(2007). http://www.eia.doe.gov/oiaf/archive.html (accessed 4 October 2009).

Eisenbeiß, G. (1995) Unsere solare Zukunft – Solarkraftwerke für die globale Strategie. Proceedings of the VDI Meeting "Solarthermische Kraftwerke II", October 11–12, 1995, Stuttgart, Germany, VDI-Berichte 1200 (ed. VDI), VDI-Verlag, Düsseldorf.

El-Sawy, A., Leigh, J., and Trehan, R. (1979) A comparative analysis of energy costing methodologies. MITRE Corp./METREC Div., Technical Report MTR-7689, sponsored by DOE-DGE, contract No. EG-77-C-01-4014 (see in particular Appendix (J. Leigh): Report on Levelized Busbar-Costing Workshop Held at MITRE/Metrek, June 29–30, 1978).

ENEA (2001) Solar thermal energy production: guidelines and future programmes of ENEA (ENEA working group; coordinator: Carlos Rubbia). Italian National Agency for New Technologies, Energy and the Environment, ENEA/TM/PRES/2001_07. http://www.solaritaly.enea.it (accessed 7 August 2009).

Enquete Commission (1990) Vorsorge zum Schutz der Erdatmosphäre (Protection of the earth's atmosphere). Report of the Enquete-Commission of the German Parliament (Deutscher Bundestag), Vol. 3: Erneuerbare Energien (970 p., including large-scale solar plants and import of solar energy carriers), Economica, Heidelberg.

Enquete Commission (1994) Vorsorge zum Schutz der Erdatmosphäre – Nachhaltige Energiepolitik für dauerhaften Klimaschutz. Final Report of the Enquete-Commission of the German Parliament, Deutscher Bundestag, Drucksache 12/8600, 31 October 1994.

Esquivel, R. (ca. 2007) Coal to methanol design report. University of California, San Diego. http://maecourses.ucsd.edu/ceng124/rpts/gp1_rpt2.pdf (accessed 4 October 2009).

EURACOAL (2008) Market report 1/2008. European Association for Coal and Lignite (EURACOAL). www.euracoal.be (accessed 10 May 2008).

European Commission (2003) External costs – research results on socio-environmental damages due to electricity and transport. EUR 20198. http://www.externe.info/externpr.pdf (accessed 4 August 2008).

Eurostat (2006) Gross inland energy consumption by fuel, Electricity generation by origin, http://epp.eurostat.ec.europa.eu (accessed August 2006).

EWI (2007) Energiewirtschaftliches Gesamtkonzept 2030 (Erweiterte Szenariendokumentation). Study from Institute of Energy Economics (EWI)/University of Cologne, Germany, and Energy Environment Forecast Analysis (EEFA) GmbH. http://www.ewi.uni-koeln.de/content (accessed 30 July 2007). (Complemented by personal communication EWI, 8 August 2007).

fondsexperte24 (2007) Uranpreisanstieg langsam unheimlich (Uranium Price Getting scary). 19 April 2007. http://www.fondsexperte24.de (accessed 25 August 2007).

Framatome (2005) Argumente: Wie lange reicht das Uran? Framatome/AREVA NP. www.areva-np.com (accessed August 2007).

Francia, G. (1968) Pilot plants of solar steam generation systems. *Solar Energy*, **12**, 51–64.

Frenkel, D. (2007) Going electric with the "energy tower". www.israel21c.org (accessed 12 October 2009).

Garcia, G., Egea, A., and Romero, M. (2003) First autonomous heliostat field PCHA project. Proceedings of the ISES Solar World Congress (CD-ROM), June 14–19, 2003, Göteborg, Sweden.

Garcia, G., Egea, A., and Romero, M. (2004) Performance evaluation of the first solar tower operating with autonomous heliostats: PCHA project. Proceedings of the 12th SolarPACES International Symposium (CD-ROM), October 6–8, 2004, Oaxaca, Mexico.

GEA Prospectuses (not dated) (1) Electricity Supply Commission – 6 x 665 MW Matimba, (2) Electricity Supply Commission – 6 × 660 MW Majuba. Prospectuses of the German plant construction company GEA, ca. 1984 bis 1990.

Gladen, H., Wasserscheid, P., and Medved, M. (2008) Multinary salt system for storing and transferring thermal energy. World Intellectual Property Organization (WIPO), publication date: 19 June 2008, publication No. WO/2008/071205.

Grasse, W. (1988) Solarturm-Kraftwerke des europäischen PHOEBUS Konsortiums und der amerikanischen Utility Teams – Ergebnisse aus den Phase 1 Studien. Kolloquium "Zukunft der solarthermischen Kraftwerke" ("Future of Solar Thermal Power Plants"), Ludwig Bölkow Foundation, 18 March 1988, Ottobrunn, Germany.

Gray, D. and Tomlinson, G. (2002) Hydrogen from coal. Mitretek Systems, Report prepared for US DOE NETL, Contract No. DE-AM26-99FT40465, Mitretek Technical Paper MTR 2002-31 (published July 2002). http://www.netl.doe.gov/technologies (accessed 5 October 2009) OR http://www.netl.doe.gov/technologies (accessed 12 October 2009).

Greenpeace (2006) Die Reichweite der Uran-Vorrte der Welt (Author P. Diehl). http://www.greenpeace.de/publikationen (accessed 17 September 2007).

Günther, H. (1931) *In Hundert Jahren – Die Künftige Energieversorgung Der Welt*. Kosmos Gesellschaft der Naturfreunde, Franckh'sche Verlagshandlung.

Haeberle, A., Zahler, C., de Lalaing, J., Ven, J., Sureda, M., Graf, W., Lerchenmueller, H., and Wittwer, V. (2001) The solarmundo project – advanced technology for solar thermal power generation. Proceedings of the ISES World Congress (CD-ROM), November 25–December 2, 2001, Adelaide, Australia.

Handelsblatt (2008a) 33 New power plants, in Südkurier (German Newspaper), 30 June 2008, quoting Handelsblatt, ca. June 2008 (German Newspaper).

Handelsblatt (2008b) Uranpreis steigt weiter. Handelsblatt, 5 August 2008.

Hansen, U. (1983) Reale und nominale Barwertmethode – Eine systematische Analyse der dynamischen Barwertmethode. *Energiewirtschaftliche Tagesfragen*, **33** (4), 218–223.

Hauch, A., Jensen, S., Ebbesen, S., and Mogensen, M. (2007) Durability of solid oxide electrolysis cells for hydrogen production. Proceedings of the Risø International Energy Conference, May 22–24, 2007, Risø, Denmark, Risø National Laboratory, Risø-R-1608(EN) (2007).

Hi2H2 (2004) Highly efficient, high temperature, hydrogen production by water electrolysis (Hi2H2), project abstract, August 2004. http://www.ist-world.org/ProjectDetails.asp (accessed 14 August 2008) OR www.ist-world.org (accessed 20 September 2009).

Hi2H2 (2005) www.hi2h2.com (accessed 14 August 2008).

Hillesland, T. Jr. (1988) See Utility-Studies (1988).

Hillesland, T. and De Laquil, P. (1988) Results of the U.S. solar central receiver utility studies. Proceedings of the VDI meeting "Solarthermische Kraftwerke zur Wärme- und Stromerzeugung", November 29–30, 1988, Cologne, Germany, VDI-Berichte 704 (ed. VDI), VDI-Verlag, Düsseldorf.

Hilsebein (Lurgi AG, Frankfurt) (2009) Personal communication, February 2009.

Hlubek, W. (1983) Betriebserfahrungen mit großen Dampfturbinen. *BWK*, **35** (1/2), 36–39.

Hoffschmidt, B. (2007) Personal communication, December 2007.

Hospe, J. (1996) Aktiv im Passivschutz – Neuer Siedewasserreaktor mit verbesserter Sicherheitstechnik. *Energiespektrum*, **11** (9), 37–40.

Hoyer-Klick, C., Schillings, C., Trieb, F., and Scholz, Y. (2006) Meteorologische Informationen für die Planung erneuerbarer Energiesysteme, in Workshop 2006, Energiemeteorologie (eds. D. Heinemann, C. Hoyer-Klick, G. Stadermann), FSV Workshop, 2 November 2006, Berlin, FVS (since 2009 FVEE), Berlin. http://www.fvee.de/publikationen (accessed 20 September 2009).

IAEA (2001) Analysis of uranium supply to 2050. International Atomic Energy Agency, Vienna. www.pub.iaea.org (accessed September 2007).

IEA Coal Information (2007) Coal information (2007 Edition)–documentation for beyond 2020 files. http://wds.iea.org/pdf/doc_coal.pdf (accessed February 2008).

IEA/OECD-NEA (2005) Projected costs of generating electricity-2005 Update. International Energy Agency/Nuclear Energy Agency. www.iea.org (accessed December 2007).

IECEC (1979) Proceedings of the 14th Intersociety Energy Conversion Engineering Conference (IECEC), August 5–10, 1979, Boston.

IEF-Jülich (2007a) Stand der Arbeiten im Bereich der SOFC-Brennstoffzelle am Forschungszentrum Jülich, 2 July 2007. http://www.fz-juelich.de/ief/ief-pbz (accessed 12 January 2009).

IEF-Jülich (2007b) The solid oxide fuel cell (SOFC), 1 August 2004. http://www.fz-juelich.de/ief (accessed 12 January 09).

IFM (2009) ARGO-TROPAT: Untersuchungen zu Zirkulation und Wassermassen-Anomalien mit profilierenden Messrobotern im tropischen Atlantik. Leibnitz Institute of Marine Sciences (IFM-GEOMAR), Kiel, Germany. www.ifm-geomar.de (accessed February 2009).

IMI (2002) Die USA treiben weltweite Rüstungsausgaben in die Höhe (Author: L. Henken). www.imi-online.de (accessed 5 October 2009).

ISE (2007) Solar electricity on a large scale–linear fresnel collectors for solar thermal power stations in a practice test. Press Release 05/07, 4 April 2007, Fraunhofer Institute for Solar Energy Systems (ISE), Freiburg, Germany. www.ise.fraunhofer.de (accessed 20 September 2009).

Jochem, E. (2004) Weltweite Perspektiven der Kohle–Klimabedrohung oder Entwarnung? in 13 Vorträge der Münchner Tagung (2004) (ed. M. Keilhacker), DPG-Meeting (German Physical Society), March 22–26, 2004, Munich, DPG, Bad Honnef, Germany.

Jones, S. (2000) Heliostat cost as a function of size for molten-salt power towers. Appendix A in Kolb *et al.* (2007a).

Kalb, H. and Vogel, W. (1980) Thesis, state examination. Institute for Applied Physics, University of Karlsruhe, Germany. (1) Kalb, H. The Future Solar-Hydrogen Energy System (in German), 1980, (2) Vogel, W. Prospects for the Energy Supply from Fuel Cells in a Solar-Hydrogen Economy (in German), 1979.

Kalb, H. and Vogel, W. (1986a) The Spain-Solar-Program: solar power plants in Spain–the future baseload power generation system for Germany, a system and cost study, draft of a research program (in German, 320 p.), Summary published in Kalb/Vogel 1986-b to -d, 1987, 1988).

Kalb, H. and Vogel, W. (1986b) Solar electricity via long-distance power line from Spain to Germany (in German). Sonnenenergie (DGS), 11 (1), 18-21, and 11 (2), 25-28. http://www.dgs.de/1924.0.html (Download system not yet operating (January 2010), should be available anytime soon).

Kalb, H. and Vogel, W. (1986c) Can solar power plants in Spain contribute to the German baseload power supply?–Proposal of a second development path for power plants (in German). Memorandum sent to ca. 80 leading institutions of economy, science, policy and press media in May 1986 and presented at.

Kalb, H. and Vogel, W. (1986d) Solar power plants in Spain instead of nuclear power plants in Germany? Press release, press conference, 13 November 1986, Stuttgart, with participation of Prof. W. Buckel (Physical Institute, University of Karlsruhe, President of the European Physical Society).

Kalb, H. and Vogel, W. (1987) Solarstrom und Solarwasserstoff (Solar Electricity and Solar Hydrogen), in *Die gespeicherte Sonne (The Stored Sun)*, 2nd edn (ed. H. Scheer), Piper, Munich, pp. 213–227.

Kalb, H. and Vogel, W. (1988) Press release. Federal Press Conference, 28 March 1988, Bonn. Participation: Prof. W. Buckel and H. Scheer (President of EUROSOLAR). See for example: (1) HANDELSBLATT (1988) Sonnenkraftwerke sollen Atomenergie ablösen (Solar Power Plants Shall Replace Nuclear Power). HANDELSBLATT (a German economics newspaper), 29 March 1988. (2) DIE ZEIT, Kein Platz fuer die Sonne, April 29 1988 http://www.zeit.de/1988/18/Kein-Platz-fuer-die-Sonne.

Kalb, H. and Vogel, W. (1993) The sunshine-project–solar energy for the electricity supply of Germany (in German, 46 p.). Study presented on the colloquium "Prospects for the Restructuring of the German Energy System in view of the CO_2-Problem", Wuppertal Institute for Climate, Environment and Energy, 16 September 1993, Wuppertal, Germany. Also presented to the Enquete Commission "Protection of the Earth's Atmosphere" of the German Parliament, Arbeitsunterlage 12/303, 3 December 1993.

Kalb, H. and Vogel, W. (1994) The sunshine-project–solar energy for the electricity supply of Germany. (in German). Meeting of the German Physical Society (DPG), 14–18 March 1994, Hamburg, in Vorträge des Arbeitskreises Energie (ed. K. Schultze), DPG, Bad Honnef, pp. 82–112.

Kalb, H. and Vogel, W. (1998) Solar thermal power plants in a South-North Linkage–a development program (in German). Study commissioned by the European Association for Renewable Energy (EUROSOLAR), sponsored by a private funding group and the federal state of Hamburg. Study presented: Kalb, H., Vogel, W. (1998) Press release (no title). Press conference, Presseclub Bonn, 23 June 1998, with participation of Prof. W. Buckel.

Kalb, H. and Vogel, W. (2004) Europäische Sonnenenergie–Thermische Solarkraftwerke in Spanien zur Stromversorgung Europas (European Solar Energy–Solar Thermal Power Plants in Spain for the European Electricity Supply). Workshop of the Energy Working Group (AKE) of the German Physical Society (DPG), 15–16 April 2004, Bad Honnef. http://www.uni-saarland.de/fak7/fze/AKE_Archiv (accessed 11 October 2009) OR http://www.dpg-physik.de/dpg/gliederung (accessed 4 June 2009).

Kanngießer, K.-W. (1998) Nutzung regenerativer Energiequellen Afrikas zur Stromversorgung Europas durch Kombination von Wasserkraft und Solarenergie, in Regenerativer Strom für Europa durch Fernübertragung elektrischer Energie (eds. H.-G. Brauch, G. Czisch and G. Knies), Proceedings of the HKF workshop (Hamburger Klimaschutzfonds), 30 September 1998, Bad Honnef, Germany, published 1999, AFES-Press, Mosbach.

Karg, J. (2008) Personal communication (Siemens, Dept. I-4 Gas turbines), 25 January 2008.

KDM (2009) Research project Euro-Argo. German Marine Research Consortium (KDM). www.deutsche-meeresforschung.de (accessed 12 February 2009).

Kearney, D., Herrmann, U., Nava, P., Kelly, B., Mahoney, R., Pacheco, J., Cable, R., Potrovitza, N., Blake, D., and Price, H. (2003) Assessment of a molten salt heat transfer fluid in a parabolic trough solar field. *Journal of Solar Energy Engineering*, **125** (2), 170–176.

Kelly, B. (2005) Nexant parabolic trough solar power plant systems analysis–Task 2: comparison of wet and dry rankine cycle heat rejection. Subcontract Report NREL/SR-550-40163, prep. 2005, publ. 2006. http://www.nrel.gov/publications (accessed 25 July 2007).

Kelly, B. (2006) Heat rejection for trough rankine cycles. Parabolic Trough Technology Workshop, February 14–16, 2006, Incline Village, Nevada. http://www.nrel.gov/csp/troughnet (accessed 25 July 2007).

Kelly, B., Barth, D., Brosseau, D., Konig, S., and Fabrizi, F. (2007) Nitrate and nitrite/nitrate salt heat transport fluids. Parabolic Trough Technology Workshop, March 8–9, 2007, Golden, Colorado. http://www.

nrel.gov/csp/troughnet (accessed 5 October 2008).

Kennedy, C. (2002) Review of mid- to high-temperature solar selective absorber materials. NREL Report NREL/TP-520-31267.

KJCOC (1999) See Cohen et al. 1999.

Klaiß, H. and Staiß, F. (eds.) (1992) *Solarthermische Kraftwerke Für Den Mittelmeerraum*, Springer, Berlin.

Kohlenstatistik (2008) Entwicklung ausgewählter Energiepreise. Statistik der Kohlenwirtschaft. www.kohlenstatistik.de (accessed 7 November 2008).

Kolb, G. (1996a) Economic evaluation of solar-only and hybrid power towers using molten salt technology. Proceedings of the 8th SolarPACES International Symposium on Solar Thermal Concentrating Technologies, 6–11 October 1996, Cologne. C. F. Müller, Heidelberg, Germany.

Kolb, G. (1996b) Personal communication, December 1996.

Kolb, G., Jones, S., Donnelly, M., Gorman, D., Thomas, R., Davenport, R., and Lumia, R. (2007a) Heliostat cost reduction study. Sandia Report SAND2007 3293. http://prod.sandia.gov/techlib/access-control.cgi/2007/073293.pdf (accessed November 2008).

Kolb, G., Diver, R., and Siegel, N. (2007b) Central-station solar hydrogen power plant. *Journal of Solar Energy Engineering*, **129** (2), 179–183.

Kreetz, H. (1997) Theoretische Untersuchungen und Auslegung eines temporären Wasserspeichers für das Aufwindkraftwerk, Diploma Thesis, Technical University Berlin.

Krewitt, W. (2002a) Externe Kosten der Stromerzeugung, in *Energie–Handbuch für Wissenschaftler, Ingenieure und Entscheidungsträger* (ed. E. Rebhan), Springer, Berlin.

Krewitt, W. (2002b) External cost of energy – do the answers match the questions? Looking back at 10 years of ExternE. *Energy Policy*, **30** (10), 839–848.

Krewitt, W. and Schlomann, B. (2006) Externe Kosten der Stromerzeugung aus Erneuerbaren Energien im Vergleich zur Stromerzeugung aus fossilen Energieträgern. German Aerospace Center (DLR), Stuttgart, and Fraunhofer Institute for Systems and Innovation Research (ISI), Karlsruhe, Report commissioned by the Center for Solar Energy and Hydrogen Research (ZSW), Stuttgart, Germany. http://www.dlr.de/tt (accessed 28 September 2009).

Kutscher, C., Buys, A., and Gladden, C. (2006) Hybrid wet/dry cooling for power plants. Parabolic Trough Technology Workshop, February 14–16, 2006, Incline Village, Nevada. http://www.nrel.gov/csp/troughnet (accessed 20 May 2008).

Laing, D. (2007) Storage development for direct steam generation power plants. Parabolic Trough Technology Workshop, March 8–9, 2007, Golden, Colorado. http://www.nrel.gov/csp/troughnet (accessed 3 May 2007).

Lata, J., Rodríguez, M., and Álvarez de Lara, M. (2006) High flux central receivers of molten salts for the new generation of commercial stand-alone solar power plants. Proceedings of the 13th SolarPACES International Symposium, June 20–23, 2006, Seville, Spain. See also *Journal of Solar Energy Engineering*, **130** (2), 021002.

Lerchenmueller, H., Mertins, M., Morin, G., Haeberle, A., Bockamp, S., Ewert, M., Fruth, M., Griestop, T., and Dersch, J. (2004a) Fresnel-collectors in hybrid solar thermal power plants with high solar shares. Proceedings of the EuroSun 2004 (14th International Sonnenforum), June 20–23, 2004, Freiburg, Germany (CD-ROM). www.ise.fraunhofer.de.

Lerchenmueller, H., Mertins, M., Morin, G., Haeberle, A., Zahler, C., Ewert, M., Fruth, M., Griestop, T., Bockamp, S., Dersch, J., and Eck, M. (2004b) Technische und wirtschaftliche Machbarkeits-Studie zu horizontalen Fresnel-Kollektoren. Study for the Federal Ministry for the Environment, Nature Conservation and Nuclear Safety (Germany), Final Report, TOS1-LERCH-0402-E1.

Litwin, R. (2002) Receiver system: lessons learned from solar two. Sandia Report, SAND2002-0084.

Litwin, R., Delgado, A., Moriarty, M., and Jones, C. (2005) Systems and Methods for Generating Electrical Power from Solar Energy. US Patent No. 6,957,536.

Lorenzo, E. (2002) Las chimeneas solares: De una propuesta espanida en 1903 a la Central de Manzanares. De los archivos historicos de la énergia solar. http://www.fotovoltaica.com/chimenea.pdf (accessed 23 January 2007).

Lovegrove, K., Luzzi, A., Soldiani, I., and Kreetz, H. (2004) Developing ammonia based thermochemical energy storage for dish power plants. Solar Energy, 76 (1–3), 331–337.

Lovegrove, K., Zavadski, A., and Coventy, J. (2006) Taking the ANU big dish to commercialization. Proceedings of Solar 2006, Annual Conference of the Australian and New Zealand Solar Energy Society (ANZSES), 13 September 2006, Canberra.

Lübbert, D. and Lange, F. (2006) Uran als Kernbrennstoff: Vorräte und Reichweite. Deutscher Bundestag, Wissenschaftliche Dienste, Info-Brief WF VIII G–069/06. http://www.bundestag.de/dokumente/analysen/index.html (accessed September 2009).

Lucier, R. (1981) System for Converting Solar Heat to Electrical Energy. US Patent No. 4,275,309.

Ludwig-Bölkow-Systemtechnik (1994) Wasserstoff im Gasnetz. (Author W. Zittel). Energiespektrum, 9 (9), 69–73.

Ludwig-Bölkow-Systemtechnik (2007) Coal: resources and future production (Authors H. Zittel, J. Schindler). Energy Watch Group (EWG) background paper (March 2007 version), EWG series No. 1/2007. www.energywatchgroup.org (accessed 11 October 2009).

Luxa, A. (2007) Chancen und Risiken von Ultra High Voltage Anwendungen für die Versorgungssicherheit. Report on the Siemens HVDC Technology, Energy Forum "Life Needs Power", Hannover Messe, 16–20 April 2007. www.life-needs-power.de (accessed 25 September 2007).

Luz II (2008) www.luz2.com (accessed 4 November 2008).

Maccari, A. (2006) Innovative heat transfer concepts in concentrating solar fields. Consultative Seminar "Concentrating Solar Power–Towards the 7 European RTD Framework Programme", 27 June 2006, Brussels. http://ec.europa.eu/energy/res/events/doc/maccari_enea.pdf (accessed 23 September 2009).

MAN corporation (1986) Personal communication.

Marcos, M., Romero, M., and Palero, S. (2004) Analysis of air return alternatives for CRS-type open volumetric receiver. Energy, 29 (5–6), 677–686.

Marko, M. (2004) Thin wall header for use in molten salt solar absorption panels. US Patent No. 6,736,134.

Martin Marietta Corp (1978) Conceptual design of advanced central receiver power systems, Final Report (Author T. Tracey). US Department of Energy, DOE/ET/20314-1/2.

Martin Marietta (1979) Solar central receiver hybrid power systems conceptual design–molten salt receiver (Author C. Bolton). Department of Energy Large Solar Central Power Systems Semiannual Review, Sandia Report SAND-79-8508.

Maulbetsch, J. (2006) Water conserving cooling–status and needs. Energy-Water, Western Region Needs Workshop, January 10–11, 2006, Salt Lake City. http://www.sandia.gov/energy-water/western.htm (accessed 23 September 2009).

Maurstad, O., Herzog, H., Bolland, O., and Beér, J. (2006) Impact of coal quality and gasifier technology on IGCC performance. Proceedings of the 8th International Conference on Greenhouse Gas Control Technologies (GHGT-8), June 19–22, 2006, Trondheim, Norway, CD-ROM, Elsevier, Amsterdam. http://sequestration.mit.edu/pdf (accessed 26 September 2009).

Mavis, C. (1989) A description and assessment of heliostat technology. Sandia Report SAND87-8025.

Maxwell, E., George, R., and Wilcox, S. (1998) A climatological solar radiation model. Proceedings of the 1998 ASES Conference, June 14–17, Albuquerque, New Mexico, 1998, American Solar Energy Society, Boulder, CO.

May, N. (2005) Eco-balance of a solar electricity transmission from North Africa to Europe. Diploma Thesis, Faculty for Physics and Geological Sciences, Technical University of Braunschweig, Germany. http://www.dlr.de/tt/en/Portaldata/41/ (accessed 1 October 2009) OR http://www.dlr.de/tt (accessed 1 October 2009)

(Department of Systems Analysis – Systems Analysis – 2006 Projects – TRANS-CSP – Additional Reports).

Mazzotti, M., Storti, G., and Cremer, C. (2004) CO_2-Emissionen vermeiden – Das Abscheiden von CO_2 aus Punktquellen oder Luft. *ETH-Bulletin*, (293), 44–47.

McDonnell Douglas Astronautics Company (1981) Second generation heliostat with high volume manufacturing facility defined by general motors – final report. Sandia Report SAND81-8177.

Mener, G. (1998) War die Energiewende zu Beginn des 20. Jahrhunderts möglich? (Was the energy turnaround feasible at the beginning of the 20th century?). *Sonnenenergie (DGS)*, (5), 40–43.

Meteotest (database Meteonorm) (2009) beam irradiation map of Spain. http://www.meteonorm.com/pages/en/downloads/maps.php (accessed 20 July 2009).

Meyer, B. (2003) Fossile Kraftwerkstechnik – Technologische Entwicklungsperspektiven incl. CO_2-Abtrennung und -Deponierung für die Verstromung von Braunkohle. Energieprogramm Sachsen, Expertise No. 6, presentation 11 July 2003 (Prof. B. Meyer, Institute for Energy Process Engineering and Chemical Engineering, Technical University Bergakademie Freiberg, Germany. http://www.ier.uni-stuttgart.de/forschung (accessed 12 October 2009).

Meyer, B. and Lorenz, K. (2004) Studie zentrale Vergasungssysteme (Study on Centralized Gasification Systems) (short title) – Final Report. TU Freiberg, Germany, study commissioned by the Federal Ministry of Economics and Labor, FKZ 0327118. http://edok01.tib.uni-hannover.de/edoks (accessed 5 October 2009).

Michels, H. and Pitz-Paal, R. (2007) Cascaded latent heat storage for parabolic trough solar power plants. *Solar Energy*, **81** (6), 829–837.

Mills, D. and Le Lièvre, P. (2004) Competitive solar electricity. Proceedings of "Solar 2004", Annual Conference of ANZSES, November 30–3 December 2004, Perth, Australia.

Mills, D., Le Lièvre, P., and Morrison, G. (2004) First results from compact linear fresnel reflector installation. Proceedings of "Solar 2004", Annual Conference of ANZSES, November 30–December 3, 2004, Perth, Australia.

Moens, L., Blake, D., Rudnicki, D., and Hale, M. (2003) Advanced thermal storage fluids for solar parabolic trough systems. *Journal of Solar Energy Engineering*, **125** (1), 112–116.

Mohr, M., Swoboda, P., and Unger, H. (1999) *Praxis Solarthermischer Kraftwerke*, Springer, Berlin.

Müller-Syring, G. (2008) Personal communication (DBI-GUT Leipzig), 16 November 2008.

Murphy, L., Simms, D., and Sallis, D. (1986) Structural design considerations for stretched-membrane heliostat reflector modules with stability and initial imperfection considerations. Solar Energy Research Institute, Report No. SERI/TR-253-2338.

Murphy, T., Otting, W., and Frye, P. (2002) Solar Dish Concentrator with a Molten Salt Receiver Incorporating Thermal Energy Storage, US Patent 7,299,633, issued on 27 November 2007, (Application 2002, Appl. No. 10324510).

NAE BEES (2004) The hydrogen economy – opportunities, costs, barriers, and R&D needs, A Report of the National Academy of Engineering (NAE), Board on Energy and Environmental Systems (BEES), The National Academies Press, Washington, D.C.

NASA (2007) Fiscal year 2008 budget estimates. NASA Budget Request. http://www.nasa.gov/news/budget/index.html (accessed 29 September 2009).

NaturalHy (2004) EU project NaturalHy. http://www.naturalhy.net/start.htm (accessed 29 September 2008).

NEA (Nuclear Energy Agency) (2006) URANIUN 2005: Resources, Production and Demand, OECD, Paris.

Neumann, T. and Riedel, V. (2006) FINO 1 platform: update of the offshore wind statistics. *DEWI Magazin*, (28), 60.

Nitsch, J. (2002) Potenziale der Wasserstoffwirtschaft. Expert's Study for the WBGU Flagship Report (2003) "World in Transition – Towards Sustainable Energy Systems". http://www.wbgu.de/wbgu_jg2003_engl.html (accessed 29 September 2009).

Nitsch, J. and Trieb, F. (2000) Potentiale und Perspektiven regenerativer Energieträger. DLR Report Commissioned by the Office of Technology Assessment (TAB) at the German Parliament. http://www.dlr.de/tt (accessed 26 September 2009).

NRAW (2006) Die Energieversorgung sichern – Politische, technologische und wirtschaftliche Implikationen. A Memorandum of the North Rhine-Westphalian Academy of Sciences (Germany). www.awk.nrw.de (accessed 24 September 2009).

NRC (2000) *Renewable Power Pathways: A Review of the Department of Energy's Renewable Energy Programs*, National Research Council, Committee on Programmatic Review of the U.S. Department of Energy's Office of Power Technologies (Chair H. M. Hubbard), National Academies Press, Washington, D.C..

NRC (2002) Critique of the sargent and lundy assessment of cost and performance forecasts for concentrating solar power. US National Research Council, BEES, Committee on Review of a Recent Technology Assessment of Concentrating Solar Power Energy Systems (Chair G. Kulcinski). www.nap.edu (accessed 23 September 2009).

NREL (2004) Solar maps (USA). National Renewable Energy Laboratory. http://www.nrel.gov/gis/solar.html (accessed 1 August 2008).

NREL (2005) Solar maps (World). Solar and Wind Energy Resource Assessment (SWERA), maps for Direct Normal Insolation (as of 2005). Africa (and Spain), China, East Asia, South America. swera.unep.net (accessed 5 August 2008).

NREL (2006) Cooling for parabolic trough power plants – overview (Author H. Price). Parabolic Trough Technology Workshop, February 14–16, 2006, Incline Village, Nevada http://www.nrel.gov/csp/troughnet (accessed 24 September 2009).

NREL (2007) CSP resource maps. http://www.nrel.gov/csp/maps.html (accessed 15 October 2009) For the DNI map without slope-filter (2007) click "PDF 2.2 MB".

NREL (2008) Solar power prospector tool. http://mercator.nrel.gov/csp/ (accessed October 2008).

NTC (2008) Engines /gas-turbine prices. Nye Thermodynamics Corp. http://www.nyethermodynamics.com/engines (accessed 21 September 2008).

OECD (2008) OECD statistics – prices and purchasing power parities http://www.oecd.org/topicstatsportal (accessed 10 November 2008).

OECD StatExtracts (2009) Exchange rates (USD monthly averages). http://stats.oecd.org/index (accessed 11 October 2009).

Olah, G., Goeppert, A., and Prakash, S. (2006) *Beyond Oil and Gas: The Methanol Economy*, Wiley-VCH Verlag GmbH, Weinheim, Germany.

Ortega, J., Burgaleta, J., and Téllez, F. (2008) Central receiver system solar power plant using molten salt as heat transfer fluid. Proceedings of the 13th SolarPACES International Symposium, June 20–23, 2006, Seville, Spain. See also *Journal of Solar Energy Engineering*, 130 (2), 024501.

Osuna, R., Fernandez, V., Romero, M., and Marcos, M. (2000) PS 10: a 10 MW solar tower power plant for Southern Spain. Proceedings of the 8th International Energy Forum, July 23–28, Las Vegas.

Pacheco, J., Showalter, S., and Kolb, W. (2001) Development of a molten-salt thermocline thermal storage system for parabolic trough plants. Proceedings of the Solar Forum 2001, April 21–25, 2001, Washington. Technomic, Lancaster, USA.

Peter, U. and Ständer, K. (1995) Ist Kernenergie in Zukunft wettbewerbsfähig? *Energiewirtschaftliche Tagesfragen*, 45 (4), 213–217.

PIER (1999) Solar two central receiver. California Energy Commission, Public Interest Energy Research (PIER), Contract No. 500-97-012, prepared by W. Stoke (Southern California Edison).

Pilkington (1994) *Prefeasibility Study of A Solar Thermal Trough Power Plant for Crete, Final Report*, Pilkington Solar International, Cologne, Germany.

Pilkington (1996) *Status Report on Solar Thermal Power Plants*. Pilkington Solar International, Cologne, sponsored by the German Federal Minister for Education, Science, Research and Technology under contract No. 0329660. http://www.solarpaces.org/Library/csp_docs.htm (accessed July 2007).

Pitz-Paal, R. (2003) Solarturmkraftwerke mit Hochtemperatur-Wärmespeicher. 300th WE-Heraeus-Seminar, Energy Working Group (AKE) of the German Physical Society, May 26–29, 2003, Bad Honnef, Germany. http://www.uni-saarland.de/fak7/fze (accessed 20 September 2009).

Pitz-Paal, R. (2004) Wie die Sonne ins Kraftwerk kommt. *Physik in unserer Zeit*, **35** (1), 12–19.

Pitz-Paal, R. and Hoffschmidt, B. (2003) Ziele bei der Entwicklung von solarthermischen Kraftwerken, in *Optionen Für Die Energie Der Zukunft* (ed. M. Keilhacker), Meeting of the German Physical Society, March 24–28, 2003, Hannover, DPG, Bad Honnef. 17–33. http://www.dpg-physik.de/gliederung/ak (accessed 2 July 2009).

Ploetz, C. (2003) Sequestrierung von CO_2: Technologien, Potenziale, Kosten und Umweltauswirkungen. Expert's Study for the WBGU Flagship Report "World in Transition – Towards Sustainable Energy Systems" (2004), Earthscan, London. http://www.wbgu.de/wbgu_jg2003_engl.html (accessed 29 September 2009).

Price, H. and Kearney, D. (2003) Reducing the cost of energy from parabolic trough solar power plants. Proceedings of the ASME 2003 International Solar Energy Conference, March 15–18, Kohala Coast, Hawaii.

PSA (1997) Personal communication, Plataforma Solar de Almería, Spain, April 1997.

PSA (2006) Annual report 2006. Plataforma Solar de Almeria, CIEMAT.

PSA (2008) Personal communication, C. Richter (DLR), May 2008.

Quaschning, V. and Geyer, M. (2000) Einsatzmöglichkeiten regenerativer Energien für eine klimaverträgliche Elektrizitätsversorgung in Deutschland. Proceedings of the DGS – 12. Internationales Sonnenforum, July 5–7, 2000, Freiburg, Germany, Solar Promotion, Munich. http://www.volker-quaschning.de/downloads (accessed 1 July 2009). See also http://www.volker-quaschning.de/publis/klima2000/index.php.

Rabl, A. (1976) Tower reflector for solar power plants. *Solar Energy*, **18** (3), 269–271.

Reilly, H. and Kolb, G. (2001) An evaluation of molten-salt power towers including results of the solar two project. SNL Report SAND2001-3674.

Richter, C. (2007) Water efficient cooling of solar thermal power plants. http://www.solarpaces.org/Tasks/Task3/EFCOOL.HTM (accessed 30 September 2009).

Roan, V., Betts, D., Twining, A., Dinh, K., Wassink, P., and Simmons, T. (2004) An Investigation of the feasibility of coal-based methanol for application in transportation fuel cell systems. Final Report, University of Florida, Gainesville. http://fuelcellbus.georgetown.edu/files (accessed 24 September 2009).

Rockwell International Corp (1979) Solar central receiver hybrid power system, in Department of Energy Solar Central Receiver Semiannual Review, Sandia Report SAND-79-8073.

Romero, M., Buck, R., and Pacheco, J. (2002) An update on solar central receiver systems, projects, and technologies. *Journal of Solar Energy Engineering*, **124** (2), 98–108.

Rosenthal, A. and Roberg, J. (1994) Twelve month performance evalutation for the rotating shadowband radiometer. Sandia Report SAND94-1248 UC-1303, DOE Contract No. DE-AC04-94AL85000.

Ruiz, H. (1989) Personal communication, Departamento de Ingeniero Energetica y Fluidomechanica, Universidad de Sevilla, 7 July 1989.

RWE (2005) Weltenergiereport 2005 – Bestimmungsgrössen der Energiepreise.

S&L (2003) Assessment of parabolic trough and power tower solar technology cost and performance forecasts. Sargent & Lundy LLC Consulting Group, Chicago, Illinois. SL-5641, NREL Report NREL/SR-550-34440.

Salt Institute (2009) Salt Institute, Alexandria, VA, USA. www.saltinstitute.org (accessed September 2009).

San Diego Renewable Energy Group (2005) Potential for renewable energy in the San Diego Region (Authors S. Anders, T. Bialek, D. Geier, D. Jackson, M. Quintero-

Núnez, R. Resley, D. Rohy, A. Sweedler, S. Tanaka, C. Winn, K. Zeng). Published by the San Diego Regional Renewable Energy Group. www.renewablesg.org (accessed 29 September 2009).

Sanchez, M., Romero, M., and Ajona, J. (1996) Sensitivity analysis of a hybrid solar tower plant for Southern Spain. Proceedings of the 8th SolarPACES International Symposium, October 6–11, 1996, Cologne, C. F. Müller, Heidelberg, Germany.

Sargent and Lundy, L.L.C. (2003) See S&L (2003).

Schiel, W. (1997) Personal communication (SBP), August 1997.

Schillings, C. (2008) Personal communication (DLR), 26 August 2008.

Schillings, C., Tereira, E., Perez, R., Meyer, R., Trieb, F., and Renne, D. (2002) High resolution solar energy resource assessment within the UNEP-Project SWERA. Proceedings of the World Renewable Energy Congress VII, June 29–July 5, 2002, Cologne, Germany, Pergamon–Elsevier Science, Oxford, UK. http://www.dlr.de/tt (accessed 3 October 2009).

Schillings, C., Meyer, R., and Trieb, F. (2004a) High resolution (10 km) solar radiation assessment for China. DLR, Solar and Wind Energy Resource Assessment (SWERA). swera.unep.net (accessed August 2008).

Schillings, C., Meyer, R., and Mannstein, H. (2004b) Kartierung der Solarenergieressource für Marokko, Work package 3.2 of SOKRATES, DLR Report, Stuttgart, Germany. http://www.dlr.de/tt (accessed 4 October 2009).

Schindler, J. and Zittel, W. (2007) Beitrag der Urankosten zu den Stromerzeugungskosten der Kernkraftwerke. http://www.energiekrise.de/uran/docs2007/Urankosten-1.pdf (accessed 30 August 2007).

Schlaich, J. (1995) *The Solar Chimney*, Edition Axel Menges, Stuttgart.

Schlaich, J. (1997) Personal communication.

Schlaich, Schiel, Scherer, Desai, and Müller (1996) Technische und wirtschaftliche Aspekte zur Beurteilung der Chancen von Aufwindkraftwerken. Working paper, study group on updraft power plants (SBP, Badenwerk, EVS).

Schlaich, J., Bergermann, R., Schiel, W., and Weinrebe, G. (2005) Design of commercial solar updraft tower systems – utilization of solar induced convection flows for power generation. *Journal of Solar Energy Engineering*, **127** (1), 117–124.

Schmitt, D. (1989) Kosten und Kostenstrukturen in der Elektrizitätswirtschaft der Bundesrepublik Deutschland. *Elektrizitätswirtschaft*, **88** (16/17), 1090–1099.

Schwaegerl, E.F. and Thieme, H. (1987) Dieselgeneratoranlagen – Wirtschaftliche Alternative zum Strombezug im Spitzenlastbereich. *BWK*, **39** (1/2), 15–20.

Sempra Energy (2005) CPUC approves SDG&E contract for solar power. Press release, December 20, San Diego, California. http://www.sempra.com/news/releases.htm (accessed 2 September 2009).

SFOE (2007) Energieperspektiven 2035. Swiss Federal Office of Energy. http://www.bfe.admin.ch (accessed March 2008).

Short, W., Packey, D.J., and Holt, T. (1995) A manual for the economic evaluation of energy efficiency and renewable energy technologies. NREL/TP-462-5173, National Renewable Energy Laboratory, Golden, Colorado. http://www.nrel.gov/docs (accessed 1 October 2009).

Smith, C. (1995) Revisiting solar power's past. *Technology Review*, **98** (5), 38–47.

Solúcar (2006) 10 MW solar thermal power plant for Southern Spain. Final Technical Progress Report, EU project NNE5-1999-356, project coordinator Solúcar Energía S.A.

Staege, H. (1980) Mittel- und hochkaloriges Gas aus Kohle unter Anwendung der Flugstromvergasung. *GWF-Gas/Erdgas*, **121** (12), 543–548.

STE (2006) Zukünftige Energieversorgung unter den Randbedingungen einer großtechnischen CO2-Abscheidung und Speicherung (Authors J. Linßen, P. Markewitz, D. Martinsen, M. Walbeck). STE Report 1/2006, Systems Analysis and Technology Evaluation (STE) Program Group, Research Center Jülich, Germany. www.fz-juelich.de/ief (accessed 20 September 2009).

Steag, R.A.G. (ca 2001) Energien für das neue Jahrtausend. www.cp-compartner.de (accessed January 2008).

Stiegel, G. (2008) US-gasification experience – yesterday, today and tomorrow. Workshop on IGCC Technology, 20 February 2008, New Delhi. http://www.indiapower.org/igcc/garyj.pdf (accessed February 2009).

Stiegel, G. (NETL) and Ramezan, M. (Science Appl. Int. Corp.) (2006) Hydrogen from coal gasification: an economical pathway to a sustainable energy future. *International Journal of Coal Geology*, 65 (3/4), 173–190. http://linkinghub.elsevier.com/retrieve/pii/S0166516205001217 (accessed September 2009). See also: Gray, Tomlinson 2002.

Stoffel, T. and George, R. (2007) Solar resource data. NREL Parabolic Trough Technology Workshop, March 8–9, 2007, Golden, Nevada. http://www.nrel.gov/csp/troughnet (accessed 1 October 2009).

Stolzenburg, K. (PLANET, Oldenburg) (2008) Personal communication, 16 November 2008.

Strachan, J. and Van Der Geest, J. (1993) Operational experience and evaluation of a dual-element stretched-membrane heliostat. Sandia Report SAND93-2453 UC-235.

Suri, M., Remund, J., Cebcauer, T., Hoyer-Klick, C., Dumortier, D., Huld, T., Stackhouse, P., and Ineichen, P. (2009) Comparison of Direct Normal Irradiation Spatial Products for Europe. http://www.unige.ch (accessed 20 July 2009).

Tamaura, Y., Utamura, M., Kaneko, H., Hasuike, H., Domingo, M., and Relloso, S. (2006) A novel beam-down system for solar power generation with multi-ring central reflectors and molten salt thermal storage. Proceedings of the 13th SolarPACES International Symposium, June 20–23, 2006, Seville, Spain.

Tamme, R. (2006) Thermal energy storage – concrete & phase change TES. Parabolic Trough Technology Workshop, February 14–16, 2006, Incline Village, Nevada. http://www.nrel.gov/csp/troughnet (accessed 1 October 2009).

Tamme, R., Bauer, T., Buschle, J., Laing, D., Müller-Steinhagen, H., and Steinmann, W.-D. (2007) Latent heat storage above 120 °C for applications in the industrial process heat sector and solar power generation. *International Journal of Energy Research*, 32 (3), 264–271. (Published online 11 July 2007).

Terres, P. (2000) Kaufkraftparitäten als Kursziel von Devisenmarktinterventionen? *Konjunktur, Zinsen, Währungen*, (6), 16–22. http://www.dekabank.de/db/de/economics (accessed 2 October 2009).

Tester, J., et al. (2001) According to NRC 2002 (References): J. Tester (MIT) et al. 2001. Letter Report to F. Wilkins, U.S. DOE, "CSP Program Peer Review", 2 pages, with 71-page Viewgraph Final Report, based on 37 viewgraph presentations by CSP participants, on November 7–9, 2001 (December 7).

Trage, B. and Hintzen, F.J. (1989) Planung und Bau von Anlagen mit indirekter Trockenkuhlung. *VGB Kraftwerkstechnik*, 69 (2), 183–189.

Tzima, E. and Peteves, S. (2003) Controlling carbon emmissions: the option of carbon sequestration. EUR 20752/EN (2003) European Commission, Joint Research Centre, Institute for Energy, Petten, the Netherlands.

UCS (2008) Nuclear reprocessing: dangerous, dirty, and expensive. Union of Concerned Scientists (UCS). http://www.ucsusa.org/nuclear_power (accessed 13 November 2008).

UIC (2007) The economics of nuclear power. Briefing Paper 8, June 2007, Uranium Information Centre Ldt (Melbourne), Australian Uranium Association. http://www.uic.com.au/nip08.htm (accessed 21 August 2007).

UIC – Newsletters Issues 4-7 (2007) Uranium Information Center Ldt (Melbourne). http://www.uic.com.au/news407.htm (accessed 2 March 2008).

Uni-Chicago (2004) See Chicago Study (2004).

University of Sydney (2007) Compact linear fresnel reflector (CLFR) power plant technology. http://www.physics.usyd.edu.au/app/research (accessed 12 October 2007).

Uranium Info (2008) TradeTech Uranium Spot Price Declines in 2008 – Year-end Price Settles at $52. TradeTech Press Release, 2 January 2009, Uranium Info

(News). www.uranium.info (accessed 2 October 2009).

US Energy Flow Chart (2004) Energy flow charts, lawrence livermore national laboratory. http://publicaffairs.llnl.gov/news/energy/archive.html (accessed 16 October 2009) OR http://www.llnl.gov.

Utility Studies (1988) Solar central receiver technology advancement for electric utility applications – phase I topical report (Project manager T. Hillesland Jr., PG&E). Pacific Gas&Electric Co., Arizona Public Service Co., Bechtel National Inc., Black and Veatch, Technical Report prepared for DOE and EPRI, DOE contract DE-FC04-86AL38740.

Verfondern, K. (ed.) (2007) Nuclear energy for hydrogen production. Schriften des Forschungszentrums Jülich, Energy Technology Series, Vol. 58, Research Center Jülich, Germany.

VGB-PowerTech (2005) Facts and figures – electricity generation 2005. http://www.vgb.org/en/data_powergeneration.html (accessed July 2007).

Vogel and Kalb See Kalb and Vogel.

Vosen, J. (1989) Wir wollen und brauchen keine Wiederaufarbeitung – Zum Kostenaspekt beim Streit um Wackersdorf. *Sozialdemokratischer Pressedienst*, **44** (77) 3–4. quoting German Atomic Forum: "Analyse 16 zur Wiederaufarbeitungsanlage Wackersdorf" (Analysis No. 16 on the Reprocessing Facility Wackersdorf', (in German)) (accessed August 1985). http://library.fes.de/spdpd/1989/890421.pdf (accessed 13 November 2008).

Warerkar, S., Schmitz, S., Göttsche, J., and Hoffschmidt, B. (2007) Wirtschaftlich und flexibel – Luft-Sand-Wärmeübertrager für die Energiespeicherung bei Temperaturen bis 800 °C. *Verfahrenstechnik*, **41** (4), 2–3.

WBGU (2006) The future oceans – warming up, rising high, turning sour (Authors R. Schubert, H.-J. Schellnhuber, N. Buchmann, A. Epiney, R. Grießhammer, M. Kulessa, D. Messner, S. Rahmstorf, J. Schmid). Special Report, German Advisory Council on Global Change (WBGU). www.wbgu.de (accessed 2 October 2009).

WEC (2007) *2007 Survey of Energy Resources*, World Energy Council, London.

Weinrebe, G. (2000) Technische, ökologische und ökonomische Analyse von solarthermischen Turmkraftwerke. Dissertation, University of Stuttgart, Germany, Institute of Energy Economics and the Rational Use of Energy (EIR), Report No. 68. elib.uni-stuttgart.de (accessed 26 September 2009).

Weinrebe, G. (2008) Personal communication.

Weinrebe, G., Bergermann, R., Schlaich, J., Schiel, W., and Hornidge, D. (2006) Commercial aspects of solar updraft towers. Proceedings of the 13th Solar-PACES International Symposium, June 20–23, 2006), Seville, Spain.

Weinrebe, G. and Schiel, W. (2001) Up-draught solar chimney and down-draught energy tower – a comparison. Proceedings of the ISES Solar World Congress (CD-ROM), November 25–December 2, 2001, Adelaide, Australia.

Weltalmanach (2006) – *Der Fischer Weltalmanach 2006 – Zahlen, Daten, Fakten*, (ed. Weltalmanach editorial team), Fischer Taschenbuch, Frankfurt, Germany.

Wikipedia (2007) Fig.: Uranpreis.svg. Diagram (Sources: UXC und Uranium.info) http://de.wikipedia.org/wiki/Bild:Uranpreis.svg (accessed August 2007). (See http://de.wikipedia.org/wiki/Datei:Uranpreis.svg (accessed 6 October 2009).

Wilcox, S., Anderberg, M., George, R., Marion, W., Myers, D., Renne, D., Beckman, W., DeGaetano, A., Gueymard, C., Perez, R., Plantico, M., Stackhouse, P. and Vignola, F. (2005) Progress on an updated national solar radiation data base for the United States. National Renewable Energy Laboratory, NREL/CP-560-37956, Conference Paper Prepared for ISES 2005 Solar World Congress, August 6–12, 2005, Orlando, FL. http://www.nrel.gov/csp/troughnet/publications.html (accessed 7 October 2009).

Winter, C.-J., Sizmann, R., and Vant-Hull, L. (eds) (1991) *Solar Power Plants*, Springer, Berlin.

WISE (2006) Nuclear fuel production chain (Update 21 March 2006). WISE Uranium Project, World Information Service on

Energy. http://www.wise-uranium.org/nfp.html (accessed 1 December 2008).

WNA Report (2005) The new economics of nuclear power. World Nuclear Association. www.world-nuclear.org (accessed 3 October 2009).

World Military Spending (2007) World military spending (Author A. Shah). http://www.globalissues.org/Geopolitics/ArmsTrade/Spending.asp (accessed December 2007).

Wu, B., Reddy, R., and Rogers, R. (2001) Novel ionic liquid thermal storage for solar thermal electric power systems. Proceedings of "Solar Forum 2001 – Solar Energy: The Power to Choose", April 21–25, 2001, Washington D.C. http://www.nrel.gov/csp/troughnet/workshops.html (accessed 3 October 2009).

Zarza, E., Valenzuela, L., León, J., Hennecke, K., Eck, M., Weyers, H.-D., and Eickhoff, M. (2004) Direct steam generation in parabolic troughs: final results and conclusions of the DISS project. *Energy*, **29** (5/6), 635–644.

Zarza, E., Rojas, M., González.L., Caballero, J., and Rueda, F. (2006) INDITEP: the first pre-commercial DSG solar power plant. *Solar Energy*, **80** (10), 1270–1276.

Zaslavsky, D. (1999) Energy towers for producing electricity and desalinated water without a collector. Proceedings of the ISES Solar World Congress, July 4–9, 1999, Jerusalem, Israel, Elsevier Science, Amsterdam, the Netherlands.

Zaslavsky, D. (2008) Tagesthemen, 13 January 2008, ARD-News (German TV). http://www.tagesschau.de/archiv (accessed 3 October 2009).

Zaslavsky, D. and Glubrecht, H. (2000) Das solarthermische Kraftwerk ohne Kollektor. Proceedings of the DGS–12. Internationales Sonnenforum, July 5–7, 2000, Freiburg, Germany, Solar Promotion, Munich.

Zaslavsky, D., Wurzburger, U., Einav, A., and Setty, S. (2001) Energy towers for producing electricity and desalinated water without a collector. Israel–India Steering Committee (Chairman U. Wurzburger), December 2001. http://www.transnational-renewables.org/Gregor_Czisch/projekte/new_et-brochure_zaslavsky.pdf (accessed 25 February 2008).

Zavoico, A. (2001) Solar power tower design basis document (revision 0). Sandia Report SAND2001-2100.

Zhang, J. and Lin, Z. (1992) *Climate of China*, John Wiley & Sons, Inc., New York.

Zhou, A., Tian, J., Zhang, X., and Wei, J. (2005) China's renewable energy policies and potentials (Supervisors J. Byrne and Y-D Wang) Center for Energy and Environmental Policy, University of Delaware, Newark, USA. http://ceep.udel.edu/energy/publications (accessed 29 September 2009).

Zimmerman, M. (1982) Learning effects and the commercialization of new energy technologies: the case of nuclear power. *Bell Journal of Economics*, **13** (2), 297–310.